Practical Engineering Geology

Practical Engineering Geology provides an introduction to the way projects are managed, designed and constructed, and how the engineering geologist can contribute to cost-effective and safe project achievement. The need for a holistic view of geological materials, from soil to rock, and of geological history is emphasised. Chapters address key aspects of

- Geology for engineering and ground modelling
- Site investigation and testing of geological materials
- Geotechnical parameters
- Design of slopes, tunnels, foundations, and other engineering structures
- Identifying hazards
- Avoiding unexpected ground conditions

This second edition includes a new chapter on environmental issues covering hydrogeology, considerations of climate change, earthquakes, and more. All chapters have been updated, with extensively revised figures throughout and several new case studies of unexpected ground conditions. The book will support practising engineering geologists and geotechnical engineers, as well as MSc level students of engineering geology and other geotechnical subjects.

Applied Geotechnics series

William Powrie (ed.)

Practical Rock Mechanics
Steve Hencher

Soil Liquefaction, 2nd ed.
Mike Jefferies et al.

Drystone Retaining Walls: Design, Construction and Assessment
Paul McCombie et al.

Fundamentals of Shield Tunnelling
Zixin Zhang et al.

Centrifuge Modelling in Geotechnics
Christoph Gaudin et al.

Weak Rock Engineering Geology and Geotechnics
Kevin Stone et al.

Introduction to Tunnel Construction, 2nd ed.
David Chapman et al.

Soil Nailing: A Practical Guide
Raymond Cheung et al.

High Resolution Pressuremeters and Geotechnical Engineering
John Hughes et al.

Practical Engineering Geology, 2nd ed.
Steve Hencher

For more information about this series, please visit: www.routledge.com/Applied-Geotechnics/book-series/APPGEOT

Practical Engineering Geology

Second Edition

Steve Hencher

CRC Press is an imprint of the
Taylor & Francis Group, an **informa** business

Cover image: Steve Hencher

Second edition published 2024
by CRC Press
4 Park Square, Milton Park, Abingdon, Oxon, OX14 4RN

and by CRC Press
2385 NW Executive Center Drive, Suite 320, Boca Raton FL 33431

First edition published by CRC Press 2012

CRC Press is an imprint of Informa UK Limited

British Library Cataloguing-in-Publication Data
A catalogue record for this book is available from the British Library

Library of Congress Cataloging-in-Publication Data
Names: Hencher, Steve, author.
Title: Practical engineering geology / Steve Hencher.
Description: Second edition. | Abingdon, Oxon ; Boca Raton, FL : CRC Press, 2024. |
Series: Applied geotechnics | Includes bibliographical references and index.
Identifiers: LCCN 2023045940 | ISBN 9781032392257 (hbk) |
ISBN 9781032392240 (pbk) | ISBN 9781003348894 (ebk)
Subjects: LCSH: Engineering geology.
Classification: LCC TA705 .H44 2024 |
DDC 624.1/51–dc23/eng/20240130
LC record available at https://lccn.loc.gov/2023045940

ISBN: 978-1-032-39225-7 (hbk)
ISBN: 978-1-032-39224-0 (pbk)
ISBN: 978-1-003-34889-4 (ebk)

DOI: 10.1201/9781003348894

Typeset in Sabon
by Newgen Publishing UK

Contents

About the author

Steve Hencher is now Emeritus Professor at Leeds and Honorary Professor in the Department of Earth Sciences at Hong Kong University. He is also Director of Hencher Associates Limited where he acts as technical advisor and expert witness on various geotechnical matters. More information is at www.hencherassociates.com.

He is a geologist by first degree and earned his PhD from Imperial College, London on the shear strength of rock joints under dynamic loading. He then joined Sir W S Atkins & Partners where he was one of only nine, yes nine geotechnical employees servicing what, even then, were the largest consultants in Europe. Atkins gave him wide experience in a very short term. This included the opportunity to investigate the ground for and supervise the construction and installation of piles at Drax Power Station, which gave a sharp insight into how large civil engineering projects work. Since then he has worked with the Hong Kong Government for 5 years where he investigated major landslides, worked on shear strength of rock and first became involved in mapping and describing thick weathered profiles. Other major experience includes being part of the Bechtel design team for the High Speed Rail in Korea, working specifically on the design of very large span tunnels and underground stations. He taught the MSc in Engineering Geology at Leeds University full time from 1984 to 1996 and supervised a large number of research students. Since 1997 he has headed geotechnics in the Hong Kong office of Halcrow and was Regional Director of the Korean office for 7 years. He has worked on various national and international committees in geotechnical engineering, in particular, on weathered rocks, piling, landslides, rock slopes and rock mass characterisation. He has acted and continues to act as an expert advisor and expert witness in numerous legal cases, including aspects of foundation design and construction, tunnelling, landslides and site formation.

Preface to 2nd edition of *Practical Engineering Geology*

The book has been revised to deal with the errors and omissions, mostly picked up by my good friend Professor Andrew Malone ex Head of GEO, Hong Kong and now at Hong Kong University. Thanks Andrew.

It includes a new chapter, Chapter 5, dealing with environmental issues, not least. It has revised figures, many in colour! It deals with "basic friction" following confirmation of data from 1976 in 2014.

It includes several more case studies in Chapter 8 whilst keeping to the general length of the book.

COVER

The cover photo is looking down from Staffa, on to Am Buachaille (the Herdsman). The north-facing hills of Mull can be seen in the distance across the sea. The hexagonal columns of basalt, in the centre of the island, rise almost vertically but those to the north and south are bent and are have all the appearance of drifting into the sea.

The upper surface of the island is a ragged series of steps leading to the top, rather than a truncated, striated surface following stripping of the rock during the last glaciation, as seen at the Devil's Postpile in the USA, (see figure below). Still lots to wonder and think about.

Figure P1 Striated surface above Devil's Postpile, California, USA

Preface to 1st edition of *Practical Engineering Geology*

The genesis of this book lies in a wet, miserable tomato field in Algeria. I was sitting on a wooden orange box, next to a large, green, Russian well-boring rig with a blunt bit. I was 3 weeks out of University. The Algerian driller hit the core barrel with a sledge hammer and a hot, steaming black sausage of wet soil and rock wrapped itself around my hands. A Belgian contractor walked up and said to me (in French) "What do you think? Four–six?". I looked at the steaming mass thoughtfully and said, "*Maybe about five*". He nodded approvingly. To this day I don't know what he was talking about or in what units.

I went to see the "Chef de Zone" for this new steelworks, Roger Payne, who seemed totally in control and mature but was probably about 28 and suggested that we should write a book on engineering and geology. He, as a civil engineer, should write the geology bits and I should write the civil engineering bits as a geologist. That way we would see what we both considered important. We would edit each others work. Well we didn't do it but this book follows the blueprint. It includes aspects of geology that I consider most relevant to civil engineering, including many things that most earth science students will not have been taught in their undergraduate courses. It also provides an introduction into the parlance of civil engineering that should help engineering geologists starting out. It is an attempt to set out the things that I wish I had known when I started my career.

Many have helped with this book mainly by reviewing parts, providing information and agreeing use of their data, figures and photos. These include Des Andrews, Ian Askew, John Burland, Jonathan Choo, Chris Clayton, Gerry Daughton, Bill Dershowitz, Steve Doran, Francois Dudouit, Ilidio Ferreira, Chris Fletcher, John Gallerani, Graham Garrard, Robert Hack, Trevor Hardie, Roger Hart, Evert Hoek, Jean Hutchinson, Justyn Jagger, Jason Lau, Qui-Hong Liao, David Liu, Karim Khalaf, Mike King, Andy Malone, Dick Martin, Dennis McNicholl, David Norbury, Don Pan, Chris Parks, Andy Pickles, Malcolm Reeves, David Starr, Doug Stead, Nick Shirlaw, Kevin Styles, Nick Swannell, Leonard Tang, Len Threadgold, Roger Thompson and Derek Williams. I would also like to acknowledge the guidance of friends and mentors including Bob Courtier, Mike deFreitas, Richard Hart, Su Gon Lee, Alastair Lumsden and Laurie Richards plus my research students whose work I have relied on throughout. Ada Li has drawn some of the figures and Jenny Fok has done some of the tricky bits of typing. Thanks to all.

Finally thanks to my long-suffering wife Marji – it has been a hard slog, glued to the computer and surrounded by piles of paper whilst the garden reverts to something resembling the Carboniferous rain forests. Sam Hencher has drawn some excellent cartoons and Kate and Jess have helped in their own sweet ways.

Chapter 1

Introduction to engineering geology

1.1 WHAT IS ENGINEERING GEOLOGY?

I remember that when I joined the Geotechnical Control Office of the Hong Kong Government in January 1980, I was designated as a "geotechnical engineer" (despite being a geologist by first degree). Geotechnical engineers of whatever background were placed into "divisions" – some into "Existing Slopes", some into "New Works" and some into "Special Projects". Those in "New Slopes Division" were to check the designs of new slopes being proposed around the territory.

"Hang on for a moment", I thought, "You are expecting me to check the design of a 30-metre-high, strutted, anchored wall, with all of those pesky bending moments". "Whilst you expect that bright-eyed civil engineer, sitting next to me, to check the geology of this major slope, with all its vagaries, as if the ground was made of cheddar cheese?"

It was early days, and I suspect that I, faced with the calculations associated with the strutted retaining wall, was far quicker to cry "Help!" than the bushy-toed engineer, thumb in mouth, happily drawing straight lines between bands of fill shown on borehole logs, overlain by decomposed granite (I kid you not) ...

So, that is a quick introduction into what distinguishes an engineering geologist from a geotechnical engineer, at least for 10 years of so... To quote from Fookes, who is often regarded as the father of British engineering geology:

> It takes a minimum of three years to formally educate a geologist and a lifetime to gain experience. This is also true of engineers and it is what brings engineers and geologists together and also keeps them apart.
>
> *Fookes (1997)*

1.2 DEFINITIONS

Many authors have attempted to define engineering geology (e.g. Morgenstern, 2000; Knill, 2003; Bock, 2006) but some of the best definitions come earlier, from Burwell & Roberts (1950), who made the following statements with respect to engineering geologists, which still ring true:

1. **"Obviously, the first requirement of the engineering geologist is that he shall be a competent geologist. ......Against this background of knowledge, he will discover the major geologic factors in advance of construction and recognize the more**

DOI: 10.1201/9781003348894-1

obscure minor details that so often exert a major influence on location, design and construction problems".

Burwell and Roberts emphasized both the "major geological factors" at a site, and the need to look for "obscure minor details". Soil and rock description for engineering purposes and site characterisation are dealt with in Chapter 4 and Appendix B. Examples of where such minor details, e.g. clay-filled, adversely-oriented joints caused failures in dams and slopes, are given in Chapter 8.

2. **"The second requirement is that he shall be able to translate his discoveries and deductions into terms of practical application. This qualification is not obtained as a result of better knowledge of geology, but of better knowledge of engineering".**

This is one of the main aims of this book, to explain how structures are designed and how the engineering geologist's geological knowledge can help the engineer, to ensure the safety of the structure – see Chapters 2 and 3.

3. **"The third requirement is dual in character. It is the ability to render sound judgements and make important decisions. Sound judgment is a priceless faculty of the geologist who is frequently called on to make decisions without all the factual data necessary to guarantee the results. It is not always economically practicable to eliminate the element of uncertainty and not infrequently his advice has to be based on few and scattered evidences in the field".**

The need to prepare ground models on the basis of "few and scattered evidences" is a key task of the engineering geologist. This is a matter of modelling (Chapter 4) as well as assessing and dealing with risk (Chapter 4 but also Appendix D).

4. **"The fourth requirement relates to the temperamental make-up or personal qualities of the engineering geologist. He should not be an alarmist. Neither faults, nor earthquakes, nor cavernous limestones, nor pervious basalts, nor low water tables should deter him from rationalizing the field evidences and proceeding to logical conclusions based on due consideration of both facts and influences".**

This calls for judgement of the risks that arise from his observations. He[1] should be prepared to judge the risk of the inherent uncertainty of his models and to warn the engineer where there is a real problem or risk with his models. Chapter 8 deals with case histories of where things went wrong; Chapters 7, 6, 5, 4 and 3 tell you how to get the project right in the first place!

Engineering Geology is defined by these early observations but there are many other tasks that he should aspire to and Table 1.1 is my attempt to set out some of the skills that, in my view, a well-rounded and competent engineering geologist should attempt to obtain and develop.

Firstly, he needs to be familiar with the nature of civil engineering[2] projects, contracts and risk, and how those structures are designed and constructed and this is the subject of Chapters 2 and 3.

Table 1.1 Basic skills and knowledge for engineering geologists

It is difficult to define engineering geology as a separate discipline but easier to define the subject areas with which an engineering geologist needs to be familiar. These include the following:

1. **GEOLOGY**
   The engineering geologist needs a good understanding of basic geology, as taught in a traditional UK geology degree, but will need a far broader perspective including methods of investigation, and degree of weathering (see Section 2, below and Appendix A.1.1). In particular, he needs to be aware that conditions are not and have never been "static" but are changing by the second, by the minute, by the decade and by the millennium. He needs to be able to interpret the geological history of a site in terms of a geological model as it is now, how it got to be in the state it is, and in a predictive manner with respect to changes that might occur, both natural changes and due to the works themselves.
2. **ENGINEERING GEOLOGY AND HYDROGEOLOGY**
   Aspects of geology and geological processes that are not normally covered well in an undergraduate geological degree syllabus need to be learned through advanced study (MSc and continuing education) or during employment. These include the following:
   Methods and techniques for sub-surface investigation.
   - Properties of soil and rock, such as strength, permeability and deformability-how to measure these in the laboratory (material scale) and in the field and how to apply these at the large scale (mass scale) to geological models.
   - Methods for soil and rock description and classification for engineering purposes.
   - Weathering processes and the nature of weathered rocks.
   - Principle of effective stress.
   - Quaternary history, deposits and sea level changes.
   - Nature, origins and physical properties of discontinuities.
   - Hydrogeology - infiltration of water, hydraulic conductivity and controlling factors. Water pressure in the ground, drainage techniques, Darcy's Law.
   - Key factors that will affect engineering projects, such as forces and stresses, earthquakes, blast vibrations, chemical reactions and deterioration.
   - Numerical characterisation, modelling and analysis.

   These are dealt with primarily in Chapters 4, 5, 6, 7 and 8.
3. **GEOMORPHOLOGY**
   Most engineering projects are constructed close to the land surface and therefore geomorphology is very important. An engineer might consider a site in an analytical way, for example, using predicted 100-year rainfall and catchment analysis to predict flood levels and carrying out stability analysis to determine the hazard from natural slope landslides. The recognition of past landslides through air photo interpretation and lidar is a fundamental part of "desk study" for many hilly sites. This is dealt with in Chapter 4.
4. **CIVIL ENGINEERING DESIGN AND PRACTICE**
   An engineering geologist must be familiar with the principles of the design of structures and the options, say for founding a building or for constructing a tunnel or a dam. He must be able to work in a team of civil and structural engineers providing adequate "ground models" that can be analysed to predict project performance and this requires some considerable knowledge of engineering practice and terminology. The geological ground conditions need to be modelled mechanically and the engineering geologist needs to be aware of how this is done and, better still, able to do so himself. This is covered mainly in Chapters 2, 3, 4 and Chapter 7.

(*continued*)

Table 1.1 (Cont.)

5. SOIL AND ROCK MECHANICS

Engineering geology requires quantification of geological models. Hoek (1999) described the process as "putting numbers to geology". That is not to say that "pure" geologists do not take a quantitative approach - they do, for example, in analysing sedimentary processes, in structural geology and in geochronology. However a geologist is usually concerned with relatively slow processes and very high stress levels at great depths. The behaviour of rock and soil in the shorter term (days and months) and at relatively low stresses are the province of Soil Mechanics and Rock Mechanics. Knowledge of the principles and practice of soil and rock mechanics is important for the engineering geologist. This includes strength, compressibility and permeability at "material" and "mass" scales, the principle of "effective stresses", strain-induced changes, "critical states" and dilation in rock masses. Chapter 7 presents key aspects of soil and rock mechanics.

6. HYDROGEOLOGY

Hydrogeology is important to many sites; groundwater changes induced by extreme rainfall leads to many landslides. Tunnels below the water table have to deal with inflows, which can lead to delays, but can also lead to settlement of houses at the surface by dropping the water table. The engineering geologist needs to be aware of these practical difficulties as well as the principles of effective stress, which controls soil and rock behaviour in many instances. This is addressed in Chapter 6.

7. UNCERTAINTIES AND RISKS

Finally, the engineering geologist needs to understand that when he presents a model of the ground, defined by mapping and sub-surface investigation, there are uncertainties, which leads to risk. He must be prepared to identify the risks clearly and forcibly to his engineering colleagues, bearing in mind the fourth attribute of an engineering geologist according to Burwell & Roberts (1950). Balance is required before shouting alarm. An example is given of the 2nd site investigation for the major suspension bridge in Turkey over the Izmit, carried out to test a structural theory, missed on the 1st investigation in Chapter 4. Other examples are given in Chapter 8 of where projects have failed, with the concept of the three verbal equations used to focus them. Finally, a practical risk assessment is given for a tunnel in South Korea in Appendix D and a risk register example provided from the Channel Tunnel in the same Appendix.

The engineering geologist should be able to identify soil and rocks by visual examination and to interpret the geological history and structure of a site as set out in Chapter 4. He needs to be able to describe these rocks and soils both geologically and, in an unambiguous way, in terms of their relative strengths. He also must have knowledge of geomorphological processes, be able to interpret terrain features and hydrogeological conditions. Principles of hydrogeology are dealt with in Chapter 5.

Engineering geologists can often make important contributions at the beginning of a project in outline planning and design of the investigation for a site. He should be familiar with the ground investigation techniques that are available and that suit the ground conditions, so that a site can be characterised cost-effectively and thoroughly, as set out in Chapter 6. He also needs to understand the way that soils and rocks behave mechanically under load, in response to fluid pressures and chemically and how to investigate their properties (Chapter 7). Much of this will not be taught in an undergraduate degree and needs to be learnt through MSc studies or through Continuing Professional Development (CPD), including self-study and from experience gained "on the job".

The better-trained the engineering geologist and the more he has learnt through experience, the more he will be able to contribute to a project as illustrated schematically in Figure 1.1. At the top of the central arrow, interpreting the geology at a site in

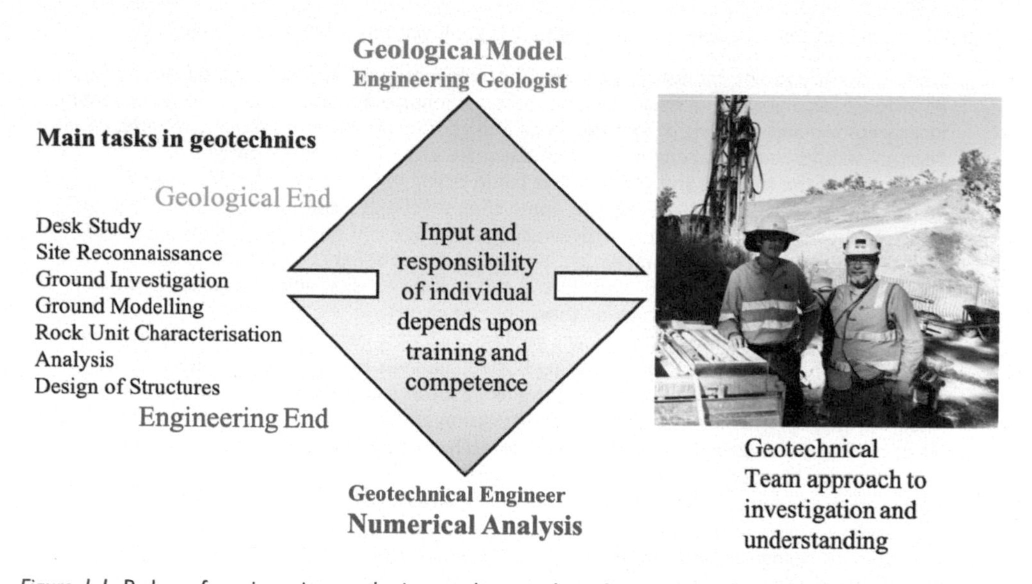

Figure 1.1 Roles of engineering geologists and geotechnical engineers. The prime responsibilities of the engineering geologist are "getting the geology right" (according to Fookes, 1997) and "assessing the adequacy of investigation and its reporting" (according to Knill, 2003) but an experienced engineering geologist with proper training can go much further, right through to the full design of geotechnical structures. Similarly some geotechnical engineers become highly knowledgeable about geology and geological processes through training, study and experience and could truly call themselves engineering geologists. The photo shows engineering geologist, David Starr and geotechnical engineer and numerical modeler, Benoit Wentzinger of Golder Associates, Australia, working in a team to investigate a major landslide west of Brisbane, which is described in Chapter 8.

terms of its geological history and distribution of strata is a job best done by a trained geologist. At the bottom end of the arrow, numerical analysis of the ground-structure interaction is usually the province of a geotechnical engineer – a trained civil engineer who has specialised in the area of ground engineering. There are however many other tasks, such as design of ground investigations and numerical modelling, that could be done by either an experienced engineering geologist or a geotechnical engineer. Many professional engineering geologists contribute in a major way to the detailed design and construction of prestigious projects, such as dams, bridges and tunnels, and have risen to positions of high responsibility within private companies and government agencies.

1.3 THE ROLE OF AN ENGINEERING GEOLOGIST IN A PROJECT

1.3.1 General

As discussed and illustrated later, some sites pose major challenges because of adverse and difficult geological conditions but the majority do not. This leads to a quandary. If a one-size-fits-all, standardised approach is taken to site characterisation and more

Table 1.2 Particular contributions that an engineering geologist might bring to a project

1. Unravelling the geological history at a site. This will come initially from regional and local knowledge, examination of existing documents, including maps and aerial photographs and the interpretation and mapping of exposed rock and geomorphologic expression. In most cases, geology should be the starting point of an adequate ground model for design.
2. Prediction of the changes and impacts that could occur in the engineering life time of a structure (perhaps 50 to 100 years). At some sites severe deterioration can be anticipated due to exposure to the elements with swelling, shrinkage and ravelling of materials. Sites may be subject to environmental hazards, including exceptional rainfall, earthquake, tsunami, subsidence, settlement, flooding, surface and sub-surface erosion and landsliding (with predictions running into hundreds or thousands of years).
3. Recognising the influence of Quaternary geology, including recent glaciations and rises and falls in sea level; the potential for encountering buried channels beneath rivers and estuaries.
4. Identifying past weathering patterns and the likely locality and extent of weathered zones.
5. Ensuring appropriate and cost-effective investigation and testing that focuses on the important features that are specific to the site and project. These should include aspects that may be important to the contractor for his construction.
6. Preparation of adequate ground models, including groundwater conditions to allow appropriate analysis and prediction of project performance.
7. An ability to recognise potential hazards and "residual risks"[4] even following high quality ground investigation.
8. Identification of aggregates and other construction materials.
9. Safe disposal of tailings and wastes, including nuclear waste (over many thousands of years).
10. Regarding project management, he should be able to foresee the difficulties with inadequate contracts that do not allow flexibility to deal with poor ground conditions if they are encountered.

particularly to ground investigation (Chapter 6) then much time and money will be wasted on sites that do not need it but, where there are real hazards, then the same, routine approach might not allow the problems to be identified and dealt with. That is when things can go seriously wrong. Civil engineering projects sometimes fail physically (such as the collapse of a dam, a landslide or unacceptable settlement of a building) or cost far more than they should because of time over-runs or litigation. Often, in hindsight, the root of the problem turns out to be essentially geological. It is also commonly found that whilst the difficult conditions were not particularly obvious, they were not unforeseeable or really unpredictable. It was the approach and management that was wrong (Baynes, 2007). Chapter 8 provides a number of case studies with the causes of the problems set out systematically.

A skilful and experienced engineering geologist should be able to judge from early on what the crucial unknowns for a project are and how they should be investigated. Typical examples of the contributions that he might make are set out in Table 1.2.

1.3.2 Communication within the geotechnical team

The engineering geologist will almost always work in a team and needs to take responsibility for his role within that team. If there are geological unknowns and significant hazards he needs to make himself heard using terminology that is understood by his engineering colleagues; the danger of not doing so is illustrated by the case example of a slope failure in Box 1.1.

BOX 1.1 CASE EXAMPLE OF POOR COMMUNICATION WITH ENGINEERS

The investigations into a rock slope failure at South Bay Close, Repulse Bay, Hong Kong are reported by Hencher (1983a), Hencher et al. (1985) and by Clover (1986). During site formation works, almost 4,000 m^3 of rock slid during heavy rainfall as shown in Figure B1.1.1. If it had happened some years later, after construction of the planned high-rise apartment blocks at the toe, then fatalities would have been highly likely. The slide occurred on a well-defined and very persistent discontinuity dipping out of the slope at about 28 degrees. The lateral continuity of the wavy feature is evident to the left of the photograph, beneath the shotcrete cover, marked by a slight depression and a line of seepage points.

A series of boreholes had been put down prior to excavation and the orientation of discontinuities had been measured using impression packers (Chapter 6). Statistical analysis of potential failure mechanisms involving the most frequent joint sets led to a design against shallow rock failures by installation of rock bolts and some drains. The proposed design was for a steep cutting with the apartment blocks to be sited even closer to the slope face than would normally be allowed. Unfortunately, the standard method of discontinuity analysis had eliminated an infrequent series of discontinuities, daylighting out of the slope and on one of which the failure eventually occurred. Pitfalls of stereographic analysis in rock slope design was written by Hencher (1985) to illustrate the problem following this near-disaster.

Examination of the failure surface showed it to be a major, persistent fault infilled with clay-bounded rock breccia about 700 mm thick and dipping out of the slope (Figures B1.1.2 and B1.1.3). In the pre-failure borehole logs the fault could be identified as zones

Figure B1.1.1 View of large rock slope failure in 1982 at South Bay Close, Hong Kong. Man, for scale, to the right-hand side. Note water flowing across the wavy surface of the fault.

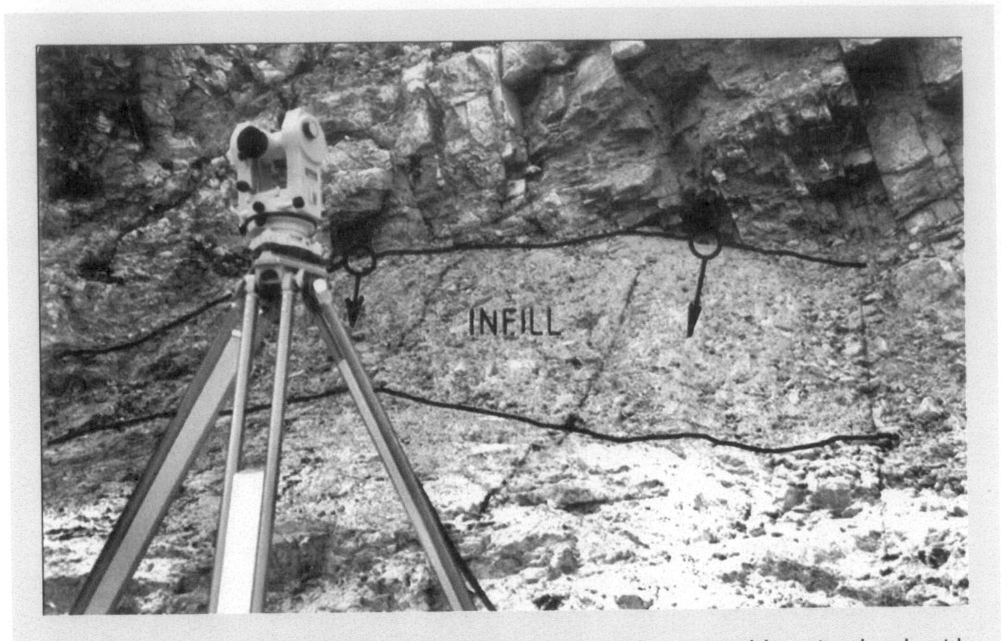

Figure B1.1.2 Exposure of up to 700 mm vertical thickness of compacted, brecciated rock, with clay infill, on the exposed thrust fault. Natural water pipes can be seen issuing above the feature.

of particularly poor core recovery; the rock in these zones was described as "tectonically influenced" at several locations. In hindsight the fault had been overlooked for the design and this can be attributed to poor quality of ground investigation and statistical elimination of rare but important discontinuities from analysis as discussed earlier but exacerbated by poor communication. The design engineers and checkers might not have been alerted by the unfamiliar terminology "tectonically influenced" used by the logging geologist; they should have been more concerned if they had been warned directly that there was an adversely-oriented fault dipping out of the slope. The feature was identified during construction but failure occurred before remedial measures could be implemented (Clover, op cit).

The slope required extensive stabilisation with cutting back and installation of many ground anchors through concrete beams across the upper part of the slope and through the fault zones (Figures B1.1.4 & B1.1.5). The site as in 2010 is shown in Figure B1.1.6. These anchors will need to be monitored and maintained continuously for the lifetime of the apartments.

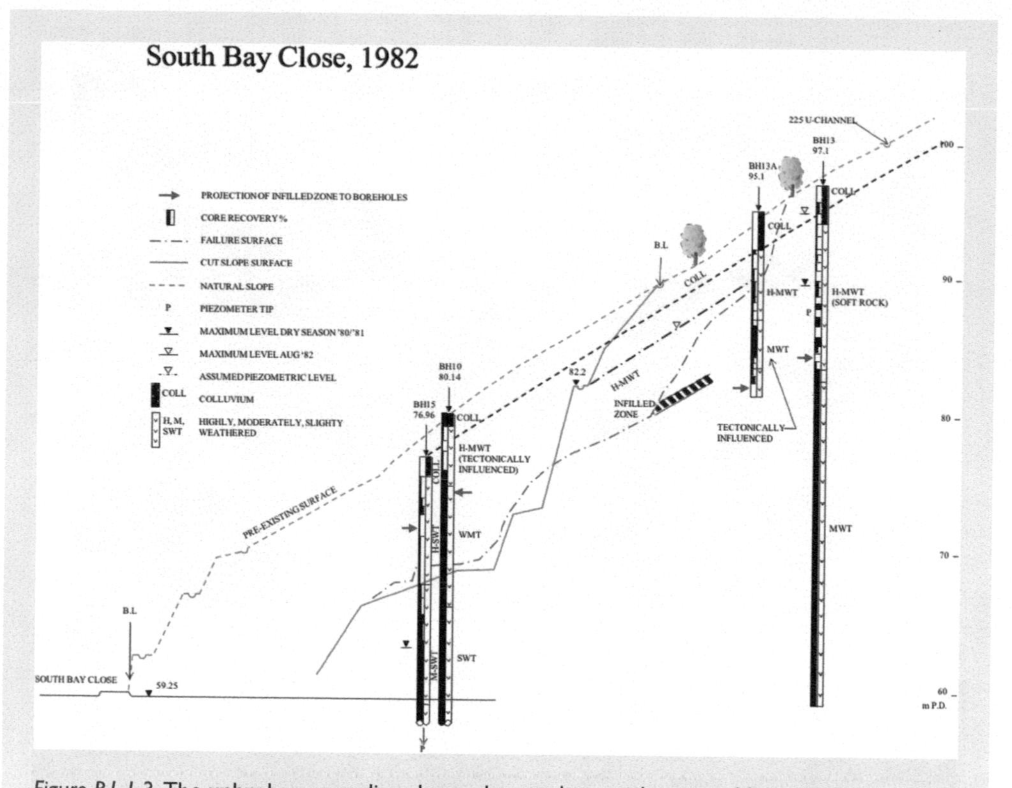

Figure B1.1.3 The unbroken green line shows the cut slope at the time of failure; the red dashed line shows the profile after failure. The likely water table is marked as a dark blue dashed line about 5 metres above the failure surface. A zone of infill made up of compacted brecciated rock is shown running along the thrust fault, rising to the right. Four boreholes are indicated, with the fault shown by red arrows (described as "tectonically influenced"). As discussed in the text, geotechnical engineers, following practice set out in Hoek & Bray (1975), contoured the discontinuity data on a stereographic projection, and statistically removed the few points depicting the fault, on which the failure later happened.

Figure B1.1.4 Engineers standing on upper berm where concrete beams have been erected with anchors, some of which have load cells to allow remote monitoring of the force applied.

Figure B1.1.5 Aerial shot showing the extent of the anchored beams, designed to support the rock above the fault.

Figure B1.1.6 View of the finished apartment blocks, with the slope and anchored beams above, taken in 2010, almost thirty years after the failure. The anchors will need to be maintained and replaced when necessary for the lifetime of the apartments.

Inadequate site investigation that fails to identify the true nature of a site and its hazards can result in huge losses and failure of projects. Similarly poorly directed or unfocussed site investigation can be a total waste of time and money whilst allowing an unfounded complacency that a proper site investigation has been achieved (box ticked). The engineering geologist needs to work to avoid these occurrences. He needs to be able to communicate with the engineers and to do that he needs to understand the engineering priorities and risks associated with a project. Those risks include cost and time for completion. This book should help.

1.4 ROCK AND SOIL AS ENGINEERING MATERIALS

Geologists are taught to describe, classify and interpret rocks as aggregates of different minerals in terms of their geological origin – sedimentary, igneous and metamorphic (Chapter 4). All naturally occurring materials are called "rocks" by pure geologists, whatever their state of consolidation, origins or degree of weathering (Whitten & Brooks, 1972). In civil engineering things are very different. Geological materials need to be split into soils and rocks essentially on differences in mechanical behaviour and especially on strength and deformability. To make it more difficult, the definitions of what is "soil" and what is "rock" may vary according to the nature of the project. For example in many senses, a "soil" may be defined primarily as material that falls apart (disaggregates) in water or can be broken down by hand but, for a large earth-moving contract, materials may be split into "soil" and "rock" for payment purposes according to how easy or otherwise the material is to excavate; "rock" might be defined as material that needs to be blasted or that cannot be "ripped" using a particular excavating machine. For engineering design, the distinctions are often pragmatic and there may be fundamental differences in approach for investigation and analysis. This is illustrated for slope stability assessment in Figure 1.2. In the left-hand diagram, the "soil", which might be stiff clay or completely weathered rock, is taken as having isotropic strength (no preferential weakness directions) albeit that geological units are rarely so simple. To assess stability, the slope is searched numerically

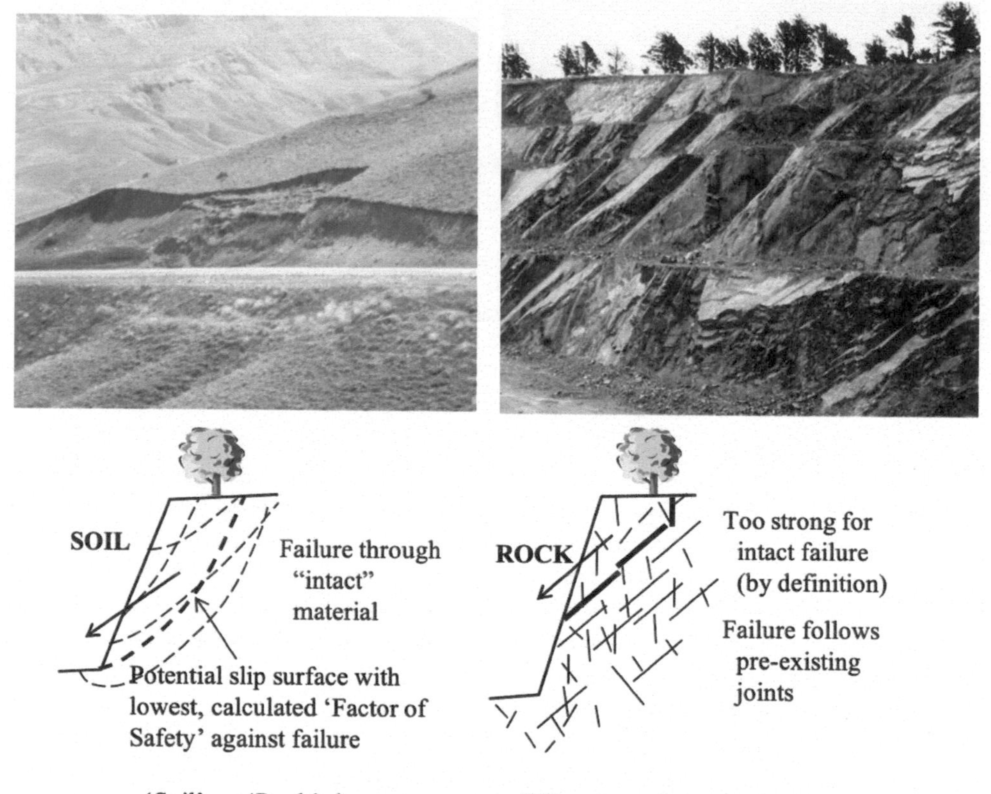

Figure 1.2 Distinction between "soil" and "rock" at a practical level for slope stability analysis. In Hong Kong, the distinction is at Grade IV to Grade III level, which is essentially the point at which a sample can be broken by hand. The soil failure illustrated is near Erzincan, Turkey. Analysis for design (or back-analysis) involves searching for the slip plane that gives the lowest FoS for the given strength profile. The rock slope is totally controlled by pre-existing geological structures (bedding planes and joints and the shear strength along them.

to find the critical potential slip surface, as explained in Chapter 4. In contrast the "rock" slope to the right is, by definition, made up of material that is too strong to fail through the intact material given the geometry of the slope and stress levels. In this case, site investigation would be targeted at establishing the geometry and strength characteristics of any weak discontinuities (such as faults and joints) along which sliding might occur. If an adverse structure is identified then the failure mechanism is analysed directly. This conceptual split is fundamental to all branches of geotechnical engineering, including foundations, tunnels and slopes, and it is important that the engineering geologist is able quickly to adapt to seeing and describing rocks and soils in this way.

The compartmentalisation of soil and rock mechanics is quite distinct in geotechnics with separate International Societies, which have their own memberships, their own publications and organise their own conferences. Text books deal with Soil Mechanics or Rock Mechanics but not the two together. There are separate methods of analysis for sites underlain by soil rather than rock. In reality this is a false or at least very fuzzy distinction and an unsatisfactory situation. Everyone working in geotechnical engineering (geologists and engineers) needs to appreciate that in nature there is actually a continuum from soil to rock and from rock to soil. Soils deposited as soft sediments in an estuary or offshore in a subsiding basin are gradually buried by younger sediments and change from soil to rock as they are compressed by the weight of overlying sediment and strong bonds are formed by cementation between mineral grains as illustrated in Figure 1.3. Conversely whilst rock such as granite is strong in its "fresh" state it can

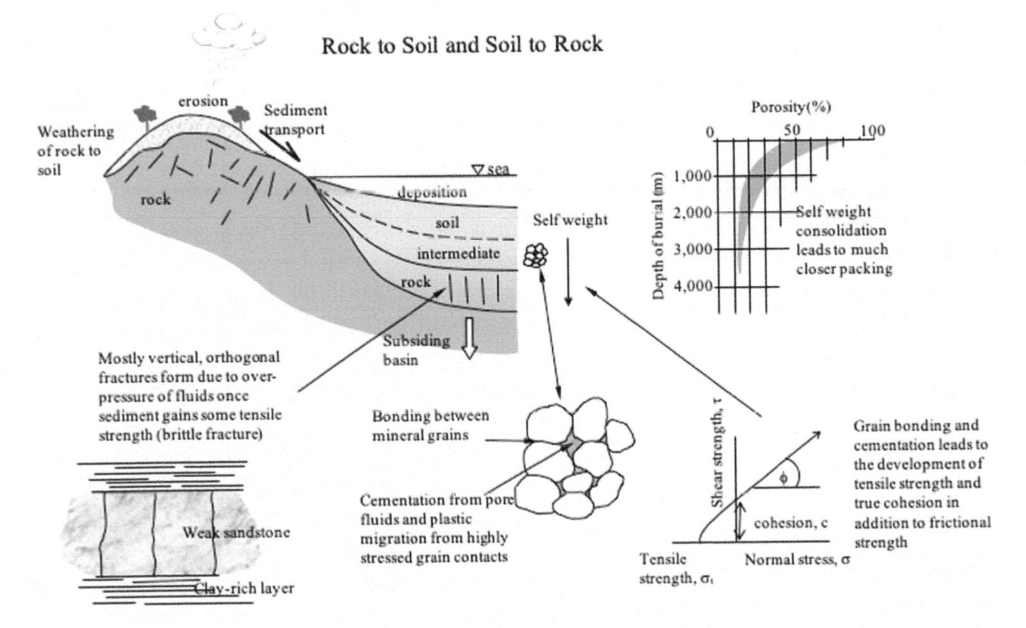

Figure 1.3 The cycle of soil to rock and rock to soil. Diagenetic and lithification processes cause soft sediment to transform into strong, cemented rock during burial. Exposed rock breaks down to soil by weathering.

be severely weakened by weathering as depicted in Figure 1.4 so that it might collapse or even flow into tunnels below the water table during excavation.

There are unifying concepts that apply to rock and soil as geotechnical materials (Morgenstern, 2000), which can be summarised as follows:

(i) All rocks and soils are porous and permeable and the concept of effective stress provides the fundamental basis for quantitative characterization.
(ii) All rocks and soils are normal stress dependent; strength increases with normal stress, stiffness increases with normal stress and permeability generally decreases with normal stress.
(iii) All rocks and soils are structure-dependent; for some, like homogeneous uniform clays, the structure is at a scale that can be characterized by the process of sampling and testing; for others, like a jointed, hard rock mass, the discontinuity fabric dominates behaviour and scale effects limit the role of sampling and testing.

An engineering geologist must be familiar with the full range of geological materials and understand the principles and methods of both soil and rock mechanics, which are tools to be adopted as appropriate, in design and construction and understanding behaviour.

1.5 PHILOSOPHICAL APPROACHES

One approach for thinking about ground engineering problems was proposed by Burland (2007) in which he emphasises the role of empiricism and "well-winnowed experience" feeding back from either laboratory testing or numerical modelling to the concept of the geological model. This concept is illustrated by the triangle in Figure 1.5. It describes a process by which we move from collecting data through ground investigation and testing (backwards and forwards), through to modelling the ground or construction, be it numerically or conceptually. This is all based on experience and precedent.

It is true that much of geotechnics must be considered on a precedent basis. I remember Professor Evert Hoek discussing the relationship between Rock Mechanics Rating[3] (which contains no estimate for Intact Rock Modulus, E_i) and the Rock Mass Modulus value (E_m) with Professor John Hudson at Leeds University. They agreed rather that "it all comes out in the wash".

Precedent is really useful throughout geotechnics, for example, for predicting settlement of buildings based on average Standard Penetration Test (SPT) value (Burbidge & Burland, 1985) but the outcome really depends upon *whose* experience is used to make the judgements, particularly over the ground model, and the testing adopted. As is ever the case, the more experience one has the better. My advice to an engineering geologist, even an experienced engineering geologist, is not to cut corners, but think through the problem, step by step, however tempting it might be to rely on your experience, which may not be all-encompassing.

The fundamental theme used throughout this book, encouraging thinking about a site holistically, is by using **three verbal equations** as outlined by Knill (1976). These equations provide a logical process, which is inclusive. These are essentially as follows:

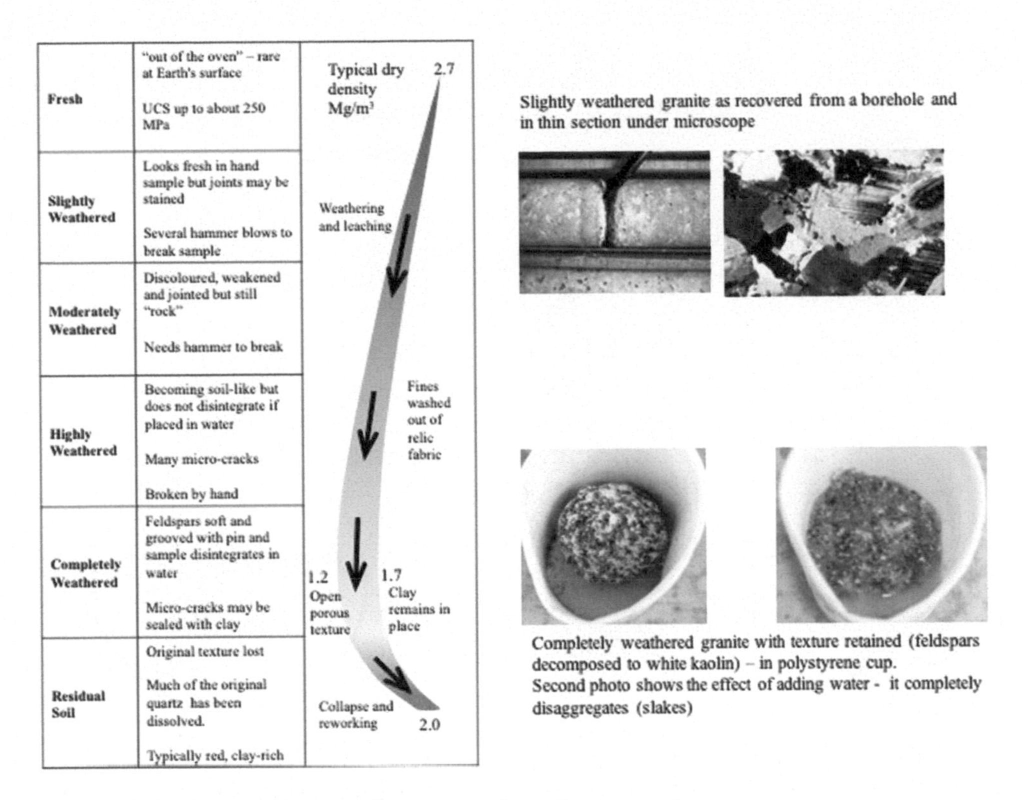

Figure 1.4 Typical stages of chemical weathering for an igneous rock.

The "geology" of a site + the environment including rainfall + the effects of the works themselves leads to the performance of the project.

These equations are set out in Chapter 2 and elsewhere in this book, in much more detail, and the consequences of missing something in the checklist, when things go wrong, are given in Chapter 8.

1.6 GEOLOGICAL MODELS

One of the important tasks of an engineering geologist is to present ground conditions as a simplified ground model that can be expressed either verbally or pictorially, or as a series of models (Fookes, 1997). Models should contain and characterise all the important geotechnical elements of a site that can affect construction or operations over future years (Baynes & Parry, 2022). Primary geological soil and rock units are usually further subdivided on the basis of factors such as degree of consolidation and strength, fracture spacing and style, hydrogeological conditions or some combination. Models must attempt to identify and account for all the natural hazards that might impact the site as illustrated schematically in Figure 1.6 for a new high-rise structure to be sited in a valley threatened by a nearby natural hillside. A simplified version of the ground model, integrated with the civil engineering structure, is used in calculations

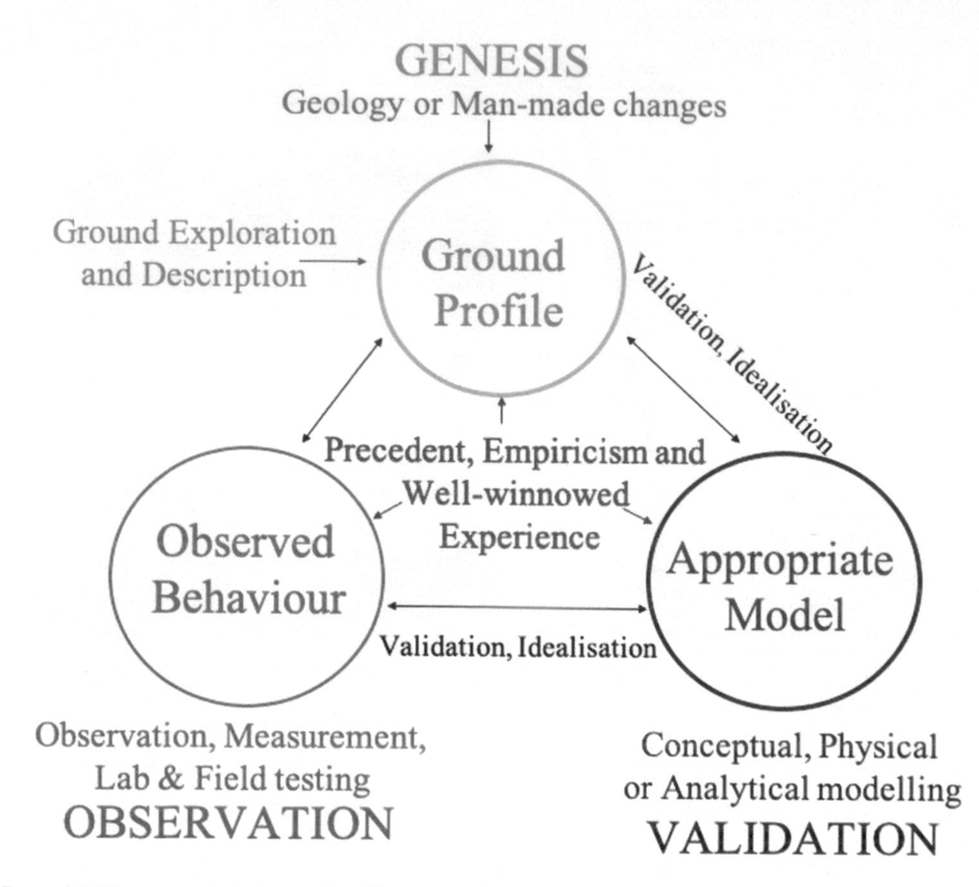

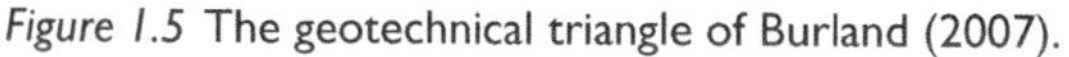

Figure 1.5 The geotechnical triangle of Burland (2007).

to ensure that the tolerance criteria for a project are achieved. For most structures the design criteria will be that the structure does not fail and that any settlement or deformation will be tolerable; for a dam the design criteria might include acceptable leakage from the impounded reservoir; for a nuclear waste repository it would be to prevent the escape of contaminated fluids to the bio-sphere for many thousands of years.

1.7 UNCERTAINTIES AND RESIDUAL RISKS

The problem with ground models, however apparently sophisticated, is that they are only models, with emphasis on some aspects or other. There are inherent uncertainties associated with them, as explained by Morgenstern (2000) and Hack et al. (2006). Fookes (1997) provides examples of ground models, which show how the same borehole information can be interpreted differently depending upon the person carrying out the work (quarry manager vs. mapping engineering geologist, who has just missed his plane, and worrying about the presence or absence of smectite). Furthermore, as more of a face is exposed and more or deeper or better boreholes are put down, so the models will (or should) change, inevitably, because more information is available.

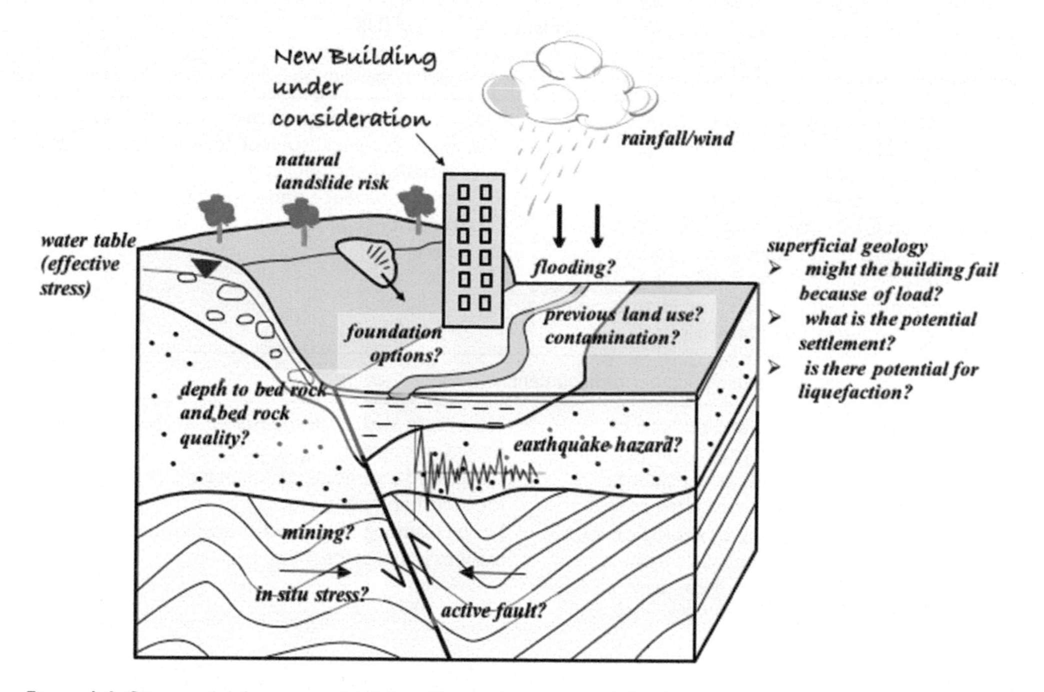

Figure 1.6 Site model for a new building illustrating some of the factors and hazards that need to be considered by the engineering geologist in preparing his model(s).

Inherently, the model will be uncertain at any stage, and the inherent risks need to be addressed by the engineer, in the knowledge that a model is only the best guess of the engineering geologist at a particular time and, to quote Hack et al. (2006), "if the engineer/geologist is good there will be a good model, or a poor model will result if the engineer/geologist is not so good".

One way of dealing with the inherent risks of the uncertainty is to over-design the works as described for a tunnel through the Himalayas (Hencher, 2015). As explained there, Jacob's ground model was necessarily patchy and knowingly limited, so the design of the Tunnel Boring Machine was to cope with the *worst* conditions that the Himalayas could offer (squeezing ground, collapsing ground at faults and so on). The tunnel was completed successfully by a very competent TBM contractor to everyone's satisfaction. The consequences of the engineers believing what the geologists tell them (inspired guess-work) about stability of a tunnel (in a very complex fault zone, which was not lined, and later collapsed, in Scotland) are described in Chapter 8 and in Hencher (2019).

1.8 QUALIFICATIONS AND TRAINING

Engineering geologists are most commonly firstly educated as geologists, later becoming engineering geologists through post-graduate training at MSc level together with experience. Within civil engineering, in the UK, Hong Kong, China and many other countries, there is a career pathway that is measured through achievement of "chartered status" or "registration" as a professional as summarised in Table 1.3.

Table 1.3 Typical routes for a career in geotechnical engineering (UK)

Engineering Geologist	*Geotechnical Engineer*
• First degree geology or other earth sciences (BSc) • MSc in engineering geology • 5+ years experience and training • Chartered Geologist (straight-forward route) – Geological Society of London • Chartered Engineer (more difficult route) – Institution of Civil Engineers or Institution of Mining, Metallurgy and Materials	• First degree civil engineering (BEng or MEng) MSc in geotechnical subject (e.g. soil mechanics or foundation engineering) • 5+ years experience and training • Chartered Engineer (Institution of Civil Engineers)
Distinctive skills at an early stage in career development	
• Knowledge of the fabric and texture of geological materials and geological structures and how these will influence mechanical properties (more so for rock than soil) • Observation and mapping of geological data • Interpreting 3-D ground models from limited information following geological principles • Identifying critical geological features for a ground model	• Numerate with sound basis for analysis and the design of engineering structures • Good understanding of mechanics (more so for soil than rock) • Understanding of project management and business principles

The aim is that works should only be conducted by those with adequate training, knowledge and experience – this is termed "competence". This is usually achieved through a first, recognised degree followed by a period of training under the supervision of a senior person within a company. Chartered or registered status is often a requirement for advancement of individuals and the practice of engineering is often legally defined and protected by government regulatory bodies. In some countries only registered or chartered engineers are permitted to use the title and to sign engineering documents (reports, drawings and calculations) thus taking legal responsibility. Details for career routes for various countries are set out in Appendix A together with details of Professional Institutions that an engineering geologist might aspire to join and links to a number of learned societies.

Notes

1 For "he" read "he/she" throughout the book.
2 I know that there are many engineering geologists who work in other industries, such as mining or landfill design, and this book is aimed equally at them.
3 There is a classic, much used empirical relationship between Rock Mass Rating of Bieniawski (1989) and Rock Mass Modulus, E_m derived originally from observation of the performance of dams by Serafim & Pereira (1983).
4 Any investigation will only allow sampling and testing of a small selection of a site.

Possible features should be considered, especially where recovery in boreholes is poor, depending on the geological and anthropogenic history. The question should always be "What if?" (de Freitas, 1993).

Chapter 2

Introduction to the management of projects

2.1 MANAGEMENT: PARTIES AND RESPONSIBILITIES

2.1.1 The Owner/Client/Employer

All civil engineering projects have owners – otherwise known as the "Client" or the "Employer" (to the Engineer) because the Owner ultimately pays for all the works and employs the various parties involved in design and construction. The Owner normally engages architectural and engineering companies to advise him and to manage, design and construct the project in a cost-effective manner. Most projects are designed by a consulting engineer and built by a contractor. Under such "Engineer's designs" the design responsibility rests with the project designer. Other projects are described as "turnkey" or "design & build" where a contractor is commissioned to deliver the whole project or part of a project as a complete package. Such arrangements–"Contractor's design" – often apply to specialist parts of projects, such as piled foundations for a building or boring a tunnel. The typical relationships and tasks for a project, designed by a consulting engineer, are illustrated in Figure 2.1.

Sometimes the Owner may have in-house technical expertise sufficient to overview the project (as in a Government Department or large energy company) but rarely will he have the staff or experience to design, construct and/or supervise all aspects of a large civil engineering project, which might require a huge range of skills – from site formation through numerical analysis to mechanical and electrical fitting out.

2.1.2 The Architect and Engineer

Engagement of an Architect and Engineer may be through competitive tender whereby several capable consulting companies are invited to make proposals for design and possibly supervision and for the cost control of construction and to give a price for carrying out this work. The Owner will select and contract with one party or with a consortium of consultants known as a Joint Venture (JV), which might be a grouping of specialist architectural, structural, mechanical and civil/geotechnical companies, which have joined together specifically to win and work on the project. The JV will need well-organised internal management to ensure that roles, responsibilities and payments are all clear and adhered to. The price paid by the Owner may be a fixed lump sum (usually with different rates quoted for engineers of different seniority and

DOI: 10.1201/9781003348894-2

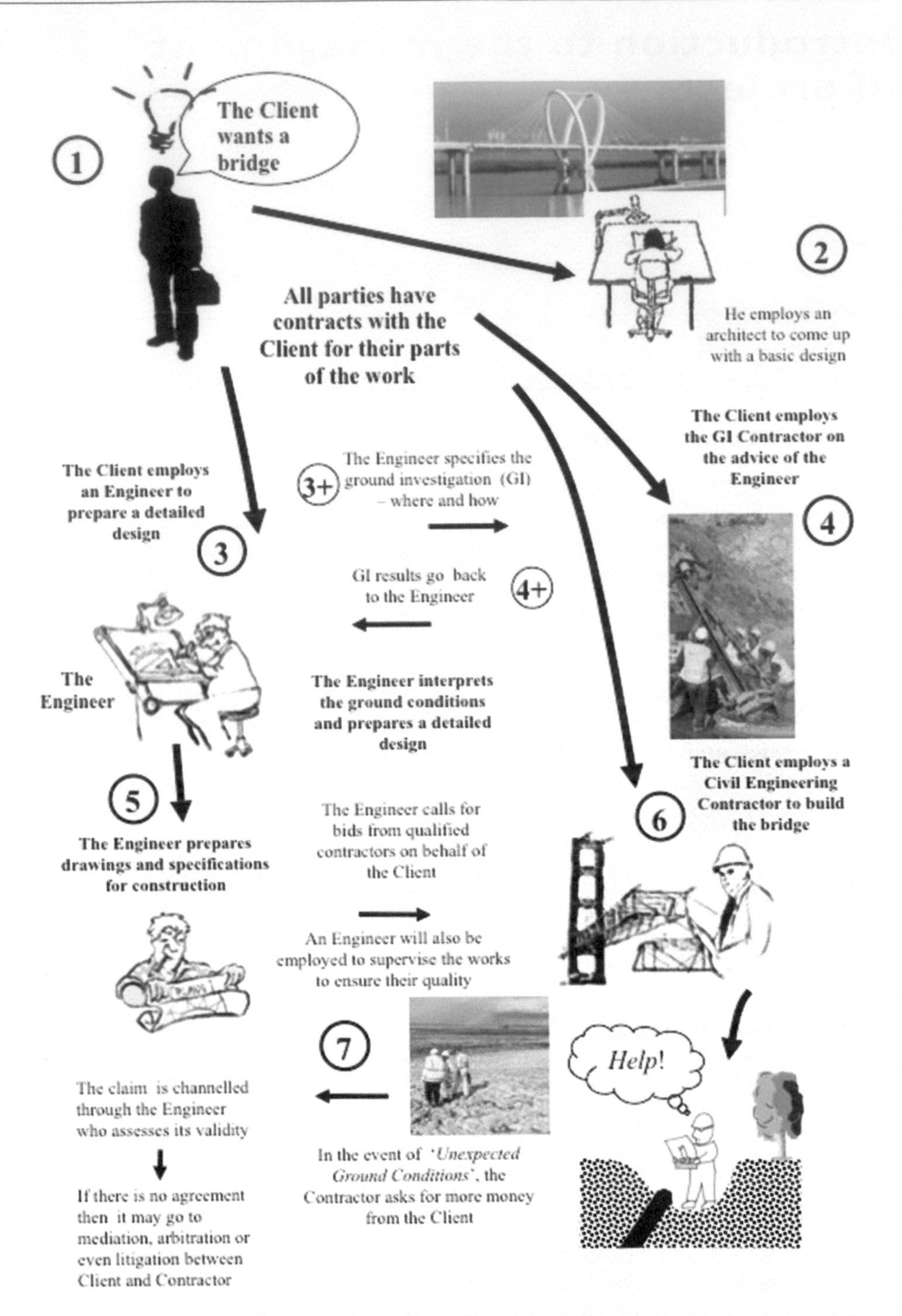

Figure 2.1 The Client wants a bridge. This figure illustrates various contractual arrangements and relationships between the main parties in an "engineered design" – one where the project is designed by a specialist design engineer and built by a specialist contractor.

expertise within the consultant organisation) or on a time charge with an agreed ceiling estimate. The roles of Architect and Engineer are legal entities with responsibilities often defined by Building Regulations within the country where the project is to be constructed. An individual within the company responsible for design may be named as an Approved Person, Architect or Structural Engineer and may be required to sign drawings and formal submissions to Government or other checking organisations.

2.1.3 The Project Design

The Engineer (and architect) plans the works, specifies investigations and designs the structure. The design is usually presented as a series of drawings, including plans and cross sections (elevations) to scale with details of what the Contractor is to construct and where. This will normally include an overall site plan showing, for example, the location of all foundation works – piles, pads or other features. Drawings are accompanied by Specifications for how the construction is to be carried out – for example, the strength of concrete to be used and any restrictions, such a prohibition on blasting because of proximity to buildings. Method Statements will be submitted later (explaining how the Contractor proposes to carry out parts of the work) and Programmes (dates for completion of the various activities making up the works) submitted by the Contractor commissioned to construct the works (see below) to the Designer for his approval.

A Project Director and Project Manager will be identified, within the consulting engineers, who will take responsibility for the successful management and "delivery" of the project. The measures of success are not only delivery of the project to the satisfaction of the Owner but also to make a profit for the design company and to meet internal requirements of the company, which includes staff development and training. When making a bid to the Owner, the price is normally estimated by calculating the staff cost input to produce the design, and then adding a margin, which might be 100 to 200%. This margin would cover overheads such as office support and infrastructure, general company costs plus actual profit for the shareholders in the company. Whereas the mark-up on staff costs might seem high, actual profit margins for most UK design consultants, once all costs are taken into account are often less than 10%.

The Engineer is in a very responsible position, as he will plan any site investigation, seek tenders from contractors to carry out all tasks and works and make recommendations to the Owner regarding which contractors he should employ. He will take the site investigation data, design the works and probably supervise the works although sometimes this is let as a separate contract or conducted in-house for consistency between separate sections of an on-going project. This is the practice of the Mass Transit Railway Corporation in Hong Kong, for example. During construction, the Engineer will usually employ or nominate a Resident Engineer and other resident site staff who will deal with the construction on site, on a day-to-day basis. The site staff will refer any needs for design changes as the works progress back to the Engineer's design office for resolution.

2.1.4 The Contractor

Various contractors may be employed for the works. Contractors are usually invited to bid to carry out works as set out in drawings, specifications and a Bill of Quantities (BOQ), which lists the works to be done and estimated amounts (e.g. volume of material to excavate). The Contractor puts a price against each item in the BOQ and the sum of all the itemised costs will constitute his offer to the Owner for completing the works. Generally a specialist ground investigation contractor will be employed to carry out sub-surface investigation of the site following a specification for those works by the Engineer. That specification will include locations and depths of sampling, types of testing and the equipment to be used (Chapters 4 & 6). Other contractors will be used to conduct and construct the various facets of a project.

Contractors, like engineers, need to ensure that they allow for some degree of profit. When the Engineer assesses the various tenders, on behalf of the Owner, he needs to be cautious that any particularly low bid is not unrealistic (which he would normally do by comparing with his own broad estimate of what the cost might be). A particularly low bid might mean that the Contractor has misunderstood the scope of the works and whilst the low price might be attractive to the Owner, quite often such situations end up in conflict or dispute with the Contractor desperately trying to compensate for his underestimation of the costs involved. Alternatively the Contractor might be trying to win or maintain market share at a time of high competition, so his bid has a deliberately low profit margin. A third possibility is that the Contractor already has in mind ways to make claims for additional payment, especially if the contracts are not well drafted as discussed later. The Engineer may recommend that the Owner does not accept the lowest tendered offer because of these various concerns and some countries and Governments have rules and methods in place for trying to eliminate unrealistic bids and ensuring that the most suitable contractor is employed.

Sometimes the Contractor might identify some better or more cost-effective way of carrying out part or all of the works and can offer this as an "alternative design" to that presented in the tender documents (the "conforming design"). The Owner might accept this proposal because of price, programme or for quality reasons. The Contractor (and his designer) might then take over responsibility for future design works and the Owner may employ another engineer to check these designs.

The Contractor may sub-contract parts of the works – for example, by employing a specialist piling sub-contractor to construct that element of the foundations. Whilst for a normal, engineer design project the consulting engineer is responsible for overall design, the Contractor may need to design "temporary works" necessary as intermediate measures in achieving the final design intent. For example, to construct a deep basement, the Contractor may have to design some shoring system to support the excavation until the final walls and bracing slabs of the final structure have been completed. Temporary works should normally be designed to the approval of the Engineer. In some instances some of the Permanent Works are designed by the Contractor or the Temporary Works somehow incorporated within the Permanent Works because to remove them might be too difficult or it is otherwise beneficial to do so.

Contractor's designs are sometimes adopted for parts of a project because of his local and specialist technical experience together with his knowledge of the costs of material, plant and labour. Another advantage is that there may be less ambiguity in terms of who is responsible for the performance of the works and in particular dealing with problems posed by difficult ground conditions. When it comes to foundations or tunnels, the Contractor should be in a position to accept the risk of any "unforeseeable ground conditions" – providing he is allowed to design and conduct an adequate ground investigation to his own specification.

2.1.5 Independent Checking Engineer

For many large projects, an independent checker is employed by the Owner to give added confidence that the design of permanent and temporary works is correct. The checker is usually a similar type of company to the design company, i.e. an engineering consultant. The check could be confined to a simple review of design assumptions and calculations but, in some instances, might involve a comprehensive and separate analysis of all aspects of a project.

2.2 MANAGEMENT: CONTRACTS

Civil engineering is a commercial business and the engineering geologist needs to understand how it works. The relations between all parties are governed by contracts. A contract is a legal document between the Owner and each of the other parties involved with a project and defines the scope and specification of works, including payment schedules and responsibilities. Contracts also need to be made between consultants and specialist sub-consultants or JV partners and between a contractor and specialist sub-contractors. It is very wise to use lawyers at this stage to ensure that contracts are well written to minimise the risk of later dispute although "standard forms of contract" are often used and large companies tend to have internal documents. The experienced engineering geologist can help ensure that contracts are reasonable, realistic and fair with respect to their treatment of ground conditions, which is where many problems arise during construction. These problems need to be resolved in a pragmatic manner and quickly during construction but there is often some dispute at a later stage over which party should pay for changes, additional costs and delays.

2.2.1 Risk allocation for geotechnical conditions

As discussed later, sites vary geotechnically, from those that are extremely difficult to understand and characterise to those that are simple and straightforward. In a similar fashion, site investigations vary in quality from focused, excellent and insightful to downright useless depending on the experience, capability and insight of the Engineer and his team planning and interpreting the investigation and the skill and quality of equipment of the ground investigation contractor. As a result there are always risks involved in projects especially where these involve substantial ground works, for example, in tunnelling or deep foundations. The risks need to be assigned under a contract and there are few mandatory rules. Each contract should state how "variations" are to be dealt with in the event of unforeseen ground conditions such as stronger or weaker rock (requiring different excavation techniques) or more water flow into a tunnel (requiring additional ground treatment works) than had been anticipated. This is a large and important subject and guidance on how to identify critical ground conditions through a systematic approach for addressing hazards and risks using focussed site investigation is presented in Chapters 4 & 6 and Appendix D. Chapter 7 takes this further and provides case examples of projects where things went wrong for some reason or other.

Some of the background and options for preparing a contract with respect to ground hazards are illustrated in Figure 2.2. Mostly projects use standard contract forms such as the New Engineering Contract (NEC) (ICE, 2005) or FIDIC (discussed by Tottergill, 2006). Some contractual forms are suitable to Engineer-design contracts and others to design-build situations.

In some forms of contract the Owner accepts all the ground risks and that makes some sense in that it is his site with all its inherent geological and environmental conditions. This kind of contract works quite well for simple sites and structures, for example, the cutting of a slope with the installation of soil nails, where the work done by the Contractor is routine and can be simply "re-measured" against the provisional BOQ priced by the Contractor when he tendered to do the work. If he excavates 2,300 m^3 of soil and 52,050 m^3 of rock during the contract, then that is what he will be paid, at the prices he originally quoted for each type of excavation, although there might be some disagreement over the definition of "soil" and "rock" by the parties. Specialist quantity surveyors (QS) assess and recommend approval of such payments to the Engineer and then on to the Owner.

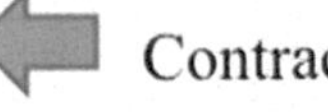

Contractor is paid the cost of completing works in full. Disadvantage is that there is no incentive for the contractor to resolve problems cost-effectively when they arise.
Client takes all risks
Some compromise alternatives:
✓ agreed reference ground conditions at start
✓ clause allowing additional payment
✓ partnering (open book) allowing both gain or loss to both parties if conditions are better or worse than anticipated
CONTRACTOR'S RISK DUE TO UNEXPECTED GROUND CONDITIONS
A disadvantage is that the contractor is unable (or unwilling) to price the risks with any certainty. Can go badly wrong (see Chapter 7).
Contractor takes all risks

Figure 2.2 The main options for forming a contract to deal with the risk of unexpectedly difficult ground conditions.

In an attempt to make it clear-cut where the responsibilities lie, some owners use contracts that place all the risks for ground conditions solely on the Contractor but this is inflexible and offers no way out when things go wrong. In practice, depending on commercial pressures, the Contractor may take a serious gamble (sometimes without fully weighing up the risks) and it is then, when things start becoming difficult such as when the ground conditions are worse than expected, that claims begin to be made and disputes can follow. Even where all the risk has been accepted by the Contractor, when things become very difficult, he and his lawyers may try to use clauses in the contract such as claiming that the works were physically or commercially impossible or just give up on the project. The arguments can be long and extremely costly for all parties. Such contractual arrangements are rarely used these days for major projects but still crop up from time to time.

For more complex projects and especially for constructions underground, the Contractor usually accepts some of the risk. The Contractor, unlike the Owner, is in the construction business, is a specialist in the particular type of works he is to undertake and may be able to spread the risk over a number of contracts to some degree (Walton, 2007). In order to get the Contractor to accept some of the risk of encountering difficult conditions however the Owner must expect to pay some additional sum to cover that insurance element through a higher contract price; if the risks do not materialise he will have wasted money but that is the nature of insurance.

In shared risk contracts the Contractor is expected to accept and cope with generally variable but predictable conditions but allowed to claim for additional money where something unpredictable and highly adverse is encountered. Despite the "pressure release valve" of old ICE Conditions of Contract Clause 12 (payment for unexpected ground conditions) and similar clauses in other standard forms of contract, it is in all parties' interests that all hazards and risks are foreseen and priced for by the Contractor in terms of the extra work and delay which will occur if the risk materialises. This is definitely the province where the engineering geologist can play a major role and in particular by engineering geologists working within the Engineer's consulting team, which is responsible for investigating the site and designing and specifying the works. There is a similarly important role for engineering geologists within the tendering contracting company who must anticipate hazards and price the job sensibly.

Unfortunately, contractors sometimes fail to take account of all the perceived risks (even where aware) partly because they know that the Owner (advised by the Engineer) will be tempted to employ the contractor offering the lowest price. There are Machiavellian aspects to all this in that each party is trying to minimise its costs and risks whilst maximising profit. Contract writing and interpretation are key parts of this. For example, a contractor will try to predict where extra quantities might be required during construction compared to the estimates by the Engineer that will form part of the contract in the BOQ (for example, in the proportion or rock vs. soil to be excavated) and quote unit prices appropriately to maximise his profits. He might include high mobilisation charges whilst trimming prices of other items on the bill to improve the payment schedule and his cash flow without jeopardising his chance of winning the contact in competition with other invited tendering contractors. This is all fair and above board but it does mean that the conduct of a civil engineering contract can be rather fraught at times.

2.2.2 Reference ground conditions

It is now common for tunnelling works especially to try to set out some Reference Ground Conditions (presented in "Geotechnical Baseline Reports") that all parties buy into for contractual purposes before the works actually begin. For larger tunnelling contracts in the UK and increasingly elsewhere (Hong Kong, Singapore, Australia, New Zealand, Europe), it is now mandatory that the hazards and risks are assessed and managed in a consistent manner (British Tunnelling Society, 2003; Davis et al., 2023). This is also the general case for some standard contracts (FIDIC). This was introduced largely because insurance companies were receiving an increasing number of claims because of tunnelling projects going seriously wrong and were threatening simply to withhold contractor's insurance on such risky, poorly investigated, poorly thought-through and mismanaged projects (e.g. Muir Wood, 2000).

Unfortunately, in practice it is often not that simple to define engineering geological conditions in a distinct and unambiguous manner. If one tries to be very specific (say on the rock type to be encountered) then it would be relatively easy for the Contractor to employ an expert geologist at a later stage to dispute that rock description in detail and then to allege that the slight difference in rock type caused all the difficulties that followed (excess wear, higher clay content etc. etc. plus delays and general loss of productivity). Drafters of Reference Conditions sometimes resort instead to broad characterisation, perhaps using rock mass classifications such as Q or RMR (good rock, poor rock and so on) as introduced in Chapters 4 and 6 and Appendix B. The problem there is that each such classification is made up of a range of parameters, such as strength and fracture spacing, each of which can be disputed because geology is never that simple (or uniform). Furthermore, experienced persons can often draw very different conclusions from the same data set. Fookes (1997) reports an exercise where he asked two engineering geologists, familiar with rock mass classifications, to interpret the same sets of boreholes and exposures for a particular tunnel in terms of rock mass rating (RMR) and Q value. One came up with an RMR = 11 (extremely poor rock: immediate collapse); the other RMR = 62 (fair rock: no support required). The Q value interpretations were similarly quite different (extremely poor vs. fair rock). In this particular case the rock, slate, had incipient cleavage and the different opinions on classifications mostly hinged upon whether that cleavage was considered a joint set or not – the standards and guidance documents do not help very much in this regard as discussed in Chapters 4 and 6. The main point is that despite Reference Conditions being set out with good intentions of helping the Contractor price the job and avoiding dispute, there is no guarantee that this will be achieved.

It is the normal case that the extent of geological/geotechnical units and position and nature of faults, for example, are uncertain. The Geotechnical Baseline Report should present the best interpretation of the ground conditions by the designers and state any limitations and reservations (Essex, 2023). In doing so, the rationale should not be, somehow to outwit the Contractor contractually, but to allow the Contractor to select the right methods for construction, and to price and to programme his works adequately. Contractually the Reference Conditions should be just that – something to refer to when considering whether some adverse ground was anticipated or anticipatable by an experienced contractor given the available information. The Contractor will have been expected to consider the site in a professional manner that would include examining any relevant rock exposures, say in quarries, adjacent to the route. Many contracts require the Contractor to satisfy himself of the ground conditions at a site or along the route but it would rarely be practical for him to carry out his own ground investigation at tender stage (with no guarantee of winning the work) and often that constraint is accepted by an Arbitrator in any subsequent dispute. This is not always the case. For major projects, bidding contractors might do their own investigation. For example, for the tender competition for the Young Dong Mountain Loop tunnel in Korea, each bidding contracting consortium carried out extensive independent ground investigation before submitting their tender documents and price (Kim et al., 2001).

One point that follows is that it is very important for engineering geologists to keep good records throughout construction. These should be factual, with measurements, sketches and photographs using standard terminology for description and classification as introduced in Chapter 4. Quite often, especially for tunnels, the engineering geologist representing the Contractor will prepare sketches of ground conditions encountered together with engineering works installed (such as locations of rock bolts and instruments) and seek to get this agreed by the supervising team on a daily basis. This means that the basis for payment is clarified and, in the event of some contractual dispute later, there are clear records for all parties to review.

2.2.3 Claims procedures

Interestingly, when things become difficult during the works, because of poor ground conditions, the Contractor has to apply through the Engineer for extra money (ultimately to be paid for by the Owner). Now it is the Engineer's responsibility to act impartially within the terms of the contract having regard to all the circumstances. In like manner the Engineer's Representative on site and any person exercising delegated duties and authorities should also act impartially (ICE Conditions of Contract). In other standard contracts, in recognition that the Engineer is employed by the Owner, the Engineer is expected to act "reasonably" rather than impartially but nevertheless he is clearly expected to treat Contractor's claims in a proper manner with due regard to the contract and the actual situation. The Engineer can however find himself in a position of conflicting interest where the ground conditions that are causing the difficulty to the Contractor might and perhaps should have been recognised and dealt with by the Engineer's investigation, design and specification for the works (Dering, 2003). He might have to allow a claim by the Contractor in the knowledge that he himself is culpable because of poor ground investigation, modelling or design. Conversely he might resist a claim from the Contractor that later proves valid. In either case he may make himself liable to legal action from the Owner.

2.2.4 Dispute resolution

If a claim cannot be resolved between the Contractor and the Owner then the claim might be passed to a third party. The two parties can jointly appoint a technical expert to help resolve the issues through a process of "adjudication". It is a far less formal process than the alternatives. The appointment of an Adjudicator might be written into the original contract (as specified in the New Engineering Contract of ICE) and his decisions should be complied with. For larger projects, the parties might appoint an agreed panel of experts at the outset (the Dispute Board). The Contractor and Owner can call upon that panel to adjudicate on the validity of any claim that conditions were not as anticipated and that they had the consequences claimed. That at least leaves the decisions in the hands of experienced professionals acting in a reasonable way rather than lawyers whose knowledge of ground conditions and ground behaviour can be rather limited at the start of litigation – but whom can learn very fast in the author's experience. However, also from personal experience, it might prove extremely difficult or even impossible, within the allowable time, to educate a lawyer regarding the meaning and relevance of geological descriptive terms. One forgets sometimes that geology is a subject that is taught at GCSE, "A" and "S" levels as well as at degree and post-graduate levels, let alone all the experience that an individual must gain professionally. Box 2.1 provides an example.

BOX 2.1 THE PROBLEMS OF EXPLAINING GEOLOGICAL TERMS TO LAWYERS

There is a concept that all lawyers are extremely intelligent – they must be, to be paid so much, for what they do?

Well, yes and no. Very many barristers – KCs – are very perceptive and pick up on the jargon of geotechnical engineers and engineering geologists very quickly. But, they have not been trained in geology and geological engineering. They are learning it as they go. They would not immediately understand the difference between granite and granodiorite or between highly and moderately decomposed rock. They would not understand the principle of effective stress without it being drummed into them.

This example concerns the collapse of the Glendoe tunnel. A 71 metre length of tunnel, that had been driven through a fault zone, using a TBM, collapsed during the tunnel operations in a hydroelectric scheme (Hencher, 2019). The case was first adjudicated, and then it was heard by a judge, in a Scottish court and later, appealed by three further judges. The case and the findings from the appeals were in open court rather than in arbitration, and therefore the findings were published.

One particular observation in this highly technical case was that the appeal judges concentrated on a rock type described as "kakerite"[1] without questioning what the term meant. The engineering geologist in the field had been encouraged to distinguish fault rocks as either mylonite or kakerite according to the proforma for rock mass classification, prescribed for use in the tunnel. I didn't recognise the term without researching it, neither did a former structural geologist in the British Geological Survey, nor did a previous Chairman of the Geological Society Tectonic Group.

A simplified classification of fault rocks in the United Kingdom is typically based on grain size and on whether or not the rock shows foliation or not, as in Table B2.1.1.

Table B2.1.1 Fault rock classification based on Woodcock & Mort (2008)

Non-foliated rock	Foliated rock
Fault Breccia (broken and disrupted, angular, > 30% fragments of rock > 2 mm); by default, cohesive, i.e. not broken down readily by hand.	
Fault Gouge "Incohesive"* – can be broken down by fingers	
Pseudo-tachylyte (glass or devitrified glass)	–
Cataclasite (classified according to grain size)	**Mylonite** (classified according to grain size)

* Brodie et al. (2007) define "incohesive" in a structural geological context rather than a soil mechanical sense.

The term "kakerite" is not included in this classification of Scottish fault rocks, its origin being rather obscure in the USA and Swiss tunnelling literature. As far as I understand it "kakerite" can be readily erodible (as in Fault Gouge) or might not be erodible at all (as in Fault Breccia). The option for the engineering geologist was to describe any rock that was not a mylonite as "kakerite". I understand that the appeal judges considered the term to have been used to describe erodible rock (whereas I do not think that interpretation was correct because non-mylonite, fault rocks might be cemented).

Please read the paper to understand (Hencher, 2019), but in my view this is an example of legally trained people, trying to understand dubious geological terms in a partial classification that did not deal with all the encountered rock types, and unfortunately coming to incorrect conclusions. In this case, they translated the term "kakerite" to mean a rock that would be erodible by water and that should have been protected by a concrete liner. As stated in the paper, it exposes the difficulties in approaching geological matters and terms as if they were points of law and highlights the advantages of a dispute board, comprising three experienced professionals, appointed as agreed by the parties.

Mediation is an option where the parties to a dispute will plead their cases to an independent mediator (who might be a lawyer rather than a technical expert). He will try to get the parties to reach an agreement and will also provide an opinion as to the likely outcome if the matter is taken to the next, more expensive level. If a party (either the Owner or Contractor) is told by an independent mediator that their position over a claim is weak, then they may be more willing to reach an agreement with the other side.

Arbitration is a higher-level process and is generally written into contracts as a way of having disputes resolved. Both parties agree at the outset that this should be so and the location where any arbitration should be conducted. Arbitration takes place in a court-like setting and there may be up to three arbitrators – perhaps one agreed between both parties and the second and third chosen by each party independently. The cost, with lawyers (probably several on both sides), barristers, independent experts (see next section) and the court expenses, can be very high. In a recent case that the author was involved with, the final award to the "winning" party was essentially the same as had been previously offered in settlement, prior to arbitration, and was far exceeded by the cost of the legal proceedings. Over recent times, because of the infectious Covid virus, disputes have been settled remotely, with witnesses and experts being questioned online, without the benefit of any "bundle" of evidence to consult, which can be difficult where there is a considerable body of evidential documents. No doubt there are cost-benefits to the remote working, but there are difficulties where dealing with reams of minutes of meetings and design reports without ready access to them.

Arbitration decisions are generally accepted as final – however disgruntled one party might feel at the result. Arbitration reports and outcomes are generally kept confidential to the parties – unfortunately this means that the geotechnical profession does not learn the lessons, which is a great pity.

In some cases however, where some party considers the results of arbitration were clearly incorrect, then they may apply for the case to be appealed through higher courts, although this should not be allowed unless the arbitrator is considered to have been evidently wrong in his interpretation of the case. If an appeal is allowed there is again the problem that legally trained rather than technically trained people may have to try to interpret civil engineering descriptions and definitions that might well have been flawed in the first place (see Box 2.2). This can lead to the situation where the appeal judges are persuaded, by some superficially reasonable argument (following some standard), whilst not being fully aware of other background issues that should have been considered.

BOX 2.2 DIFFICULTIES WITH STANDARDS

GENERAL COMMENTS ON STANDARDS

Governments and associated standards offices (such as the British Standards Institution) prepare "Standards". These are produced to a perceived high level ("they are written by experts") that set out standards to be employed and that will hopefully be helpful to the practitioners. The geotechnical standards attempt to deal with geological and geotechnical variability, how things are to be described and classified and how properties might be measured. Unfortunately the Standards change from time to time as industry develops (as per BS 5930 and Eurocode 7 – see see Appendix B, Table B4 to see the difficulties). Again, unfortunately, these are set out as the "gold standard" and beware anyone who uses the terms (current) incorrectly (Norbury, 2017). This leads to the difficult situation where an author may have been using a particular standard (of a particular year), to describe and classify the nature of the rock encountered, as beaten into his psyche on an MSc course of a particular era, only for the logs to be interpreted by another engineering geologist (working on either an earlier, or later, or just different standard). I did actually warn of exactly such a situation in the first edition of this book (see page 366 of Edition 1) and have sadly seen the consequences of such misinterpretation.

The UK-based engineer working in a Middle Eastern country, had been given a set of borehole logs showing "weak" rock (logged according to ISRM, 1981). He and his companions interpreted them, however, as if they had been logged by someone using BS 5930:1999 (following his experience).[2] He thereby *underestimated* the strength of the rock by 400 to 500%. He then analysed the stability of slopes in the excavations. This led to the ridiculous and unnerving result that he found that the slopes were marginally safe. In truth the Factors of Safety were in excess of 5.0 because his assumed strengths were far too low.[3] The resultant slopes included anchors and diaphragm walls unnecessarily costing the contractor large sums of money.

HONG KONG PHOOEY

A second example is from Hong Kong. In Hong Kong ground investigations are (should be) carried out following Geoguide 2 (Geotechnical Control Office, 1987) and the description and classifications of rocks and soil from Geoguide 3 (Geotechnical Control Office, 1988).

The ground investigation industry in Hong Kong (HK) has developed to the point where, there is now considerable skill in recovering 100% of core in weak and weathered rocks and colluvium, by using a combination of tools such as double and triple tube barrels, Mazier sampling and foam flush. That said, in some situations there is core that is lost. This should simply be noted as core loss, but in Hong Kong there has developed the practice of interpreting (wrongly), the loss as "soil" – i.e. either Grade IV or V materials, whereas the length of core loss could just be broken rock or due to poor drilling practice. The geotechnical logs that are produced look impressive and comprehensive, but there are these flaws. Things are made even more difficult where the local Standards (or Code of Practice) for Foundations define core recovery of particular weathering grades as representative of particular bearing stresses.

Furthermore, there are particular problems regarding Standard Penetration Tests (as discused more fully in Chapter 6).

WEATHERING GRADE CLASSIFICATION BY SPT

Firstly it needs to be stated that there is **no** guidance regarding SPT "*N*" values of different weathering grades in Geoguide 3 and very little reference to SPT "*N*" values in the Engineering Geology of Hong Kong (Geotechnical Engineering Office, 2007). UK practice regarding SPT "*N*" values is presented in Clayton (1995) but there are no reliable measures to link ranges of SPT values to weathering grade descriptions.

There was, however, a comprehensive study carried out on SPT values in granite by the Geotechnical Engineering Office (Irfan, 1996). In this study he recorded completed SPT values (after full 450 mm penetration into the ground, the last 300 mm counting for the "*N*" value). He also had the advantage of high quality Mazier samples above and below the SPT test horizon, from which he measured dry density of the samples. This is about as good data as you can have, and Dr Irfan is a very good engineering geologist whose main research is on weathered rocks. The results were very revealing and are reproduced here as Figure B2.2.1.

It can be seen that Completely Decomposed Granite (Grade V which slakes in water) has approximate dry density between 1.2 and 1.8 Mg/m^3 and SPT "*N*" values ranging between 10 and 120. Highly Decomposed Granite (Grade IV in which samples can still be broken by hand but do not slake in water), showed SPT "*N*" values ranging from about 110 to about 150. Dry densities were about 1.6 to 1.85 Mg/m^3.

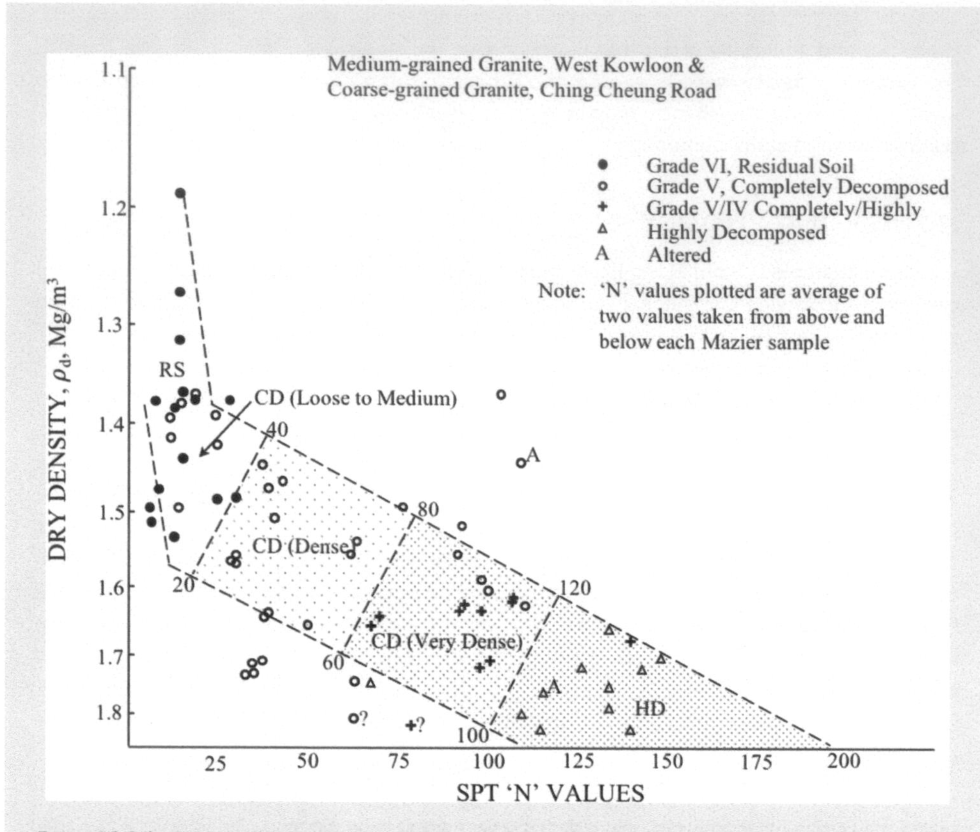

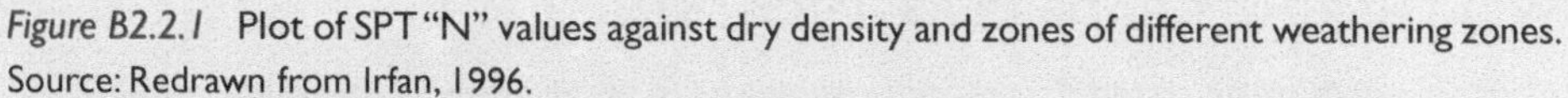

Figure B2.2.1 Plot of SPT "N" values against dry density and zones of different weathering zones. Source: Redrawn from Irfan, 1996.

FOUNDATION GUIDELINES (HKIE, 2004 AND 2017)

The geotechnical fraternity in Hong Kong clearly overlooked Dr Irfan's results and data in preparing their guidance document for Foundations. Part of Table 2.1 of HKIE, 2004, which gives allowable vertical bearing pressures for different categories of rock is reproduced here as Table B2.2.1.

There are concerns regarding definitions of strength and recovery in all categories, for example, in 1(d) does it mean that all of the rock in the ground, or only that recovered as core, must be of Grade III or better or only the 50% that is recovered? These definitions are tricky, have been rewritten in HKIE 2017, but the issues are still to be debated. Probably best to leave them alone for the moment.

What is important, and needs correcting immediately however, is Category 2 – "Intermediate soil". This is described as "Highly to completely decomposed, moderately weak to weak rock of material weathering Grade V or better with SPT N-value ≥200". This has been reproduced, word for word, in the 2017 version but as Category 3 rather than 2.

Now from Irfan's findings, Grade V is typified by SPT "N" values up to 120 and this is confirmed in Engineering Geological Practice in HK (Geotechnical Engineering Office, 2007). Grade IV has a range up to 150. There is absolutely no support for the statement in

Table B2.2.1 Presumed allowable vertical bearing pressure under foundations on horizontal ground

Category	*Description of rock or soil*	*Presumed allowable bearing pressure (kPa)*
1(a)	*Rock (granite and volcanic)* Fresh strong to very strong of material weathering grade 1, with 100% total core recovery and no weathered joints, and minimum uniaxial compressive strength of rock material (UCS) not less than 75 MPa (equivalent point load index strength PLI_{50} not less than 3 MPa)	10,000
1(b)	Fresh to slightly decomposed strong rock of material weathering Grade II or better, with a total core recovery of more than 95% of the grade and minimum uniaxial compressive strength of rock material (UCS) not less than 50 MPa (equivalent point load index strength PLI_{50} not less than 2 MPa)	7,500
1(c)	Slightly to moderately decomposed moderately strong rock of material weathering Grade III or better, with a total core recovery of more than 85% of the grade and minimum uniaxial compressive strength of rock material (UCS) not less than 25 MPa (equivalent point load index strength PLI_{50} not less than 1 MPa)	5,000
1(d)	Moderately decomposed, moderately strong to moderately weak rock of material weathering Grade better than IV, with a total core recovery of more than 50% of the grade	3,000
2	*Intermediate soil (decomposed granite and decomposed volcanic)* Highly to completely decomposed, moderately weak to weak rock of material weathering Grade V or better, with SPT N-value ≥200	1,000

Source: After Hong Kong Institution of Engineers 2004 Code of Practice for Foundations.

HKIE's table of "Completely to Highly Decomposed rock with SPT ≥200". SPT's ≥200 are typical of rock of Grades III and better.

My difficulty is that this contradiction needs to be explained to lawyers – one of the two interpretations is correct. Which one? In my view Irfan's is correct and the table (official production of the Hong Kong Building Regulations) needs to be revised.

QUALITY OF SPT'S AND THEIR INTERPRETATION, IN HONG KONG

Unfortunately, the situation is not made any better by the standards of ground investigation description of rock and soils in Hong Kong. SPT tests are designed for "soils" to give some rough idea of density of sand, for example. They are not designed for rock where they would just bounce off. For this reason, and to protect the equipment, they are specified to be stopped if they do not penetrate the full 150 mm + 300 mm after 50, 100 blows or

Figure B2.2.1a Core obtained from a ground investigation carried out in the Northern New Territories, Hong Kong.

even 200 blows (according to particular specifications) and instead to record the distance penetrated. The contractor falsifies many such records. The supervising engineer wrongly instructs many others, as shown in the following example.

Just imagine the rock face represented by the core shown in Figure B2.2.1a. A sheer cliff of excellent rock with RQD = 100%. Fresh to slightly decomposed granite with occasional joints.

Unfortunately the supervising engineer (I pray that it was not an engineering geologist) at a depth of 34.42 metres obviously noticed a change in flush returns. He dutifully instructed the contractor to pull up his core barrel and, evidently without inspecting the core, instructed an SPT test to be conducted (within the rock). The SPT test recovered 5 cm of pegmatite rock after 100 blows (see Figure B2.2.1b), without actually entering the 300 mm over which the "*N*" value should be measured.

Following the SPT test, the rock was reamed out to 34.82 m depth, presumably using a rock roller bit, with no recovery for the 0.4 metres. As a consequence of this action, the SPT test horizon was regarded as "probably Grade V", rock head was forced lower (ridiculously), and the borehole trundled on, with depth to pile to be installed later, increased further (again ridiculously).

Figure B2.2.1b Detail, showing recovered rock from SPT test.

2.2.5 Legal process and role of expert witness

When disputes reach the stage of either arbitration or civil court, where one party sues another, it is usual for the parties to employ experts to advise them on the validity or otherwise of their case and, if they agree with their client's position, to make a report stating the reasons why. Because many ground condition claims are fundamentally linked to a poor appreciation of geology, engineering geologists often become involved in disputes as experts.

Initially the expert may be advising his client on the rights of the claim "without prejudice". If the expert disagrees with his client's position, he must tell him as soon as he recognises that situation. It will then be up to the party and his legal advisors to decide how to proceed.

If the expert thinks his client's case is valid and he writes a report that may be used in evidence it is important that he recognises that his overriding duty is to the court rather than to the party that is paying for his services. The expert needs to understand that in complex, technical cases his evidence can often be pivotal. Questions put to an expert and his replies to them are treated as part of his evidence. Experts are required to make some "statement of truth" that the expert believes that the facts stated in his report are true and that the opinions expressed by him are correct and his own. He will also need to make an oath in court.

The court will often request that the experts employed by the two sides hold meetings and prepare joint statements, identifying where matters are agreed and where matters are in disagreement. In principle this sounds straightforward but sometimes instructing lawyers will prevent or limit the agreement of experts, partly because whilst experts may agree broadly or compromise over some technical issue (as they would if they were working on an engineering project) there may be subtleties in the legal considerations and case law that might hold sway. That said technical experts must beware being led by lawyers in preparing their reports to argue points that they are not happy with or that fall outside their knowledge and expertise. Such a position does not survive in court where barristers, judges and arbitrators will question experts very thoroughly and a tenuous and weakly argued position will usually be exposed for what it is.

Geotechnical experts need to recognise that they are not legal experts. The author was involved in a case where the Contractor had accepted all ground risks. The situation appeared clear cut and hopeless to a layman but a barrister educated me that the Owner and his design engineers had made "representations", which would affect things legally. Quite often a party is clearly at fault in some way but there may be some question over their legal responsibility. To be held negligent the party need not have done everything right, just at least as well as his peers (on average). I have heard an expert say in court "I have seen worse" as an excuse for poor practice. It is then up to the arbitrator or judge to decide whether that excuse is persuasive.

2.2.6 Final word on contracts: Attitudes of parties

In practice, much depends upon the attitudes of the various parties. Even a poorly drafted contract can be made to work so that the Owner gets his project constructed within his budget and the Contractor makes a profit but this requires co-operative and non-adversarial attitudes. To foster this attitude formal "partnering sessions" are commonly used where everyone is asked to agree some set of rules of behaviour and professional dealings.

Whether or not this works is often down to individuals – especially the Resident Engineer and the Contractor's Site Agent. The author has experience of a large project involving several different contracts where the Resident Engineer had a high regard for one contractor but mistrusted another because of previous encounters on other projects. He was of the opinion that the second contractor had "bought" the contract for an unrealistically low price and therefore would be out to make their profits through claims. The first contract went very well despite many technical problems, which were overcome in a pragmatic manner, working as a team. Reasonable claims were dealt with expediently and everything was completed on time, to the required technical standards and with the Contractor leaving the site a happy man. The second contract was a direct contrast. All site supervisory staff were instructed by the Resident Engineer not to give him any advice, help or site instructions to avoid chinks in the contractual armour that the Resident Engineer thought the Contractor might exploit through spurious claims. The contract went badly wrong; there were technical difficulties, delays to all later works, financial losses and bad feelings all around.

The Mass Transit Railway Corporation (MTRC) in Hong Kong have run several very challenging projects in Hong Kong recently with the construction of underground structures in heavily congested urban areas with all sorts of problems to be overcome. As an MTRC spokesman put it verbally – "conditions are tough enough anyway without contractual difficulties on top". They therefore try to agree Target Cost Contracts on a "cost-plus" basis for complex projects. The Contractor does what he has to and gets paid accordingly. If the contract is brought to completion below target price then the Contractor takes a bonus, if not he "shares the pain" with the MTRC. During one particularly challenging contract at Tsim Sha Tsui in Hong Kong, more than 600 ideas for better working practices were presented during the works together with 371 value-engineering proposals (ways to do things more cost-effectively). As a consequence the programme was reduced by 5 months with significant savings in costs to everyone's benefit.

Notes

1 "Kakerite" (or kakirite) was one of two terms used for describing all the fault rocks encountered in the tunnels, the other being "mylonite". Mylonite is well defined as a re-crystallised rock with flow structures. This leaves the term kakerite to describe all other fault rocks, including cataclasite, fault gouge and fault breccia – whether or not they can be broken down by hand.
2 Differences are shown in Appendix B at Table B4. Weak according to ISRM (1981) is used to describe rock with UCS ranging from 5 to 25 MPa. Weak rock according to BS5930 (1999) includes rock with UCS ranging from 1.25 to 5 MPa.
3 The general standard for high-risk slopes, with strengths correctly assigned, is 1.4 (Geotechnical Control Office, 1984a).

Chapter 3

The design process, analysis and construction

3.1 INTRODUCTION

The main parties involved in the development of a structure, the Owner, the Architect, the Engineer, the Contractors and sub-Contractors, were introduced in Chapter 2 and the procedures behind the design and construction of a bridge under a design-build contract are illustrated in Figure 2.1. The design process involves a number of stages:

1. A conceptual phase, in which the need for a structure is identified and costed by the potential Owner as explained in Chapter 6 (Box 6.5 summarises the process and the hazards to be considered for a new hydroelectric scheme).
2. The locations where the structure might be built are identified and the Engineer is appointed.
3. A desk study is carried out including the use of maps, aerial photographs, satellite images and perhaps airborne "light detection and ranging" (LiDAR) and drone flights.
4. The location may be essentially fixed or flexible and optimised through sorting, as site data become available.[1]
5. The site investigation is initiated, ideally carried out in a number of phases as detailed in Chapter 6, gradually building up a "ground model" and a "geotechnical model" of the site, which meets the safety requirements of the structure be it a road, a tunnel, a dam or a bridge, for example.
6. The structure (conceptual) is analysed. There may be some formal requirement to establish that the design "works" by mathematical calculation. Numerical analytical methods such as Plaxis or FLAC might be employed and the calculations required to prove a Factor of Safety by limit equilibrium (or equivalent methods), although, as ever, if the geological model is incorrect, so will be the analytical results.
7. The ground model might be "almost right" but still failing in some small way (in terms of the overall model), but in a highly *significant* way, in terms of the potential failure mechanism.[2] This is a problem in difficult, "unforgiving" sites and an area where an engineering geologist can play a key role in identifying potential hazards whilst avoiding "crying WOLF" – the fourth attribute of a competent engineering geologist according to Burwell & Roberts (1950) as outlined in Chapter 1.
8. Alternatives to the calculation method known as limit equilibrium which is used by geotechnical engineers, include designing the structure using "characteristic values and partial factors" to Eurocode 7, reliability-based design methods (Low

DOI: 10.1201/9781003348894-3

& Phoon, 2015), load and resistance factor design (LRFD) or allowable stress factors (ASD).[3]

9. Instrumentation such as piezometers and strain monitoring devices, might be installed early on, and monitoring then continued through the construction phase, and into operation, for the structure's lifetime (at least 100 years in the case of a dam).
10. Construction is carried out, paying attention to ground conditions as encountered, fed back into the ground model in an iterative way. The ground conditions need to be interpreted with feedback from instrument monitoring, especially where the observation method is used for design (Peck, 1969).
11. Sometimes the instrumentation feedback is directly interpreted and ground improvement carried out to limit settlement by TAM grouting as at Waterloo Station where the engineers managed to keep settlement down to 10 mm in the case of the Victory Arch and 14 mm in the case of Waterloo and City Line stations (Harris et al., 1994). In the case of the Heathrow Express, however the instrumentation was not reacted to quickly enough and grouting may have contributed to the tunnel collapse as explained in Chapter 8.
12. Eventually when construction is complete and the structure is handed over to the Owner, his engineers will need to monitor behaviour for years and decades and sometimes even longer.
13. The case of the Glendoe tunnel is dealt with in Chapters 2 & 8 where a long length of the handed-over tunnel, constructed by tunnel boring machine (TBM) through fault rocks was left with minimal support, and soon collapsed (Hencher, 2019).

This chapter introduces the concept of the design process, analysis and construction so that the engineering geologist can understand better the requirements of projects in terms of his involvement throughout the development of a structure. The design process involves the overall conceptual development of a structure, be it a dam, a bridge or a high-rise building. A difficulty (and this is where the engineering geologist's skills really come in to play), is that the structure is to be built, at a real site, where the geomorphology, geological make-up and conditions affecting the site are uncertain for the moment.

Analysis, which follows the conceptual phase requires testing the various ways in which the structure might fail, using a range of geotechnical parameters, and detailing the structure, the reinforcement that might be required, and any ground improvement methodology. The design process overlaps with that of site investigation as dealt with in Chapter 6. As data are obtained from the investigation, so there is a progressive development of the design and feedback of additional requirements from the investigation.

3.2 DESIGN

3.2.1 Design codes

Building works throughout the world are generally covered by local regulations, which are mandatory together with Codes of Practice and Standards. Such documents cover most aspects of works including ground engineering and sometimes aspects of engineering geological practice. Some of the key documents that the UK engineering geologist needs to be aware of are listed in Table 3.1. Similar codes and standards exist for many other countries.

Table 3.1 Selected codes and standards that the engineering geologist needs to be familiar with in the UK

Table 3.1 Selected codes and standards that are useful or essential references for the engineering geologist. The bias here is towards UK practice.

CODES FOR SITE INVESTIGATION AND TESTING INVESTIGATION

UK: BS 5930 +A1: 2020 Code of practice for site investigations

BS 5930 (1st edition 1981; revised 1999, 2015, 2020) deals with the investigation of sites for civil engineering and building works in the UK; parts of the 1999 version were superseded by Eurocode 7 (BSI, 2004b) but these two codes now are a "pair of complementary documents without conflict between the two" according to Norbury (2017) although I have my doubts about the validity of this statement.

The first issue in 1981 was a brave attempt at standardisation of ground investigation practice and terminology, yet attracted 423 pages of constructive criticism at a conference at Surrey University in 1984 (Hawkins, 1986). It was revised extensively in 1999, largely incorporating the comments from the conference as well as the findings from Anon (1995), by the Engineering Group of the Geological Society on weathered rocks. Eurocode 2007 and associated documents were supposed to supersede the various British Standards but had controversial changes – some of which literally did not make any sense at all (Hencher, 2008). In 2020 the "new" BS 5930 was presented. This incorporated some of the criticisms, for example on describing the degree of weathering and strength, but made mistakes in copying tables from one document to another, changed the definitions of rock strength with no scientific support that I am aware of to justify these changes, and incorporated some of the old, poor definitions from Eurocode 2007 that had been previously criticised.

BS 5930 is supposed to encourage good practice (as determined by a group of individuals plus some obscure method of commenting from interested parties) and gives sources of information and references to original literature. In-depth guidance is given on a wide range of techniques in ground investigation including drilling, boring, *in situ* testing and geophysical works. The code is almost a textbook in its own right and provides some advice on designing and managing site investigations in the UK. The engineering geologist in the UK needs to be very familiar with this code of practice because it defines geotechnical descriptive terms (in a "should" style). There is no equivalent advisory style European document. Beware in applying this code in other countries where terminology is different, and even in the UK, where the standard is incorrect, questionable or wrong!

The term Site Investigation is used in the code in its broad sense including desk study; the narrower subject of sub-surface exploration is termed Ground Investigation. Whilst common practice in the UK is covered in detail, some techniques that are used in ground investigation in other countries are dealt with only briefly or not mentioned. For example, it makes no mention of Mazier sampling, which has been standard practice in Hong Kong for over 30 years, with excellent recovery in weathered rocks of all grades.

BS 5930 gives guidance on standard rock and soil description for civil engineering purposes and the terminology in the BS is apparently used routinely in the GI industry in the UK. As discussed in Chapters 4 and 5 different schemes are used in other countries and important subjects such as rock mass classification are not covered in BS 5930. Amendment 1 to BS 5930 was revised to comply with BS EN ISO 14688 1:2002, BS EN ISO 14688-2:2004 and BS EN ISO 14689-1:2003, which apply in Europe generally. These changes in terminology are not universally accepted as improvements [see discussion by Hencher (2008), Chapter 4 and Appendix B].

There remain fundamental errors in the 2020 version, for example, in weathering zone description, and in the description of slaking, which actually renders the BS rather dangerous as it will be referenced in courts in the UK (who will, as engineering lay people, rely on the BS as being fundamentally correct)! This is a generic problem with standardisation committees as referenced in Box 2.2 and in Chapter 5, see Boxes 5.5 and 5.6 and in Appendix B. It is also important to recognise that the BS 5930 is NOT applicable in various countries (including most of the Far East, Australia and New Zealand, for example) where they have their own codes, or other countries such as the USA where they rely mainly on ISRM definitions, for example.

Table 3.1 (Cont.)

One chief concern that I have is in the description of compressive strength where the SAME terms are used widely, but differently, and these are quite fundamental to description. Appendix B shows how strength "descriptors" have changed from revised version of BS5930 to revised version of BS5930, leaving the industry gasping for breath (and subject to litigation)!

In summary, it is very important that the relevant and applicable codes (dated) are used in whichever country the engineering geologist is working. The MSc student cannot claim the defence of having been trained to describe according to BS 5930 1999; or to the 2020 version, which has different definitions and terminology neither of which are definitive!! It is also important to recognise the fundamental limitations of standards committees who might be off the mark in making their recommendations, practically and scientifically, as referenced in Box 2.2.

TESTING

Standard UK methods for some laboratory and *in situ* soil testing are given in BS 1377 (British Standards Institution, 1990), which has been partly superseded by Eurocode 7 Part 2 (Ground Investigation and Testing) (British Standards Institution, 2004). Internationally, reference is often made to American Standards (ASTM) or to methods recommended by ISRM (Ulusay & Hudson, 2007) and others. The recommended methods sometimes differ in detail (such as dimensions of samples) and care must be taken to ensure that appropriate guides are being adopted according to the nature of the project and location. Modern, sophisticated and relatively uncommon testing practice is generally not dealt with in country standards and codes of practice and reference must be made to the scientific literature (see also Chapter 6).

The ISRM standard (2014) for direct shear testing is an example of the "state-of-art" in geotechnics. Keeping it simple rather than sorting out the problem. As explained later in Chapter 7 there are a range of "basic frictions" that might be defined for a rock joint (not *one* as Barton & Bandis, 1990, suggest). The value depends on the surface finish and debris on the surface. A coarse sandpaper surface, which does not dilate during shear might well show higher "friction" than a finer sandpaper surface. It is the same for saw-cut surfaces of rocks as demonstrated by Coulson (1971).

You might think that ISRM would seek to incorporate this truth into their standard tests, but no, they sought to standardise the tilt test so that it gave reliable numbers to slip into Barton's equation, and ignored the evidence of variability (too difficult ...).

CODES FOR GEOTECHNICAL DESIGN

The British codes of practice discussed below are now generally withdrawn and replaced by Eurocode 7 Part 1 (Geotechnical Design) (British Standards Institution, 2004) for geotechnical design purposes. Nevertheless they provide general advice and guidance and therefore remain useful references on good practice based on "well-winnowed experience" (Burland, 2007).

FOUNDATIONS

BS 8004:1986 Code of practice for foundations

As with BS5930, this code of practice (British Standards Institution, 1986) gives general guidance and background information that is very useful in guiding the geotechnical engineer and engineering geologist. The code provides recommendations for the design and construction of foundations for buildings and engineering structures. It introduces general principles of design as well as detailed consideration of the design and installation of the foundations. The code also discusses site operations and construction processes in foundation engineering and the durability of the various materials used in foundation structures. Section 11 deals with safety issues.

In the UK, BS 8004 is superseded by BS EN 1997-1 Eurocode 7 Part 1 "Geotechnical Design", which adopts a limit state approach to design rather than a lumped safety factor approach as discussed later.

(*continued*)

Table 3.1 (Cont.)

Other codes and standards

Whereas BS 8004 has been superseded in the UK, similar codes are still used internationally. For example CP4: 2003, the Code of Practice for Foundations in Singapore (Singapore Standard, 2003) provides general guidance on foundation design, specific to local ground conditions. In Hong Kong comprehensive guidance is given in GEO Publication No. 1/2006 (Foundation Design and Construction).

EARTHWORKS AND RETAINING STRUCTURES

BS 6031:1981 Code of practice for earthworks

British Standards Institution (1981a) gives advice on formation of earthworks for civil engineering projects such as highways, railways and airfields and on bulk excavations for foundations, pipelines and drainage works. It gives some UK-focused advice on design and construction of cuttings and embankments. Advice is also given on methods of excavating trenches, pits and temporary support to the sides including timbering, sheet piling, diaphragm walls and contiguous bored piled walls.

BS 8002:1994 Code of practice for earth retaining structures

BSI (1994) is aimed at UK practitioners and provides guidance on the design and construction of retaining structures up to about 15 m high.

More detailed guidance on retaining wall design, especially where dealing with weathered rocks, is given in Geotechnical Engineering Office (1993) Geoguide 1: Guide for Retaining Wall Design which, like many other Hong Kong guides and publications, is downloadable from the Hong Kong Government, Civil Engineering Design and Development website: www.cedd.gov.hk.

EUROCODE 7: GEOTECHNICAL DESIGN

The Eurocodes comprise a suite of ten standards, now adopted as British Standards that have replaced the majority of older national codes of practice as the basis for designing buildings and civil engineering structures in the UK and in most member states of the European Community. As commented above the superseded codes of practice still contain very useful guidance on good practice albeit that there has been a fundamental shift in design concept from "lumped Factor of Safety" to a "partial factors" approach.

Fundamentally the concepts used in the earlier codes of practice and the Eurocodes are the same: under extreme loading conditions structures must not fail catastrophically and in "day to day service" structures should not suffer deformations that would (a) render the structure incapable of achieving the use for which it was designed or (b) suffer deformations that take the structure beyond its aesthetic appearance requirements: these are different examples of "limit states".

Ultimate limit state failure might include the collapse of a slope, bearing capacity failure of a building or blocks of rock falling out of the roof of a tunnel or might be identified as piping failure through the foundations of a dam. Serviceability limit state failure could be defined as excessive settlement; the classic example being the Leaning Tower of Pisa, which has settled dramatically but not collapsed.

In traditional design, uncertainties are dealt with by adopting Factors of Safety. These give a broad protection against the inherent uncertainty in models, calculations, loads, strengths, workmanship and so on. If the site conditions such as geological model and geotechnical parameters are understood well or if potential consequences are minor then a low FoS might be adopted. Where less certain or the risk is greater then a higher FoS is adopted.

In the Eurocode approach, rather than assuming a global Factor of Safety, it has been taken as fundamental that different parts of the calculation are known with different certainties; which is certainly true in many situations and is a refinement to design philosophy. "Partial Factors" are then applied to material properties, resistances and/or actions (loads) according to the level of uncertainty. The Eurocode clauses are written as "Principles" and "Application Rules". Principles.

Table 3.1 (Cont.)

use the word "shall" and are mandatory, whereas, "Application Rules" use the words "may", "should" and "can" and allow more judgement. Although this use of language suggests a more prescriptive approach than in earlier codes, in practice, the Eurocodes provide a similar level of latitude for the designer. For example, in assessing geotechnical risk, Eurocode 7 contains Application Rules that define three geotechnical categories and alternative methods are allowed for assessing geotechnical risk. For routine design cases, the geotechnical design may be assessed by reference to past experience or qualitative assessment. For complex or high risk situations, e.g. weak/complex ground conditions or very sensitive structures, Eurocode 7 allows the use of alternative provisions and rules to those within the Eurocode. In such situations, rational design based on site-specific testing and numerical modelling might be more appropriate. Detailed guidance is given in Bond & Harris (2008).

Limit state design approaches are used elsewhere similar to current European practice. For example, Canadian practice has moved in that direction and AASHTO (2007) is used in the USA and internationally for the design of major projects such as the 2nd Incheon Bridge in Korea completed in 2009 (Cho et al., 2009a, 2009b).

Sometimes local mandatory codes or guidelines conflict with others prepared in other countries or by international learned societies, not least in terminology for soil and rock description and classification with the same words used in different codes to mean different things. The engineering geologist who wishes to work in different countries needs to be aware that the standards and terms that he will need to use may change from country to country. He also needs to be aware that the advice given in codes and working party reports regarding geological matters is often generalised and sometimes difficult to adopt; for example, guidance prepared for temperate zones may not be applied readily in tropical areas and vice versa. This is addressed in more detail in Chapter 6 and Appendix B when discussing soil and rock description for engineering purposes.

3.2.2 Application of engineering geological principles

Despite codes and standards, ground conditions continue to be the major source of failure in civil engineering projects – through catastrophic failure or unacceptable performance and even more commonly due to claims, delay and litigation. In hindsight the problems can often be attributed to a poor and unperceptive site investigation, incorrect interpretation of the geological conditions or inadequate design and analysis. Poor management and contractual arrangements often contribute to the problems. Ways of avoiding "unexpected ground conditions" are presented in Chapter 6 and case examples, mostly of projects where failures occurred, are presented in Chapter 8.

3.2.3 Keeping records

Engineering geologists on site should keep careful records as works advance using daily notebooks. Excavations should be examined, described and photographed as necessary. It is often useful to take photographs and use these as the base for overlays on which to record features such as geological boundaries, strength of materials, discontinuity orientations and style and locations of seepage. Such records will be very helpful in the case of any future disputes over payment or if anything goes wrong. Pairs of photographs taken some distance apart can be used to allow a 3D image to be viewed and this is particularly useful where access is difficult or hazardous. Where discontinuities are measured it is important to record the location. In drill and blasted tunnels, description proformas are commonly used as permanent records after each pull, agreed and signed by the contractor and engineer as illustrated in Figure 3.1.

3.2.4 Checking the ground model and design assumptions

It is fundamentally important that design predictions are checked during construction. Design is usually based on widely spaced boreholes and the interpretation will certainly have been oversimplified. Often this does not have any major consequence but sometimes it does, and the engineering geologist on site should be alert to any indications that the ground model is incorrect or inadequate. Any changed conditions should be flagged up quickly to the designers so that necessary rectifications can be made.

In rock slope design and construction, the fundamentally important features of discontinuity orientation, lateral persistence, roughness and infill can only be surmised from typical ground investigation data. Ground models and assumptions need to be checked as the rock is exposed. There are many cases of rock slope failure where sliding occurred on features that were exposed during construction but were either not mapped by the site staff or the significance not recognised. This can be a very difficult situation in that even where rock is well exposed, the presence or otherwise of rock bridges or steps along adversely oriented discontinuities can only be guessed at. Care should be taken to note how easily the rock is being excavated and how stable or otherwise temporary slopes are as such information will be helpful in judging the hazards.

3.3 LOADING

Most civil engineering projects involve either loading the ground say from the weight of a new building, or unloading because of the excavation of a slope (see calculation example in Box 3.1). In the case of a tunnel or underground excavation, the ground is unloaded locally beneath the upper point of excavation (the crest). Stresses in the side walls however, become concentrated as the weight of the ground essentially flows around, and is carried by, the arch created by the tunnel. Tunnels where drained (water flows into the tunnel) will also result in the effective stress increasing at some distance from the tunnel, which can result in settlement of the ground. Load changes can be permanent or temporary, static (due to weight) or dynamic (due to blasting, for example). A further important consideration for most geotechnical problems is *in situ* stresses.

3.3.1 *In situ* stresses

At any point in the Earth's crust the stresses can be resolved into three orthogonal directions and these are termed the maximum, intermediate and minimum principal

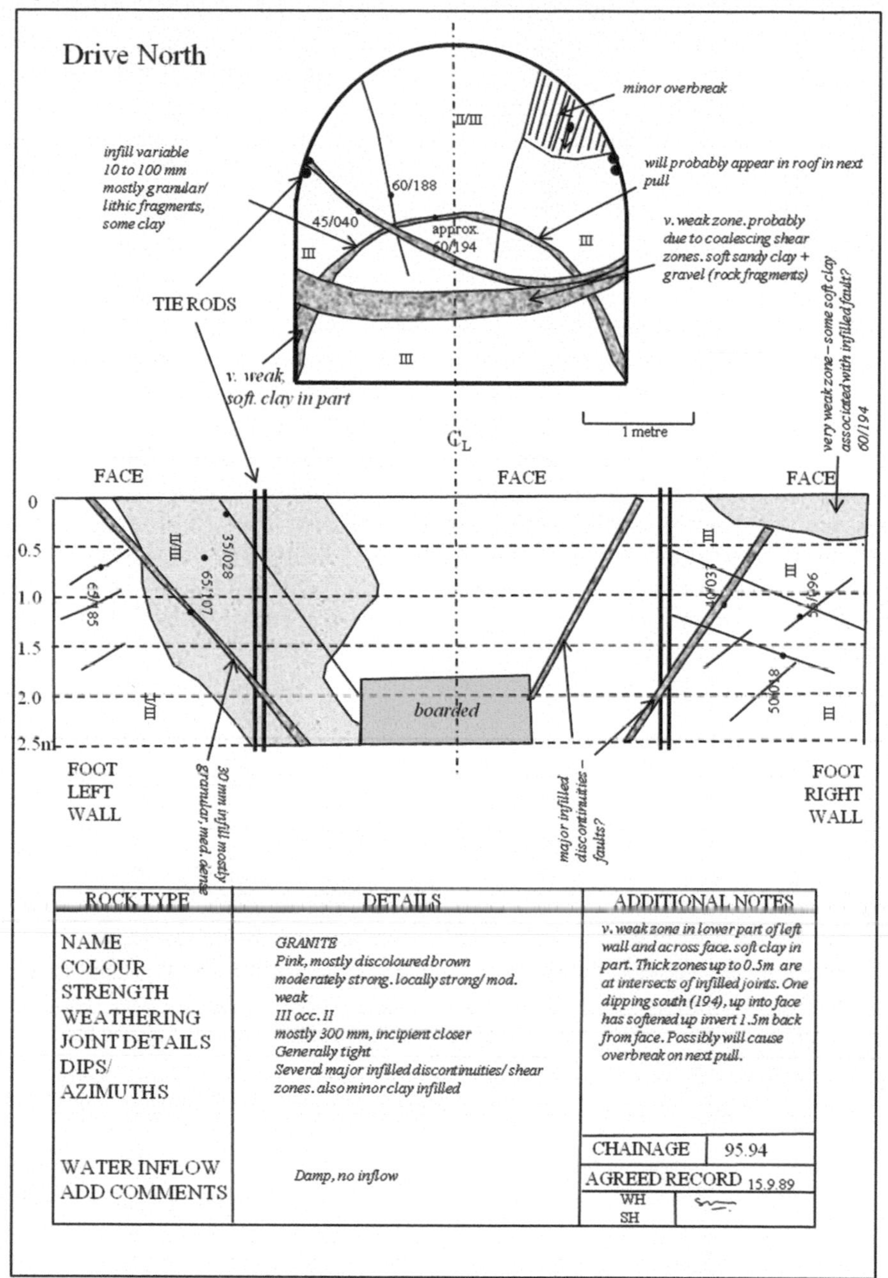

Figure 3.1 Example of agreed record of ground conditions in a tunnel. Queens Valley Reservoir diversion tunnel, Jersey, UK.

stresses and depicted σ_1, σ_2 and σ_3, respectively. By definition, the planes to which the principal stresses are normal are called principal planes and the shear stresses on these planes are zero. An important point regarding rock engineering is that all unsupported excavation surfaces are principal stress planes because there are no shear stresses acting on them (Hudson, 1989). One of the principal stresses will always be perpendicular to the Earth's surface (Anderson, 1951) and this is generally assumed to be vertical ($\sigma_1 = \sigma_v$) although that is not the case at shallow depths or in hilly terrains (Goodman, 1989).

3.3.1.1 *Lithostatic stress*

For projects close to the Earth's surface such as cut slopes or foundations, natural stresses are important but at most locations these stresses are only due to self-weight, weight of included water and buoyancy effects below the water table, which reduces the total stress to an effective stress (weight of soil minus water pressure) as illustrated in Box 3.2, Figure B3.2.1. As the rock or soil is compressed under self-weight, it tries to expand laterally and a horizontal stress is exerted. This is termed the Poisson effect. Typically, in a soil profile at shallow depths, the *in situ* horizontal stress (σ_h) due to self-weight will be between about 0.3 (in loose sand) and 0.6 times (in dense sand) times the vertical gravitational stress. The ratio 0.3 to 0.6 is called the coefficient of earth pressure at rest. In normally consolidated clay the value is about the same as for dense sand, 0.6. For most rocks Poisson's ratio is slightly less than 0.3. Most continental rocks weigh about 27 kN /m^3 so at a depth of 500 metres the total vertical stress can be anticipated to be about 13.5 MPa and horizontal stresses (σ_h) about 4 MPa.

BOX 3.2 EXAMPLE STRESS CALCULATION

VARIATIONS FROM LITHOSTATIC STRESS CONDITIONS

Whereas in many areas of the Earth's crust stress conditions can be estimated reasonably well by calculating the weight of the soil/rock overburden to give vertical stress and taking account of Poisson's effect for horizontal stress, considerable variation is found (Hoek & Brown, 1980). In particular, horizontal stresses can be higher or lower than anticipated.

EXAMPLE 1 OVERCONSOLIDATED SOILS

Soils that have gone through a cycle of burial, partial lithification and then uplift and erosion are termed "over-consolidated". They typically have lower void ratios (percentage of pores) and are stiffer than would be expected for "normally consolidated" soils at similar depths of occurrence. They are also sometimes partially cemented as described in Chapters 1, 4 and 6. Under compression they show high moduli up until the original maximum burial stress at which point they revert to the normally consolidated stress curve as described in soil mechanics textbooks (e.g. Craig, 1992). Because the stress level has been much higher in

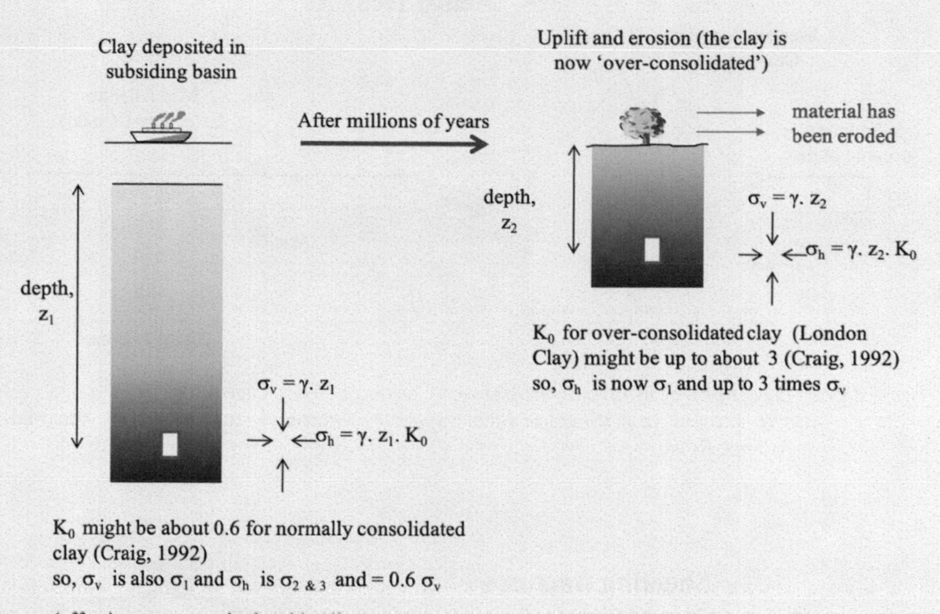
Clay deposited in
subsiding basin
After millions of years
Uplift and erosion (the clay is
now 'over-consolidated')
material has
been eroded
depth,
z_1
depth,
z_2
$\sigma_v = \gamma . z_2$
$\sigma_h = \gamma . z_2 . K_0$
K_0 for over-consolidated clay (London
Clay) might be up to about 3 (Craig, 1992)
so, σ_h is now σ_1 and up to 3 times σ_v
$\sigma_v = \gamma . z_1$
$\sigma_h = \gamma . z_1 . K_0$
K_0 might be about 0.6 for normally consolidated
clay (Craig, 1992)
so, σ_v is also σ_1 and σ_h is $\sigma_{2 \& 3}$ and = 0.6 σ_v
(effective stresses calculated in all cases,
allowing for depth of water)

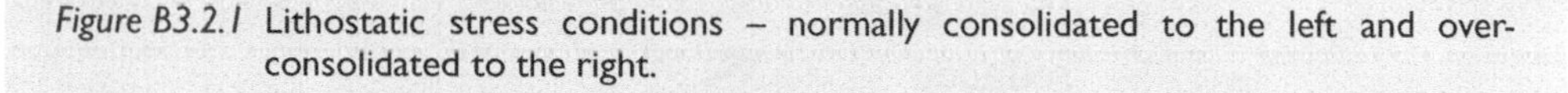

Figure B3.2.1 Lithostatic stress conditions – normally consolidated to the left and over-consolidated to the right.

geological history, the horizontal stress may have become locked-in as a residual stress and may be much higher than the vertical principal stress as illustrated in Figure B3.2.1. Craig quotes "earth pressure at rest" K_0 values up to 2.8 for heavily overconsolidated London Clay. Further discussion of earth pressures and how they relate to geological history is given by Schmidt (1966).

EXAMPLE 2 ACTIVE AND ANCIENT TECTONIC REGIONS

Deviations from lithostatic stress conditions can be anticipated at destructive plate margins as along the western margins of North and South America where high horizontal stresses are to be expected. Conversely in extensional tectonic zones the horizontal stresses can be anticipated to be tensile. Variations can also be expected in ancient mountain chains or areas of igneous intrusion where relict horizontal stresses can be very high resulting in rock bursts and large deformations of structures (e.g. Holzhausen, 1989)

EXAMPLE 3 TOPOGRAPHIC STRESSES

Stress conditions may be strongly affected by local topography exacerbated by geological conditions. At an extreme scale large scale mountain structures are ascribed to gravity

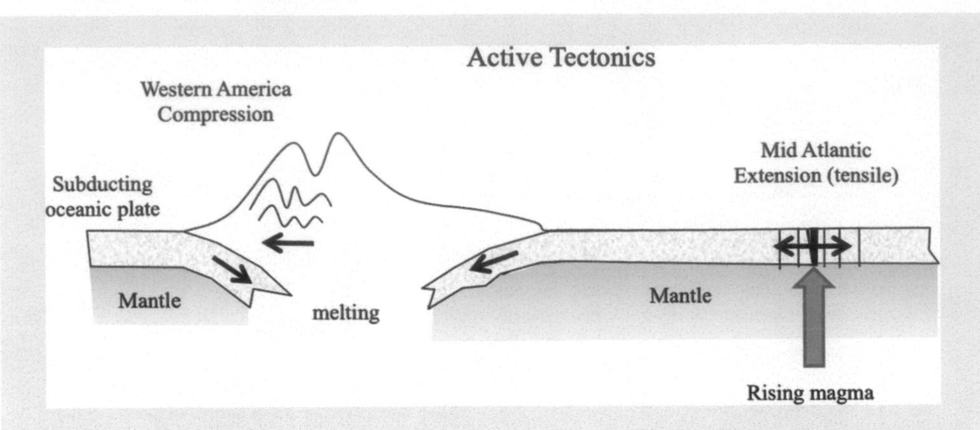

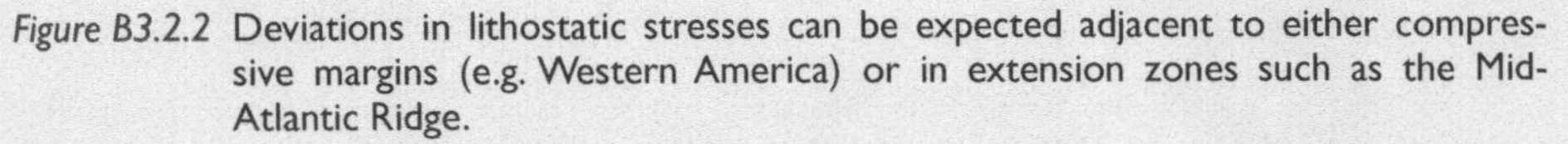

Figure B3.2.2 Deviations in lithostatic stresses can be expected adjacent to either compressive margins (e.g. Western America) or in extension zones such as the Mid-Atlantic Ridge.

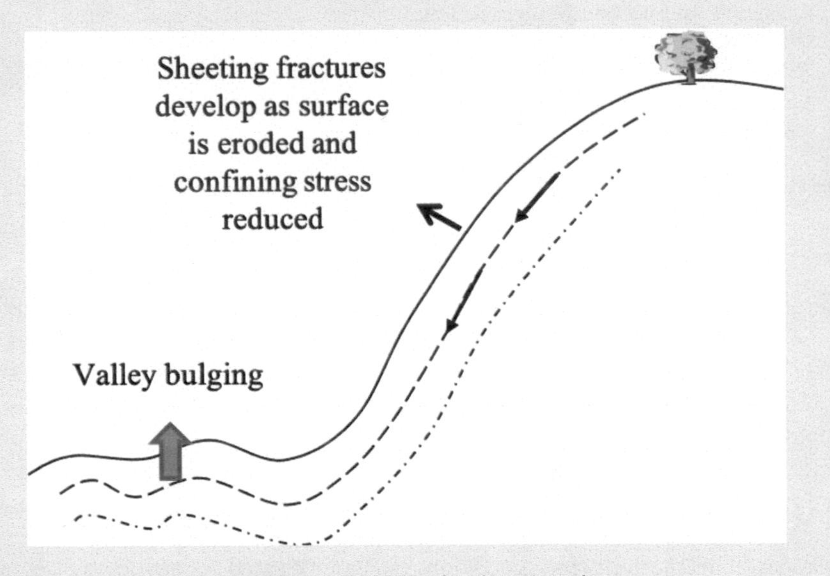

Figure B3.2.3 Natural stress conditions affected by local topography.

"gliding" (e.g. Graham, 1981) and certainly large landslides have ample evidence of compression and tensile zones. Other key examples of the effect of localised topographic stress are sheeting joints where the minimum principle stress runs at right angles to the Earth's surface (Figure B3.2.3) and equals zero (Hencher et al., 2011) and valley bulging (Parks, 1991).

3.3.1.2 Overconsolidated clay

At some locations tectonic or topographic stresses can be dominant even very close to the Earth's surface, with horizontal stresses, sometimes locked in from a previous geological event and far in excess of that due to gravity and the Poisson effect. As illustrated in Figure B3.2.1 in over-consolidated clays such as London Clay, where the rock has been buried to considerable depth before uplift, erosion and unloading then the earth pressure at rest can be up to 3 times the vertical stress.

3.3.1.3 Active plate margins

As illustrated in Figure B3.2.2, in tectonically active regions stresses can be higher or lower than lithostatic. Horizontal: vertical stress ratios as high as 15 have been measured in areas where tectonic or thermal stress has been locked in as the overburden has been eroded (Hoek & Brown, 1980). These stresses can adversely affect engineering projects resulting in deformation in tunnels, rock bursts and propagation of fractures (e.g. Karrow & White, 2002).

3.3.1.4 Topographic effects

In mountainous terrain as illustrated in Figure B3.2.3, principal stress trajectories will follow the topography so that the maximum principal stress runs parallel to steep natural slopes, with the minimum stress perpendicular to the slope and this leads to spalling off of the rock parallel to the natural slope as sheeting joints and valley bulging at the toe of the slope.

Stress conditions have been measured across the world from instruments, by interpretations of "breakouts" in deep drillholes for oil and gas exploration or analysis of earthquakes and many such data are compiled centrally and are freely available at www.world-stress-map. *In situ* stresses are sometimes investigated specifically for projects (Chapter 6) but this is expensive and can be inconclusive because of the small scale and localised nature of tests.

Where stress assumptions have proved wrong the consequences can be severe as at Pergau Dam, Malaysia where it had been anticipated that stresses would be lithostatic (i.e. caused by self-weight). During construction open joints and voids were encountered in tunnels together with high inflow of water (Murray and Gray, 1997). It was established that horizontal stresses were much lower than had been predicted and this necessitated a complete re-design of shafts and high-pressure tunnels and their linings at considerable cost. Further examples are given later in this chapter.

3.4 TEMPORARY AND PERMANENT WORKS

The Engineer's design generally concerns the permanent works – the long-term stability and performance of the finished project. Performance is measured by criteria specific to a project such as settlement, leakage, durability and long-term maintenance requirements. During construction there will usually be other design considerations including stability of temporary excavations, disturbance to the groundwater conditions and water inflow to the works. Temporary work design is generally the

Figure 3.2 Temporary works for an underground station construction in Singapore. Piles were excavated by a large diameter drilling rig and then concreted. As excavation has proceeded the piles have been anchored back into the ground and strutted using systems of waling beams (horizontal, along the face of the piles) and struts, supported where necessary by additional king posts.

responsibility of the Contractor and his design engineers, perhaps checked by an independent engineer to Category 3[4] (BSI, 2019). The design of deep temporary excavations can be just as demanding as for permanent works as illustrated in Figure 3.2. Catastrophic failure of such works is unfortunately common – in recent years affecting such high-profile projects as the International Finance Centre in Seoul, Korea and the Nicoll Highway subway works in Singapore (Chapter 8). In both cases the strutted excavations collapsed. Guidance on the design of such structures is given in Puller (2003) and Geotechnical Control Office (1990).

In tunnels, during construction there may be a need to stabilise the walls and possibly the working face using rapidly applied techniques including shotcrete with mesh or steel fibres, steel arches or lattice girders and rock bolts (Hoek et al., 1995). Such measures are generally specified and installed by the Contractor, typically agreed with a supervising engineer who may well be an engineering geologist. The engineering geologist will probably be involved in identifying the rock mass conditions and identifying any geological structures that might need specific attention as discussed later. The decisions taken will often have cost as well as safety implications. Usually measures installed to allow safe working will be ignored when designing and constructing permanent liner support but, in some tunnels, there is no permanent lining so the temporary measures become permanent works. In the latter case, the materials and workmanship will be specified accordingly and appropriate to the design life of the project. Close supervision will be required on site to ensure that the specified requirements are met and the quality of the works is not compromised.

3.5 FOUNDATIONS

3.5.1 Loading from a building

A structure will change the stresses in the ground and, in turn be acted upon by stresses from the ground due to gravity and tectonic forces, wind, snow, earthquakes and perhaps from anthropogenic sources including blasting and traffic. The loading condition for a high rise building constructed on piles is illustrated in Figure 3.3. It is the task of the geotechnical team, given the loading conditions from other members of the design team, to ensure that there is an adequate "Factor of Safety" for the foundations against failure and that settlement is within the tolerance of the structure. The traditional, permissible stress approach involving a lumped Factor of Safety to cover all uncertainties has been replaced in Europe and some other countries and design codes by a limit state approach, which ostensibly encourages more rigorous consideration of different modes of failure and uncertainties in each parameter and in the calculation processes itself.

The load from a building is passed to the ground through the foundations. This will include the vertical, dead weight of the building – its walls and fixed fittings and a "live" load which arises for the occupants and other transient loads such as from snow, wind or earthquake loading. The stress from a building, if placed directly on to flat, essentially isotropic ground, will decrease with depth and can be expressed as a bulb of pressure as illustrated in Figure 3.4. At a depth of perhaps 1.5 to 2.0 times the diameter of a building, the stress level can be anticipated to reduce to 10% of the stress immediately beneath the foundation (Tomlinson, 2001). This is an important rule-of-thumb

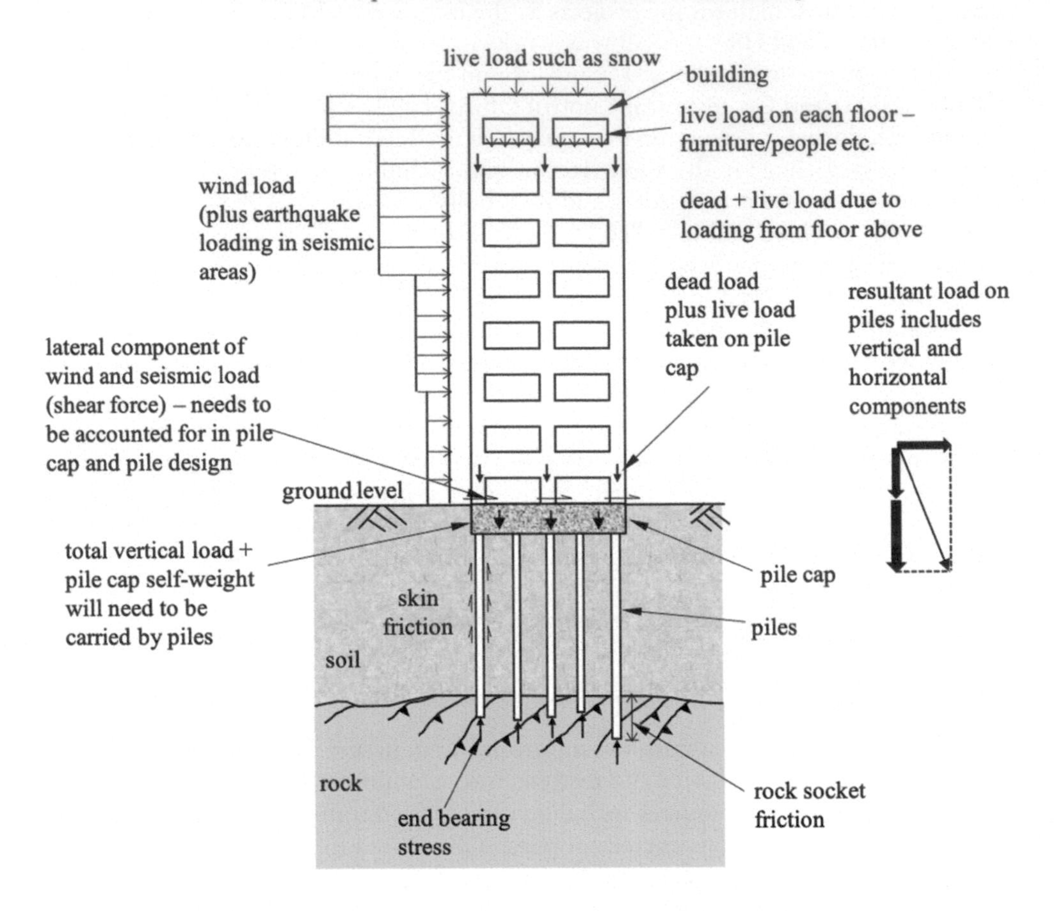

Figure 3.3 Typical loading conditions for a high-rise building to be founded on piles.

for the engineering geologist to keep in mind because it gives an indication of the minimum depth of ground to be investigated as discussed in Chapter 6.

For a small house, the overall load is not very great (imagine the structure collapsed as a pile of bricks and concrete, which might only be a metre or two in height), so the weight of the house can usually be carried safely by narrow "strip footings", which run beneath the "load-bearing" walls. For a footing of about 0.5 m width, the typical load from a two-storey house might be about 50 kN per metre length of wall so the bearing pressure on the foundation would only be about 100 kN/m^2 (= 100 kPa), which would be safely carried by a stiff clay or dense sand. For such a narrow strip footing, the

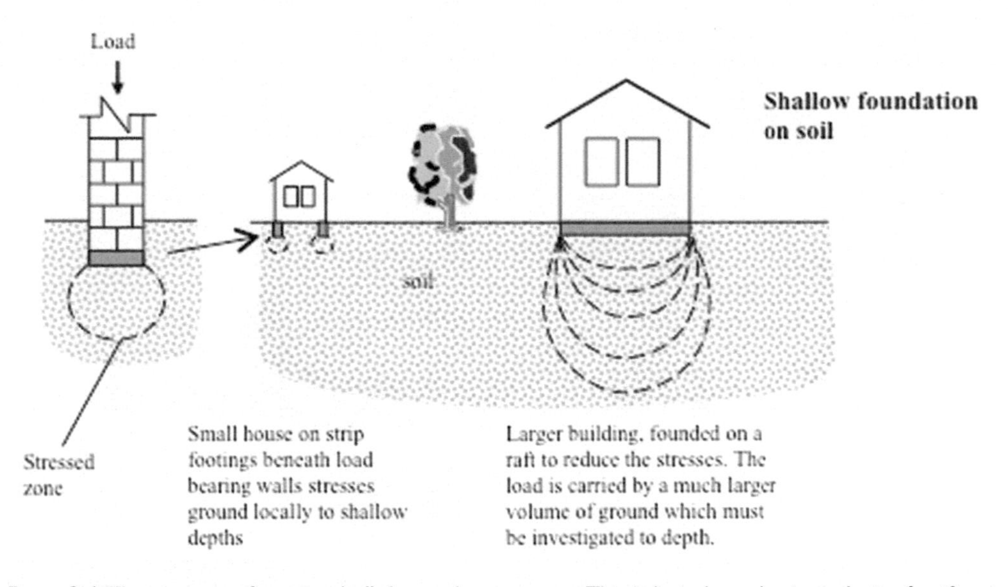
Load
Shallow foundation on soil
soil
Stressed zone
Small house on strip footings beneath load bearing walls stresses ground locally to shallow depths
Larger building, founded on a raft to reduce the stresses. The load is carried by a much larger volume of ground which must be investigated to depth.

Figure 3.4 The concept of a stress bulb beneath a structure. This is based on elastic analysis of uniform materials but is indicative and helpful. The wider the structure, the greater the volume and depth of ground that will be stressed. This has implications for the ground investigations that must be carried out.

appropriate depth of ground investigation for assessing potential settlement might be 1 to 2 metres.

3.5.2 Options for founding structures

There are two key considerations for foundation design. Firstly, there should be a check against bearing capacity failure or the "ultimate limit state" of the underlying soil or rock. This involves analysis of the various loads and calculating the strength of the supporting ground. In traditional design calculations, a "Factor of Safety" (FoS) of between 2.5 and 3 is adopted against ultimate bearing capacity failure to allow for the uncertainties. This means that the allowable bearing stress should be at least 2 or 3 times lower than the load that the ground could theoretically support without failing catastrophically.

The second check is for settlement (otherwise known as a "serviceability state"). Settlement is inevitable as a building is constructed and the ground loaded but there are certain tolerances that the designer needs to be aware of. Many structures can cope with perhaps 25 mm of vertical (total) settlement and some, such as an earth embankment dam (constructed from soil and rock fill), may settle by metres without distress; the key question is usually the tolerance of a structure to differential settlement whereby some parts of the structure settle more than others causing shear stress between different parts. This may happen where the ground is not uniform – perhaps one corner of the building "footprint" is less weathered and therefore stronger and

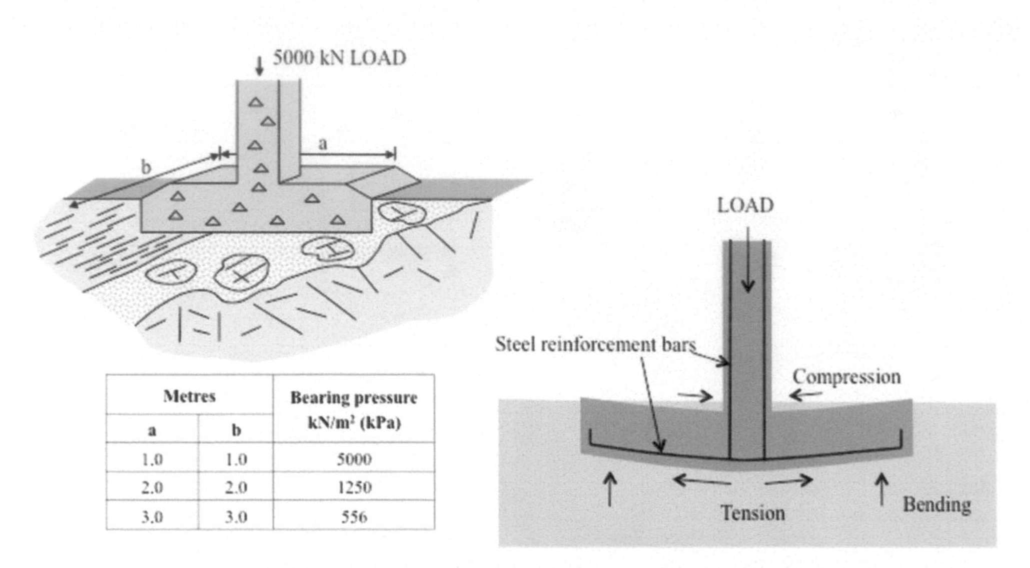

Metres		Bearing pressure kN/m² (kPa)
a	b	
1.0	1.0	5000
2.0	2.0	1250
3.0	3.0	556

Figure 3.5 Demonstration of how the bearing pressure on the ground can be reduced by increasing the dimensions of the footing whilst carrying the same building load. The weaker the ground, the larger the foundation will need to be, or some other solution might be necessary such as piling to stronger material at depth. Steel reinforcement will be necessary to resist tensile stresses throughout the structure and to also to resist buckling in the column.

less compressible than the rest. If the design of the foundations and/or load distribution of the building does not properly account for this variability, the building will settle more towards one end. Usually the limiting relative rotation for a framed structure is taken as about 1 in 500 to 1 in 300 to avoid cracking in walls and partitions (Skempton & Macdonald, 1956; Burland & Wroth, 1975); for a high-rise building the tolerance may be lower. Some structures may be even more sensitive and have special requirements for restricting settlement. The structural designers will need to tell the geotechnical engineer and engineering geologist what is the tolerance for the project so that foundations can be designed accordingly.

It is often cost-effective to design shallow foundations for structures. For framed structures comprising columns and beams of concrete or steel, the load is carried on the columns, which are then founded on "pads" of reinforced concrete. The size of pad will control the bearing pressure on the underlying soil or rock. If the ground cannot carry the applied stress from the building without unacceptable settlement, then the pad size may be increased as illustrated in Figure 3.5. For a concrete frame structure, steel reinforcement would be placed in the columns and towards the base of the foundation where the concrete may be subject to tensile stress by bending; concrete (and rock) is relatively weak in tension – typically about one tenth of its strength in compression. Examples of design calculations are given in Tomlinson (2001). Because of the variable ground conditions below the pad, great care would be needed not to overstress any weaker zones and it might be necessary to carry out local "dentition" to excavate pockets of soil or weak rock with structural concrete. Excavations should be examined and logged by an experienced engineering geologist or geotechnical engineer to check that the conditions are as good as was assumed for the design, before concreting. Such checks and approvals should be well-documented. As the space between the required pads becomes smaller, it may make sense to combine the foundations in a single "raft" over the full building footprint. It must be remembered however that the wider the foundation, the greater the volume of ground stressed as shown in Figure 3.4 and the ground investigation must establish the nature of ground over that full depth. There are many cases where weak compressible material at depth has caused problems for foundations (e.g. Poulos, 2005).

Instead of using a raft, it is often cost-effective to take the foundations deeper using piles, which might be made of steel, concrete or in some countries, timber or even bamboo. These may be used to transfer the entire building load to some stronger stratum at depth and this is called "end bearing".

BOX 3.1 VARIATIONS FROM LITHOSTATIC STRESS CONDITIONS

EXAMPLE STRESS CALCULATIONS

Generally, stresses are estimated by calculating the total stress beneath a vertical column of soil based on unit weight measurements. Effective stress is estimated by subtracting measured or estimated water pressure from the total stress due to the bulk weight of the soil or rock (including contained water).

In Figure B3.1.1 a ground profile is shown with sand overlying clay. The water table (upper surface of saturated ground) is 4 m below ground level (mbGL).

The unit weight (γ) of the damp sand above the water table is 16 kN/m^3; the unit weight below the water table, sand plus pores full of water (γ_{sat}), is 19 kN/m^3. The underlying saturated clay has unit weight, γ_{sat} = 21 kN/m^3. The unit weight of fresh water, γ_w, is about 9.81 kN/m^3 (10 is generally a near-enough approximation given other assumptions).

We wish to estimate the vertical stress at the crown of a tunnel to be constructed at a depth of 10 mbGL.

As shown in Figure B3.1.2

At depth	*Total vertical stress (σ_v) kN/m^2*	*Water pressure (u) kN/m^2*	*Effective vertical stress (σ') kN/m^2*
4 m	4 × 16 = 64	0	64
8 m	64 + 4 × 19 = 140	4 × 10 = 40	140–40 = 100
10 m, at tunnel crown	140 + 2 × 21 = 182	6 × 10 = 60	182–60 = 122

Therefore, before tunnel construction, the estimated, vertical effective stress at the tunnel crown is 122 kN/m^2. During construction, due to seepage into the tunnel the water table would be lowered or this might be done deliberately to excavate "in the dry" to avoid flowing or ravelling of the soil into the tunnel. If the water pressure dropped so the effective stress would increase. If the water table was lowered so that water pressure was zero at tunnel crown level then the effective stress would equal the total stress (= 182 kN/m^2).[5]

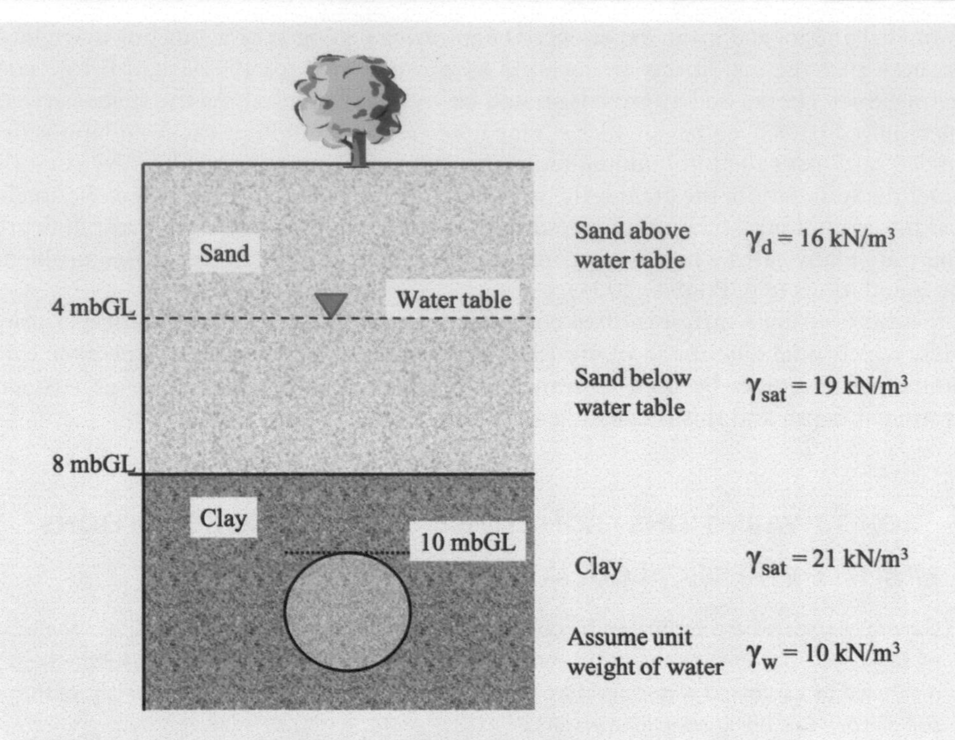

Figure B3.1.1 Example ground profile with tunnel passing through it.

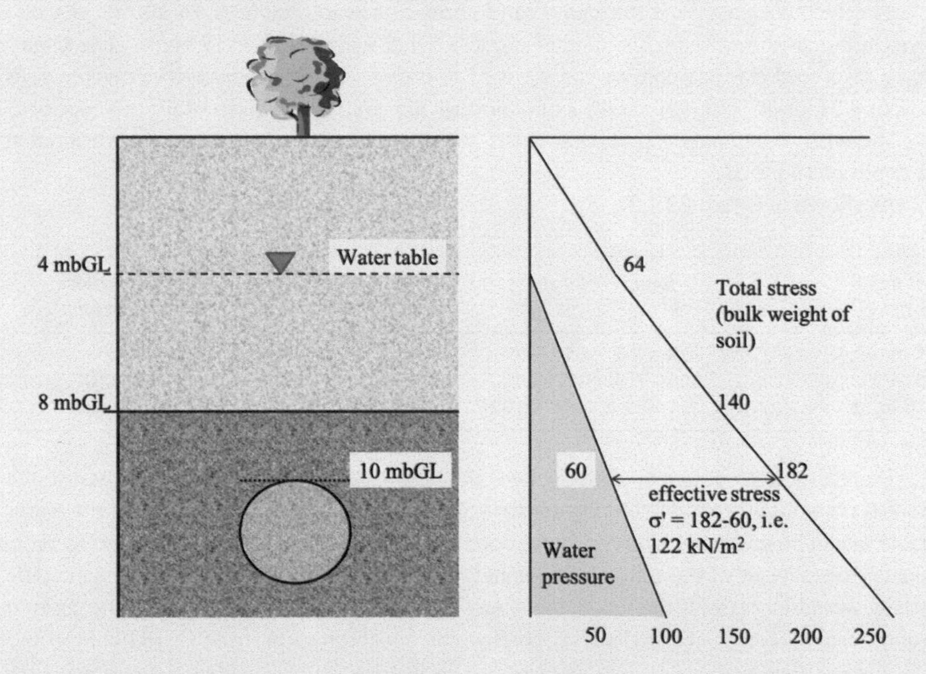

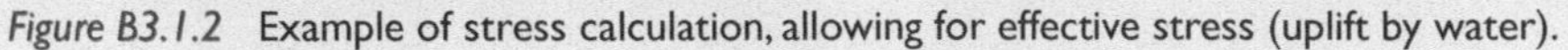

Figure B3.1.2 Example of stress calculation, allowing for effective stress (uplift by water).

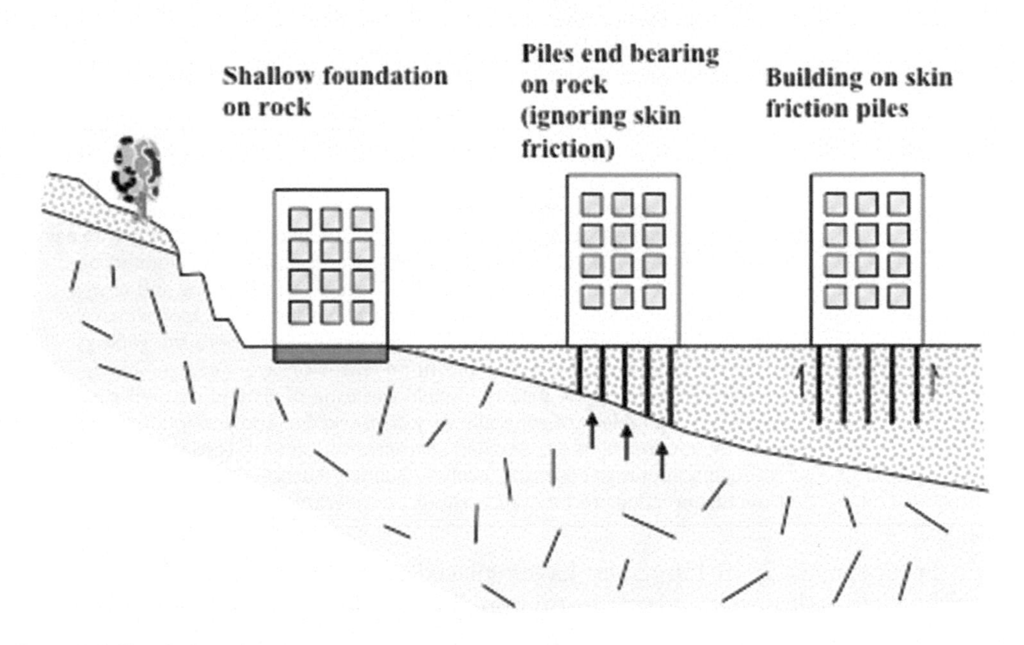

Figure 3.6 Simple foundation options.

Other piles rely on support gained from the underling ground profile by "skin friction", for example, by driving a pile into sand until it can be driven no further. Many piles are designed to be part end bearing and partly relying on skin friction (Figure 3.6).

Foundations are the interface between a building and the ground and transfer loads from the building to the underlying soil and rock. Detailed and practical guidance on foundation design and construction issues is given by Tomlinson (2001). Wyllie (1999) deals specifically with foundations on rock. If ground conditions are suitable then shallow foundations are used because of cost considerations. These include strip footings beneath the walls of a house, pads beneath columns for a steel or concrete-framed structure or a raft supporting several loading columns and walls.

3.5.3 Shallow foundations

For traditional design involving a single Factor of Safety, which is probably the easiest to understand and still employed as the standard approach to design in many parts of the world, the following definitions are used:

Bearing Pressure	The net loading pressure: load from structure, divided by the area of the foundation, minus the weight of material removed from the excavation.
Ultimate Bearing Capacity	The loading pressure at which the ground fails. This is the same as the "ultimate limit state" in the limit state approach (Eurocode 7).

Allowable Bearing Pressure	The maximum loading pressure that meets two criteria: 1. An adequate Factor of Safety against failure, and 2. Settlement within tolerance of the structure (specific to the particular structure).
Presumed Bearing Pressure	A net loading pressure considered appropriate for a given ground condition based usually on local experience and incorporated in Building Regulations or Codes of Practice such as BS 8004 (UK) (British Standards Institution, 1986) and CP4 (Singapore Standard, 2003). Typical bearing values are presented in Table 3.2 and can be used for preliminary design purposes. They allow the practicability of foundation options to be assessed and to select appropriate ground investigation, testing and design methods. Presumed values are only appropriate if the site is approximately level (not for example at the top of a steep slope) and where the geology is relatively uniform and isotropic with no lenses or layers of significantly weaker or compressible material within the zone of ground that will be stressed. Such tables are generally very conservative and economies can be made by conducting more detailed characterisation with testing and analysis although sometimes regulating bodies (Building Authorities) may be loath to allow higher values to be used without considerable justification.

In Europe, since 2010 Eurocodes have replaced national standards and should be used for design (British Standards Institution, 2004). The Ultimate Limit State (ULS) is essentially the same as Ultimate Bearing Capacity but with possible failure modes spelt out, including sliding resistance and structural capacity, heave, piping and so on, which were implicit in BS 8004 approach as factors that a responsible geotechnical engineer should consider. The Serviceability Limit State (SLS) of Eurocode 7 is defined as "States that correspond to conditions beyond which specified service requirements for a structure or structural member are no longer met" and this equates effectively to the idea of Allowable Bearing Pressure as far as settlement is concerned but includes other considerations such as vibration annoyance to neighbours, and so on – again factors that would usually be considered automatically by experienced and responsible geotechnical engineers when adopting a BS-type approach.

From Table 3.2 it can be seen that, for rock, the two governing parameters are generally taken to be uniaxial compressive strength and degree of fracturing and this is expressed in charts presented in BS 8004 and similar standards world-wide. For rock such as sandstone or granite with an intact compressive strength of 12.5 MPa (just broken by hand) the allowable bearing pressure would also be 12.5 MPa provided discontinuities were to be widely spaced apart, reducing to about 10 MPa as discontinuity spacing increases to about 0.5 m and reducing to 2.5 MPa when discontinuity spacing is 150 mm. If the fracturing is particularly adverse or includes discontinuities with low shear strength that could combine to form a failing wedge, then this needs specific consideration and analysis as dealt with in Goodman (1989, 1993) and Wyllie (1999).

Variability across the foundation footprint may also be an issue. If there are soft or weathered pockets, these may need to be excavated and replaced with concrete or other suitable material. Karstic conditions with voids at depth that may be particularly difficult to investigate comprehensively can pose particular difficulties for foundation design and construction as illustrated by a case example in Chapter 8 and discussed

Table 3.2 Examples of presumed bearing pressures for foundations

	Examples of rock and soil type (indicative only)		*Presumed bearing value (MPa)*	
ROCK	Bearing on surface of rock		Strip Footings < 3 m wide. Length not more than ten times width	
	Strong. Discontinuity spacing more than 200 mm		10 to 12.5	
	Strong. Discontinuity spacing 60–200 mm		5 to 10	
	Moderately strong. Discontinuity spacing 60–200 mm		1 to 5	
	Notes: Figures given are for igneous rocks, well cemented sandstone, mudstone and schist/slate with flat lying cleavage/foliation. For other rock types see references quoted. Strength definitions are from BS 5930:1999 and Hong Kong's Geoguide 3 Strong rock (σ_c = 50 to 100 MPa) requires more than one hammer blow to break Moderately strong rock (σ_c = 12.5 to 50 MPa) core cannot be broken by hand			
SOIL	*Sand & Gravel: foundations at least 0.75 m below ground level*	*SPT N-value*	*Foundation width*	
			<1 m	*<2 m*
	Very dense	> 50	0.8	0.6
	Dense	30–50	0.5–0.8	0.4–0.6
	Medium dense	10–30	0.15–0.5	0.1–0.4
	Loose	5–10	0.05–0.15	0.05–0.1
	Clay: foundations at least 1 m below ground level	*Undrained Shear Strength (MPa)*	*Foundation width*	
			<1m	*<2m*
	Hard	> 0.30	0.8	0.6
	Very stiff	0.15–0.30	0.4–0.8	0.3–0.5
	Stiff	0.075–0.15	0.2–0.4	0.15–0.25
	Firm	0.04–0.075	0.1–0.2	0.075–0.1
	Soft	0.02–0.04	0.05–0.1	0.025–0.05

by Houghten & Wong (1990). Conversely if there are particularly strong areas – for example, an igneous dyke through otherwise weak rock, in a pad foundation – then this must be accounted for – otherwise the foundation may fail structurally. In all cases it is essential to check any assumptions from preliminary design as the foundation excavation is exposed. If the ground is worse than anticipated then re-design may be required. In severe cases where, for example, a major fault is exposed unexpectedly, the required change in design may be drastic, but that is the price paid for an inadequate site investigation. Time must be allowed for checking during construction and taking any actions that prove necessary.

For soils, compressibility and settlement is often the main concern. The presumed values given in Table 3.2 should restrict settlement to less than 50 mm in the long term, but estimates may be widely in error and even supposedly sophisticated methods of prediction are often inaccurate. For foundations on granular soils empirical methods relying on SPT or CPT data tend to be used for predicting settlement. Burland & Burbidge (1985) compiled data for sand and gravel and showed that predictions of settlement are often in error by factors of 2 or more. Das & Sivakugan (2007) provide an updated review.

For cohesive soil, where relatively undisturbed samples can be taken to the laboratory, oedometer tests are used to determine settlement potential and to predict rate of consolidation (Chapter 7). Estimates of settlement can be made given the thicknesses of the various strata in the ground profile, their compressibility and the stress changes. Details are given in many references including Tomlinson (2001) and Bowles (1996). For major structures, engineers will often carry out numerical modelling using software such as Plaxis or FLAC, which can be used for sensitivity studies. Such software is also used to predict deformations during different stages of excavation and construction and to determine support requirements.

3.5.4 Hazards for buildings

A normal ground investigation might not provide warning of the many potential hazards that might affect buildings. These include factors such as general site instability leading to landslides or coastal sites where the coast line is retreating. The former is best investigated by expert geomorphologists who can identify hummocks and other ground features as signs of ongoing movement. The latter is just a matter of common sense and reading the literature. Remember that many former villages are simply under the sea now; nothing really to be done about it (despite the super-engineering efforts in Holland where a third of the country is below sea level). Other hazards to low-rise buildings include expansive clays; beware trees planted too close to the building leading to extraction of water by the roots, and inevitable settlement (Jones & Jefferson, 2012). This is not just a problem with recent clays but also with older mudstones, certainly back to the Carboniferous. Movement of foundations can also be the result of gypsum growth as iron pyrites in the clay or mudstone is oxidised and the acidic groundwater reacts with Ca^+ ions to produce gypsum, jarosite and other minerals the growth of which can cause heave of buildings. Gypsum, conversely, can be washed out below foundations, resulting in sinkholes. Fortunately, such occurrence is relatively rare (Building Research Establishment, 1993). Other hazards to be considered at a site include rockfall, flooding and seismicity as set out in Table 3.3. Further examples of hazards in the UK are discussed by Giles (2020).

3.5.5 Buoyant foundations

If the weight of the soil removed from an excavation is the same as the building constructed within the excavation then theoretically, no settlement should occur, as illustrated schematically in Figure 3.7, and this design concept has been used for many major structures incorporating deep basements, which can be utilised for parking spaces. There may be a need to include holding down piles or anchors in the design to combat any uplift forces. Construction of deep foundation boxes often involves the construction of diaphragm walls using the same techniques as for barrettes discussed below. Once the walls are in place, excavation is conducted inside the walls with either bracing and/or anchorages used to stabilise the works.

Table 3.3 Examples of hazards that might need to be considered, even for low rise buildings

Building on unstable ground	The bearing pressure from Table 3.2 would not deal adequately with sites that are broadly unstable, as in the example of a house in Portugal, built on the edge of a cliff in schist that is degrading. Instability hazards might be recognised by experts who can recognise the signs of ongoing landslides, often by using air photograph interpretation.	 *Figure T3.3.1* Failure in schist slope, adjacent to house foundation, Portugal.
Coastal erosion	Other sites to avoid include retreating coastal areas such as Holderness in Yorkshire, UK, where cliffs are retreating at more than 1 metre per year. If you wish to pass on your house to your grandchildren you should buy at least 200 metres inland.	 *Figure T3.3.2* Coastal erosion leading to undermining of house foundations (photo courtesy of BBC, UK).

(*continued*)

Table 3.3 (*Cont.*)

Old mine workings	Old mine workings can collapse causing a subsidence trough that travels gradually across the countryside. Old shafts can open up in gardens and directly beneath buildings and other structures–best investigated through intelligent drilling (based on type of mining) and geophysics although this is not easy (Clayton et al., 1982)	 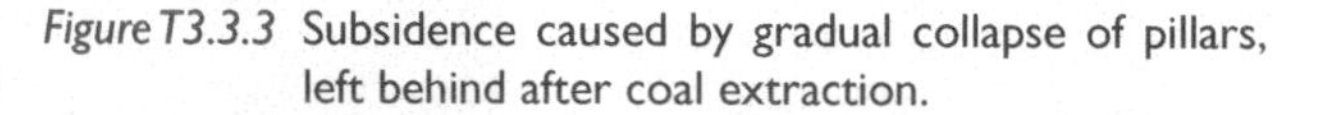 *Figure T3.3.3* Subsidence caused by gradual collapse of pillars, left behind after coal extraction.
Expansive clays	Trees such as willows can extract water beneath foundations especially in times of drought causing expansive clay to shrink with consequent damage. Other expansive soils, which readily change volume with moisture content are a hazard to housing and slopes world-wide. The most common problems are associated with smectites, especially montmorillonite, for example, in India, Africa, Australia and the USA. Residual friction angle can be as low as 7 degrees.	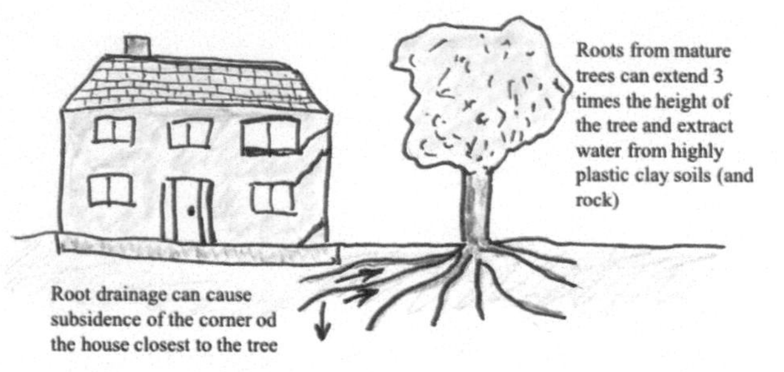  *Figure T3.3.4* Tree too close to building causing settlement.  *Figure T3.3.5* Rapidly degrading slopes in smectite-rich, weak rocks derived from volcanic ash, Turkey.

Table 3.3 (*Cont.*)

Adverse materials	Sites underlain by salt and karstic deposits can cause sinkhole development with obvious consequences (20% of the Earth's surface according to Waltham, 2016)	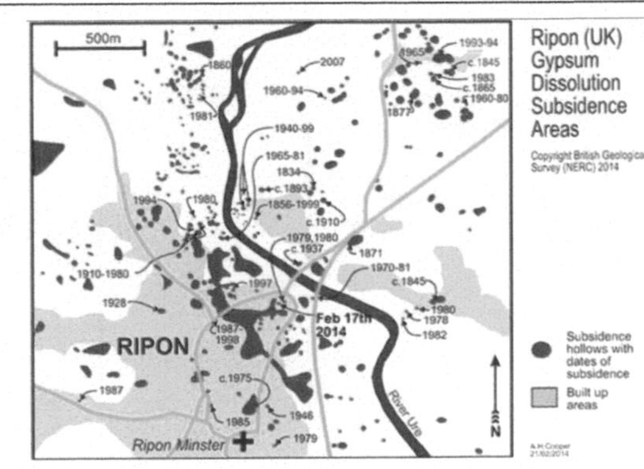 *Figure T3.3.6* Map of Ripon, UK, showing development of sink holes (courtesy of BGS).
Seismicity	In seismic zones, special care is required in the design of foundations. In particular, saturated silt and sand underlying a structure can turn to a liquid during a strong earthquake (liquefaction).	 *Figure T3.3.7* Collapsed house in new development at Erzincan, Turkey, 1992 quake.
Other hazards	Other hazards to consider include flooding (don't buy a property within the *floodplain* of a river!!), settlement due to water and oil extraction and tunnelling, impact by boulders and falling trees from adjacent ground, frost heave and thaw in permafrost areas and other environmental hazards including hazardous gases especially from previous land use.	 *Figure T3.3.8* Planners oblivious to boulder fall risk, South Africa.

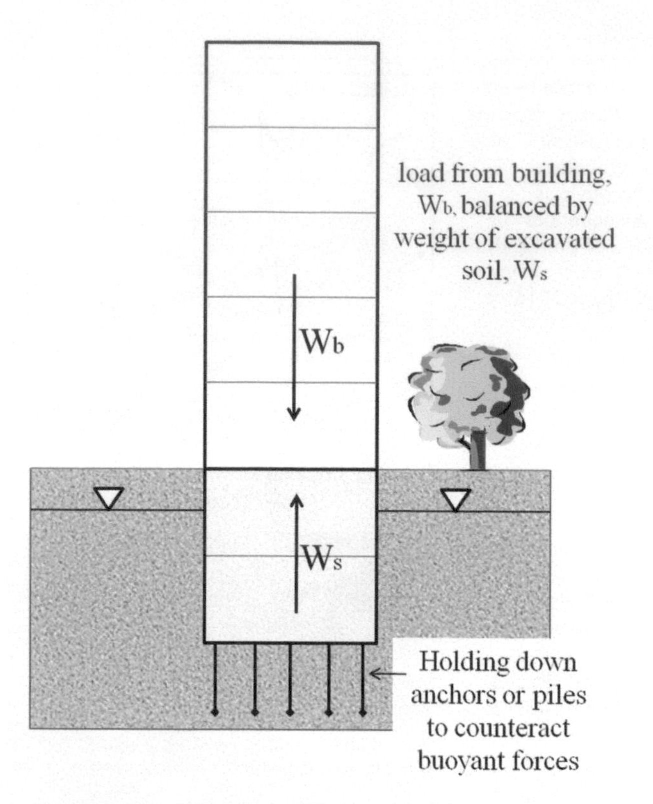

Figure 3.7 Concept of buoyant foundation design. The weight of the building balances the excavated soil so that the net increase or decrease in pressure is minimised.

3.5.6 Deep foundations

3.5.6.1 Piled foundations

Piles are used to transfer building loads, via pile caps, to deeper levels in the ground profile. There are two main types driven and bored. Driven piles are hammered into the ground and are also termed displacement piles. Hammering is sometimes done by dropping a large weight on the top of the pile from a crane or by using a diesel or hydraulic hammer (Figure 3.8). Bored piles are generally constructed using bucket augers, soil grabs and rock roller bits with heavy-duty rock cutting tools used to grind their way into the underlying rock and to form rock sockets as necessary. Even using the most powerful equipment formation of sockets can take a very long time, advancing perhaps only 100 mm per hour in strong rock, and therefore can be relatively expensive so designers should be wary of being ultra conservative in their specification of socket length.

Figure 3.8 Diesel hammer (centre of photo) being used to drive pre-stressed concrete piles, Drax Power Station, UK. Elsewhere piles are being "pitched" into holes formed by auger. In the background the kentledge can be seen for a proof test on a working pile.

3.5.6.2 Driven piles

Driven piles are generally made of timber, steel or concrete. Figure 3.9 shows concrete piles being manufactured on site in a factory-type operation to allow 20,000 piles to be driven in just 18 months for Drax power station completion (Hencher & Mallard, 1989). The purpose-made pile beds were heated to allow rapid curing of concrete and the piles were pre-stressed to improve their resistance to tensile stresses and to allow the piles to be lifted from their forms quickly. For most sites, piles will be manufactured off site sometimes as different lengths that are joined together on site to suit requirements. One of the advantages of using driven piles is that an estimate can be made of the driving resistance given the known energy being used to drive the pile and the penetration into the ground per blow of the hammer. Piles are therefore driven to a "set", which is a pre-defined advance rate (such as 25 mm for 10 blows by the hammer). However, resistance during driving may not always give a very good indication of how the pile will behave under working conditions because of "false sets", generally due to water pressure effects as described for the Drax operation in Chapter 8. Driving resistance and "set" are however part of the process of quality control during construction. Pile driving analysers (PDAs) using accelerometers and other instruments attached to the pile can be used to estimate driving resistance in a more sophisticated way than the traditional method of measuring the "quake" with a pencil although the same limitations apply regarding whether or not dynamic behaviour is a

Figure 3.9 Piles being cast in formers, Drax Power Station, UK. Note lifting eyes cast into the concrete piles, steel plates at end of piles (trapezoidal) and pre-stressing cables, which are to be cut before lifting piles from the casting beds.

reliable indicator of future performance. PDAs are sometimes used after the pile has been installed (both driven and bored piles) to test its capacity but this can be somewhat of a black art with many assumptions being made and the method is certainly not fool-proof or as reliable as full static load tests as discussed below.

3.5.6.3 *Bored piles*

Bored piles are excavated as described earlier. Temporary or permanent steel tubes (casing) may be used to prevent collapse of the hole and, if the hole is formed below the water table, often bentonite or some other mud or polymer is used to support the sides of the hole. Once the hole has been completed and cleaned out then a steel reinforcing cage is introduced and finally concreting carried out. Concrete needs to be "tremied" by a pipe from the surface to the bottom of the hole. This avoids the concrete disaggregating and the concrete will hopefully displace soft sediment that might have accumulated at the bottom of the bored hole after the final clean out. It will also displace the bentonite slurry or water from the bored pile excavation so this can be a very messy operation. Despite best efforts

Figure 3.10 Reinforcing cage for bored pile with included tubes to allow proof drilling through the toes of the completed pile and cross-hole geophysical testing to prove integrity, Hong Kong.

"soft toes" of sediment will still sometimes occur (perhaps associated with the removal of temporary casing) and sometimes ground movements occur causing "necking" of piles. Clearly there is a need for high quality work and for close supervision. Currently in Hong Kong all bored piles are installed with steel tubes attached to the reinforcing cage (Figure 3.10). After concreting, rotary drilling is carried out down one of the tubes, through the concrete and into the underlying natural ground to prove that the pile is founded as designed and that there is no soft ground beneath the pile. If there is, then remedial measures such as pressure grouting might be needed. Other tubes installed through the concrete are used to carry out geophysical cross-hole tests (seismic) to check for concrete "necking" and other construction defects. In severe cases, piles may prove inadequate to carry the loads and remedial works are required. This might not be discovered until the superstructure is constructed. In one extreme case in Hong Kong, two 44-storey tower blocks had to be demolished. Such problems may be put down to workmanship, the inherent difficulties of the operation, poor investigation and design and sometimes fraud (Hencher et al., 2005).

Once the piling is completed a pile cap is constructed as a reinforced box of concrete that bridges between several piles to support major columns in the superstructure.

3.5.6.4 Design

Piles are designed to suit the ground profile. If rockhead is at relatively shallow depth and the overlying soil does not contain boulders that could cause difficulties, then driven piles might be adopted, end bearing onto the rock. At Drax Power Station, the piles were driven to found several metres into dense sand overlying sandstone, thereby picking up some skin friction as well as end bearing. If there is no rock then the piles will need to gain their resistance mostly from skin friction in the soil. For example, the Sutong Bridge across the Yangtze River, China which was (in 2011) the longest cable stay bridge in the world with a main span of 1088 m is founded on bored piles taken to 117 m and relying upon skin friction from alluvial sediments as shown in Figure 3.11.

Ways to estimate skin friction parameters and end bearing resistance are given in textbooks such as Tomlinson (2001) and might be governed by standards such as AASHTO (2007) used as the basis for design of the 2nd Incheon crossing completed in 2009. The principles are quite simple; skin friction is calculated as soil shear strength multiplied by an adhesion factor, then multiplied by the surface area of the pile shaft. End bearing is often calculated as an empirical value for the soil or rock quality multiplied by the basal area of the pile. At some sites the bottom end of the pile is enlarged by under-reaming to increase the end bearing contribution although sometimes the difficulty of this operation is hardly justified by the increase in pile capacity that might ensue.

A worked example of pile design to Eurocode 7 using partial factors specified uniquely for the UK (to correlate with traditional design experience) is presented in Box 3.3 based on one presented by Bond & Simpson (2010). Other countries might use different partial factors and other approaches as allowed in the Eurocode. In the example presented, the main unknowns – variable live load, shaft resistance and base resistance are factored

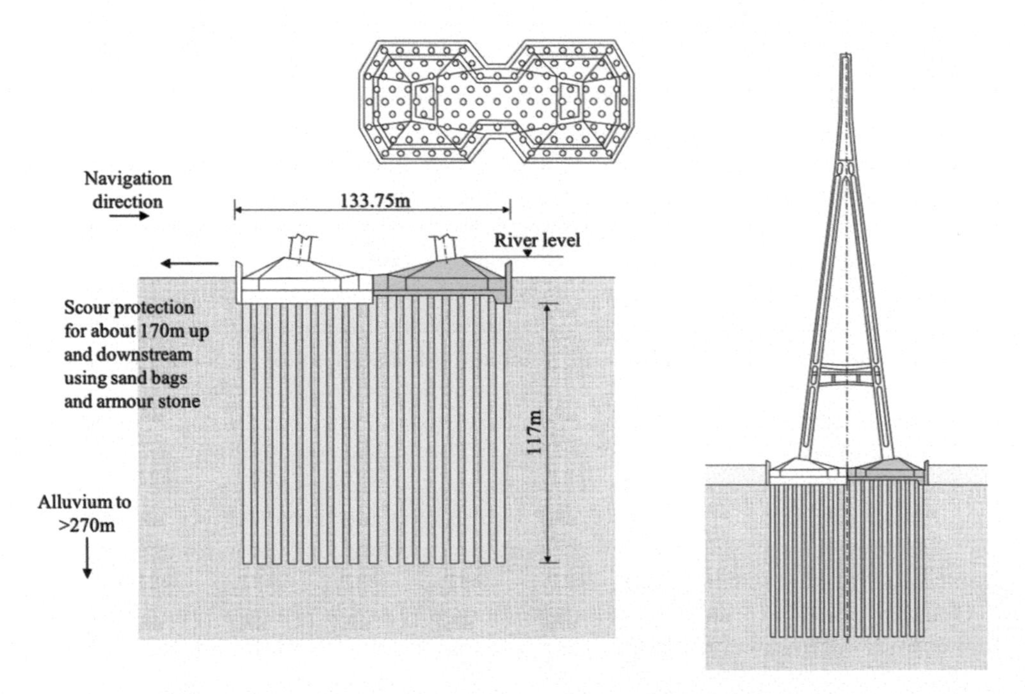

Figure 3.11 Sutong Bridge, China, designed to be carried by piles relying on skin friction.

up and down as appropriate towards a safe solution – the results are compared to the FoS as determined using a traditional approach – best estimate of strength divided by best estimate of loading. It is to be noted from this example that whichever approach, there is considerable judgement and approximation involved. Shear strength is taken as "undrained" in this example, which is conceptually questionable for the long-term; adhesion factor estimates range from 0.3 to 0.9 for different soils. If an effective stress approach was adopted – as would generally be done for sand and weathered rock – then estimates would be needed of stress conditions and "shaft resistance coefficients", which also requires estimation and judgement. Workmanship may also play a key role in whether or not shaft friction will be mobilised and whether the base of a bored pile excavation is properly cleaned out prior to concreting. The use of a partial factors' approach does concentrate attention on where the key unknowns are (rather than just on geometry and fixed loads) but doesn't take away the need for proper ground characterisation, analysis and design judgement. The fixed nature of the partial factors might seem rather prescriptive to cover all soil, rock and founding situations. Selection of parameters, adhesion and shaft resistance factors are reviewed well in GEO (2006) and the use of Eurocode 7 for design is summarised by Bond & Simpson (op cit).

BOX 3.3 EXAMPLE CALCULATIONS FOR PILE DESIGN

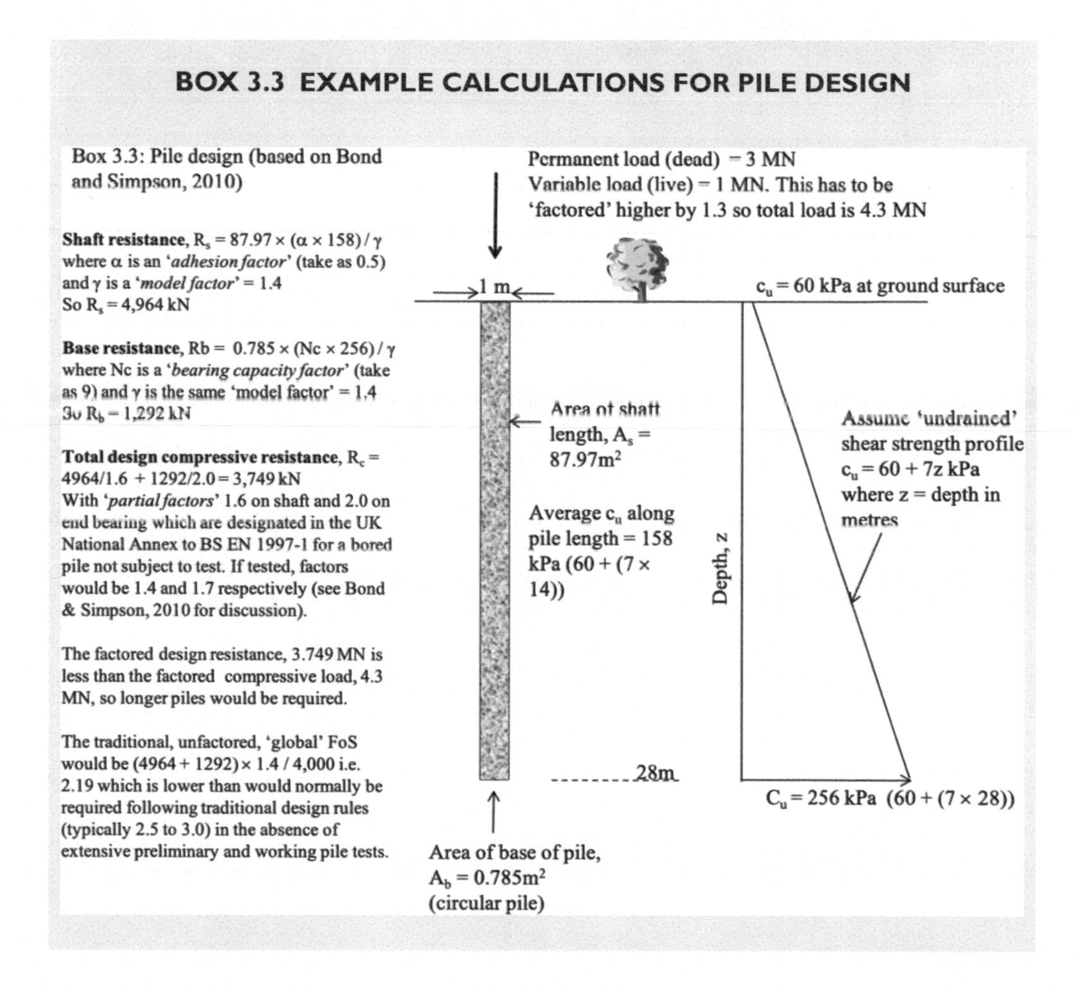

Figure 3.12 Pile test set up with kentledge, Singapore.

A site-specific way to obtain design parameter values, especially for large projects, is to install test piles and measure their performance up to perhaps 2.5 times the design load of the working piles. Test piles are often instrumented along their length using strain gauges so that the actual resistance being provided by the ground can be measured throughout the full profile and these parameters can be used in the design of other piles. Traditionally piles are loaded from the top using "kentledge" of concrete blocks or steel (Figure 3.12). Jacks are used to push the pile into the ground whilst the kentledge provides the reaction. One of the difficulties of this is that much of the support comes from the upper soil at early stages of the test and there is little idea of how the toe is performing until a test approaches failure (Figure 3.13). A system has been introduced where "Osterberg" cells are incorporated into the pile construction at depth and then expanded against the test pile both upwards and downwards (Figure 3.14). The end bearing resistance below the cell is balanced by the skin friction from the soil above the cell. This system was used for the Incheon Bridge design, using up to five cells in a single 3 metre diameter pile to generate forces of over 30,000 tonnes (Cho et al., 2009b). The obvious advantages include the fact that no reaction is required at the ground surface but a limitation is that the forces upwards must be balanced by those downwards, which would be difficult to achieve where the pile is mostly end bearing.

An additional aspect to be considered in pile design is possible future settlement of the ground around the pile due to self-weight, earthquake liquefaction or perhaps

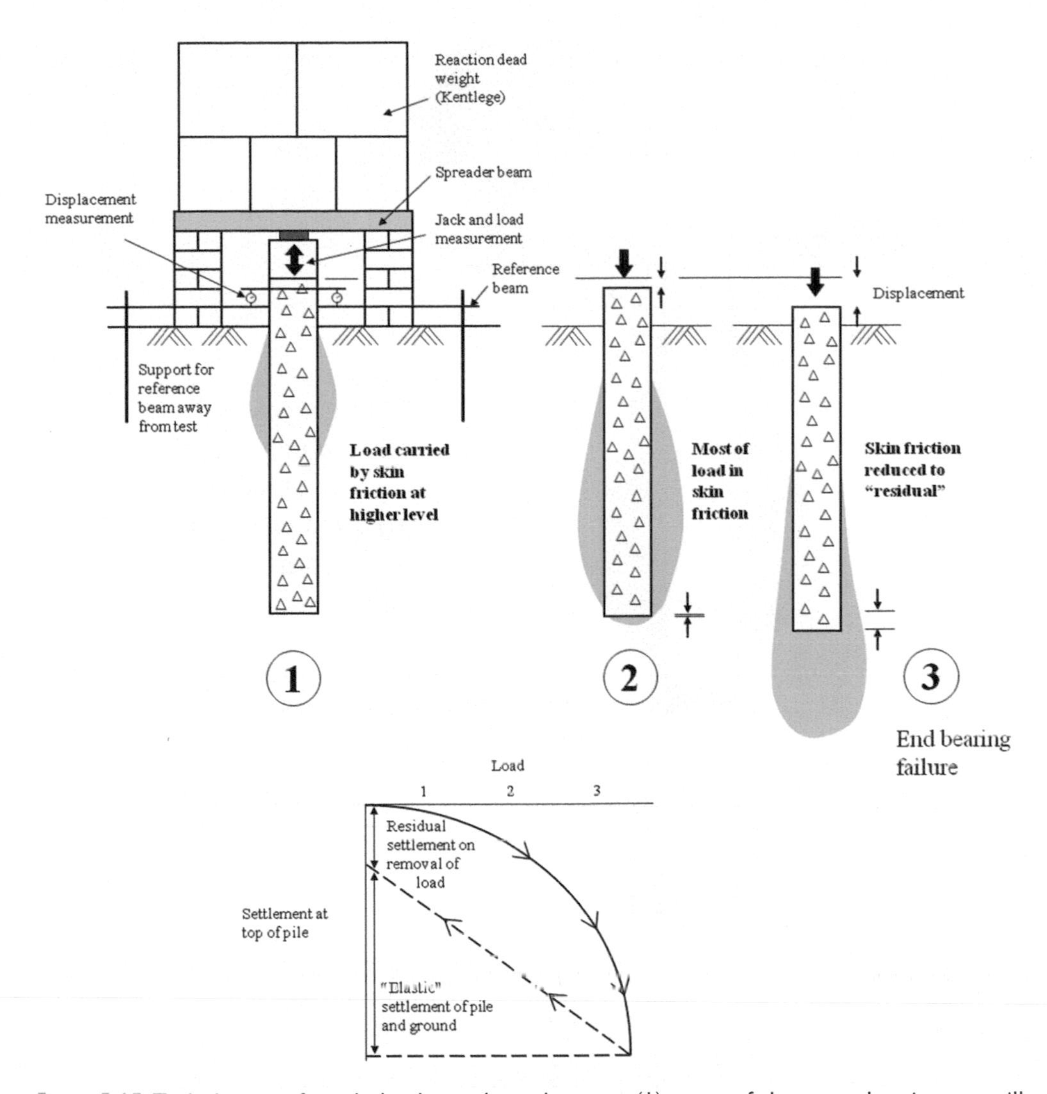

Figure 3.13 Typical set up for pile load test. At early stages (1), most of the ground resistance will come from skin friction at shallow depths. End bearing is not mobilised until later stages (2) and (3) of the test (depending on the configuration of the pile and ground profile). The rate of settlement increases as the ground resistance becomes fully mobilised and there will be some permanent displacement (residual settlement) once the pile is unloaded.

groundwater extraction, which can result in a drag down force on the pile known as negative skin friction and illustrated in Figure 3.15. The potential for negative skin friction is generally a matter of engineering judgement based on the ground profile and perceived future usage of the site and applied as a nominal additional load to be carried by the piles.

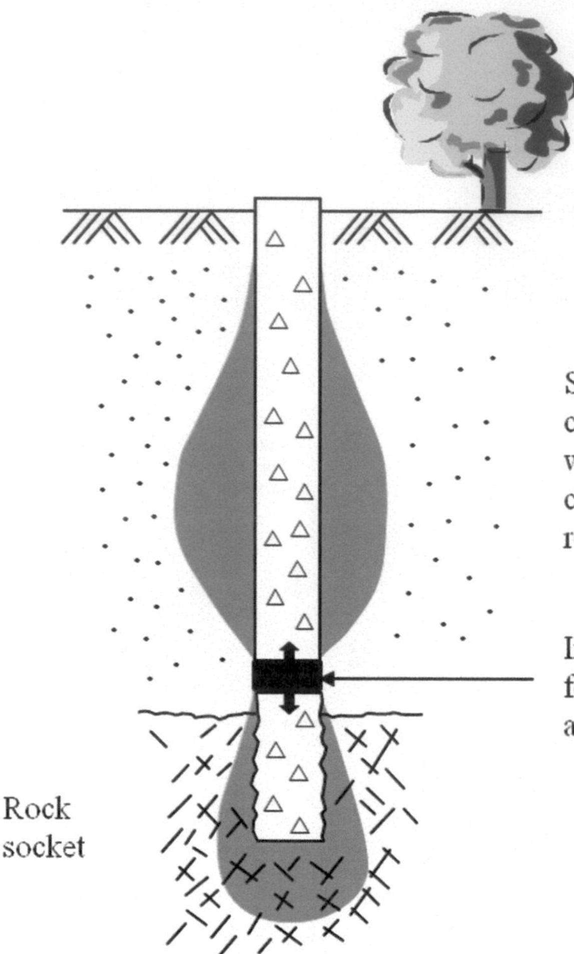
Strain gauges and load cells included in pile tell where the load is being carried (a check on soil and rock parameters)
Included jack expands, forcing upper part of pile against lower part
Rock socket

Figure 3.14 Pile test using an integral jack or set of jacks. This set up allows the end bearing part of the pile to be jacked against the upper parts (skin friction). If strain gauges are built into the pile then a good interpretation can be made of ground parameters.

3.5.6.5 *Proof testing*

Proof tests are typically carried out on one in a hundred piles or so. The test pile should be selected by the supervising engineer, after construction and with no pre-warning to the Contractor so that he does not exercise special care in its construction. Full loading tests are carried out with kentledge or some other reaction system such as ground anchors and should be taken up to loads of perhaps 1.5 times the working load for the pile. The displacement during the test (partly elastic deformation of the pile) and residual settlement after the test is completed are used as criteria of whether the tested pile and its neighbours are acceptable (Figure 3.8). If not, then additional piles may need to be installed and the existing piles down rated. There may be time or space restrictions (such tests are very expensive and time consuming) and the Contractor

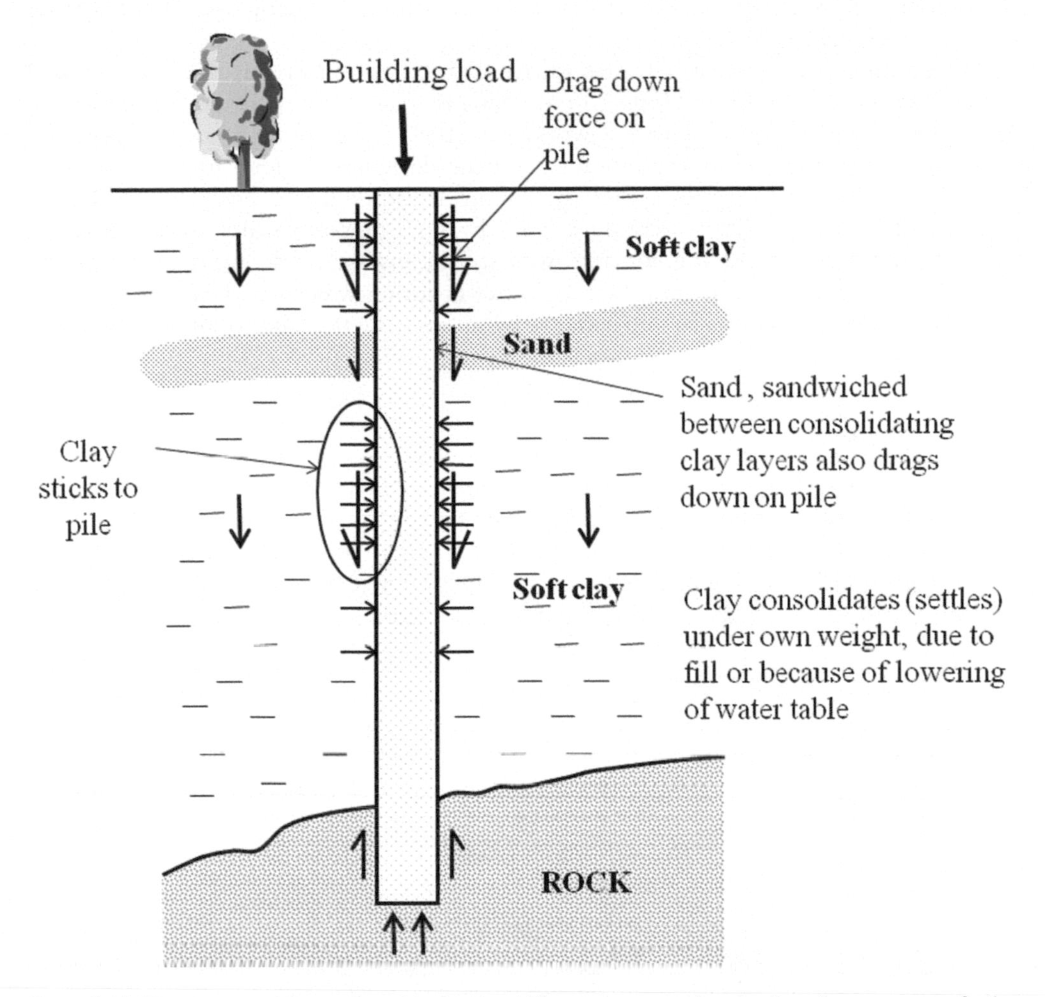
Building load
Drag down force on pile
Soft clay
Sand
Sand, sandwiched between consolidating clay layers also drags down on pile
Clay sticks to pile
Soft clay
Clay consolidates (settles) under own weight, due to fill or because of lowering of water table
ROCK

Figure 3.15 The concept of negative skin friction. Where the ground around a pile or group of piles settles significantly (cm), then the ground will cause a drag down force on the pile. At the same time, the upper parts of the pile cannot provide positive skin resistance.

might urge the use of dynamic pile analysers as an alternative way of proving acceptability. As noted earlier such tests can be unreliable and may give no measure of end bearing resistance. Specialist tests are used to determine pile integrity, for example, by using a vibrator to take the pile through a series of frequencies so that its response can be measured. Resonance will indicate the length of the responding section, which will help in deciding whether or not the pile is broken.

3.5.6.6 Barrettes

Barrettes, like piles, are deep foundations but constructed in excavated trenches using special tools called hydro-fraises, often under bentonite to support the sides of the trench. Otherwise construction is similar to a bored pile with a steel cage inserted into the trench prior to concreting. Barrette shapes can follow the geometry of load bearing walls in the finished structure.

An example of the use of barrettes rather than bored piles is for the International Commerce Centre (ICC) in Hong Kong. The 118-storey building was the tallest in Hong Kong and fifth tallest in the world (in 2011). Granite bedrock is reportedly 60–130 metres deep below the building and the designers decided to use 241 post-grouted rectangular barrettes rather than more traditional end bearing bored piles (Tam, 2010). The barrettes were "cracked" by high-pressure water injection down pre-installed pipes whilst the concrete was still at low strength. Once the concrete had reached its 21-day strength, high-pressure grouting was carried out through the "cracked" path around the barrettes metre by metre from the base to improve the skin frictional resistance.

3.5.6.7 *Caissons*

Caissons are large box structures formed of steel or concrete and are used as a common solution for bridge foundations off shore. The box is typically constructed on shore then floated and towed to its location where it is sunk. Sometimes caissons are sunk into the ground by driving and digging, elsewhere they just sit on a prepared surface on the sea floor. Different types are illustrated schematically in Figure 3.16. Once the

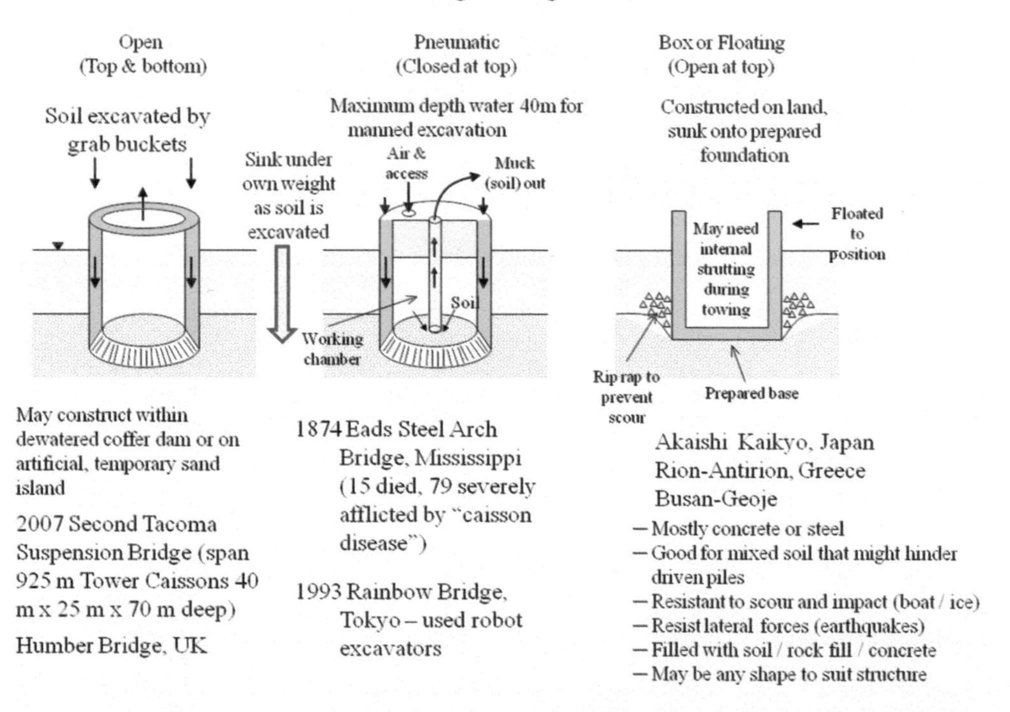

Figure 3.16 Different types of caisson commonly used for large bridge foundations with examples.

caisson is in place/sunk to the required depth, then it is backfilled with rock and concrete. Caissons are also often used to form sea walls for reclamation schemes, the boxes formed on land then floated and towed to position where they are sunk onto prepared foundations and then backfilled.

3.6 TUNNELS AND CAVERNS

3.6.1 Tunnels

Engineering geologists (EG) are often closely involved in the investigation of tunnel routes, preparation of Reference Ground Conditions for contracts, tunnel design and during construction. Tunnelling has been carried out since ancient times; originally probably making use of techniques developed in mining, which date back many thousands of years. Aqueduct tunnels for water supply were constructed in ancient Rome, Greece and the Middle East. Modern tunnelling really started with the development of extensive canal systems in the UK and mainland Europe in the 16th and 17th centuries, where the alternatives to tunnelling were either long detours around hills or deep cuttings. Originally tunnels were hand-dug, using gunpowder where necessary. Many modern tunnels are constructed in similar fashion; improvements include computer-controlled drilling for blast holes, rapid and sophisticated support methods and much better ventilation and safety systems. Generally, "drill and blast" tunnels involve a cycle of drilling, blasting, "mucking out" and support, generally advancing a few metres per "pull". "Hand"-excavation is also sometimes used employing essentially mining techniques, perhaps using powerful "road headers" moved across the face to dislodge soil and rock but not taking out the full tunnel profile at one time. It is often the EG who, during this type of tunnelling will examine the exposed ground and make decisions on the degree of temporary support that is required together with any special requirements for further investigation or ground treatment before the tunnel is advanced. One advantage of drill and blast is that methods can be modified rapidly to suit rapidly changing and difficult ground conditions. Another advantage is that mobilisation is fast – tunnelling can be begun quickly and carried out in remote areas of the world. The disadvantage is that it is often much slower than by using a tunnel boring machine (TBM). The options as to whether to drill and blast or to use a TBM are complex and a matter of the availability of workers, length of tunnel and difficulty of conditions. Most tunnels in Scandinavia, for example, use drill and blast techniques because of familiarity. Both major types of tunnel can be risky and it is often considered rather better to use drill and blast techniques, though slower, as these allow adaption to changing geological conditions. A TBM, provided it is designed properly and is staffed by competent persons, can generally succeed even in harsh environments such as the faulted and squeezing ground below the Himalayas (Hencher, 2015).

TBMs were gradually introduced for excavation in particular for the underground railway tunnels in London. In an early attempt at tunnelling beneath the Channel a 2.13 metre diameter boring machine tunnelled 1,893 metres in 1881 from the UK side and a similar machine advanced 1,669 metres from its portal in France. Today there is a huge range of TBMs available ranging from ones specifically designed for hard rock through to ones that tunnel through soft water-logged sediments using pressurised slurry in front of the machine to support the soil. Pre-cast segmental tunnel liners can be erected directly following the cutting part of the machine and bolted together with gaskets to form a watertight tube. TBMs can be highly successful with hundreds of metres advance in a single month compared perhaps to a hundred metres using drill and blast methods so for long tunnels, the cost and possible delays in manufacturing a TBM for the job may be justified. Quite often however TBMs run into difficulties from ground conditions that can slow them down or even stop them completely despite huge sophistication in their design. It is very important in tunnelling to consider all the potential hazards that might be encountered and to make sure that the TBM can cope as addressed in Chapters 4, 6 and Appendix D. This is even more important for TBMs than drill and blast tunnels because it may be very difficult to modify the method of working and ground support. Whereas in a drill and blast tunnel the engineering geologist can examine the face and tunnel walls before and after a blast, in a tunnel excavated by TBM all that can often be seen is the spoil being excavated (often contaminated with drilling mud) so it is rather difficult to confirm that the ground conditions are as anticipated. Engineering geologists are therefore relatively little used during TBM construction, until something goes wrong and needs investigation.

The author has experience of tunnelling through weathered rock in Singapore using a very sophisticated slurry machine where, in one section of the tunnel, the TBM was slowed to a standstill because of the high strength of the rock and the lack of natural discontinuities. Elsewhere on the same drive, the rock was weathered to a residual soil that was so clay-rich that the slurry treatment plant could not cope, again causing delays and necessitating re-design of the treatment plant. For the same TBM machine the machine operators had difficulty in selecting the pressures to adopt in the slurry. If the pressure was too low, the ground collapsed, if too high, slurry was ejected into the street above. Shirlaw et al. (2000) give examples of problems in tunnelling through weathered rock terrain and other examples, especially in "squeezing ground" and zones of high stress are given by Barla & Pelizza (2000). Further case examples are given in Chapter 8.

In tunnels, key aspects to consider are safety for the workers and public above the tunnel, the feasibility of different excavation techniques, limiting water ingress, stability of the face and side walls during construction and in the longer-term, effect on surrounding structures (mainly settlement, undermining, vibration and noise) and cost.

Shallow tunnels and other excavations such as underground railway stations may be constructed as concrete boxes in open excavations from the ground surface (maybe 50 metres deep). "Immersed tube tunnels" are formed from boxes of concrete constructed on land and then floated by barge to position where they are sunk to the prepared river or sea bed and bolted together.

In bored tunnels in soil at relatively shallow depth, the main concerns will usually be stability of the soil in the face, inflow of water and settlement of adjacent structures. If necessary the ground can be pre-treated by using grout, frozen or by using compressed air to restrict water inflow and stabilise the ground. When using compressed air there are considerable health considerations and regulations and there is danger of a blow-out occurring, particularly where running close to some pre-existing structure such as a well or borehole (Muir Wood, 2000).

In shallow tunnels in rock, the prime considerations will be blocks of rock failing into the opening or encountering faults, which may be full of water. Small failures generally are to be expected in drill and blast tunnels or are protected by the shield in TBM operations. Generally, in fractured rock, shotcrete is applied quickly together with steel mesh and rock bolts as necessary to stabilise zones of potentially unstable rock. Similar temporary support systems are sometimes used in "soil" tunnels although the principles are different as discussed in Chapter 6 and there have been many failures in soil tunnels when trying to adopt an observational approach to temporary support (essentially NATM).

When a tunnel is formed, the existing ground stresses have to "flow" around the created void (Figure 3.17). Depending on the ratio of σ_1 (maximum principal stress) to σ_3 (minimum principal stress) tensile zones will develop although this may be less than the available rock strength. Generally, these will not cause any great problem other than some minor cracking and possibly some water ingress at shallower depths. In deep tunnels however the concentration of compressive stress in side walls to a level of the intact rock strength, combined with lack of confining stress can lead to spalling off and rock bursts (Carter et al., 2008). Such phenomena are not really a problem for most tunnels but can be significant for those constructed deep through a mountain chain or for deep mining. Hoek (2000) reports particular problems for tunnel stability where the *in situ* stress was 5 times the rock mass strength.

3.6.2 General considerations for tunnelling

Tunnels will be constructed as part of an overall project, for example, water supply, drainage, rail, road, or in connection with power generation. As a result, there may be little flexibility over route and therefore geological and hydrogeological conditions and size and shape of tunnel. It is up to the engineering team to come up with a cost-effective solution.

One factor that will influence the chosen method of construction and lining (or not) are the final finish requirements for road and rail tunnels and whether or not it might carry water under pressure in hydraulic tunnels. The main issues for the engineering geologist and design team are likely to be the following:

- The geology along the route, how this will affect the selected method of tunnelling and any particular hazards, such as natural caverns, mining or major faults.
- Stress levels and ratio of vertical to horizontal stress. High stress at depth and the concentrations in stress resulting from perturbation of the stress field by the construction can result in failure of the rock, which might result in spalling in brittle rocks or squeezing in generally weaker rocks (Hoek & Brown, 1980; Hoek et al., 1995).
- Hydrogeological conditions, especially possible confined aquifers, and the risk of unacceptable water inflows with extensive ground settlement resulting, and possible flooding. This is always a major issue for undersea tunnels, but can also be a concern for tunnels constructed under land.
- Existing structures that might be adversely affected by the tunnel during construction, for example, by blast vibrations or due to undermining as the tunnel

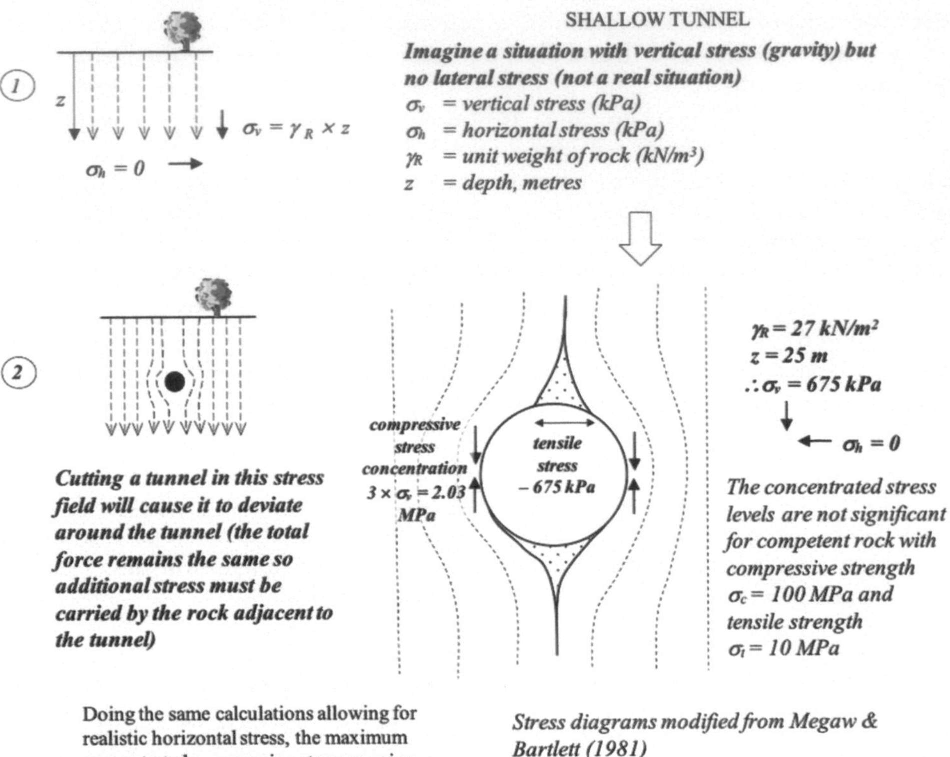

SHALLOW TUNNEL
Imagine a situation with vertical stress (gravity) but no lateral stress (not a real situation)
σ_v = vertical stress (kPa)
σ_h = horizontal stress (kPa)
γ_R = unit weight of rock (kN/m³)
z = depth, metres
1
z
$\sigma_v = \gamma_R \times z$
$\sigma_h = 0$
2
compressive stress concentration $3 \times \sigma_v = 2.03$ MPa
tensile stress − 675 kPa
$\gamma_R = 27$ kN/m²
z = 25 m
$\therefore \sigma_v = 675$ kPa
$\sigma_h = 0$
Cutting a tunnel in this stress field will cause it to deviate around the tunnel (the total force remains the same so additional stress must be carried by the rock adjacent to the tunnel)
The concentrated stress levels are not significant for competent rock with compressive strength $\sigma_c = 100$ MPa and tensile strength $\sigma_t = 10$ MPa

Doing the same calculations allowing for realistic horizontal stress, the maximum concentrated compression stress remains below 3 MPa with lower tensile stress in the tunnel crown. Stress concentrations would not cause cracking and tunnel stability would depend solely on geological structures allowing collapse.

Stress diagrams modified from Megaw & Bartlett (1981)

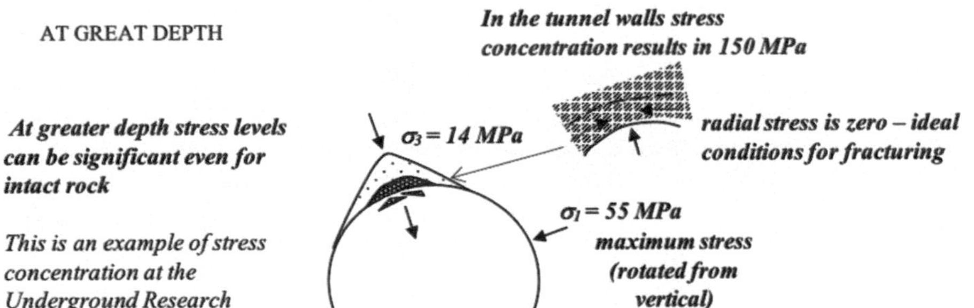

Figure 3.17 Stresses in a shallow and deep tunnel.

passes by. In the longer-term, lowering of groundwater may cause settlement and/or affect water supply boreholes.

As for all geotechnical work, one needs a ground model for design. Because tunnels are often long and may be at great depth it may be impractical to do more than a rather superficial investigation, relying largely on geological mapping and extrapolation of data although if a serious obstacle is anticipated, such as a major fault zone, then boreholes might be targeted at that particular feature, using inclined boreholes or even drilled along the line of the tunnel. An example of this is given in Hencher (2015), where Halcrow designed a ground investigation in the Himalayas, designed to test the qualities of particular rock mass items, rather than the particular tunnel route. The ground investigation unknowns were matched by the design of the TBM, which could and did cater for any particular ground conditions that might be met. Alternatively, a small diameter pilot tunnel might be constructed before the main tunnel – possibly for later use as a drainage or service tunnel – because small diameter tunnels tend to have fewer difficulties (Hoek, 2000). The pilot tunnel essentially works as a large diameter exploratory borehole.

The ground model needs to include estimates of rock or soil quality along the tunnel drive. For rock this is often done using rock mass classifications (RMCs), such as Q, RMR or GSI described in Chapter 6 and Stille & Palmström (2003). This will allow some estimation of support requirements and allow a contractor to choose his method of working and type of machine if a tunnel boring machine (TBM) option is selected (Barton, 2003). The ground model will also be used for hazard and risk analysis as discussed later, and may sometimes be used as the basis for Reference Ground Conditions in geotechnical baseline reports (Chapters 2 & 6) against which any claims for unexpected or differing ground conditions can be judged. As noted in Chapter 2 however, RGCs may be too coarse to represent geological conditions realistically. They may also be open to different interpretations, leading to disputes that are difficult to resolve.

3.6.3 Options for construction

Up to about a century ago, all tunnels in soil or rock were excavated by hand, using explosives where necessary to break up the rock in advance of mucking out. Now many are excavated using powerful machines. The main options generally adopted in modern tunnelling and typical support measures are set out in Table 3.4. The method of tunnelling will often be decided on factors including length of tunnel, availability of TBM, local experience and expertise. In South Korea, for example, most rock tunnels, including very long ones, have been constructed in preference by "drill and blast" rather than TBM. There is a wide variety of tunnel boring machines designed for all kinds of conditions from rock to soft soil. The engineering geologist needs to be able to predict the ground conditions so that the tunnel designers and tendering contractors can select the correct machine. It usually takes a long time to manufacture and launch a TBM with a whole series of ancillary equipment in the following "train" and if the machine proves unsuitable for any reason it can be a costly mistake. Some machines are designed to be able to cope with "mixed ground" conditions but can still run into

Table 3.4 Options for tunnelling

Ground Type	*Excavation*	*Support*
Strong rock	Drill & blast or TBM	Nil or rockbolts
Weak rock	TBM or roadheader	Rockbolts, shotcrete, etc.
Squeezing rock	Roadheader	Variety depending on conditions
Overconsolidated clay	Open-face shielded TBM or roadheader	Segmental lining or shotcrete etc.
Weak clay, silty clay	EPB closed-face machine	Segmental lining
Sands, gravel	Closed-face slurry machine	Segmental lining

Source: After Muir-Wood, 2000.

difficulties. Nevertheless, many TBM tunnels proceed well and at much faster rates than hand dug/drill and blast tunnels. The adoption of hazard and risk analysis (British Tunnelling Society, 2003) will help reduce incidents but will not necessarily eliminate hazards entirely.

3.6.4 Soft ground tunnelling

Soft ground including severely weathered rock may be excavated by hand or by tunnel boring machine. For open face excavation, behaviour can be predicted using classification such as the "Tunnelman's Classification" of Heuer (1974), which allows prediction of whether the soil will stand firmly whilst the liner is put in place or is likely to "ravel, run, flow, squeeze or swell". Behaviour depends on the nature of soil, water conditions and stress levels. For example, uncemented sand might be expected to flow below the water table especially at depth. Such empirical predictions are also useful for weathered rocks where the application of conventional soil mechanics principles is questionable (Shirlaw et al., 2000). When tunnelling in soil or in mixed face conditions it is the behaviour of the weakest or most mobile material that generally governs the need for, and magnitude of, the support pressure that is needed at the tunnel face.

If the "soil" is stiff and cohesive then NATM methods can work successfully as has been achieved for example in the London Clay (van der Berg et al., 2003) and in the Fort Canning Boulder Bed and the Old Alluvium in Singapore (Shirlaw et al., op cit). Where soils are unstable then various options include grouting, dewatering, freezing or the use of compressed air. All of these are costly, may have severe health and safety implications and restrictions and take time to install. Nevertheless, such methods are often necessary to recover and restart a tunnel that has encountered a major problem and perhaps collapsed.

Tunnel boring machines used in soft ground are of the closed face type as illustrated in Figure 3.18a & b. Guidance on machine selection and use is given by the British Tunnelling Society (2005).

Earth Pressure Balance (EPBM) and Slurry machines use pressurised soil at the cutting face to hold up the ground as the tunnel advances. In an EPBM machine the broken-down soil remains in the plenum chamber behind the cutting head, balanced by pressure in the Archimedes screw, which removes the spoil under the control of the operators. In a slurry machine which tends to be used in higher permeability soils, bentonite slurry is introduced to the plenum chamber, mixes with excavated soil which

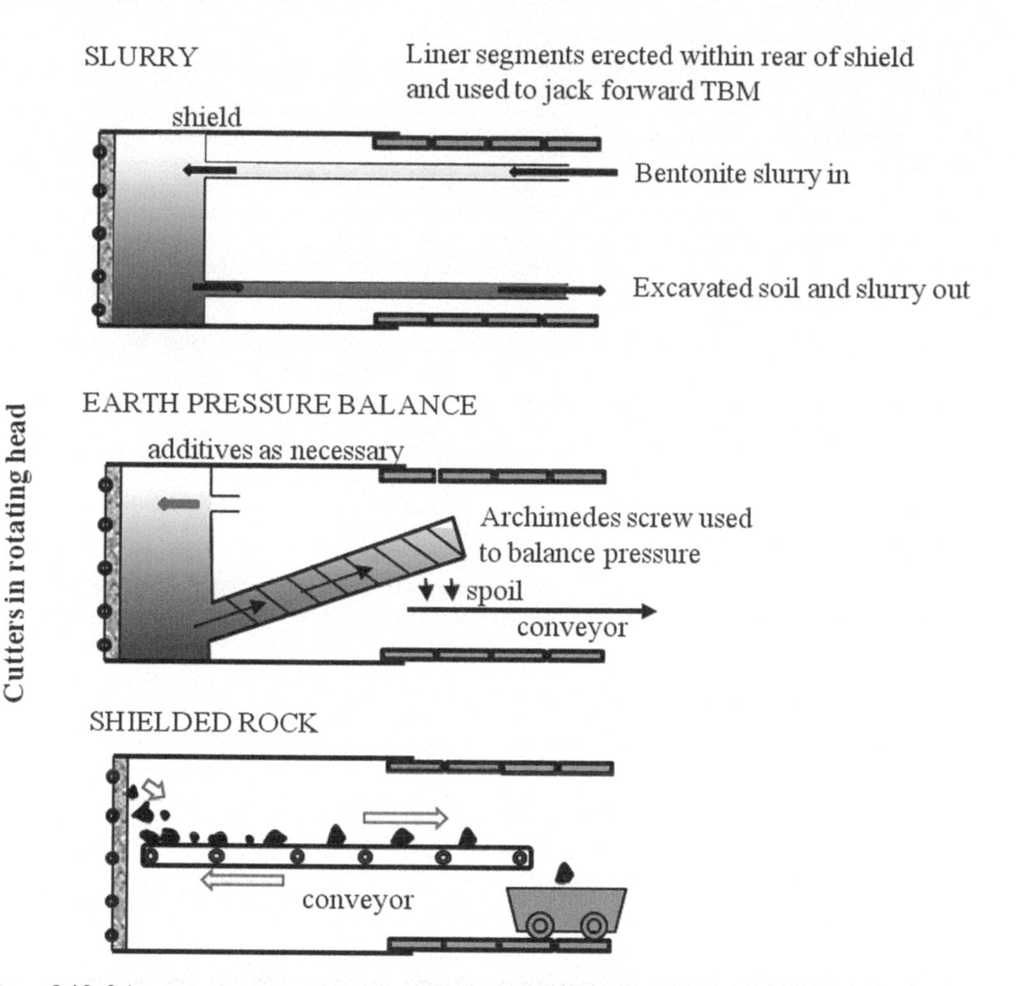

Figure 3.18 Schematic diagrams of shielded TBMs. (a) Slurry machine, Bentonite slurry is pumped to plenum chamber and mixes with spoil cut at the face. Mixture is removed for separation and treatment before recycling. (b) Principles of EPBM. Cut soil (with additives as necessary) is removed by a screw device with the pressures monitored and maintained. (c) Single shield rock TBM. Rock cut from the face is mucked out and TBM pushes forward against the liner erected to the rear of the shield. Other rock TBMs use grippers pushed against the walls of the tunnel and use this as the reaction force for advancing the TBM.

is then removed for separation, disposal, and re-use (bentonite) by pipes rather than on a muck conveyor. Permanent concrete lining is formed from pre-cast segments, directly behind the machine and this liner is used as a reaction to push the TBM forward. TBMs often work well for the specific conditions for which they are designed but also commonly run into problems with the machine getting stuck or running into rock that is either too hard or too soft or too wet for the type of machine (see Table 3.5). Shirlaw et al. (2003) report cases of settlement and collapse in Singapore even using sophisticated EPBMs. Similarly, an EPBM machine was recently stopped by silt

Table 3.5 Common difficulties with TBM tunnels

Problem	*Mitigation*
Ground too strong (intact strength and/or lack of discontinuities)	May need to pull TBM back and advance with drill and blast
Ground too weak and collapsing (should have been an earth balance or slurry machine perhaps)	Ground improvement might be necessary in advance of tunnel drive – grouting or freezing
Major faults	Collapse of ground and TBM gets stuck. May need to sink a shaft in front of machine and construct a tunnel back to and around the TBM to free it up. Ground treatment and possible hand construction through fault zone may be required to get the TBM going again.
Weak ground and high *in situ* stresses leading to squeezing action on TBM	Can cause huge delays. Ground improvement to strengthen the ground and resist the squeezing pressures.
Too much clay for slurry treatment	Can cause delay and necessitate installation of additional treatment plant – extra hydrocyclones etc.
Ground abrasive because of high silica content causing too much wear on teeth leading to cost and delay.	Cost maybe prohibitive necessitating a change of excavation method.
Too much water and TBM electrics not protected	Drilling and grouting in advance of machine or possibly ground freezing or compressed air working. Possible change of method to drill and blast or employ different machine with suitable spec.
Excess tunnel slurry pressure cause blow out at ground surface	Lower pressure.
Pressure too low causes face collapse	Reverse of the above.

"breaching the tunnel liner" on a contract in the UK and a further example (Heathrow Express Tunnel collapse) is discussed in Chapter 8. Recovery options include freezing the ground and grouting the ground to stabilise it to allow the TBM to be withdrawn (NCE, 19th Jan 2011).

Where the materials to be excavated include strong and weaker material, this is known as mixed face conditions. For stability, the major issue concerns relative mobility of the materials rather than just strength. A mixed face of strong boulders and hard clay presents problems in terms of rate of excavation, but generally not in terms of heading stability. However, a combination of strong, stable rock with a more mobile material, such as flowing, rapidly squeezing or fast ravelling material provides conditions where the overall stability of the heading can be very difficult to control as well as difficult to excavate. Shirlaw et al. (op cit) provide examples of major inflows resulting from the use of conventional rock tunnelling methods too close to the transition from rock-like to soil-like conditions. Ironically, this particular type of mixed face condition has become even more problematic with the introduction of modern tunnelling technology.

3.6.5 Hard rock tunnelling

The main options are drill and blast, a road header excavating machine or to use a TBM that may be either open (without a protective shield) or shielded (Figure 3.18c).

3.6.5.1 Drill and blast/road headers

Drill and blast tunnels are usually more flexible to variable ground conditions than TBM tunnels and allow difficult ground conditions to be understood and overcome, but they may be much more time consuming unless a number of access points can be found to allow operations to proceed from several faces at the same time.

Holes are drilled in the face, and explosives placed in the holes. Issues of tunnel blast design are addressed by Zare & Bruland (2006). The holes are detonated sequentially to break to a free face over micro seconds. The aim is to break the rock to manageable size so it can be mucked out readily with machines without further blasting or hammering. Other aims may be to keep blast vibrations to a minimum and not cause damage or offence to nearby residents, and usually to keep as closely as possible to the excavation shape prescribed by the designers, i.e. minimising "overbreak". Typical advances per round are 3 to 3.5 m, sometimes up to 5 m in very good rock conditions. Depending on the size of tunnel and ground conditions the full face may be blasted in one "round" or may be taken out as a series of smaller headings – top, or side, that may be supported by sprayed concrete with steel mesh or steel/carbon fibres, rock bolts, and/or steel arches or lattice girders, before the tunnel is advanced. Figure 3.19 shows the tunnel portal for a diversion tunnel at Queen's valley reservoir, following the first blast, with steel arches being erected to protect the tunnel access.

After blasting, and dust and gases have dissipated and safety checks made (e.g. for methane or radon), the broken rock is excavated ("mucked out") and it is the engineering geologist's unenviable task to examine and map the geological conditions exposed. The freshly blasted rock may well be unstable and the geologist should not approach the face until the Contractor has carried out all necessary scaling and/or rock support work to make the tunnel safe. The Contractor has overall responsibility for site safety and his instructions should be followed at all times in this respect. A decision will then be taken on whether the ground is as expected, if the ground is changing (and probing ahead is required), and the support requirements. Any potential for deteriorating conditions or, for example, a major potential wedge failure, need to be identified quickly so that support measures can be installed. As illustrated in Figure 3.20, often the rock mass is self-supporting. As the tunnel is excavated the tunnel walls move inwards and the rock mass locks up as it dilates. If there is an inherent weakness such as a free wedge of rock or a fault zone then local collapse can be followed by ravelling failure, which could "chimney" to the ground surface. In two of the examples discussed in Chapter 8 the situation deteriorated quickly. If conditions are poor and getting worse then the ground might be supported in advance of the tunnel by an umbrella of "spiles" or "canopy tubes", and/or by pressure grouting.

In suitable rock, other mining approaches may be used including the use of large "road headers" that cut their way into the rock but do not excavate the full-face profile in one operation unlike a TBM. In a tunnel formed by drill and blast or road header it is possible to examine and record the ground conditions throughout construction and make decisions as to the support required. In a TBM tunnel little can be told about the ground ahead of the machine without stopping and drilling in front of the face, which disrupts operations and is therefore to be avoided.

Figure 3.19 After first blast and mucking out, construction of temporary steel arches to protect tunnel portal, Queens Valley Reservoir, Jersey, UK in 1993 (Full protective clothing would now be worn, although this would not protect from rock fall from the recently blasted face behind the men).

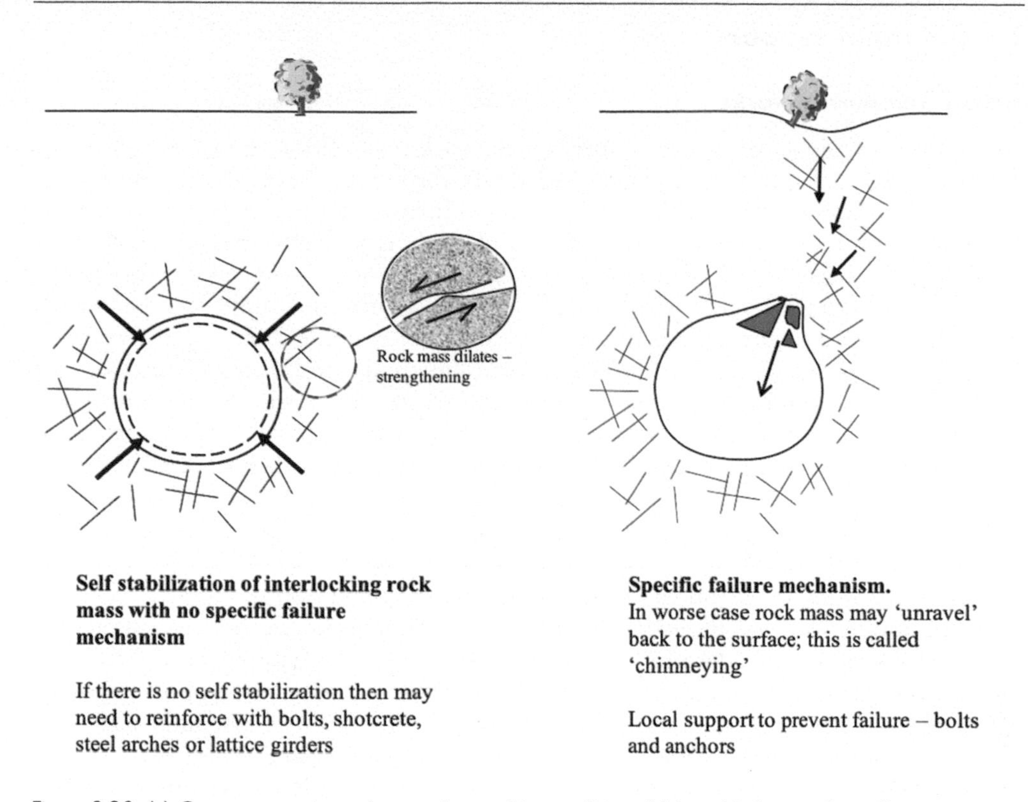

Rock mass dilates –
strengthening
Self stabilization of interlocking rock mass with no specific failure mechanism
If there is no self stabilization then may need to reinforce with bolts, shotcrete, steel arches or lattice girders
Specific failure mechanism.
In worse case rock mass may 'unravel' back to the surface; this is called 'chimneying'
Local support to prevent failure – bolts and anchors

Figure 3.20 (a) Convergence in rock tunnel to stable condition. (b) Local failure and ravelling to ground surface.

3.6.5.2 TBM tunnels in rock

The design and use of modern hard rock TBMs are covered comprehensively by Maidl et al. (2008). In good rock with high RQD, open TBMs are sometimes used but generally only for relatively small diameter tunnels. The tunnel advances by jacking forward against grippers that are extended laterally against the tunnel walls. Clearly if the rock becomes poor quality then there may be problems with the grippers. There is also no way of preventing groundwater ingress other than by grouting, preferably in advance of the machine. In Chapter 8 a case (SSDS) is presented where open rock TBMs were selected anticipating good rock conditions with low water inflows and the operations were halted when inflows became too great and grouting in advance was extremely difficult.

In poorer quality rock generally, shielded TBMs are used. A single shield machine pushes against the liner as for soil TBMs (Figure 3.18c). In other set ups there are two shields; the rear shield has grippers and provides the reaction against which the front shield can push forward. The cutter head has discs that rotate as the cutter head itself rotates. The thrust of the machine may cause the rock to fail, mainly in tension. A major consideration is the life time of the cutting discs before they need to be replaced as addressed by Maidl et al. (op cit). A case example in Chapter 8 describes considerable wear in an EPBM used to tunnel through abrasive sandstone.

3.6.6 Tunnel support

3.6.6.1 Temporary works

Rock tunnelling in general relies largely on the rock mass "locking up" as joints and interlocking blocks of rock interact and dilate during the process of convergence towards the excavation. Good quality rock often forms a natural arch and no or little support is needed. However, in weaker ground, such as in fault zones, the rock mass cannot support itself even with re-inforcement, and requires artificial support in the form of steel arch ribs, typically encased in shotcrete. Optimising support requirements in weaker ground requires prediction of likely convergence rates, making observations as excavation is undertaken, i.e. "observational" methods, and then applying support such as rock bolts and/or shotcrete and/or steel arch ribs to control the movement and prevent excessive loosening (Powderham, 1994). In stronger, blocky rock masses rock movement will be much less and the purpose of the support is then to prevent loss of loose blocks and wedges, which would destabilise the arch and maybe lead to ravelling failure.

Rock mass classification systems introduced in Chapter 4 are linked to charts allowing decisions to be taken as to the immediate (temporary) support measures required and these are reviewed by Hoek et al. (1995). In practice, decisions may often be biased by other considerations such as the materials and equipment at hand and the workers perceptions of the degree of risk and how well previous support measures have worked. This may of course have cost implications and may also later become a matter of dispute as to what was really necessary, as discussed and analysed by Tarkoy (1991). The importance of good engineering geological records during construction is emphasised. In severe situations, such as high stress or intense water inflow, steel lining may be used but even then this sometimes proves inadequate as happened during the construction of the Tai Po to Butterfly Valley water supply tunnel in Hong Kong where unexpectedly high water pressures buckled the liners (Buckingham, 2003; Robertshaw & Tam, 1999).

3.6.6.2 Permanent design

There are two main areas for consideration, firstly the area around the portal, especially for tunnels that are part of a road or rail system, and secondly the need for a permanent liner.

3.6.6.3 Portal design

The area above the entrance to a tunnel often requires careful engineering to make it safe, both during construction and during operation. The problems are essentially the same as for general slope stability design as discussed later in this chapter but the need for long-term inspection and maintenance whilst maintaining tunnel usage sets portal design in a rather special category. A canopy is often constructed to protect the portal area from falling rock and other debris as illustrated in Figure 3.21. Catch nets, barriers (such as gabion walls) and *in situ* stabilisation can be used to prevent debris impacting the portal area. Rock and soil masses immediately above the portal area are

Figure 3.21 Canopy extending out from tunnel liner (being waterproofed), to protect portal area. A55, North Wales.

often covered with steel mesh and shotcrete or similar hard covering and dowelled, nailed or anchored back using post-tensioned bolts and cable anchors (Figure 3.22). The requirements for designing, protecting and maintaining ground anchorages are set out in national standards and codes of practice such as BS 8081 (BSI, 1989) and BS EN 1537 (BSI, 2000). Despite such standards things occasionally go wrong either because of ground conditions or flaws in the anchorage itself and designers must appreciate the practical difficulties that might be associated with maintenance programmes whilst ensuring safety for the road user. If a major problem is found then the tunnel might need to be closed or restricted in use whilst the problems are rectified. Several cases of the failure of rock anchorages, even in projects post-dating BS 8081, are discussed in Chapter 8.

3.6.6.4 Permanent liners

The options for permanent tunnel liner design include the following:

- Unlined (ignoring temporary support measures).
- Unreinforced concrete.
- Reinforced concrete.
- Steel.

Figure 3.22 Rather a cautious approach to rock hazard at a portal in North Wales.

Lined tunnels can be designed to be undrained in which case the permanent lining must withstand the full groundwater pressure as well as rock loads. Other tunnels are designed to be drained whereby the outer surface (or extrados) of the arch of the liner is lined with a waterproofing membrane, but which is placed on geotextile sheets, which carry water down to drains and sumps below the tunnel invert. The sumps may need continual pumping and the whole drainage system needs maintenance over the life of the project. Figure 3.23 shows details of a design as used in some recent rail

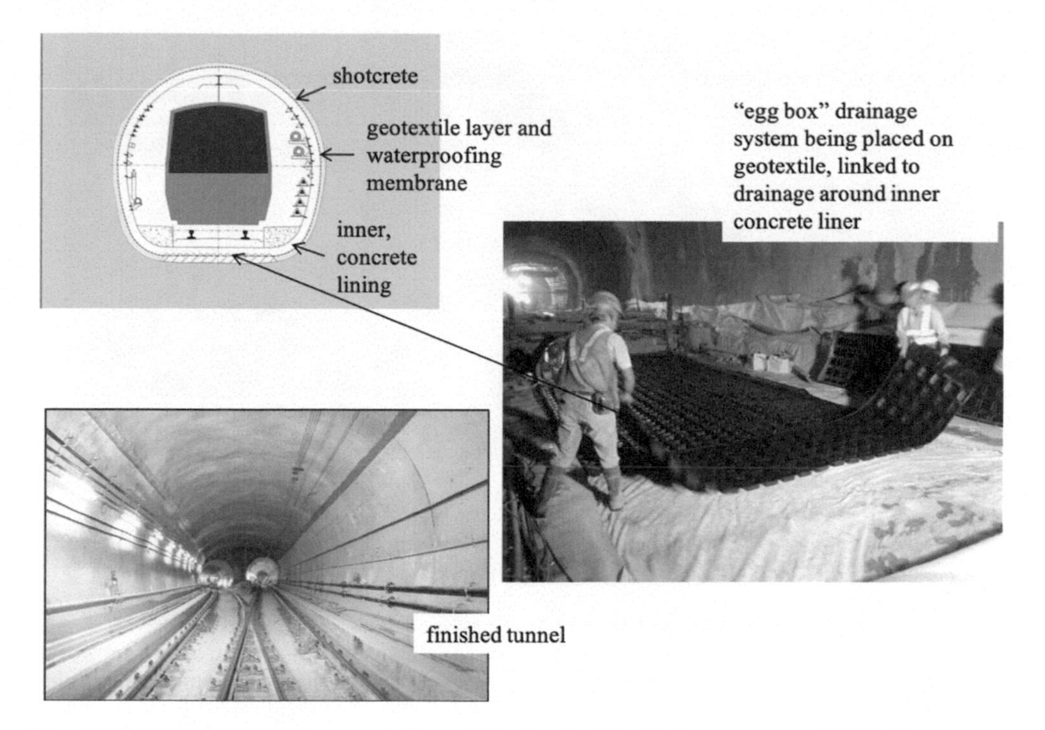

Figure 3.23 "Egg box" drainage system for a drained tunnel (by kind permission of MTR Corporation, Hong Kong).

tunnels in Hong Kong. After shotcreting the tunnel walls, layers of geotextile (outer) and waterproof membrane (inner) are placed followed by an inner concrete liner (250 mm thick). Groundwater is thereby channelled via the geotextile to an "egg box" drainage system in the invert. For any drained lining design, care must be taken that any permanent drawdown in the water table has no adverse effects on structures above the tunnel or on water supply from groundwater sources.

Pre-cast concrete segments are commonly erected as part of a TBM excavation and support process, mainly in soft ground tunnels, but also in some hard rock applications. The segments are manufactured externally and then erected within the shield surrounding the advancing machine and bolted together. If required, segments can be fitted with gaskets to form fully waterproof concrete liners (Figure 3.24). As noted earlier, the installed liner can be designed to provide a reaction to push the TBM forward.

3.6.6.5 Pressure tunnels

One of the most severe design situations is in high-pressure water supply tunnels associated with hydropower constructions where for some operational periods the tunnel carries water under high pressure, but at other times the same tunnels are empty and have to withstand significant external water and rock pressures.

Figure 3.25 Rock core from vertical borehole in strong mudstone, Taegu, South Korea. Note near-vertical persistent joint infilled with calcite. This network of joints (two sets orthogonal to bedding) were encountered frequently in preliminary boreholes and appreciation of their significance led to (a) reconsideration of the potential rock loads on the permanent liners and (b) additional ground investigation using inclined rather than vertical holes to characterise the rock mass better.

The main concerns with pressure tunnels are as follows:

- Potential damage by hydraulic fracturing (formation of new fractures) or jacking (opening of existing fractures) within the rock mass.
- Stability, durability and low maintenance.

To avoid hydraulic fracturing an empirical rule is sometimes used:

$$D\gamma_R / H\gamma_W > 1.25 \text{ (Haimson, 1992)} \quad (3.1)$$

where γ_R is unit weight of rock and γ_W is unit weight of water, D is rock overburden at tunnel location and H is the water head. However, it is important to recognise that this formula only considers vertical *in situ* stress. Horizontal stress can be very low in some situations, for example, close to valley sides and this will control the risk of hydraulic fracture or jacking if water from the tunnel can reach the excavated rock surface at sufficiently high pressure.

Where the confining rock stress, vertical and/or horizontal, is too low, fully welded continuous steel liners are generally used to prevent the high-pressure water from reaching the rock mass. Concrete liners may be used in competent rock but might crack under high internal water pressure if the confining stresses are too low. In such cases there is a risk of leakage to surrounding ground (with a risk of causing landslides in some situations) and/or water flow into other underground openings. Haimson (1992) presents examples of schemes where the importance of stress conditions and the correct choice of lining only became evident late in the design process with "unpleasant consequences". An important task of the engineering geologist is to ensure that the *in situ* stress conditions along the route of a pressure tunnel are evaluated fully and reported to the design team, preferably at an early stage in project planning.

In certain situations, typically in low-pressure headrace tunnels, a concrete liner can be designed with drainage holes to relieve water pressure on the tunnel lining. Consolidation grouting is usually carried out around the tunnel to reduce leakage out of or into the tunnel (depending on the relative internal and external water pressures). Unlined tunnels can be used in good rock conditions and with favourable *in situ* stresses, but there may be higher maintenance requirements and the need to construct rock traps to catch any fallen debris. The proper design of hydraulic pressure tunnels is particularly important as the consequences of failure are usually very severe and costly to repair. A comprehensive summary of the principal design and construction considerations is presented by Benson (1989).

3.6.7 Cavern design

Caverns are large span underground openings and these are used for many purposes including sports halls, power stations and oil and gas storage (Sterling, 1993). Hydroelectric power caverns and large three-lane road tunnels are typically 20 to 25 m span, but caverns have been constructed successfully in good quality rock with spans in excess of 60 metres (Broch et al., 1996) and natural caves are found with much larger spans.

There is considerable guidance in the literature on approaches to their design and construction (e.g. Geotechnical Engineering Office, 1992; Hoek & Brown, 1980). Many design issues are similar to tunnels but because they are at fixed locations, ground investigation decisions are more straightforward. The other major difference is scale. Whereas many tunnel walls "lock up" as the rock mass dilates and need little support to ensure stability, in a cavern there is more potential for large scale strain and failure mechanisms to develop. For example, large caverns were required for a proposed high-speed rail station at Taegu, Korea in strong mudstone. Preliminary numerical analyses were carried out to design permanent concrete liners and bolting support assuming essentially isotropic rock mass parameters. The design had to be revised when it was realised that the rock structure was strongly anisotropic with bedding mostly horizontal and many near-vertical tensile joints infilled with calcite (Figure 3.25). These joints could allow discrete failure into the crown of the openings as illustrated by Maury for mine workings (1993).

Hoek & Moy (1993) and Cheng & Liu (1993) describe different aspects of the design and construction of the Mingtan pumped storage project in Taiwan and illustrate the need for an integrated approach of geological investigation, numerical modelling, design, observation, construction and instrumentation. An exploration/drainage gallery and two other galleries were used to install corrosion-protected, permanent cable anchors to reinforce the roof arch of the main cavern 10 metres below, prior to

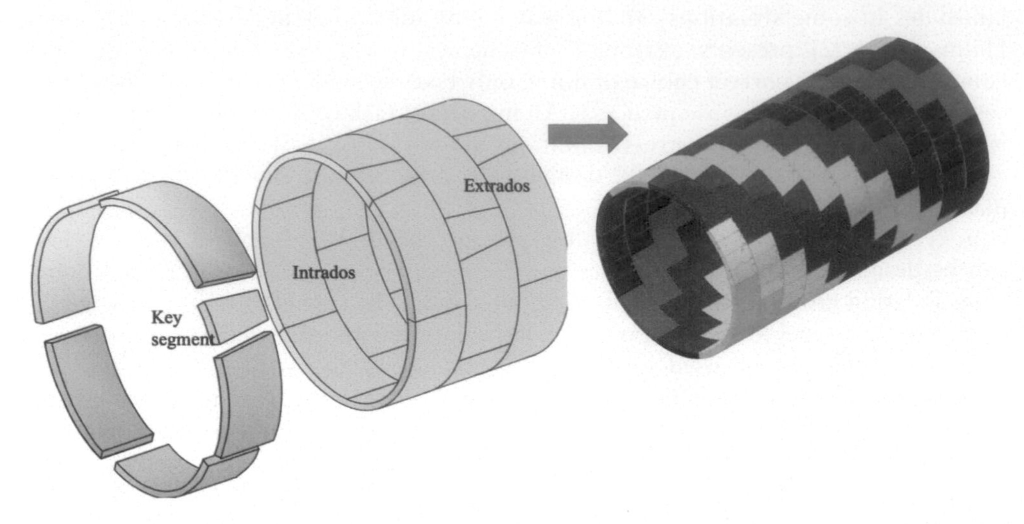

Figure 3.24 Interlocking tunnel segments, pre-fabricated and erected to the rear of a TBM shield. (Figure courtesy of Mike King, formerly with Halcrow.)

its excavation. Small loads were applied to the cable anchors, which only took on their full loads as the cavern was excavated.

3.6.8 Underground mining

Underground mining is quite different from the formation of caverns and tunnels for civil engineering although many of the skills required are the same. In mining the objective is to extract the ore whilst minimising waste rock production. Safety is a prime concern as it is for civil operations but mining involves the formation of non-permanent voids, many of which will be allowed to collapse or be packed loosely with waste rock so the fundamental operational concepts are obviously quite different. Rock mechanics of underground mining operations are discussed by Brady & Brown (2004). In terms of geological hazards, of particular concern are flammable and/or noxious gases including radon and the control of dust and ventilation is very important. Such matters are generally mandated by national standards on Health and Safety but still accidents occur regularly world-wide.

A general concern for construction in mining areas is continuing ground settlement or sudden collapse of old workings. These are matters to be considered at the desk study stage of site investigation as addressed in Chapter 6.

3.6.9 Risk assessments for tunnelling and underground works

In Chapter 6 a system was introduced whereby site investigation is conducted or reviewed following a checklist approach whereby firstly geological hazards are considered, then environmental factors and finally hazards associated with the specific type of project or construction method. Tunnels are often particularly risky undertakings because they are so dependent upon geotechnical conditions, which may vary considerably along their length and it is seldom feasible to carry out as comprehensive a ground investigation as it is for other types of project. Good reviews of tunnel collapse mechanisms and case histories are given by Maury (1993) and Geotechnocal Engineering Office (2009), respectively. Consequently, industry has developed several approaches whereby hazards are considered in detail so that strategies can be prepared to reduce or mitigate the risks. This can be done at the option assessment and design stages and then later as part of the management of construction.

The British Tunnelling Society Joint Code of Practice for Risk Management of Tunnel Works in the UK (British Tunnelling Society, 2016; 2017) was prepared jointly by the Association of British Insurers and the BTS and sets out requirements regarding risk assessment and management for tunnels. In effect it is mandatory in the sense that without its adoption no contractor's insurance will be forthcoming for an underground project. The Code of Practice sets out how and when risk is to be assessed and managed, and by whom. Risks are to be assessed at the project development stage (design), by the Contractor at tender stage and during construction through a risk register.

The Code also requires the ground reference conditions or geotechnical base line conditions to form part of the contract but as noted in Chapter 2, definition of such conditions is not always straightforward. Whilst the intention to avoid dispute is laudable, there may be considerable difficulty in summarising geological and geotechnical conditions succinctly and unambiguously.

3.6.9.1 Assessment at the design stage

The ways that risk can be assessed at investigation and design stages are illustrated by the example of the 16.2 km Young Dong rock tunnel in Korea as presented in Appendix D.1 and D.2. Given an appreciation of the ground conditions along the route, based on a well-conducted site investigation, the hazards associated with the various options for construction can be considered. Once these have been identified, their likelihood and seriousness can be rated in terms of potential consequence (e.g. programme, cost, health and safety) and methodologies devised for mitigation prior or during construction. Decisions can then be made on how to proceed.

3.6.9.2 Risk registers during construction

During construction, hazards that were anticipated at the design stage may prove real or illusory. New ones will be identified and need to be dealt with. The current way of so-doing is to employ a risk register in which hazards are identified and assigned to individuals in the project team to derive strategies for their avoidance or mitigation. In the BTS Code of Practice (2003) this is identified as a task for the Contractor but the register will include risks brought forward from the project development stage. In practice it may well be the Project Engineer rather than the Contractor who manages the construction risk register, perhaps at monthly meetings held to monitor progress on mitigating each of the identified risks, remove from the register those that have been dealt with, and recognise and assign to individuals any new risks identified during the course of the work. Brown (1999) outlines the risk management procedures adopted for the successfully completed Channel Tunnel Rail Link Project in the United Kingdom and a list of typical tunnelling hazards to be considered during construction is presented in a table in Appendix D.3.

3.7 SLOPES

Landslides cause major economic loss and physical damage and kill many people each year. Slopes can be split into natural and man-made. The hazards from natural terrain landslides in mountainous regions and at the coast are considered in Chapter 5.

Man-made slopes include cut slopes (cut into the natural hillside) and fill slopes. Fill slopes might simply comprise the excess debris from an adjacent cutting, dumped or compacted onto the adjacent hillside to form an extra carriageway but can also include sophisticated, high and steep slopes incorporating geotextiles or other materials to strengthen the soil (Figure 3.26). Stability of existing slopes needs to be assessed by engineers and if considered unacceptably low, measures must be taken to improve the stability to an acceptable level.

Figure 3.26 Construction of reinforced earth embankment for Castle Peak Road widening, Hong Kong.

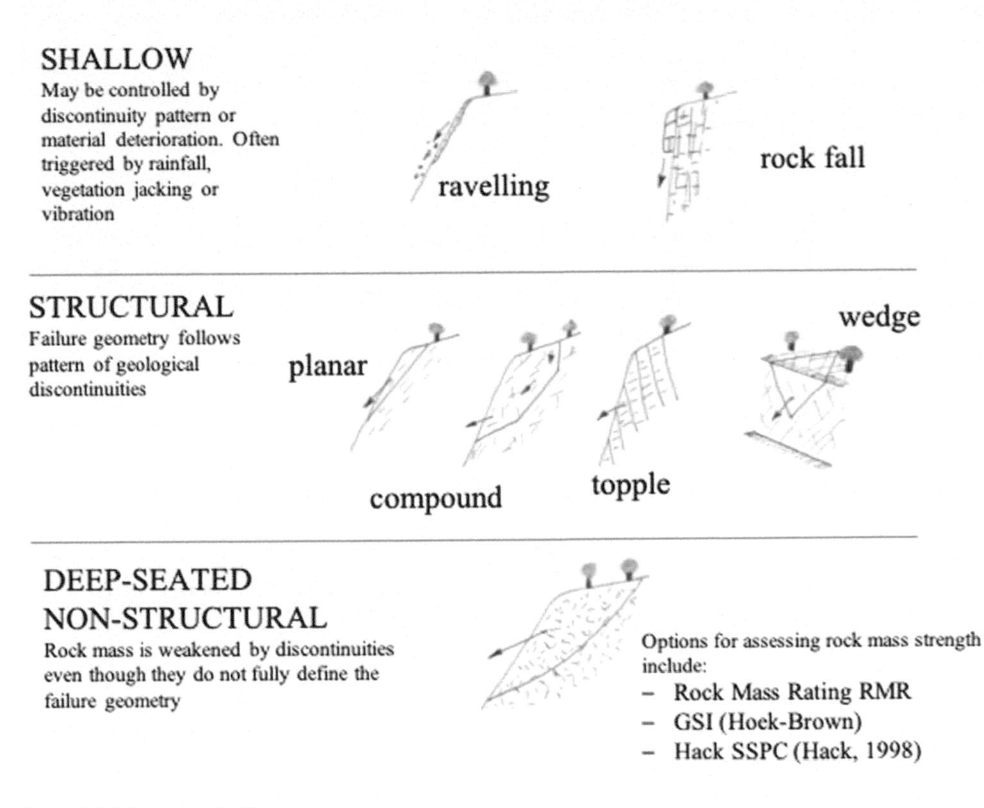
SHALLOW
May be controlled by discontinuity pattern or material deterioration. Often triggered by rainfall, vegetation jacking or vibration
ravelling
rock fall
STRUCTURAL
Failure geometry follows pattern of geological discontinuities
planar
compound
topple
wedge
DEEP-SEATED NON-STRUCTURAL
Rock mass is weakened by discontinuities even though they do not fully define the failure geometry
Options for assessing rock mass strength include:
- Rock Mass Rating RMR
- GSI (Hoek-Brown)
- Hack SSPC (Hack, 1998)

Figure 3.27 Modes of failure in rock slopes.

3.7.1 Rock slopes

Rock slope stability is generally controlled by the geometry of pre-existing, adversely oriented discontinuities, including bedding planes, faults and master joints including sheeting joints. Failure types can be grouped as illustrated in Figure 3.27 and as follows:

Shallow failures – superficial failures, generally low to medium volume.

Structurally controlled failures – sliding may occur on one or more intersecting discontinuities that are adversely oriented relative to the slope geometry. Toppling can result because of the presence of unstable columns of rock, perhaps dipping steeply back into the slope. Large, compound failures can involve sliding on a number of adverse sets of discontinuities, together with some breaking through intact rock, which allows the full mechanism to develop (Martin & Kaiser, 1984; Malone et al., 2008; Wentzinger et al., 2013).

Deep seated failures that are not-structurally controlled: the rock can be considered an interlocking mass of rock blocks lacking adverse fabric, such as bedding, schistosity or systematic joints. For such situations (rare), the Hoek–Brown GSI model provides estimates for rock strength.

3.7.1.1 Shallow failures

Steep rock slopes are sources of rock falls that can be a major risk, especially where adjacent to a busy road or railway. All rock slope surfaces deteriorate with time (Nicholson et al., 2000). Rock material weathers, vegetation grows and opens up joints and blocks get undermined by erosion (Figure 3.28). Even small blocks can cause accidents, for example, a minor block fall in Hong Kong caused the death of a van driver, which led to major litigation about working above live carriageways. Along large lengths of highway through a mountainous region there will be a need to identify where the risk is greatest so that the risks can be mitigated cost-effectively (Box 3.4). This can be done by using some rock mass rating appraisal system together with software such as Rockfall (by Rocscience) capable of predicting where falling rock might end up, as in Figure 3.29, taken from the Hong Kong Highways Manual (2017). Nevertheless, it is often a matter of engineering judgement, taking account of the history of rocks falls (Woods et al., 2022). In such an assessment it should be remembered that relatively minor rock falls may be precursors to major rock collapses. Methods of mitigating rock falls and other potential landslides are discussed below.

Figure 3.28 Dr Dick Martin close to Lantau Peak, examining rock slope to the right. Close up of the natural rock slope shows many open, near vertical fractures and a developed sliding plane, which will fail, given some time.

	Results of Rockfall Trajectory Analyses	Protective Barrier Required	Required Slope Treatment Prior to Excavation
LOW ROCKFALL HAZARD 	No rocks reach roadside barrier.	Simple fence/barrier or concrete block barrier.	No significant requirements.
MODERATE ROCKFALL HAZARD 	No rocks reach barrier in full flight. Bouncing rocks impact barrier at less than 1 m height.	Low energy requirements of up to 100 kJ. Movable barrier with posts socketed in rock and netting.	Increasing demand for slope treatment or preventive works and protective measures to control the block size and quantity of rockfall prior to excavation, with provision for standby plant and measures to break up and clear debris, repair or replace protective fence/barrier, and to implement road closure during excavation.
HIGH ROCKFALL HAZARD 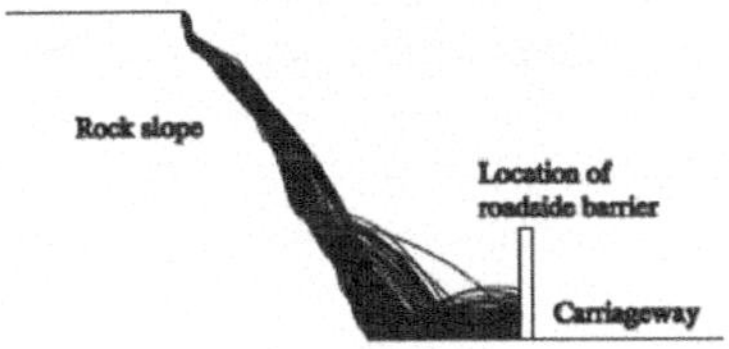 Rock slope Location of roadside barrier Carriageway	Bouncing rocks impact barrier at less than 3 m height. Rocks in full flight impact barrier at less than 1 m height.	Moderate energy requirements of up to 250 kJ. Specialised rockfall fence with adequate energy-absorbing capacity.	
VERY HIGH ROCKFALL HAZARD 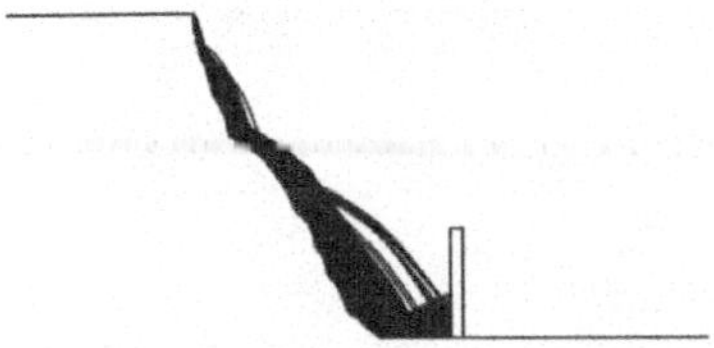	Rocks in full flight impacting barrier at heights less than 5 m.	Requires high capacity energy-absorbing rockfall fence, up to a maximum available capacity of 2 500 kJ.	
EXTREME ROCKFALL HAZARD 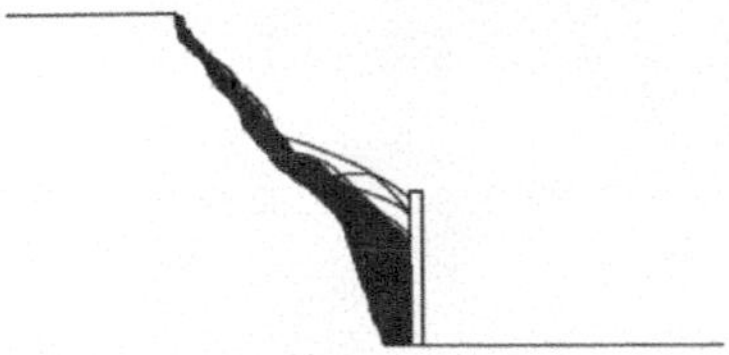	Rocks in full flight impacting barriers at heights of 5 m or more.	Requires rockfall fence of maximum available capacity of 2 500 kJ approx. The height of the barrier can become a critical consideration.	Extremely demanding requirements where it is not possible to close the road.

Figure 3.29 Rock fall hazard diagram prepared by Dr Laurie Richards and presented in Highway Slope Manual, Geotechnical Engineering Office, 2017.

BOX 3.4 JUDGING THE SEVERITY OF ROCK FALL HAZARDS AND THE ASSOCIATED RISKS

"People - even experts - rarely assess their uncertainty to be as large as it usually turns out to be."

Baecher & Christian (2003)

The assessment of hazard of rock slope failure is always rather subjective as illustrated in Figure B3.4.1 taken in the Lake District in the UK. Which engineer or engineering geologist would assess this rock as safe? Yet it is, for the moment...

The concept of hazard is similarly illustrated by the petrographs at Anhwa-ri, Goryeong, Korea in February 2008. The rock exposure shown in Figure B3.4.2, above the rock carvings, appears to be on the brink of failure and one would be tempted to fence off the area immediately followed by the removal of any blocks that cannot be stabilised by dowelling and dentition works. However, the fact that the precarious open-jointed rock is directly above the ancient rock carvings is evidence that this rock face has not retreated very far over a period of more than 2000 years. The process of deterioration and collapse is actually quite slow and judgment of the risk as immediate and obvious requiring urgent action would therefore err on the conservative side.

Conversely, the slope shown in Figure B3.4.3 is in the Cow and Calf Quarry at Ilkley, Yorkshire in the UK and was used to teach MSc Engineering Geologists to map rock discontinuities for several years. The collapse to the left of the photograph occurred unexpectedly between mapping exercises, despite its repeated examination and systematic logging on scan lines, without the failure mechanism having been identified.

Figure B3.4.1 Overhanging rock in Lake District, UK.

Figure B3.4.2 Petrographs at Anhwa-ri, Goryeong, Korea. Rock slope above petrographs (with small protective fence) shows signs of vegetation wedging with loose blocks resting against trees.

Figure B3.4.3 Unexpected rock failure in Cow and Calf Quarry, Ilkley, West Yorkshire, UK.

These examples illustrate our uncertainty and the difficulties in judging the degree of hazard by examination of exposed rock alone. It is highly likely that even after sub-surface ground investigation our ability to judge the severity of the situation is often rather poor. The conclusion must be that consequence should be the priority when assessing the risk of slope failure. If there is a major risk to life then works should be done. This is the underlying philosophy behind the Landslide Preventive Works (LPM) strategy in Hong Kong where the catalogue of tens of thousand of slopes prepared in the 70s and 80s has been compiled and ordered in terms of perceived risk (a function of height, angle and proximity to vulnerable facilities). Each slope is checked and "*upgraded*" in order of perceived risk. Most of these are dealt with using essentially "prescriptive", engineering works including soil nails and inclined drains installed to a pattern.

If there is clear danger from the hazard then it should be dealt with. In the Korean case discussed above, despite the apparently slow retreat of the rock exposure above the petrographs, visitors to the site should be protected against the evident rock fall hazards.

QUANTITATIVE RISK ASSESSMENT OF ROCK FALL TO ROADS

At the site shown in Figure B3.4.4, while it might be intuitively obvious that there is some risk to life from rock fall along the road and some history of such rock falls, the cost of preventive works may be very expensive. One way to deal with this quandary is to try to quantify the risk and compare this to the cost of reducing the risk.

Figure B3.4.4 Road cut through limestone with very little engineering support or protective measures, Tailuko Gorge, Taiwan.

To do this requires the following data to be measured or estimated:

- Frequency and size of rock fall incidents, (per day)
- Number of vehicles per day, average length and velocity
- Vulnerability of persons in vehicles to rock fall (depends on size of falls).

The annual probability of risk of death can then be calculated and compared to published guidelines on acceptable risk (e.g. Fell et al., 2005). Different sections or road will be shown to have different risk levels, which will allow decisions to be made on where to carry out mitigation works. Quite often such a calculation will show that risks are acceptable even if judgmentally the hazard is still intolerable (the situation looks very worrying). It may well be found that relatively simple and measures such as scaling off the most obvious loose rock and providing netting or cheap barriers such as gabions locally will reduce risk considerably whilst also making the situation feel safer. Further information on judging rock fall hazards and the use of rock fall rating systems is given by Bunce et al. (1997) and by Li et al. (2009) and discussed in the main text.

3.7.1.2 Structurally controlled landslides

The distinction of failure mechanisms into planar, wedge and toppling, and the discontinuity geometries and conditions responsible for each style of failure are set out clearly by Hoek & Bray (1974) and this has been updated by Wyllie & Mah (2004). The most common type of failure is sliding on a single discontinuity and this is simple to analyse. The main difficulties are in assessing shear strength of the rock discontinuities as set out in Chapter 6 and how to deal with groundwater pressures. Generally, a simple analysis is done in which it is supposed that water pressure at the slope face is zero, increasing back within the slope, to some height below ground surface at the rear of the slope (Figure 3.30). This is often a conservative assumption in that water pressure will be localised, not acting throughout the whole slope at the same time. Richards & Cowland (1986) discuss a well-investigated site where it would have been unrealistic to design the slope to withstand the maximum water pressures at each location, all acting at the same time because instruments clearly showed pulses of water pressure travelling through the slope following a rain storm.

Even small intact rock bridges can provide sufficient true cohesion to stop seemingly hazardous slopes from failing. This can be a major dilemma because the rock bridges cannot be seen or identified by any realistic investigation method. Careful geological study has failed to identify a useable link between persistence and any other measurable joint characteristic (Rawnsley, 1990) and, it must be remembered, traces exposed at the Earth's surface may be poor representations of characteristics inside the unexposed mass because of stress relief and weathering. Shang et al. (2018, 2016; 2017) successfully conducted experiments to open up rock bridges by using expansive chemicals. This methodology could be repeated for different weathered zones in an attempt to differentiate the degree of rock bridging (which would be helpful for nuclear waste studies). Until that time, designs will need to be conservative with the risk of failure minimised by incorporating toe buttresses, reinforcement with anchorages of some kind, or some other protection, possibly using an avalanche shelter as discussed later.

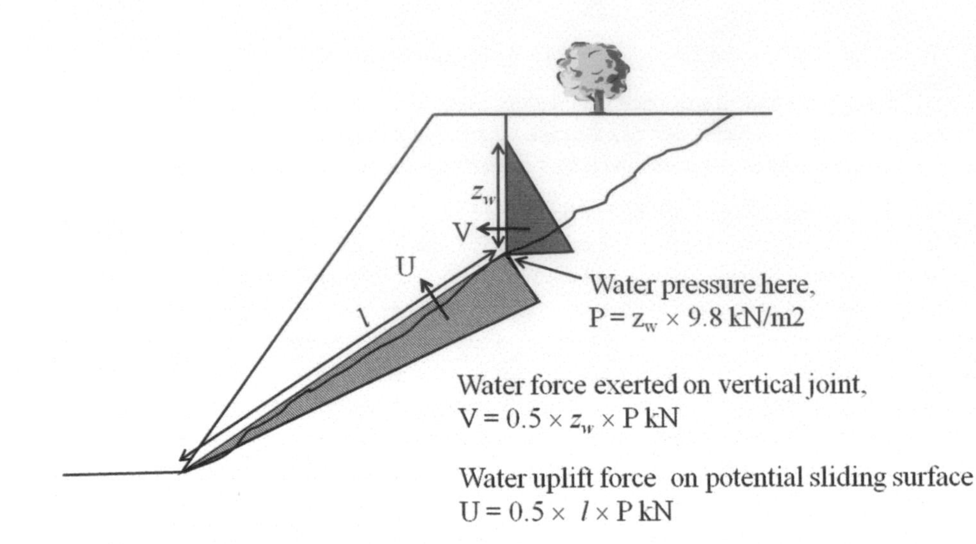

Figure 3.30 Typical model for analysing influence of water pressure on stability of a sliding rock slab.

From experience, wedge failures are relatively rare so that even where these are identified as a problem from stereographic analysis this might not develop in practice. Similarly, most slopes that appear to have a toppling problem do not do so in reality, generally because of impersistence. Care must be taken therefore to be realistic in appraising the results of any geometrical analysis that suggests there to be a problem. One factor that must be considered is risk which is the product of hazard (likelihood of a failure) and consequence (likelihood of injury or damage). One other aspect is that where major failures do occur it is often found by later inspection that the rock mass was in serious distress long before failing and this might have been discovered by carefully targeted investigation. Key factors to look for are open and infilled joints and distorted trees though again the situation might be less risky than it immediately appears (Box 3.4). There is no easy answer to this – it is a matter requiring observation, measurement, analysis, experience and judgement and consideration of consequence. Monitoring can be conducted in the real time, for example using Total Systems that record movements at short intervals automatically or vibrating wire strain gauges with data transferred to the responsible person. Alternatively, periodic examinations using inclinometers, ground radar or photogrammetry can all be effective.

3.7.1.3 Deep-seated failure

Very large rock slope failures often involve some zones where sliding on discontinuities is happening whilst elsewhere the rock mass may be acting as an isotropic fractured mass in a Hoek–Brown way and in other areas intact rock may be failing. Explaining such complex failures is a much easier task than prediction. Many large failures have been studied in detail and these cases are probably the best place to look for ideas and inspiration when dealing with large slope failures (e.g. Eberhardt et al., 2004; Bisci et al., 1996; Malone et al., 2008; Barla & Paronuzzi, 2013).

3.7.2 Soil slopes

For those "soil" slopes, where the ground mass can be regarded as essentially isotropic within each unit or layer (stratum), analysis involves searching through the slope geometry looking for the potential slip planes with the lowest FoS. This can be done easily using available software such as Slope-W and SLIDE. In weathered rock and indeed many "soils", stability might well be controlled by adverse relict joints and other weak discontinuities so that a variety of possible failure modes need to be addressed. An example of hazard models requiring particular analysis within cut slopes in Eocene mudstone at Po Chang in Korea is given in Box 3.5.

BOX 3.5 HAZARD MODELS FOR A SLOPE, PO CHANG, KOREA

The slope shown in Figure B3.5.1 is within a development site near Po Chang, South Korea. The rock comprises weak to strong, bedded mudstone containing strong rounded concretions. The slope was excavated several years prior to the photograph and in some areas is deteriorating very rapidly with large screes of disintegrated mudstone debris.

Figure B3.5.1 Steep, unprotected slope at Po Chang, South Korea.

Figure B3.5.2 Unprotected slope, adjacent to live carriageway with risk to population from several hazards.

Apart from bedding, the main discontinuities are orthogonal vertical sets of joints, probably formed during burial. There are also conjugate shear fractures, inclined at steep angles. As Figure B3.5.2 shows, the same rock is exposed in unprotected slopes adjacent to main roads on the outskirts of Po Chang. There are evident recent failure scars in some of the slopes.

This case provides an example of how a single slope or series of slopes may contribute several different hazards each of which needs to be considered in a different way as illustrated in Figure B3.5.3.

Shallow hazards include boulder fall from the concretions, undermined from the continuing ravelling deterioration of the mudstone. Trees may collapse in a similar way.

There is a risk of structurally controlled failure on the steeply inclined conjugate shear set of joints with vertical joints providing the release surfaces. There was evidence of such failures in some exposures. Finally, there is a risk of large scale landslide in these steep slopes involving a generalised slip surface through the closely fractured rock. The question is how to determine an appropriate set of strength parameters for analysis. It might be reasonable to use the GSI approach (Marinos & Hoek, 2000) but the first step would be to collect more empirical data on the way that the Po Chang mudstone behaves regionally and to identify whether there are any large scale failures that might be back-analysed.

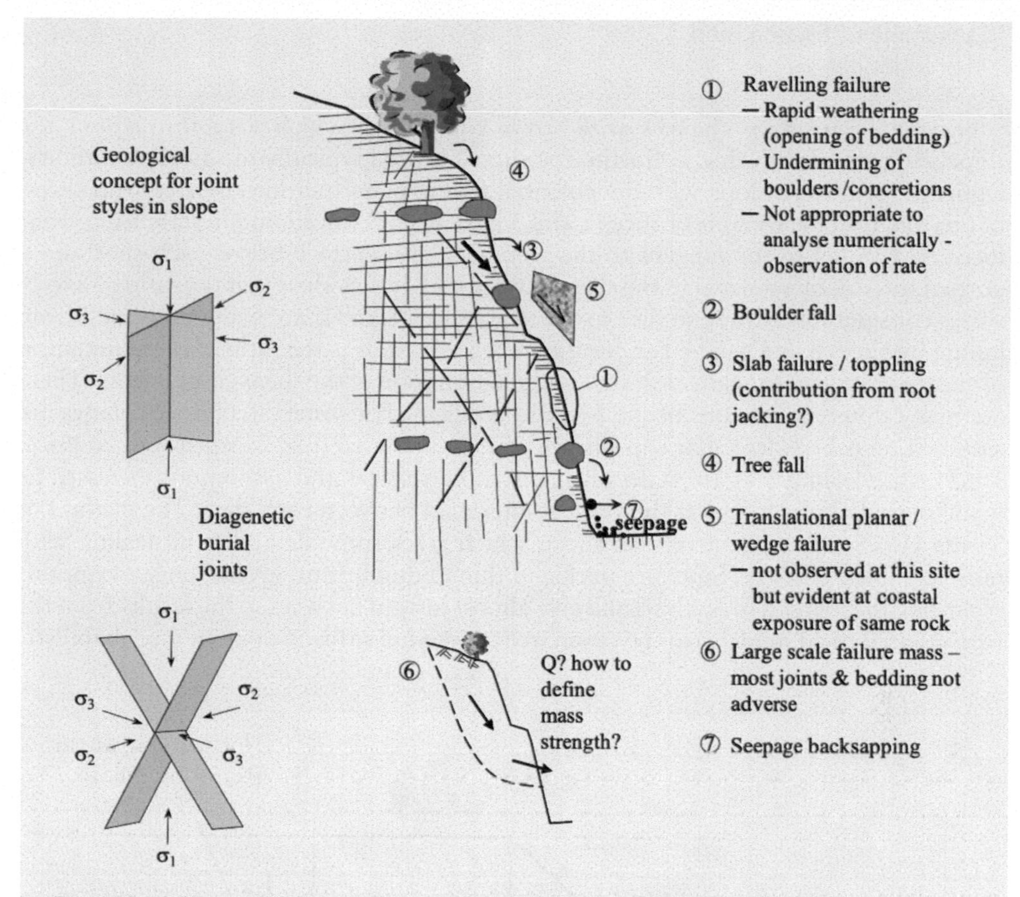

Figure B3.5.3 Various hazard models for slope at Po Chang, South Korea.

The simplest type of analysis is "undrained" in which it is assumed that the soil has a uniform strength, independent of stress level, expressed as cohesion along the potential slip plane. The logic is that any change in normal stress is matched by a change in water pressure so that change in effective stress and frictional resistance is zero. This type of analysis is only appropriate for earthworks in clay, immediately after cutting, and is not considered further here (see also Box 3.3 re pile design).

More generally an "effective stress" analysis is used where the strength of the soil (or closely fractured rock) is considered to be derived from two components – friction and cohesion as per the familiar Mohr–Coulomb expression:

$$\tau = (\sigma - u) \tan \phi + c \tag{3.2}$$

where τ is shear strength
σ is total stress (generally due to weight) normal to the failure plane
u is water pressure reducing σ to an effective stress, σ'

ϕ is angle of friction, and
c is cohesion

Frictional resistance changes with stress conditions, which vary throughout the slope and to deal with this, a "method of slices" is used typically to calculate stability. Figure 3.31 shows a slope with the potential failing mass split into four vertical slices. In this diagram the weights of slices 1 and 3 have been resolved into destabilising shear force, S, parallel to the tangent to the section of slip surface below each slice and a normal force acting normal to the shear surface (N). It is evident that the ratio of S to N varies considerably from one slice to the next. Slices 1 and 2 are being prevented from failing by Slices 3 and 4. The FoS for the slope as a whole is the ratio of the summation of shear resistances beneath each slice to the summation of the shear components. There are many different versions of the Method of Slices. For some circular slip planes are assumed, in others, irregular slip surfaces can be analysed (e.g. Morgenstern & Price, 1965). Slice boundaries are generally taken to be vertical and assumptions need to be made regarding the forces at the vertical interfaces between each slice. The method of Sarma (1975) allows non-vertical slices, which gives some flexibility in dealing with more complex geology. Software packages (limit equilibrium) give a range of options regarding the method of analysis and give almost instant answers so the results from the various analytical models can be compared. Sometimes this is done in a probabilistic

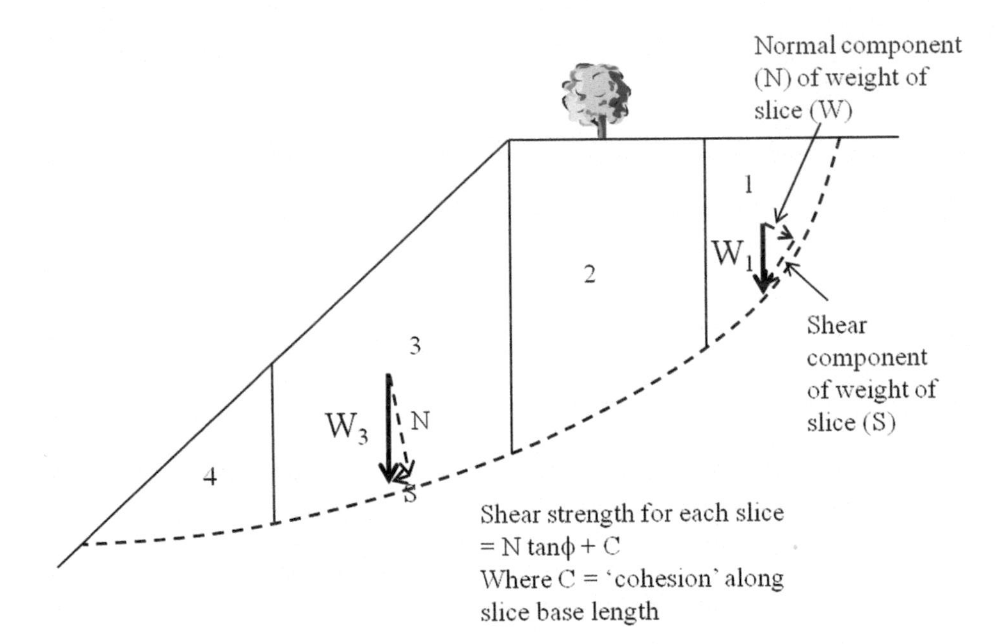

Figure 3.31 The method of slices for slope stability analysis (simplified). Note that to the rear of the slope, in slice 1, the driving force (Shear) is almost as large as the Normal Force, acting vertically, which provides the frictional resistance at the base of that slice. To this is added any cohesion along the shear path. In Slice 3, further down the slope, the Normal Force is much higher than the Shearing Force. Totting up the various forces leads to the calculation of a Factor of Safety (generally done using SlopeW or Slide, these days).

manner, varying the various strength parameters through their anticipated ranges and distributions (Priest & Brown, 1983). Generally, these analyses are carried out to try to establish that the FoS exceeds some chosen value – typically between 1.2 for a slope with low consequence of failure and 1.4 for a higher risk slope and empirically most slopes analysed with such FoS will stand safely provided that the ground model is correct – probably in part because of inherent conservatism in most assessments of mass strength parameters (see discussion on disturbance in Chapter 6 and 7). More sophisticated analyses can be carried out using generalised representations of soils and their properties in both 2 and 3 dimensions (e.g. FLAC SLOPE and FLAC3D – Itasca) and these software packages allow the engineer to see how the failure develops in a time-stepping manner, which is very helpful. The best use of stability analyses is to test the significance of the various assumptions to the outcome. Lumb (1976) addressed some of the problems of the "Factor of Safety" approach and advocated that engineers think instead in terms of probability: "*forcing the designer to consider the reliability of all his data and to face up to the consequences of his being wrong*". If for example water level is shown as critical to stability then that should lead to a careful assessment of the need to prevent infiltration and to install drainage systems.

In the partial factor approach of Eurocode 7, each part of the analysis – forces and strength parameters – are factored in a prescriptive manner. Commonly used software packages can cope with this. This approach might be regarded as rather limiting and perhaps giving an incorrect impression that everything is understood and that all factors are always the same. For example, the Eurocode partial factor for cohesion is the same as for friction (1.25), whereas it is common experience that friction can generally be measured or estimated with far more confidence than cohesion and changing assumptions on cohesion can have a disproportionate influence on calculated FoS.

All analyses are of course only as valid as the input parameters and especially the geological and hydrogeological models; if the model is wrong, so will be the analysis. In a study of the failures of several engineer-designed slopes Hencher (1983d) concluded as follows:

> Six of the eight cut slopes that failed had been investigated by drilling in recent years. In five of these cases, important aspects that controlled the failure were missed. In only one case were the true geological conditions recognised but even then, the groundwater levels were underestimated considerably. In all cases where piezometric data were available and the groundwater level was known by other means, albeit approximately (e.g. observed seepage), the piezometric data did not reflect peak water pressure at the failure surface. This was principally due to failure to observe rapid transient rises and falls in water levels. A further problem was that many of the piezometers were installed at levels where they could not detect the critical perched water tables which developed.

More recently, Lee & Hencher (2009) document a case study where a slope was subject to numerous ground investigations and analyses (often in response to some relatively minor failure) over many years before the slope finally collapsed in a disastrous manner. There were fundamental misconceptions about the geological conditions by all of the investigators. The potential for self-delusion that such methods of analysis truly represent actual stability conditions is expressed a little cynically in the song "Oh Slopey, Slopey, Slopey" in Box 3.6. Lerouiel & Tavernas (1981) used various classic examples of slope failures and their analysis to demonstrate how different assumptions can lead to different results and explanations.

BOX 3.6 OH SLOPEY, SLOPEY, SLOPEY!
– sung (and danced if you wish) to the tune of the Hokey-Cokey

You Put Your Phi Value in
You Take Your *c′* Value out
You Add a Bit of Suction
And You Shake it All About
You Do the Old Janbu[6] and You Turn Around
That's what it's All About.
Chorus:

Oh Slopey, Slopey, Slopey
Oh Slopey, Slopey, Slopey
Oh Slopey, Slopey, Slopey
It's So Easy
One–Point–Four![7]

Written and sung by the GCO Cabaret Stars, 1982

3.7.3 Risk assessment

A decision needs to be made on whether the risk from slope failure is acceptable or not and whether the cost of engineering works can be justified. A modern approach to assessing the need for preventive measures is to use quantified risk assessment as described by Pine & Roberds (2005). The project described involved remediation and stabilisation of several sections of high cut and natural slopes dominated by potential sheeting joint failures and by the potential for failure of rock blocks and boulders bouncing down exposed sheeting joints to impact the road below. Design of slope cut backs and stabilisation measures was based on a combination of reliability criteria and conventional FoS design targets aimed at achieving an ALARP (as low as reasonably practicable) risk target which, in actuarial terms, translated to less than 0.01 fatalities per year per 500 metre section of the slopes under remediation. Further examples of quantitative risk calculation are given by Fell et al. (2005).

3.7.4 General considerations

Remediation of stability hazards on slopes is often not trivial especially where the works are to be conducted close to existing infrastructure and implementation of the works can itself increase the risk levels albeit temporarily. Factors that will influence the decision on which measures to implement include the specific nature of the hazards, topographic and access constraints, locations of the facilities at risk, cost and timing. The risks associated with carrying out works next to active roads both to road users and to construction workers themselves need to be addressed (Geotechnical Engineering Office, 2000a). Pre-contract stabilisation works might be needed to allow site access and preparation. Preventive measures such as rock bolting may be carried out at an early stage to assist in the safe working of the site and designed to form part of the permanent works. Options for the use of temporary protective barriers and catch nets to minimise disruption to traffic during the works also need to be addressed, as do contractual controls and alternatives for supervision of the works. Traffic controls may be needed, and in some circumstances, it will be necessary to close roads or evacuate areas temporarily, especially where blasting is to be used. The use of a risk register, as piloted for tunnels (Brown, 1999), with clear identification of particular risks and responsible parties, helps to ensure that all hazards and consequences are adequately dealt with during construction. Decision analysis is now widely applied at an early stage to assess whether to mitigate slope hazards (e.g. by rockfall catch nets) or to remediate/resolve the problem by excavation and/or support approaches. If construction of intrusive engineering measures to stabilise hazards might be unduly risky then passive protection can be adopted instead. A hybrid solution is often the most pragmatic approach for extensive, difficult slopes where some sections might be stabilised by anchors and buttresses, with other sections protected by nets and barriers (Carter et al., 2002; Pine & Roberds, 2005).

3.7.5 Engineering options

Some of the options for improving the stability of slopes are illustrated in Figure 3.32 and listed more comprehensively in Hencher et al. (2011). These can be split into

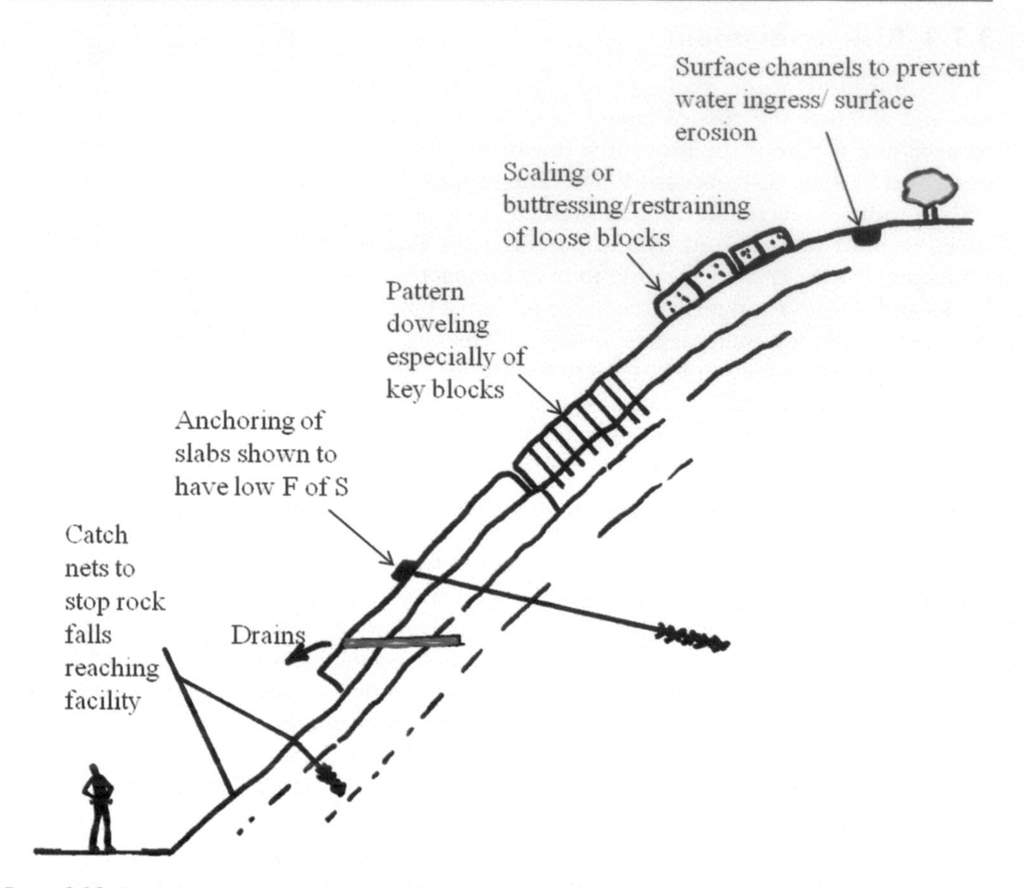

Figure 3.32 Schematic representation of various measures for stabilising rock slopes or protecting public.

passive options that either deal with the possible failure by controlling surface deterioration at source or installing preventative reinforcement to increase local factors of safety, or adding walls or buttresses to restrain detached debris before it causes injury or damage and *active* measures that enhance overall Factors of Safety of larger sections of slope by major engineering works, including cut backs or buttresses or heavy tie-back cable anchors.

3.7.5.1 *Surface treatment*

Many risks can be mitigated cost-effectively through surface treatment to stabilise or remove relatively small blocks of rock. Surface drainage is important using adequately sized concrete channels with a fall across the slope and channels down the face that may be stepped to try to reduce velocity of flow. Further guidance is given in Geotechnical Control Office (1984) and in Ho et al. (2003).

There is a temptation to use hard slope treatments such as shotcrete to constrain loose blocks at the slope surface but such measures if not properly designed can restrict drainage from the slope, hide the geological situation from future investigators and can themselves cause a hazard as the shotcrete deteriorates allowing large slabs of shotcrete to detach. Furthermore, shotcrete is increasingly an unacceptable solution for aesthetic reasons and there is a push towards landscaping high, visual slopes where safety is not compromised (Geotechnical Engineering Office, 2000b). Bio-engineering is used to improve generally the stability and reduce erosion from natural slopes. Roots bind the soil and vegetation can increase surface runoff. Most bio-engineering solutions cannot however be relied upon to improve the long-term stability in risky slopes because vegetation can rot and die or be destroyed by fire. Furthermore, root growth can lead to rock blocks becoming loosened and detached.

3.7.5.2 Rock and boulder falls

Where individual rock fall sources are identified, these can be scaled off, reinforced by dowels, bolts, cables or dentition buttresses and/or netted where the rock is in a closely jointed state. Removing large blocks can be difficult because of the inherent risks associated with breakage techniques including blasting and chemical splitting which can dislodge blocks unexpectedly. Care must be taken to protect the public and workers during such operations. The most difficult zones to deal with are those with poor access. Implementing passive or active protection needs to start from safe ground and move progressively into the areas of more hazardous stability.

Rockfall trajectory analysis using widely available software allows prediction of energy requirements and likely bounce heights and run-out damage zone extent (Figure 3.29). Where energy considerations allow, toe zone protection measures, catch benches, catch ditches, and toe fences provide the most viable mitigation approaches.

Catch nets or fences can be positioned on-slope or in the toe-zone of the slope depending on energy requirements and site restrictions as shown in Figure 3.33. Where energies computed from rockfall analyses are too high for toe zone protection alone to maintain risk levels below prescribed criteria for highway or rail users, on-slope energy protection fences become a necessity to reduce total energy impact at road level. Where the road (or railway) passes under areas prone to continuous rockfall an avalanche shelter is commonly used (Figure 3.34).

3.7.5.3 Mesh

Wire mesh is commonly used to restrict ravelling-type rock failure and can be fixed at many anchorage points or can simply hang down the face, fixed with anchors at the top and weighted with scaffold bars or similar at the toe. Mesh (varying from chain-link, triple twist, hex-mesh to ring-net in increasing order of energy capacity) can be placed by a variety of techniques ranging from climber-controlled unrolling of the mesh to the use of helicopters (Koe et al., 2018). An example demonstrating the reduced hazard is given in Figure 3.35.

3.7.5.4 Drainage

Deep drainage can be very effective in preventing the development of adverse water pressures and this is often a combination of surface protection and channelling of

Figure 3.33 Retaining structures and catch nets to stop natural terrain landslides impacting new road, Lantau Island, Hong Kong.

water away from the slope and inclined drains drilled into the slope (Figure 3.36). Regular patterns of long horizontal drain holes can be very effective, but it must never be expected that all drains will yield water flows and the effectiveness of individual drains will probably change with time as subsurface flow paths migrate. Typically drains comprise plastic tubes with slotted crests and solid inverts, threaded into pre-drilled holes. Inner geotextile liners might be used that can be withdrawn and replaced as and when they get clogged up. Drains might need to be flushed out periodically. Attention should be made to detailing the drain outlets properly otherwise erosion might result. If not maintained, vegetation can block outlets rendering the drains damaging their effectiveness.

In rock slopes there is a need to target subsurface flow channels many of which will be shallow and ephemeral. The paths may be tortuous and hard to identify and drainage measures can therefore be rather hit or miss. If the exposed joint is badly weathered the weak material may back-sap and possibly pipe leading to destabilisation, partially caused by lack of free drainage and careful detailing will be required to prevent deterioration. "No-fines" concrete whilst appearing to be suitable to protect weathered zones often ends up with lower permeability than designed and should not be relied upon without some additional drainage measures.

Figure 3.34 Rock fall nets and avalanche shelter, near Cape Town, South Africa.

As an alternative to deep drains drilled into the slope from the surface, drainage adits and tunnels are sometimes used to lower the water table generally with drainage holes drilled radially into the rock mass from the tunnel walls. Other solutions include deep caissons constructed at the rear of the slope to intercept through flow with inclined drains leading away from the slope at their base (McNicholl et al., 1986). Pumped wells are also occasionally used, pumps being activated when water levels reach critical heights within the slope.

3.7.5.5 *Reinforcement*

Stability can be improved by a variety of reinforcement options. For rough matching joints, provided there has not been previous movement, the interlocking nature provides considerable shear strength. If the joint can be prevented from movement by reinforcing at strategic locations then full advantage can be taken of the natural shear strength. Depending on configuration, rock may be stabilised by passive dowels, tensioned bolts or cable anchors. Passive dowels allow both mobilisation of a normal force (due to the resistance provided by the fully grouted dowel) plus active shear restraint provided by the steel of the dowels resisting block slide mobilisation (Spang & Egger, 1990).

Figure 3.35 Photos of slope before and after construction of drape net, anchored at the top of the slope, and reducing the hazard to the road-user.

Figure 3.36 Drilling deep water drains to reduce landslide risk in Taiwan.

The Geotechnical Engineering Office in Hong Kong has published some guidelines on prescriptive measures for rock slopes and in particular gives guidance on rock dowelling for rock blocks with volume less than 5 m^3 (Yu et al., 2005). In essence it is advised to use pattern dowels with one dowel per m^3 of rock to be supported with minimum and maximum lengths of 3 and 6 metres, respectively, and where the potential sliding plane dips at less than 60°. The dowels are to be installed at right angles to the potential sliding plane, with the key intention to allow the dowels to act in shear, whilst also enhancing the normal restraint due to asperity ride during sliding. In practice dowels frequently need to be used in more variable orientations. Designs must be checked in the field during installation to check that the perceived ground model is correct. If not, then the design must be revised.

Sub-horizontal cable anchors can be used if capacities larger than about 20 tonnes per reinforcement member are required. Great care needs to be taken to ensure that such tensioned anchors are adequately protected against corrosion and regular checking and maintenance will be required. Figure 3.37 shows a trial anchor exhumed from slope, being examined, on the A55 in North Wales. Several cases of anchors that have failed due to corrosion are discussed in Chapter 8. For weaker rock and soil, pattern soil nailing is now commonly used. The "nails", which typically comprise 50 mm or so diameter steel bars connected as necessary by couplers every 6 m, are usually installed in pre-drilled holes, held centrally by "lantern" spacers and then pressure grouted over their full-length using tubes installed with the nail (Figures 3.38 & 3.39). Soil nails are usually installed as a passive reinforcement that would only take on load if the slope began to deform prior to failure.

Figure 3.37 Exhumed, triple protected anchor at A55 works, North Wales.

Figure 3.38 Soil nail drilling in Landslide Preventive Works scheme in Lantau, Hong Kong.

Figure 3.39 Details of passive soil nails showing installed nails and head development, to the right.

3.7.5.6 Retaining walls and barriers

Retaining walls are commonly used to support steep slopes, especially where the slope comprises weak and broken rock and where space is constrained. There are many different types as illustrated in Figure 3.40.

For temporary works, corrugated steel sheets are generally driven, vibrated or pushed into soil prior to deep excavation, with each sheet linking to its neighbour. As the excavation proceeds, sheets are usually braced by a system of struts and waling beams although they may also rely on depth of embedment. Diaphragm walls formed by concreting deep trenches excavated under bentonite mud are also used as part of temporary works and then may be incorporated in the permanent structure. Permanent retaining structures are often created using piles (Figure 3.41). Alternatively, provided that the ground can be anticipated to stand steeply temporarily during construction, the full slope is cut back and then a wall of concrete constructed at some short distance in front. The space between the wall and the natural ground is back filled with granular free draining material, often with geotextile material at the interface, feeding water down to a drain. Drainage is very important if the retaining wall is not going to act as a dam. "Gabion" structures are made from galvanised steel or plastic baskets, backfilled with rock. The main advantages are that they are free draining, can be landscaped, and they can be made cheaply on site using locally derived rock to fill locally woven baskets. They are therefore very suitable for forming retaining structures or barriers in remote locations (Fookes et al., 1985). Deflection structures, impact barriers and "debris brakes" are commonly used for mitigation against channelised debris flows.

3.7.5.7 Maintenance

Whatever the engineering solutions adopted, slopes should be examined periodically for signs of distress and for maintenance such as cleaning out of drainage channels. The requirement for inspection, testing and possibly remediation works should be built into the design of any new slope with careful consideration for how this is to be achieved. In Hong Kong the current practice is to prepare "maintenance manuals" for slopes and to carry out routine engineer inspections at regular intervals. New and newly upgraded slopes are generally constructed with access ladders and often with hand rails provided along berms to allow safe inspection.

3.8 EXCAVATION, DREDGING AND COASTAL ENGINEERING

3.8.1 Excavatability

Site excavation is usually carried out by heavy machinery and the main questions for the engineering geologist are what machinery would be suitable and whether the rock would need to be blasted first.

Where blasting is restricted then the contractor might need to use some kind of chemical or hydraulic rock splitter, but the noise levels of drilling and rock breaking might still be a problem. Generally, the factors that will control whether or not blasting is needed are intact rock strength and the spacing between joints (MacGregor et al.,

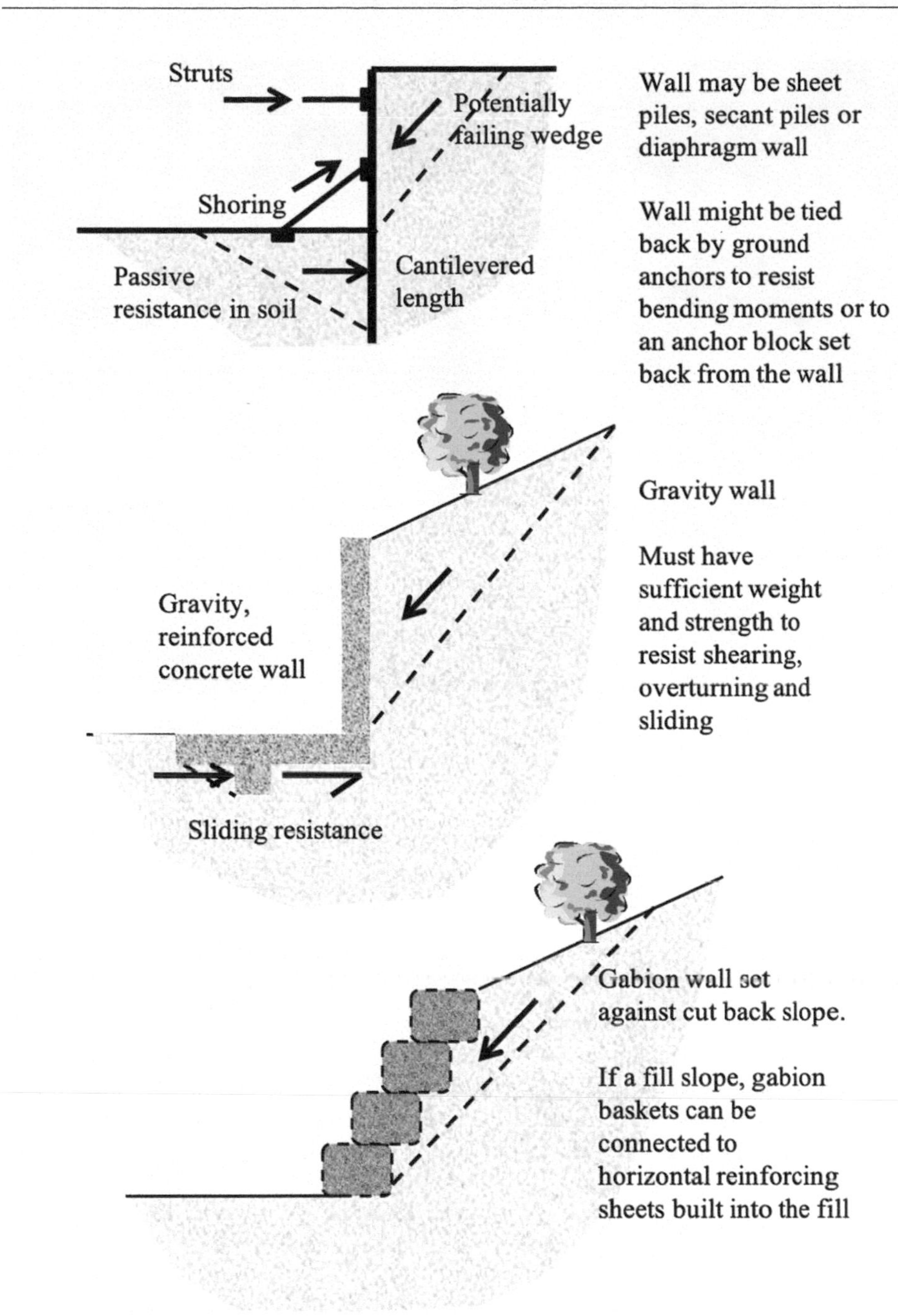
Struts
Potentially failing wedge
Wall may be sheet piles, secant piles or diaphragm wall
Shoring
Wall might be tied back by ground anchors to resist bending moments or to an anchor block set back from the wall
Passive resistance in soil
Cantilevered length
Gravity wall
Must have sufficient weight and strength to resist shearing, overturning and sliding
Gravity, reinforced concrete wall
Sliding resistance
Gabion wall set against cut back slope.
If a fill slope, gabion baskets can be connected to horizontal reinforcing sheets built into the fill

Figure 3.40 Different types of retaining wall.

Figure 3.41 Retaining wall made up of hand-dug caissons, under construction in Hong Kong. Note that workers nowadays would be expected to wear boots, hard hats and full protective clothing.

1994; Pettifer & Fookes, 1994). As emphasised elsewhere, care must be taken to differentiate between mechanical fractures with low tensile strength and incipient fractures with high strength as this will affect strongly the ability of machines to rip the rock.

3.8.2 Dredging

Dredging is commonly carried out for port works, as part of other land reclamation projects (Bray et al., 1997). There are several types of dredger and these vary in their capacity to deal with soil and weak rock. Where the rock is stronger and or the joint spacing is wide, excavations can be difficult even for the strongest suction cutter dredgers and then some pre-treatment, normally blasting, will be required. Guidance is given on borehole spacing requirements by Bates (1981) but the site-specific requirements must depend upon the geological situation and limitations and sensitivity of the available equipment. Reviews on dredging practice in various countries including the USA, UK, HK and Singapore are given in Eisma (2006).

3.8.3 Coastal engineering

Coastal engineering is an important topic dealing with protection of the seafront, and tied into our concept of geological makeup of the land, impact from the sea (risk) and obviously sea level rise in the future as dealt with in Chapter 5 and especially Box 5.1.

The methods for assessing the sea conditions and the options for protecting the coast are set out by Reeve et al. (2018).

3.9 GROUND IMPROVEMENT

3.9.1 Introduction

At many sites the ground conditions are too weak or wet to allow construction by the preferred method or even to allow access by heavy construction equipment. Ground improvement might therefore be carried out, often as an alternative to some engineering solution such as piling and the engineering geologist should be aware of the techniques that might be employed to deal with a particular site condition. Ground improvement might be used in temporary works, such as freezing the ground to allow tunnelling through saturated and potentially flowing materials or the construction of barriers to water flow or to restrict vibrations during construction. In other situations, ground improvement might provide a permanent solution, such as densification or using chemical additives to provide additional strength.

3.9.2 Dynamic compaction

One of the simplest methods is dynamic compaction, which involves dropping a large weight, up to about 30 tonnes, from a crane, over a regular pattern and then backfilling the depressions with granular material. Further drops are carried out at closer spacing. The depth of improvement depends upon the weight dropped, size of "pounder" and the height. Typically, a weight of about 15 tonnes dropped 20 m might be expected to improve ground to about 10 m deep (e.g. Bo et al., 2009). The method is most suitable for improving fills and granular soils generally but sites underlain by clay have also been improved although consideration must be given to the pore pressures that might be generated and how these dissipate. Generally, the improvement is measured by tests before and after improvement using techniques such as the SPT, CPT or the Menard pressuremeter that was developed specifically for this purpose (Menard & Broise, 1975). The technique has been applied successfully for quite prestigious projects involving large scale reclamation, such as Nice Airport. In Hong Kong, in the 1980s, the method was used to densify the upper horizons in old fill slopes and thereby improve their stability although the environmental nuisance of the method means that it is no longer acceptable (Geotechnical Engineering Office, 2005).

3.9.3 Static preloading

If time allows, then an effective way to improve the consolidation characteristics of soil at a site is to pre-load it, often by placing an embankment of fill material that can be removed again later or re-graded at site, compacted properly in thin layers. The process of consolidation is generally accelerated by introducing a series of vertical drains to increase the mass permeability and allow excess pore pressures to dissipate monitored using piezometers. The drains can be "sand wicks", which are sausages of geotextiles, filled with sand and installed in predrilled holes. Other systems include "wick" drains that are geotextile covered plastic elements pushed into the ground using a purpose-built machine.

3.9.4 Stone columns

Stone columns can be used to enhance drainage, and are installed to depths of 10 m and sometimes more. These are formed by using a vibrating poker, pushed into the soil to form a void and then filling the void with gravel and sand, which is compacted in stages using the same vibrating tool (McCabe et al., 2009). Stone columns have been used to increase mass permeability and prevent liquefaction of loose silty sand during an earthquake although in such a usage settlement will still occur but in a relatively uniform and non-catastrophic manner. Stone columns are also used generally to improve the bearing conditions at a site, the improvement depending upon the ratio of cross-sectional area of stone columns to untreated ground. Groups and lines of stone columns can be used as weak piles to provide support to structures such as oil tanks.

3.9.5 Soil mixing and jet grouted columns

Clay soils especially can be improved by mixing with lime slag and cement either at the ground surface (to prevent erosion in slopes, for example) or in columns or trenches using hollow stem augers and similar equipment. The works will improve the bearing capacity of the ground although the improvement might be difficult to quantify. Stronger columns can be formed by using jet grouted columns formed using high-pressure grout jets as a drilling string is rotated and lifted from depth. The resulting column of mixed soil and grout can be used to carry structures or to form cut off barriers to restrict water flow, for example, beneath dams. Jet grouting is sometimes used to form structural members during temporary works construction of deep excavations (Puller, 2003; also see case study of Nicoll Highway collapse in Chapter 8).

3.9.6 Drainage

For deep excavations and tunnelling it is commonly necessary to lower the groundwater during construction although there are many factors that must be considered, not least associated settlement of the ground due to increased effective stress and self-weight compaction and consolidation and drying up of land in adjacent properties (Preene & Brassington, 2003). New and steep flow paths through the soil can lead to seepage piping and liquefaction in the floor of excavations. The cheapest and simplest way to lower water is just to let it happen naturally as the excavation proceeds, to channel water inflow to collection sumps and to then pump this water away, although disposal may be an issue on environmental grounds and pumping from great depth will require a series of pumps at different levels. Active dewatering is generally conducted using well-point systems or submersible pumps in wells and details are given by Puller (2003). As noted elsewhere dewatering is often important to the stability of slopes and semi-permanent solutions include drains, drainage caissons and adits. Emergency pumping systems are sometimes set up to be triggered if piezometric levels become dangerously high.

3.9.7 Geotextiles

Geotextiles are fabric or plastic sheets that have many different uses in ground engineering and a few of these are discussed below.

3.9.7.1 *Strengthening the ground*

To improve site access sheets of plastic mesh may be laid on the ground and then a layer of gravel placed and compacted on top. The purpose of the geotextile is to prevent the gravel being pushed into and mixing with the underlying soil that may be wet and soft. In this way temporary road access can be provided. In other circumstances, more complex solutions might be designed involving elements, such as stone columns or piles together with a geotextile grid draped across and linking the structural elements.

Geotextile mats and strips are also used in the design of reinforced earth structures (as are metal grids and strips) as illustrated in Figure 3.26. Basically, the frictional resistance between the soil and grid or mats, placed horizontally and regularly within a fill structure enhances the overall strength of the soil mass and prevents it failing. Where facing walls are used or the geotextile is wrapped around at the face to prevent soil erosion the finished structure can be very steep or even vertical.

Plastic grid boxes, infilled with rock cobbles, have been used to form gabion walls as barriers. Care must be taken that the situation is not one where the finished structure can be destroyed by fire and that the deterioration rate is acceptable given the proposed lifetime of the structure.

3.9.7.2 *Drainage and barriers*

Geotextile sheets are available that are highly permeable but also designed with a mesh size that restricts soil erosion in the same way as traditional soil filter systems. Geotextiles are therefore used, for example, as part of the drainage system behind concrete retaining walls. Plastic sheets (geomembranes) are used as barriers to water flow, especially for landfill sites. Great care must be taken to ensure that sheets are welded one to the other and that those welds are tested. Membranes must be resistant to and protected from puncturing. Any leakage may be extremely difficult and expensive to rectify at a later stage. In Chapter 8 an example is given where a combination of permeable geotextiles and impermeable geomembranes were used to reduce leachate loss from a quarry used for landfill.

3.9.8 Grouting

Grouting has been referred to many times in this chapter in association with a wide range of engineering applications.

Grouting is generally used to increase strength of a rock or soil mass and to reduce permeability (Warner, 2004). In the case of drill & blast tunnels, pre-drilling is used to check flow and then the ground is pre-treated by grout before continuing. In some cases where flow develops rapidly and will not stop, hot bitumen grout may have to be applied whereby the interface between the bitumen and flowing water creates a crust, beneath which the bitumen continues to flow like a Newtonian fluid. The method has been used successfully to stop huge flows of 4000 litres/s through marble at Yung Chung Tunnel in Taiwan (Fu et al., 2007).

Grouting is used routinely below dams to provide a cut-off "curtain" to restrict seepage through the foundations. The efficacy of the grouting is assessed by pumping tests before and after grouting using the Lugeon test (Houlsby, 1976). If the foundations are still leaking water, another pattern of grout-holes is added, and the Lugeon testing repeated.

A main consideration is the type of grout – usually Portland cement is used in gravel and coarse sand but sometimes chemical grouts or resin must be used to penetrate low permeability ground such as jointed rock. Grout will not penetrate clay or even silt so where dealing with such low permeability materials, some physical barrier like sheet piles, diaphragm walls or possibly jet grouting may be needed. The grout is injected typically using tam a manchette tubes to grout over 0.5 metre lengths. Grouting might jack open existing joints in rock or form new fractures in soil and weak rock and this is called "claquage". Grouting may be used to correct settlement or other deformations caused by engineering works such as tunnelling (e.g. Harris et al., 1994) but care must be taken that the grouting itself does not make matters worse as per the Heathrow Collapse described in Chapter 8.

3.9.9 Cavities

Cavities that engineering geologists and geotechnical engineers need to contend with include natural cavities such as those often found in limestone areas (Waltham, 2016), more rarely in other rock types including unlikely candidates such as weathered granite (Hencher et al., 2008). The other main problem is mine workings as illustrated earlier in Table 3.3. Ground investigation for such voids is a matter of proper desk study (including likely mining method if that is the hazard), focused investigation possibly using geophysics such as micro-gravity and resistivity and probing, perhaps using percussive drilling to keep the costs down. If and when voids are found these can be explored and characterised in terms of size and extent using cameras, echo sounders and radar. In the case of mines and caves, inspection may be required by suitably equipped and experienced persons following proper procedures as specified in health and safety regulations. Depending on their extent, if necessary voids may be backfilled with gravel, grouted or structurally reinforced as appropriate. When extensive mine workings were encountered unexpectedly during tunnelling for the high-speed railway from Seoul to Taejon in South Korea one proposed solution was simply to construct a concrete structure through the mine but this was considered politically unacceptable because the public was already aware of the situation. Instead the route for the railway was moved several km and previous railway works abandoned.

3.10 SURFACE MINING AND QUARRYING

Surface mining and quarrying are industries that have strong demands for geotechnical expertise including engineering geology. Slope design is often very important and the design practices discussed later, used in civil engineering also apply to quarries and open pits and open cast mines. The main difference is that in such enterprises many of the slopes are always changing in geometry as the works progress. One key to success is establishing a safe layout for operations, such as crushing and processing plants and for haul roads whilst avoiding sterilising valuable resources because of the siting of infrastructure such as site offices and treatment plants. Major haul roads also need to be established in a safe manner to avoid disruption to operations if instability occurs. Other faces may well be temporary and are therefore formed at angles that would be unacceptable as permanent slopes in civil engineering. For large open pit mine operations, the scale of overall slope formation can be huge, extending hundreds of metres, and predicting stability often requires numerical modelling by experts, tied in to monitoring systems. Excavation of rock usually involves blasting and this is a specialist operation as it is for tunnelling. A good review is given in Wyllie & Mah (2004) and detailed guidance is given in Read & Stacey (2009). Key considerations for all blasting operations are fragmentation to avoid producing large blocks that cannot be handled easily and need secondary breaking operations, avoiding damage to the remaining rock, avoiding overbreak beyond the design profile, safety and risk from fly rock, gases and vibrations.

Waste from mining needs to be disposed of. In open pit coal mining the waste rock is backfilled into the void as part of the on-going operations and nowadays in the UK at least, the final re-instatement of the area is strictly controlled with every attempt made to simulate the natural countryside as it was pre-operations. Other wastes are often wet and contaminated and held behind tailings dams that should be designed and analysed with just as much care as any other civil engineering structure. Unfortunately, this is often not the case and there have been many major failures world-wide over the last 50 years, which have resulted in severe contamination and many deaths (e.g. Rico et al., 2008).

3.11 NUMERICAL MODELLING FOR ANALYSIS AND DESIGN

3.11.1 General purpose

There are two main groups of programmes commonly used – finite element (FE) and finite difference (FD), time-stepping type software. PLAXIS is a general-purpose FE package that allows geotechnical situations – foundations, slopes or tunnels, to be modelled. The model is set up and run to give a quick solution to complex equations – perhaps of deformation or calculation of Factor of Safety of a model that is split into elements – mostly triangular. It can also be used to model fluid flow. As with all sophisticated software, it should only be used by those knowledgeable of the underlying mechanics and the way these are dealt with within the computer programme. Following the Nicoll Highway collapse discussed in Chapter 8 it was established that there had been a mistake made in the manner in which the design of the diaphragm walls was carried out using an inappropriate soil model. The same problem would have arisen for any finite element package used in this incorrect manner – it is not unique to PLAXIS. The mistake resulted in excessive deformation of the walls and an under-design of their moment capacity. The Nicoll Highway Committee of Inquiry Report on the collapse includes a well-written section on the problem with the Mohr Coulomb model (Magnus et al., 2005).

The finite difference programme FLAC is probably the second most generally used software for geotechnical design.

Until recently the programme was quite daunting, requiring individual commands to be typed in but recent versions have a graphic interface, which makes things easier. As for PLAXIS and other sophisticated programmes, a great deal of knowledge and understanding is needed if reasonable results are to be achieved. For example, the model must first be set up with proper boundary conditions and brought to equilibrium as "natural" ground before any engineering works such as excavation are simulated. FLAC progressively calculates and checks solutions. Intermediate stages can be calculated, saved and expressed graphically as a movie, which can illustrate how strains are developing with time. FLAC like its sister programme UDEC can cope with large displacements, more so than typical FE analyses. FLAC is used mainly for soil and rock that can be characterised as continua. UDEC is used for fractured rock and each fracture or set of fractures can be specified individually in terms of geometry and engineering parameters. Both UDEC and FLAC can be used for foundation design, tunnels and slopes.

Other commonly used software includes the suite produced by Rocscience such as Phases and their use is discussed in Hoek et al. (1995) and at www.rocscience.com/education/hoeks_corner.

Many authors, experienced in the development and use of software have recommended that sophisticated software should be used in an investigatory way "like a detective" (Starfield & Cundall, 1988) using many simple models to check sensitivity to assumptions rather than trying to prepare a single, complex model in an attempt to simulate all aspects of a situation at the same time. Swannell & Hencher (1999) discuss the use of software specifically for cavern design.

3.11.2 Problem-specific software

Many suites of software have been developed for particular purposes. SLOPE/W and SLIDE, for example, are commonly used for routine design of slopes. The software calculates stability employing the method of slices as discussed later and gives instant solutions for FoS for a wide range of potential slip surfaces, the broad geometries of which are specified by the operator. Controlling factors such as slope geometry, strength parameters and groundwater conditions can be varied rapidly allowing sensitivity analysis. Structural elements such as soil nails and rock bolts can be included in the models. There are many similar packages available all of which are verified and validated against standard mathematical solutions. There is a danger that the ease of use of such software in sensitivity analysis, varying a range of likely parameters, can give a misplaced confidence that all possible conditions have been dealt with. If the ground model is seriously wrong the results will be meaningless.

Other specialist software packages are used for particular design tasks, such as rock fall trajectory analysis, stresses around tunnels, pile design and groundwater and contaminant migration modelling and details of many of these are reviewed at the web page maintained by Tim Spink: www.ggsd.com. Most engineering companies also have in-house spread sheets (often based on EXCEL) used to solve common analytical problems.

3.12 FRAUD AND CORRUPTION

As a final note, the engineering geologist should be aware that fraud and corruption occur in civil engineering as in other walks of life. Whole boreholes in ground investigations have been known to be fictitious, let alone individual test runs. Cases are known where core from one site is placed into core boxes at another site. Tests are sometimes not carried out as specified with sample sheets carried around in vans. Data on piles and other foundations are sometimes made up, for example, depths, materials used and test results (Hencher et al., 2005). Engineering geologists need to be aware that such practice does occur, albeit rarely, and remain alert in their supervising duties.

Regarding deaths resulting from earthquakes, fictitious accounts are commonplace, both high to attract more international aid and low as part of a coverup as reported by Ambraseys and Bilham (2011).

Notes

1 This necessitates a considerable degree of rapport between the design engineer and the engineering geologist. The primary requirement, with respect to this book, is to ensure that the engineering design is compatible with, and can be adjusted to, the geology (Knill, 1978).

2 Burland (1987) states "*it is my experience that over 90% of our design decisions are based on a knowledge of the ground profile with calculations playing the secondary role of refinement and justification... There are countless examples where minor structural features ... have dominated behaviour...*".

3 It is remarkable that, with all these methods in use, civil structures made of steel and concrete still collapse regularly. Bolton (1979) comments "*Fresh graduates are often intrigued by their employer's evident lack of confidence in clever calculations, and amazed at the ingenuity of the experienced engineer who will so design his works that calculations will be redundant*".

4 British Standard 5935 (British Standards Institution, 2019) states that a Cat 3 check is required when a temporary works design is complex, unusual or in a high hazard location, such as a nuclear environment or works adjacent to a railway.

5 Note: The actual stress conditions near a tunnel would be more complex than this calculation. The tunnel would distort the stress field – refer to Muir Wood (2000) or Hoek et al. (1995).

6 Janbu is the author of a commonly used limit equilibrium method of slices for calculating Factors of Safety of slopes and can be applied to non-circular surfaces. There are two forms – a "routine" method and a "rigorous" method (Janbu, 1973). Lumsdaine & Tang (1982) carried out an exercise comparing results of calculations by 6 Government Offices and 36 others and found a very high proportion of analytical errors and lack of documentation, which of course is over and above any uncertainty in ground model, parameters adopted and assumed groundwater conditions – either positive pore pressure or suction.

7 A Factor of Safety of 1.4 is generally regarded as an acceptable number to guard against failure in a high risk slope in Hong Kong (Geotechnical Control Office, 1979; 1984).

Chapter 4

Geology and ground models

4.1 CONCEPT OF MODELLING

4.1.1 Introduction

The geology at a site can range from apparently simple to apparently complex at scales of metres, tens or hundreds of metres (Figures 4.1 & 4.2). Geological complexity does not however always equate with difficulty in engineering terms (Morgenstern & Cruden, 1977). Conversely even where the soil or rock mass is apparently relatively uniform there may be a single feature or a single property that will cause problems. Figure 4.3 shows a specific fault, which allowed a particular mechanism of failure to develop in a huge slope in Malaysia. It might be a rather more mundane soil property such as a change in porosity leading to a different permeability, which causes the problem, or a rather more critical attribute, such as a super-transmissive joint at a nuclear repository, which leads to an unacceptable, high permeability (unknown other than as postulated). It is one of the main tasks of the engineering geologist to interpret the geology at a site and to identify those characteristics and properties that might be important to the engineering project. Much of the detail will be insignificant; the skill is in recognising what is and what is not. At some stage during the design process the geology will need to be differentiated in some way into units that can be characterised with essentially uniform mechanical properties or where the properties change in some definable way, perhaps with depth. Sometimes the way to do this is obvious – for example, a layer of fill (man-made ground) overlying alluvium which in turn overlies bedrock which will define the way foundations are designed but at other locations identification of the key attributes is more difficult. Thin layers that might be overlooked, or even lost, in logging a borehole could turn out to be the most important features at a site.

For civil engineering, ground models need to be prepared that are simplified representations of a site and that should incorporate all the important elements relevant to design and construction. The models are generally developed from a preliminary 3D interpretation of the geology based on desk study and surface mapping and then refined by further study of environmental factors, such as earthquake hazard and hydrogeology. Models will be improved by ground investigation and testing and finally presented as a design model specifically tuned to the project (Baynes et al., 2020). The process is illustrated simply in Figure 4.4 and expanded upon later in this chapter. One of the key features of many ground models is differentiating between upper, soil-like

DOI: 10.1201/9781003348894-4

Figure 4.1 Massive, horizontally bedded, Eocene conglomerate and sandstone unconformably overlying Triassic Lower Musschelkalk, Sierra de Montsant, north of Falset, Spain.

Figure 4.2 Folded and faulted extremely strong Devonian Radiolarian Chert interbedded with thin bands of extremely weak organic shale, near Cabacés, Spain.

Figure 4.3 Pre-existing fault (tight and planar) allowing accommodated movement as a counter scarp, Po Selim Landslide, Malaysia.

materials and underlying rock with the separating boundary being called rockhead or sometimes "engineering rockhead". Care must be taken in using this term because it has various definitions and connotations and is sometimes used in an over-simplistic way for what is a complex situation. The consequences of wrong perception can be severe if, for example, soil is encountered at depth and below the water table unexpectedly in a "hard rock" tunnel. Definitions of rockhead are set out in Box 4.1.

BOX 4.1 DEFINITION OF ROCKHEAD

Care must be taken in using the term rock head because it can be defined in various ways and the wrong impression may be conveyed within a geotechnical design team that things are clear cut when they are not.

GEOLOGICAL DEFINITION

Rockhead is defined in BS 3618 (British Standards Institution, 1964) as "the boundary between superficial deposits (or drift) and the underlying solid rock" and this definition is also adopted by the US Department of the Interior (Thrush et al., 1968). The term *solid rock* is defined, in turn, in Thrush et al., following Challinor (1964) as "rock which is both consolidated and in-situ". "Solid" is also generally used in a geological sense to describe formations that predate superficial deposits (Whitten & Brooks, 1972) as in "solid" vs. "drift" maps. Rockhead used in this way essentially defines a geological boundary usually marking an unconformity. The "solid" rock shown on a geological map says nothing about its strength or weathering state so "rockhead" does not necessarily mark a boundary between soil and rock in strength terms.

GEOTECHNICAL DEFINITION

The term rockhead or engineering rockhead is often used in geotechnical design to define a boundary between soil-like material and rock that is stronger and more resistant whatever the geological conditions. It is also sometimes used more generally "as the level at which the engineering parameters of the ground satisfy the design parameters for a specific project" (Geotechnical Engineering Office, 2007).

Sometimes the geological profile is simple (recent soil over rock) and rockhead is readily defined but often the situation is more complex and care must be taken not to represent a difficult and variable geological condition in over-simplified diagrams that might be misunderstood by designers. Weaker material or voids below the first occurrence of "rock" in a borehole might have a controlling influence on mass strength, compressibility and permeability and have severe affects on constructability, for example, collapse of pile borings or sudden inflow of soil into a tunnel.

It is particularly difficult to define a simple level for rockhead in regions of weathered rock. In the opinion of Knill (1978), in the case of karstic limestone, rockhead is the geological contact between *in situ* limestone and overlying superficial deposits (despite often great complexity due to dissolution features). Similarly Statham & Baker (1986) define rockhead by the top of *in situ* limestone (despite the presence of sediment – infilled voids "below rockhead"). Goodman (1993) comments that "the unevenness of the top-of rock surface (or rock head in British usage) on karstic limestone presents obstacles for the designer". He notes the many potential difficulties for design and construction and states that "unlike most other rocks, the existence of a solid-appearing outcrop right at the location of a footing or pier does not guarantee that good rock will occur below the outcrop". The same is true for other rock types with large corestones sitting on the ground surface underlain by severely weathered rock as discussed by Ruxton & Berry (1957). An example of a landslide that occurred where rock head was misinterpreted by the slope designers on the basis of boreholes that terminated 5 m in rock is described in Hencher & McNicholl (1995).

4.2 RELEVANCE OF GEOLOGY TO ENGINEERING

Attempting to form a ground model of a site based solely on descriptions from boreholes and test results without recourse to an informed geological interpretation of the data would be like trying to put together a complex jigsaw without having the picture on the box lid. Geologists are trained to examine rocks and soils at scales of a hand specimen or a quarry and to draw conclusions on the likely origins and history of the sample or exposure. Then by examining other samples and exposures in and around a site, they can start to develop a picture of how the various different components relate to one another. The conceptual model for the geology at a site can be used to extrapolate and interpolate observations to make further predictions on the basis of geological knowledge. Ground investigation can then be designed and used to target residual unknowns.

4.3 GEOLOGICAL REFERENCE MODELS

4.3.1 A holistic approach

Fookes et al. (2000) encourage a "total geological approach" whereby any site is assessed with regard to its full geological history. That history includes original formation of the soils or rock underlying the site, tectonic events, weathering, erosion, deposition of any overlying superficial deposits, geomorphological development and anthropogenic influences. The changes that have taken place at the Earth's surface to form the landscape and the extensive time involved are almost inconceivable but must be considered when interpreting the geology at any site.

Total geological analysis should allow the distribution and nature of the various strata at the site and other features such as hydrogeological conditions to be explained. The assessment should be based on the extensive literature on geological and geomorphological processes, which comprises the "toolbox" for interpreting the conditions that are encountered at any site. Useful sources of information on geology focused on civil engineering application are Blyth & de Freitas (1984) and Goodman (1993) but often, to understand what is happening at a site one has to refer to more fundamental geological literature. This chapter introduces aspects of geology that are relevant to engineering design and performance. It commences by considering the three basic rock types – igneous, sedimentary and metamorphic – focusing on their typical characteristics and associations that may be of particular importance to engineering. The next section introduces rock structures, particularly the origins and characteristics of discontinuities that tend to control rock mass properties. Towards the end of the chapter guidance is given on developing ground models for a site.

4.3.2 The need for simplification and classification

Simplified approaches are generally adopted for the description and classification of soils and rocks for engineering purposes largely because geological detail is often irrelevant. This especially applies to logging soil and rock encountered in boreholes. Nevertheless, the engineering geologist needs to be alert to situations and cases where geological detail might be important to explain the geological situation or because particular characteristics have some special significance. In the author's experience, whilst many sites are described and characterised quite adequately using shorthand terms and classifications, occasionally one meets situations where to understand what is happening, to an adequate level for an engineering project, intensive study is necessary into geological minutiae including chemical analysis, thin section examination and even radiometric dating.

4.3.3 Igneous rocks and their associations

Igneous rocks were once molten. As magma cools minerals grow with an interlocking texture. As a result, most igneous rocks are strong and sometimes extremely strong in their fresh state – several times the strength of concrete. They are primarily split into *intrusive* rocks that solidify below the Earth's surface and *extrusive* rocks that form on the Earth's surface. They are then differentiated according to chemistry. Rocks with high silica content either directly as quartz (SiO_2) or tied up in the structure of other silicate minerals are termed "acidic". "Basic" rocks have low silica content and "ultrabasic" rocks even lower. A simplified classification of igneous rocks is presented in Table 4.1.

Table 4.1 Classification of igneous rocks

IGNEOUS ROCKS: generally, have massive structure and crystalline texture. Typically, high strength in fresh state.
Volcanic rocks deposited as sediments are dealt with in Table 4.2: Sedimentary Rocks. More details are given in Streckeisen (1974; 1980) and Thorpe & Brown (1985).

Grain size	*ACID* *Much quartz*	*INTERMEDIATE* *Little quartz*	*BASIC*[1] *Little or no quartz*
	Pale colour ⟵	⟶	Dark
	Relatively light in weight		Heavy
Coarse	GRANITE[2]	DIORITE	GABBRO
>2 mm	GRANODIORITE	SYENITE	
Medium	MICRO-GRANITE[3]	MICRO-DIORITE	DOLERITE[4]
0.06–2 mm	MICRO-GRANODIORITE	MICRO-SYENITE	
Fine	RHYOLITE[5]	ANDESITE	BASALT
<0.06 mm	DACITE	TRACHYTE	
Glassy	OBSIDIAN	VOLCANIC GLASS	

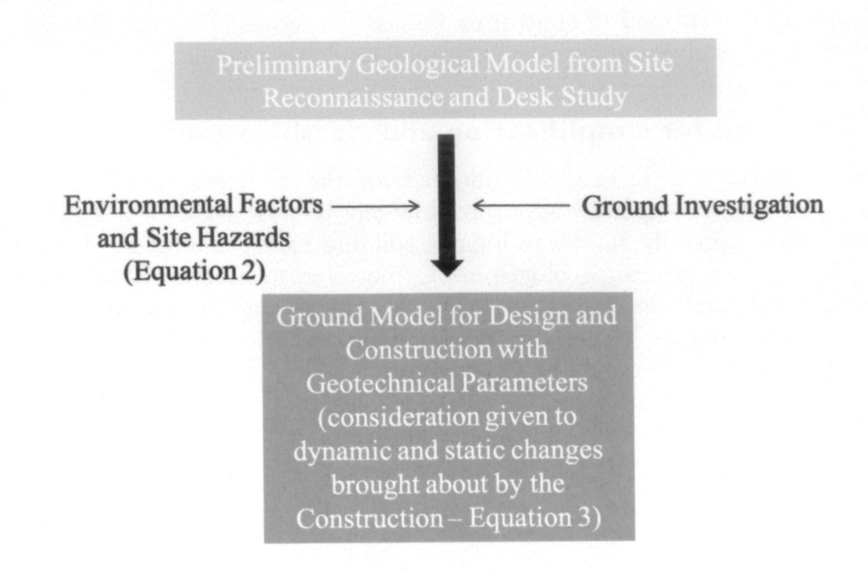

Figure 4.4 Basic process of creating a ground model for a site.

Intrusive rocks, solidified at depth, include extensive igneous bodies now exposed at the Earth's surface such as granite that makes up Dartmoor in the UK. This rock solidified very slowly from temperatures in excess of 1000°C at depths of several kilometres in the Earth's crust and the slowness of the cooling allowed large mineral grains to grow very gradually. Since formation they have been uplifted and the overlying rocks eroded away with huge consequential changes in stress and temperature conditions.

Granitic rocks are light-coloured and relatively light in weight (unit weight 27 kN/m^3 which is 2.7 times that of water). In terms of mineralogy, granite has a high proportion of quartz and feldspar. Quartz (SiO_2) is hard, has no cleavage weaknesses and is much more resistant to chemical weathering than feldspar. Feldspar (orthoclase and plagioclase) is a much more complex silicate mineral with cations of potassium, aluminium, sodium and calcium and relatively weak cleavage directions that make it prone to chemical attack. The feldspars therefore break down, primarily to form clays, which are a series of minerals of essentially the same chemical makeup as the feldspars but which are more stable at the Earth's surface temperature and chemistry. Minor minerals in many igneous rocks include biotite, hornblende and magnetite which contain iron which is released and then oxidised on decomposition hence giving the rust-red of weathered rock profiles in many sub-tropical and tropical parts of the world. Iron oxide and carbonate products play an important role in cementing recently deposited sediments as discussed later. Granitic rocks are found in continental regions and probably largely represent melted and reconstituted crust as plates are subducted beneath mountain chains or in extension zones (Davis & Reynolds, 1996).

Oceanic regions are made up of "basic" igneous rocks – mainly basalt – which is the fine-grained chemical equivalent of gabbro. Basalt erupts along extensional plate boundaries such as that running down the centre of the Atlantic Ocean as the European and African plates move away from the American plates at a rate of about 20 mm each year. Basic igneous rocks are darker coloured and heavier than granite largely because they are rich in iron and magnesium. Gabbro has a unit weight of about 30 kN/m[4]. Basic rocks are sometimes extruded in continental regions where faults allow basaltic magma to rise as molten rock from great depth to the Earth's surface. Basalt rock that originated as lava fields makes up the Giants Causeway in Ireland and much of the other Tertiary volcanics of Northern Britain. The basaltic Deccan "traps" in India cover an area of more than 50,000 km^2.

Lavas such as basalt are finer-grained than granite or gabbro because they cool relatively quickly; where cooled extremely quickly, say by extruding into water, they may form natural volcanic glass. Basaltic lava has relatively low viscosity and flows a long way unlike the more viscous pale coloured acidic lavas such as rhyolite (the fine-grained equivalent of granite). Acidic volcanoes erupt in a much more violent way than basaltic volcanoes. Other rocks associated with volcanoes are formed from the huge amounts of dust and other air-borne debris in volcanic eruptions. Volcaniclastic sediments (Table 4.2) are called tuffs and often exhibit many sedimentary features especially where deposited in water. Hot clouds of debris deposited on land become welded tuffs or ignimbrites. A welded tuff with pumice inclusions (fiamme) that were hot and plastic when deposited and then flattened by the weight of overlying material, is illustrated in Figure 4.5.

Dykes are intruded in extensional (tensile) regions and cut across other geological structures (Figure 4.6); sills follow existing geological structures such as bedding and are more concordant. Dykes can be quite local or very extensive. The Great Dyke in Zimbabwe can be traced across country for more than 450 km. Tertiary dyke swarms and associated intrusion complexes in Scotland, Ireland and Northern England are shown in Figure 4.7.

Some of the characteristics of an area intruded by igneous rocks are illustrated schematically in Figure 4.8. Hot igneous rocks affect the country rock that they intrude, giving rise to thermal metamorphism as discussed in Section 4.4.5. Some metamorphosed zones may be much weaker than the associated igneous body with obvious potential consequences if encountered in an engineering project. Figure 4.9 shows a zone of weak, hydrothermally altered rock encountered at a depth of about 200 metres in a tunnel, well below the level of anticipated weathering and therefore rather unexpectedly. It resulted in a minor collapse that delayed tunnelling. The extent of the zone was investigated with horizontal drill holes. Following forward probing, ground treatment and support, tunnelling was able to proceed.

4.3.4 Sediments and associations – soils and rocks

4.3.4.1 General nature and classification

Sediments and sedimentary rocks are derived from the breakdown of older rocks. Detritus is transported by gravity, water, wind and glaciers before it is deposited and gradually buried and transformed into rock by diagenetic or lithification processes as introduced in Chapter 1. Some weathering products are carried in solution to be deposited directly on the Earth's surface, transformed into animal shells or carried by fluids to be deposited as cements in the sediment pile as it is self-compacted. During

Table 4.2 Classification of sedimentary rocks. See Tucker (1982) for more detail

mm	*Clastic soil*[6, 7]	*Clastic rock*		*Volcaniclastic or pyroclastic rock*	*Chemical & biochemical rock*
	BOULDERS	CONGLOMERATE (rounded clasts) BRECCIA (angular)		PYROCLASTIC BRECCIA or AGGLOMERATE	LIMESTONE (examples) :
>200	COBBLES				Chalk
60	GRAVEL			LAPILLI TUFF	
2	SAND	SANDSTONE Greywacke (generally poorly sorted) Arkose (feldspathic sandstone)		COARSE ASH TUFF	Calcarenite (sand and gravel size) Calcilutite (mud size matrix)
0.06					
<0.002	SILT	SILTSTONE	MUDSTONE as general term SHALE if fissile	FINE ASH TUFF	Oolite
	CLAY	CLAYSTONE			DOLOMITE (Mg rich) EVAPORITE (salts) COAL FLINT & CHERT (Cryptocrystalline silica)

Figure 4.5 Welded tuff with flattened fiamme that were plastic when deposited sub-aerially, Ap Li Chau, Hong Kong.

transport detritus is sorted by size and density, largely in response to the velocity of water flow or wind. Sediment deposited close to its erosion source may have a wide range of grain sizes as illustrated in Figures 4.10 and 4.11. Sediments transported by water or wind tend to be winnowed to a limited range of grain sizes and are then called

Figure 4.6 Basalt dyke cutting granite, Hong Kong.

"well-sorted" (Figure 4.12); the same material would be called "poorly-graded" by an earthworks engineer – he looks for soils in embankments to have a wide range of sizes so that a dense degree of compaction can be achieved. Grains also are abraded and rounded as they are blown by the wind especially and this is an important engineering characteristic. Angular sand has an internal friction angle that can be 10 degrees higher than rounded sand due to dilatancy (Rowe, 1962) depending upon the normal stress level (Bolton, 1986).

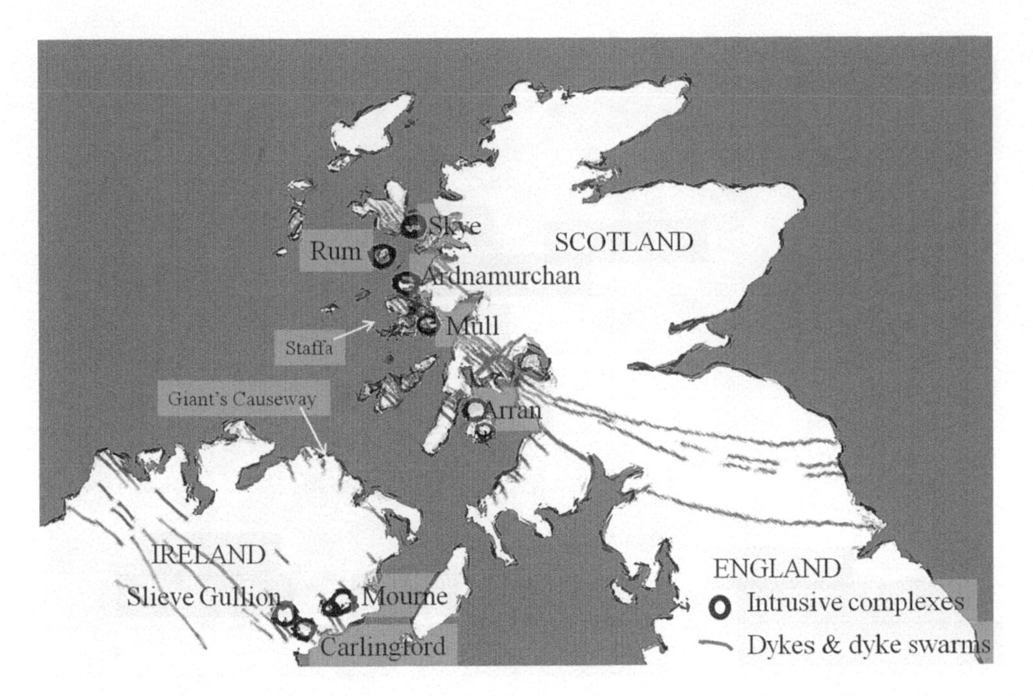

Figure 4.7 Simplified map showing Tertiary dyke swarms and igneous intrusion complexes across northern Britain. Lava flows from the volcanoes above the igneous complexes that are now exposed (e.g. Figure 4.35) included the extensive flood basalt sheets of which the Giant's Causeway and the Isle of Staffa are part (see Figures 4.46 to 4.48). Data are from Holmes (1965) and Richey (1948).

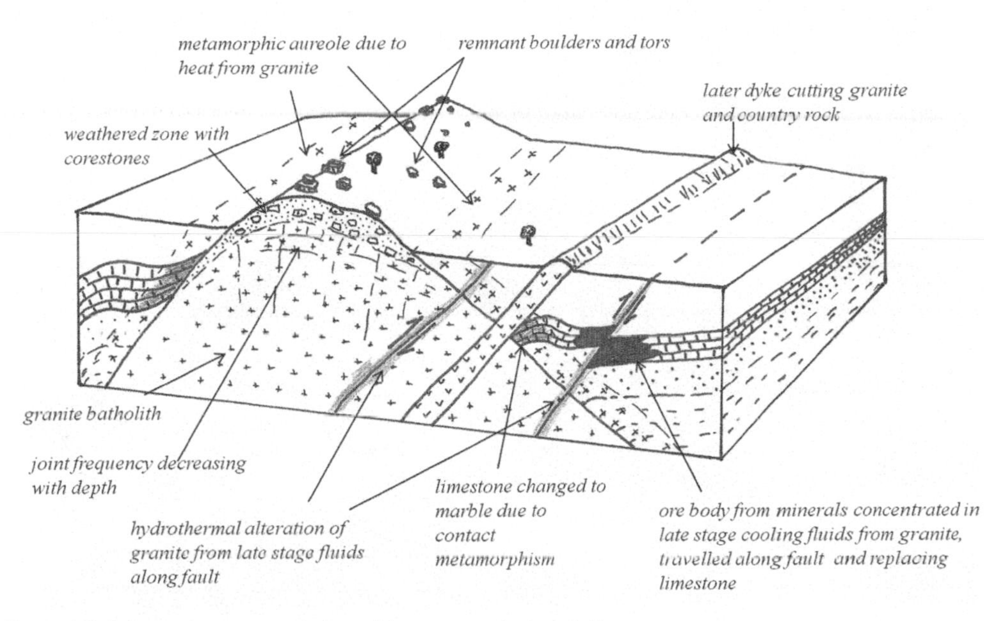

Figure 4.8 Schematic representation of igneous rock associations.

Figure 4.9 Hydrothermal zone (darker, lower part of tunnel face), Black Hills Tunnels, Hong Kong.

Figure 4.10 Slope comprising colluvium, cut almost vertically adjacent to live carriageway, South Africa.

Figure 4.11 Ancient colluvium derived from Fei Ngo Shan (in background) in cut slope on Clearwater Bay Road, Hong Kong (now hidden by slope works).

Figure 4.12 Sediments derived from Mount Cook, New Zealand representing a fairly consistent coarsely-graded soil.

Sediments and detrital/clastic sedimentary rocks are classified primarily on dominant grain size as in Table 4.2. In terms of soil mechanics and geotechnical behaviour, granular soils are generally distinguished from clays in that they are non-cohesive and are comparatively inert chemically although some authors would dispute the term cohesive being applied to clay. Granular soils include everything down to silt-size and even some clay-sized soil where this is derived from mechanical breakdown (rock flour is clay-size material, generally quartz, produced by glacial abrasion). Gravel and larger grain sizes are usually made up of rock (lithic) fragments rather than mineral grains. In continental regions where granitic rocks dominate, sand and silt are often made up predominantly from quartz, the most resistant mineral from granite. In areas where chemical decomposition is inactive feldspar might also survive (arkosic).

Many clays are distinct from granular soils not only because of their grain size and shear behaviour but because they comprise a new series of minerals with their own structure and chemistry; they are derived from other rocks but not only as detrital mineral grains but as precipitants from solution (Eberl, 1984). Other clays are primary, associated with igneous or hydrothermal activity or are formed by transformation of other minerals. Clays comprise three groups – phyllosilicates, weakly crystalline aluminosilicates and hydrous oxides of iron, aluminium and manganese. Selby (1993) provides a very useful review of clay mineralogy as well as their engineering properties.

The most common clay minerals are the phyllosilicates, which, like mica, comprise sheets or layers of silica and sheets of alumina. The various clay mineral species owe their differences to the ways that sheets are arranged, substitutions of other cations are made into the structure that produce distortion of the crystal lattice and the bonding between sheets. The type of clay that will be formed at any location depends primarily upon the source rock and climatic effects. Kaolinite and illite are relatively inactive clays and commonly produced by the weathering of granitic rocks. Montmorillonite (major member of the smectite class) is much more active and absorbs water readily thereby swelling dramatically. When it dries it shrinks and these characteristics have important consequences for engineering. Smectites also tend to have low shear strengths when wet (Chapters 6 and 7). They are commonly the result of weathering in basalt in tropical areas and form difficult but productive soils in many counties including Australia, Africa, India and the USA. Soil types include "vertisols" and "black cotton" soils.

Aluminosilicate clays such as allophone develop from volcanic ash with silica content below a critical level and are important soils in New Zealand, Japan and Indonesia (Selby, 1993). Like montmorillonite they are very responsive to changes in water content, prone to flow and liquefy when wet and to shrink and crack when dry.

"Atterberg limits" are measured routinely in laboratories to define the nature of soils and to interpret the likely clay mineralogy by index testing. If water is added to a clay or clay-silt mix one can get it to flow like a liquid. The moisture content at which it changes from a plastic state to a liquid state is known as the "liquid limit" (LL) and can be determined using standard tests where a groove cut in the surface closes as the sample is struck in a standard way or a cone is dropped onto the sample and penetration measured. If the sample is then dried out the sample gets stronger and the water content at which a sausage of 3 mm diameter can be rolled without breaking is called the "plastic limit" (PL). The difference in moisture content between the liquid state

(LL) and plastic-semi-solid condition (PL) is the plasticity index (PI). The PI is found to be a good indicator of the type of material, species of clay and thence engineering behaviour as shown in Figures 4.13a & b. It can also be indicative of the expansiveness of a clay as discussed in Chapter 3.

Clay particles are carried by rivers to estuaries where a change in salinity causes clay platelets to flocculate. Because of their shape and the distribution of excess electrical charges, they tend to link end to face and form open structures with very low density and high water content. This open structure is broken down, partly by bioturbation but also by the weight of overlying sediment. The clays develop a laminar structure and water is squeezed out. Eventually once the clay is buried by about 2 km,

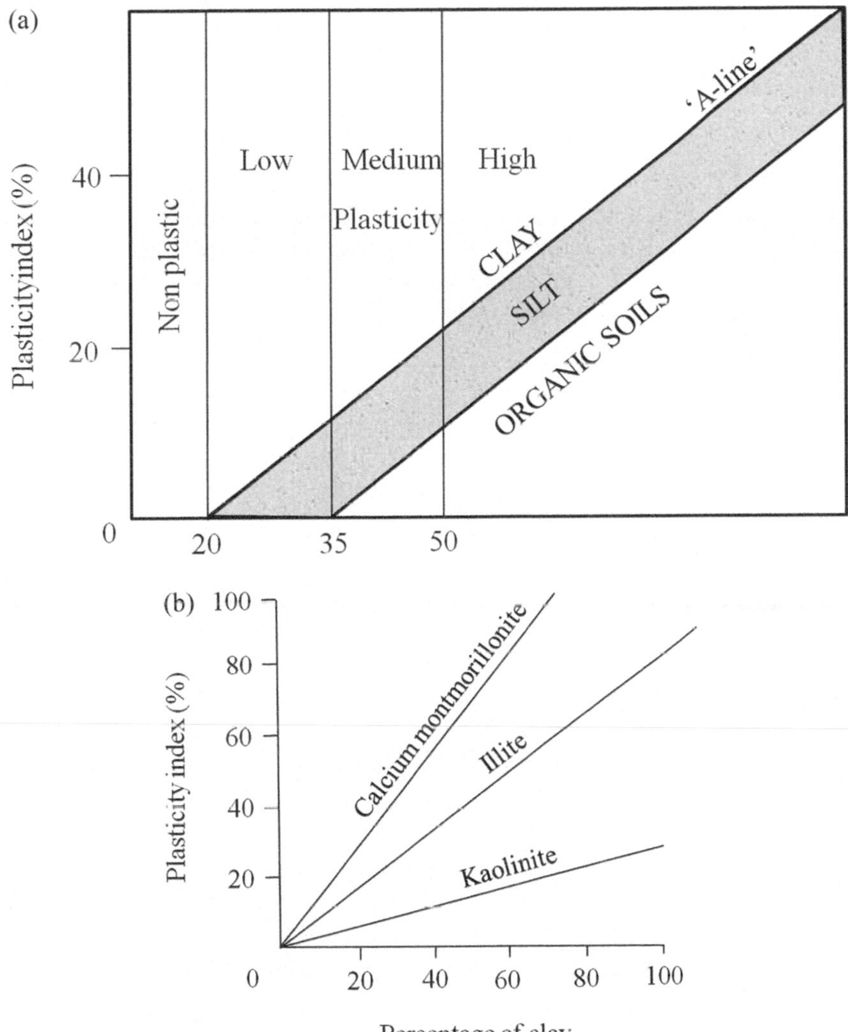

Figure 4.13 (a) Chart of plasticity index vs. moisture content used for identifying the nature of fine grained soils using simple, standard tests (Atterberg limits). (b) Use of plasticity index to indicate clay mineral type in soil sample (after Skempton, 1953).

theoretically there may be dry contact between clay crystals with the development of strong, covalent bonds with shared ions (Osipov, 1975). There would also be transformation of clay, say from smectite to illite. During burial and self-weight compaction and consolidation, water may be expelled dramatically with the formation of mud volcanoes on the sea floor. Mud volcanoes also occur on land with the mud apparently sourced by erosion of mudstones at great depth although the mechanism is uncertain. The LUSI mud volcano in East Java has had eruption rates of up to 180,000 m^3 per day continuing over several years (Davies et al., 2011).

4.3.4.2 *Sedimentary environments*

The sketch in Figure 4.14 illustrates a number of sedimentary environments. The source rock, nature of weathering and erosion and especially method by which the sediment is transported and finally deposited result in the wide range of sediments and sedimentary rocks encountered.

4.3.4.2.1 Onshore

Sediments deposited on land generally include colluvium (landslide deposits and slope wash) and glacial deposits, which are often poorly sorted. Alluvium is deposited in rivers and if transported for any distance will become sorted and coarse clasts will be rounded. Other important soils on land include organic soils such as peat which becomes coal when lithified.

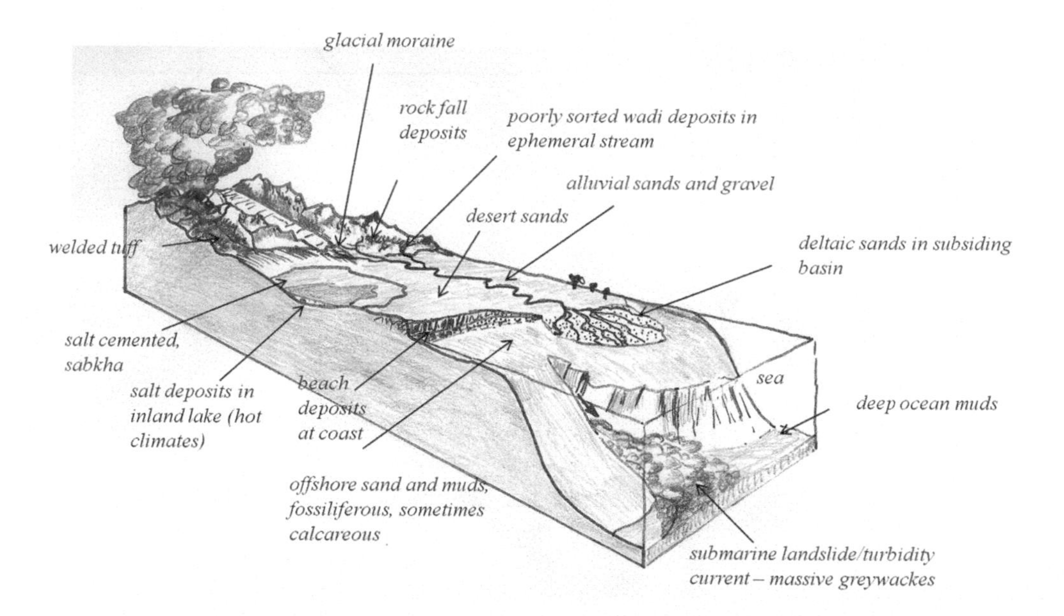

Figure 4.14 Model indicating types of sediment that might be found onshore and offshore.

Figure 4.15 Uniform Miocene weak sandstone with rounded grains, weakly cemented with iron oxide cement providing some cohesion, Algarve, Portugal.

Desert sands are distinguishable from river transported sediments by their rounded, almost single sized grains, which may become cemented especially by iron oxides giving typical red beds, such as the Permian sandstone in the UK (Figure 4.15). They also lack mica which is either abraded to dust or blown away. Finer wind-transported silt forms thick deposits called loess. Where cemented, loess has been exploited to construct settlements especially in China. This can be risky as loess is prone to collapse when wetted or disturbed by earthquakes and there are numerous instances of loss of life as a result (Derbyshire, 2001).

Glaciers have covered much of northern Europe, Canada and the northern USA several times in the last 3 million years. In Britain everywhere north of a line from London to the Bristol Channel was covered in ice sheets. The rest would have been areas of permafrost (periglacial conditions). Man has inhabited Britain on and off for at least 700,000 years but has had to abandon the country because of advances in ice several times, most recently about 13,000 years ago. Only for the last 11,500 years has Britain been occupied continuously (Stringer, 2006). As a consequence, much of northern Britain shows signs of glacial erosion and also materials deposited in association with the glaciers. Much of this is poorly sorted "boulder clay" or "till" but there are also great thickness of sorted and layered outwash sediments laid down on land or in lakes as the glaciers retreated. These materials can be very variable vertically and laterally (Fookes, 1997) so great care is needed in interpreting ground investigations. This is illustrated by one case history (piling for Drax Power Station) presented in

Chapter 8. Associated with these periods of glaciations, the sea level across the world has fluctuated widely. About 20,000 years ago the sea level was almost 150 metres lower than it is today as is discussed in some detail in Chapter 5 (Pirazzoli, 1996). Ancient river channels that ran across the land surface have been submerged to become infilled with marine sediments, which can prove hazardous for engineering projects such as tunnelling or foundations for bridges across estuaries. For example, during design of the recently constructed 4.24 km immersed tube tunnel as part of the Busan-Geoje fixed link crossing in South Korea, a depression was found in the sea floor unexpectedly underlain by thick marine sediments and this required extensive engineering works to support the tunnel involving the use of soil-cement mixed piles. Buried channels are also found on land, often with no obvious surface expression. Krynine & Judd (1957) report several case examples including a dam at Sitka, Alaska where river fill was fortuitously found by drilling, extending more than 25 metres below the dam foundations. If it had been missed then there would have been considerable water leakage beneath the dam, requiring remedial works. In Switzerland, during the construction of the Lötschberg rail tunnel, 25 miners were killed by inrush of glacial sediments when they blasted out of rock unexpectedly into an over-deepened glacially scoured valley about 180 metres below the valley floor (Waltham, 2008). A case from Hong Kong is presented in Fletcher et al. (2000) and summarised in Chapter 8 where the planned construction of a tower block had to be abandoned because of the potentially huge cost of deep bored piles. A large cavernous xenolith of marble had been found unexpectedly within granite beneath the site; the caves were partially infilled with soft sediments to depths of 150 m. The formation of the cave and sediment infill certainly occurred when sea levels were much lower.

Other poorly sorted, mixed soils include landslide colluvium. One such deposit is called the Fort Canning Boulder Bed which underlies much of the Central Business District in Singapore and has caused numerous difficulties for construction, not least because it has sometimes been misinterpreted during ground investigation as weathered rock as discussed in Box 4.2.

BOX 4.2 THE FORT CANNING BOULDER BED, SINGAPORE

The Fort Canning Boulder Bed (FCBB) underlies much of the Central Business District of Singapore and is an example of a complex mixed rock and soil deposit, the geological nature and interpretation of which has had and continues to have great consequences for civil engineering construction. When the Fullerton Building was constructed in the 1920s, it was found that "the foundations which consisted of clay and boulders, were of a dangerous character, and this great structure had to be placed on a concrete cellular raft, which is so designed as to give each superficial foot of soil not more than one ton to carry" (Straits Times, June 27, 1928). Since then in several reported cases engineers have struggled with construction on and through this material. Ground conditions have sometimes been misinterpreted following inadequate ground investigation. In 1952 the original ground investigation for the Asia Insurance Building incorrectly interpreted the site as underlain by *in situ* rock rather than FCBB. The mistake was only discovered during construction and

Figure B4.2.1 Site close to Mount Sofia, Singapore with large diameter rock augers at work. Behind the rig can be seen a pile of very large angular boulders extracted by rock coring.

necessitated total redesign of the foundations (Nowson, 1954). Shirlaw et al. (2003) provide several other examples.

The FCBB is almost 100 metres thick in places, is not exposed at the ground surface, but is commonly encountered in foundation construction and tunnels (Shirlaw et al., op cit). The deposit typically comprises a poorly sorted mixture of clay, silt and sand with boulder content between 5% and 35% (Singapore Standard, 2003). The mostly angular boulder content is probably derived from the adjacent Jurong Formation of Triassic/Jurassic age, with strong sandstone predominating. Figure B4.2.1 shows boulders recovered and partially drilled in large diameter bored piles from one site near Mount Sophia. In between the boulders, the matrix sometimes comprises hard, red clay with undrained shear strength up to 2 MPa (weak rock) but sometimes sandier and silty and of much lower strength. Broms & Lai (1995) also report encountering a highly permeable gravel layer within the FCBB at depth.

Shirlaw et al. (op cit) suggest that the FCBB is a colluvial deposit originating from Fort Canning Hill (Figure B4.2.2) but the low height of the hill relative to the thickness of the deposit makes this somewhat doubtful and the red clay is often much stronger than would be expected even for clay that has undergone quite deep burial. Furthermore, the clay is clearly not a weathered product of the sandstone and, anyway, the contained boulders in

Figure B4.2.2 View towards Fort Canning Hill from the Central Business District and a few locations where FCBB have been encountered.

Figure B4.2.3a Part of core box through Fort Canning Boulder Bed (core 60 mm diameter). Core ranges from red clay to rock fragments showing apparent structure.

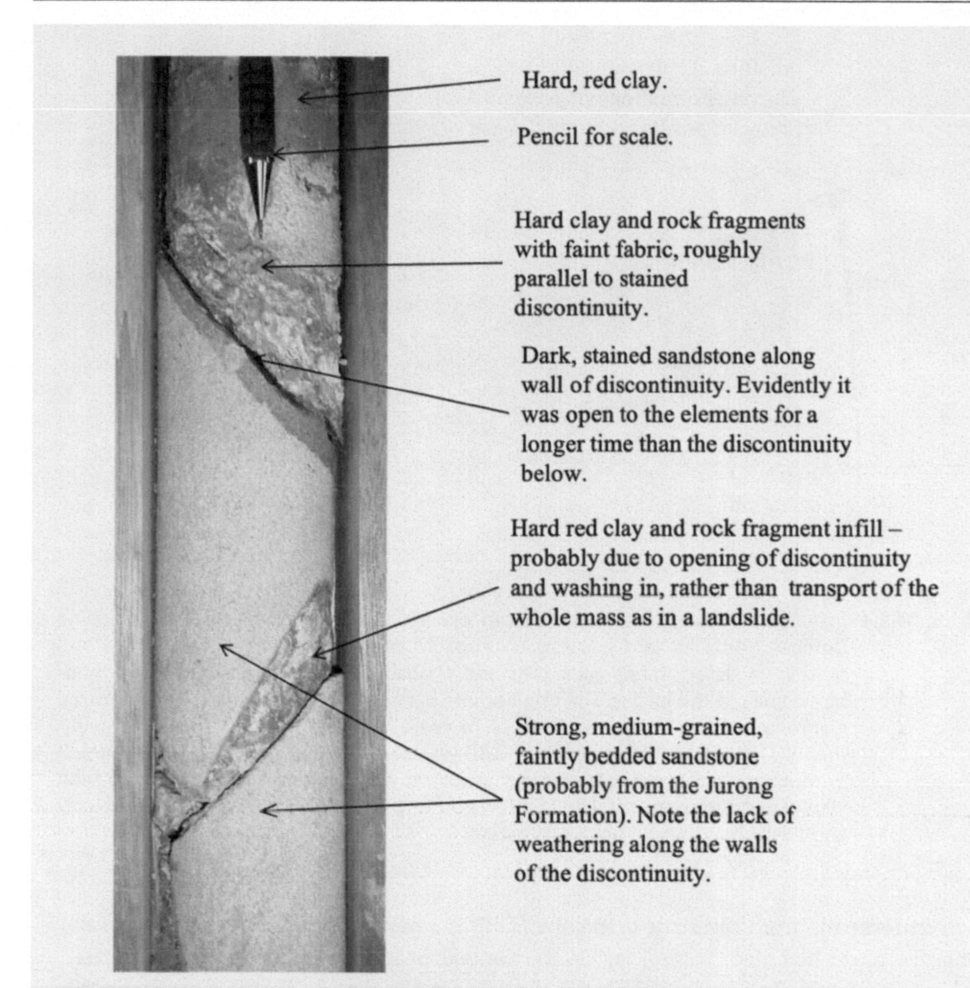
Hard, red clay.
Pencil for scale.
Hard clay and rock fragments with faint fabric, roughly parallel to stained discontinuity.
Dark, stained sandstone along wall of discontinuity. Evidently it was open to the elements for a longer time than the discontinuity below.
Hard red clay and rock fragment infill – probably due to opening of discontinuity and washing in, rather than transport of the whole mass as in a landslide.
Strong, medium-grained, faintly bedded sandstone (probably from the Jurong Formation). Note the lack of weathering along the walls of the discontinuity.

Figure B4.2.3b Detailed section of core (about 60 mm diameter) showing locally stained sandstone with apparent clay and rock fragments infill.

the FCBB generally show little weathering. The red clay sometimes appears as an infill to apparent relic joint structure in sandstone blocks (Figure B4.2.3).

Broms & Lai (op cit) present two sketches of caisson excavations for the Republic Plaza, one of which appears to have a vertical fabric. If this was a transported colluvial deposit, boulders would be expected to be essentially random or preferentially flat lying. The same sketch shows vertical zones containing few boulders with a form similar to "gulls" found in areas of cambering (Parks, 1991).

An alternative model for the origin of the FCBB therefore is rock cliff deterioration, regression and local collapse as sea levels rose from – 150 m at the end of the Holocene glacial periods. The red clay is probably partly washed in infill to the open fractures within the deteriorating rock mass, in a terrestrial environment, as illustrated schematically in Figure B4.2.4. If the model is correct, the collapsed, colluvial facies of the FCBB would grade laterally into essentially *in situ*, deteriorated Jurong Formation rocks.

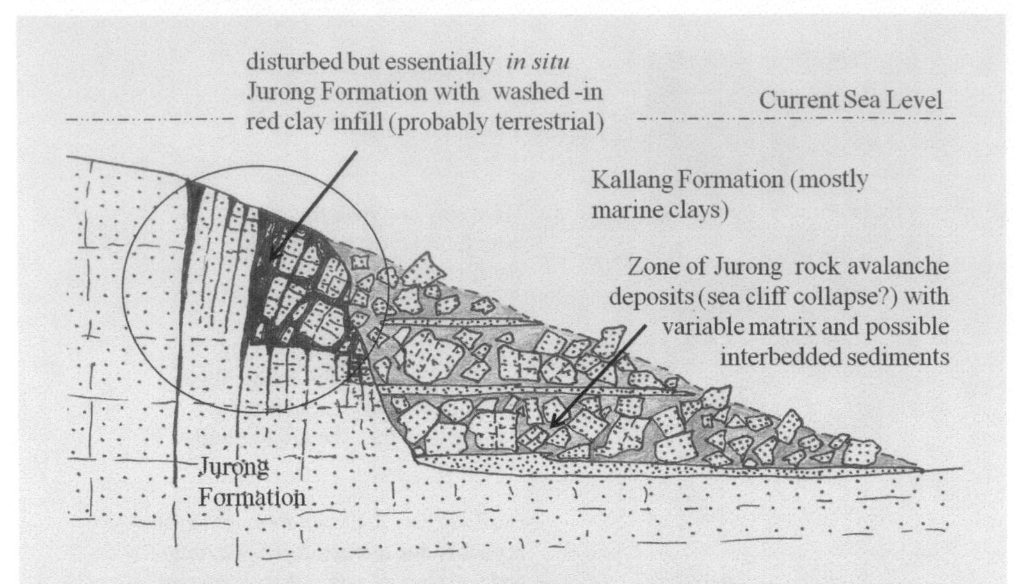

Figure B4.2.4 Schematic geological model that can explain the features of the Fort Canning Boulder Bed. The hard red clay is probably infill, washed in to open joints and fissures in deteriorating rock cliffs and probably derived from weathered mudstone units in the Jurong. The clay became hardened and cemented in a terrestrial environment in the same way as "terra rossa" in weathered limestone environments. The colluvial facies of the FCBB was probably deposited as rock and soil avalanches, at the coast. It remains unsorted although may be intercalated with alluvial or beach deposits. The FCBB was later buried locally by Kallang marine clay and fluvial horizons and by reclamation.

Whatever the truth regarding origin, the FCBB is a most difficult founding stratum and one that might be called "unforgiving" in the context proposed in Chapter 6, but situated in an area of one of the most valuable real estates in the world. Clearly if dealing with a site which comprised weathered but essentially undisturbed Jurong one might use rather higher design parameters than for a colluvial soil that might contain weaker horizons. Similarly tunnelling through the weathered Jurong the potential hazards might be different from tunnelling through a colluvial and possibly interbedded section of the FCBB with weaker horizons that could flow or ravel. The literature has many examples of problems that arise through getting the ground model wrong and as Shirlaw et al. (op cit) note, identification of the FCBB can be difficult if adopting a routine approach to ground investigation.

4.3.4.2.2 Offshore

Sediments are deposited off shore in tectonically-controlled, continually subsiding basins and can be thousands of metres thick (Leeder, 1999). Skempton (1970) compiled data on rates of deposition for argillaceous sediments (clay rich) and reported these as ranging from 0.03 m/1000 years for deep sea conditions, about 1 to 2.5 m/1000 years

Figure 4.16 Bedded, Miocene sandstone, Capela da Sra da Rocha, Nr Porches, Algarve, Portugal.

for estuaries and shallow marine conditions and up to 120 m/1000 years for deltaic situations. Sand deposited close to the shoreline tends to be quite well sorted, bedded and often fossiliferous (Figure 4.16). Off shore, below the continental shelf, thick, often relatively poorly sorted sandstones are sometimes deposited very rapidly from dense turbity currents arising from submarine landslides. Rocks formed in this way (once lithified) include greywacke. They are often bedded and cyclic with predominantly finer grained horizons followed by predominantly coarser beds (Figure 4.17). Submarine landslides are a major hazard to offshore engineering works, including pipelines, cables and oil rigs and are often triggered by earthquakes although they may simply occur as the sediment accumulation on the continental shelf reaches a critically unstable geometry.

Other sediments are deposited as a result of chemical processes. Many limestones including Chalk are essentially bio-chemical involving microplankton and algae in their formation. Other types include bio-clastic and reef limestone. Limestone can be fine grained rock, uniform and extremely strong and as per the Carboniferous Great Scar Limestone of northern England. Other limestones such as oolite can be weak and porous. Limestone is readily dissolved by acidic water leading to karstic conditions (irregular rockhead and cavernous conditions) with obvious implications for site formation and design. Rocks susceptible to dissolution, including limestone, dolomite and salts (discussed below), can constitute natural hazards due to sudden collapse of sink holes and general subsidence. Relatively minor geological differences can give rise

Figure 4.17 Carboniferous schistose turbidites, Cachopo Road, Portugal.

to very different engineering properties. Within the chalk of southern Britain there is a gradational change from the characteristic White Chalk, which is relatively strong and brittle down into grey, clay-rich rocks. The latter, Chalk Marl, intermediate between the overlying fractured, brittle White Chalk and underlying Gault Clay was recognised as the ideal tunnelling medium for constructing the Channel Tunnel (Varley & Warren, 1996). The presence of hard flint and chert bands in chalk can give rise to considerable difficulties during construction because of abrasion and wear on cutting tools as illustrated by a case example in Chapter 8.

Limestone can become dolomitized, which involves partial replacement of calcium carbonate by magnesium carbonate, leading to greater porosity. Dolomitization often occurs in the presence of evaporates (salt deposits) and dissolution of underlying salts such as gypsum can lead to the development of sinkholes and brecciation of the overlying rock followed by re-cementation. This process is described for Magnesian Limestone in the UK by Dearman & Coffey (1981). Similarly, the Miocene limestone and dolomite sequences in Qatar and Saudi Arabia are complex and extremely variable due to their post-formation dolomitization, collapse and re-cementation (Sadiq & Nasir, 2002). The rock is sometimes strong, elsewhere very weak and can be cavernous. This is important for founding engineering structures and for other activities including dredging (Vervoort & De Wit, 1997). Limestone is an important rock economically particularly as a source of cement, aggregate and, where massive and strong, is commonly used as armour-stone for breakwaters.

Deposits of salt are formed by evaporation of lakes and even seas (the Mediterranean completely dried up about 5 million years ago), and are also very important economically as source rocks for chemical industries, such as fertilisers. They are significant for the oil and gas industry because they have low density and low permeability and gradually rise through the overlying denser country rock as diapiric structures. Traps for oil and gas at the boundaries of these diapiric structures are searched for using geophysical methods and then targeted by drilling. Salt deposits are also considered to have great potential as nuclear waste repositories because of their low permeability although there are also reservations because of potential dissolution, mobility and influence of heat (e.g. Krauskopf, 1988). Salt also provides a cementing medium for sediments in some environments (sabkha) and can be very important for founding structures as in the Gatch underlying Kuwait (Al-Sanad et al., 1990).

Engineering properties of sediments and sedimentary rocks and their investigation are addressed in Chapters 6 and 7.

4.3.5 Metamorphic rocks and their associations

Metamorphic rocks are rocks that have been changed by heat, pressure or both. By definition they do not include low-temperature and low-pressure diagenesis processes affecting soils as discussed in Chapter 7.

A general classification of metamorphic rocks is presented in Table 4.3. Contact metamorphism occurs at the host boundaries of igneous bodies as illustrated in Figure 4.8. The metamorphic aureole surrounding a major granite or gabbro pluton can extend for hundreds of metres. The greatest affect is on sedimentary rocks; recrystallised, generally very strong rock found close to large plutonic igneous rock bodies is called hornfels. At greater distances the effect of the intrusion is less – often the only indication of change being the growth of new minerals, such as kyanite or cordierite in the

Table 4.3 Classification of metamorphic rocks. Refer to Fry (1984) for more detail

	Foliated	*Non-foliated*
Coarse to Medium	GNEISS Often widely spaced and irregular foliation	
	MIGMATITE Mixed schist and gneiss	MARBLE Derived from limestone/dolomite
	SCHIST Strong foliation	QUARTZITE Recrystallised sandstone
Fine <0.06mm	PHYLLITE Undulose foliation. Often micaceous, shiny	HORNFELS Generally recrystallised contact rock
	SLATE Planar cleavage	SERPENTINITE Metamorphosed peridotite/norite
	MYLONITE Fault gouge	

otherwise largely unaltered country rock. A good field example is the Skiddaw Granite in the Lake District, UK where the crystalline hornfels zone extends more than 3 km from the exposed granite outcrop (the granite shallowly underlies the ground surface). The limit of metamorphism can be traced up to about 5 km away from the exposed granite; the metamorphic aureole, measured at right angles to the granite, is about 1 km thick (Institute of Geological Sciences, 1971).

Marble forms from the metamorphism of limestone through heat and pressure often during mountain building processes. Metamorphism of sandstone can form quartzite in which the sand grains are welded together. This rock is often extremely strong and abrasive to drills and tunnel boring machines.

Minor intrusive rocks such as dykes and sills, though very hot when emplaced often cause little metamorphism as illustrated in Figure 4.18 for a basalt dyke cutting limestone because of their relatively small volumes.

Regional pressure during mountain building can impose a marked cleavage or schistosity perpendicular to the maximum compressive stress. Rock types formed in this way range from relatively low-temperature slate, through phyllite to high-temperature schist. In all cases the rocks are recrystallised and the new fabric and structure dominates their mechanical properties although original bedding may be evident. Phyllite is intermediate between slate and schist and generally has shiny, low friction foliation because of the presence of minerals such as mica and chlorite. In schist, the original bedding may still be broadly recognised by chemical layering throughout the rock mass – some zones could be richer in silica (originally sandstone) others might be graphitic (the original rock perhaps having been organic mudstone with coal). Sometimes there have been several phases of metamorphism with several different cleavage or schistosity foliations imposed on the same rock mass leading to blocky rock, which may cause difficulties for underground excavations as occurred for the power house at Kariba Dam, Zambia (Blyth & de Freitas, 1984).

Mineralogical, grain size and shear strength variability along schistose foliation can result in joint styles changing very rapidly from layer to layer as illustrated in Figure 4.19, which causes obvious difficulties for characterisation of the fracture network. Schist is sometimes associated with thin (say 100 mm) shear zones of low frictional strength (15°–25°) often running roughly parallel to foliation and sometimes extending laterally for more than one kilometre (Deere, 1971). Not surprisingly these often cause problems for engineering structures, including tunnels and slopes. Deere gives several examples particularly of tunnels running parallel to the strike of steeply dipping schistosity. There are various possible origins for these shear zones but many are probably the result of slippage along foliation during folding – similar in origin to intra-formational shear zones in folded sedimentary rock sequences as discussed below.

4.4 GEOLOGICAL STRUCTURES

4.4.1 Introduction

Where plates collide, large compressional stresses are generated as along the western coast of North and South America. The consequence is uplift of the Rocky and Andes mountains and earthquakes on active faults, such as the San Andreas in California

Figure 4.18 Tertiary dyke through Jurassic limestone, Island of Muck, Scotland.

and on the subducting plate beneath Chile and Peru. The rocks are squeezed and are either deformed plastically (see Figure 4.2) where temperatures and confining pressures are high, or fracture or both (Figure 4.20). Folding may control the disposition of the various rocks at a site and a specific geotechnical hazard involved with folding is intra-formational slip. As the rocks are folded, different layers slip relative to one another possibly resulting in high polished planes of low shear strength (Salehy

Figure 4.19 Joint system geometry varying with each stratum. About half way along Sector 9A of the Via Algarviana–São Bartolomeu de Messines to Barragem do Funcho, Algarve, Portugal.

Figure 4.20 Severely folded and thrusted sandstone of the Table Mountain Group, Cogmanskloof, South Africa.

et al., 1977; Kovacevic et al., 2007). Such highly polished intra-formational shear surfaces are common in the Coal Measures in the UK and have been responsible for large landslides.

For geotechnical engineering, the geological structures that are of prime importance are called discontinuities. These are fundamentally important to the mechanical properties of rock and some soil masses and how they perform in engineering projects.

4.4.2 Types of discontinuity

For geotechnical purposes, a discontinuity may be defined as a boundary or break within the soil or rock mass, which marks a change in engineering characteristics or which itself results in a marked change in the mass properties. At a macroscopic scale the most important discontinuities that engineering geologists need to consider are the following:

1. Geological interfaces such as bedding, pluton boundaries, dykes and sills, unconformities.
2. Faults.
3. Joints and other fractures.

4.4.3 Geological interfaces

Geological interfaces are the main boundaries mapped by geologists and therefore fundamental to unravelling the geological history at a site, interpolation between boreholes or field observations and extrapolation to some other location. The boundaries such as unconformities and dyke boundaries often represent a major gap in time, sometimes of many millions of years. There may be sudden contrasts in rock and soil type and in degree of fracturing across the boundary (Figure 4.21). Such contrast is often mapped in the field by lines of seepage and marked changes in slope. Often however rock strata boundaries, especially where unweathered, are of little engineering consequence as illustrated in Figure 4.22.

4.4.4 Faults

Faults are geological fractures on which there has been demonstrable shear displacement. They range from minor breaks with only a few mm movements to breaks in the Earth's crust extending many kilometres laterally and vertically and with cumulative displacements, over many years, also of many kilometres. Faults can form in compressive regions ("reverse" faults) and in extensional zones ("normal" faults). The term "normal" originates from coal mining in the UK because most fault blocks dropped away, down the dip of the fault, and the miners knew in which direction the productive coal seam was likely to be found. Faults in the upper 10 km or so of the Earth's crust break in a brittle manner producing fractured, brecciated rock. At greater depth, where the temperature and stress is much higher, faults occur more plastically. Typical features of brittle and plastic faults zones, exposed now at the Earth's surface,

Figure 4.21 Signal Hill, Cape Town, South Africa. Joints orthogonal to horizontal bedding in Table Mountain Sandstone, unconformably overlying Cape Granite.

Figure 4.22 Fused boundaries between volcanic rock and granitic intrusion, Anderson Road Quarry, Hong Kong.

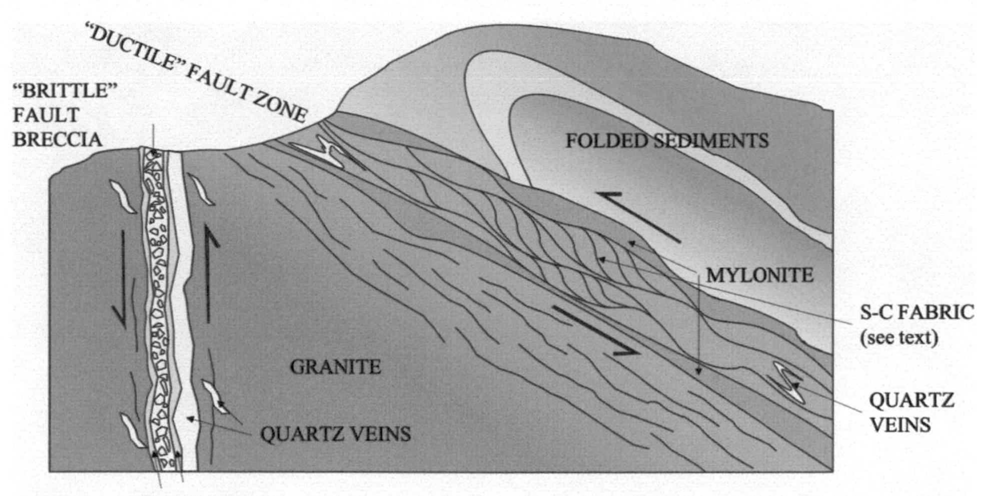

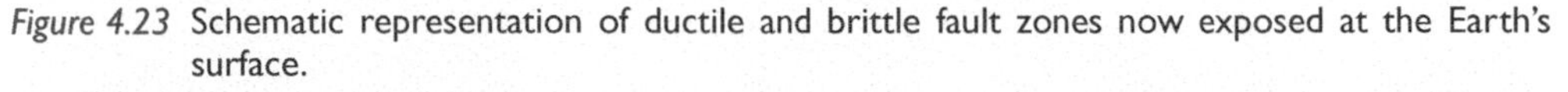
Figure 4.23 Schematic representation of ductile and brittle fault zones now exposed at the Earth's surface.

Source: After Fletcher, 2004.

are illustrated in Figure 4.23. Fault breccia (broken rock) is associated with brittle fault formation; ductile zones involve plastic movement and deformation of the rock. Typical fabric in such ductile zones is described as "s-c"; "s" comes from the parallel tilted banding or schistosity, the "c" from the French word "cisaillement", describing shearing along kink bands, roughly parallel to the shear zone walls (Cosgrove, 2007; Vernon, 2018). Fault zones can be very extensive, with many metres of broken rock and gouge between the walls of the fault (Figure 4.24) but can otherwise be represented by a single surface with very little gouge (Figure 4.25). One example presented in Chapter 8 (TBM tunnel collapse) serves as an illustration of how important it is to know the nature of any fault zone. Sometimes the fault movement results in a highly fractured shear zone rather than a discrete plane with gouge and such zones are often highly permeable.

Faults are a particular concern in geotechnical engineering in that they can be associated with sudden and often rather unexpected changes in rock quality. They may act as barriers to flow (termed fault seals in oil reservoirs) or, conversely, they can be highly permeable zones, full of water and lead to a sudden inrush of water into tunnels. Faults are also, of course, the main source of earthquakes. By definition faults disrupt the rock mass and may throw rocks of very different engineering characteristics together. As a result, a tunnel may pass from hard and good rock to extremely poor rock conditions over a very short distance and without warning. Such situations can be very difficult to deal with necessitating a change in excavation methods, support requirements and sometimes a complete re-think of a project (e.g. Ping Lin Tunnel – a case study in Chapter 8). For foundations there may be a sudden change over a few metres from simple pad foundations resting on rock to the need for deep piles to carry the load of a structure.

Figure 4.24 Thick fault zone, about 3 metres across – note discontinuous pale stratum, Po Selim Landslide, Malaysia.

Not all faults cause problems for projects so there is a danger of being over-cautious leading to over-expensive investigation and unnecessary allowance in design for potential poor ground. Furthermore, faults shown on geological maps are sometimes conjectural, inferred by the mapping geologist on some topographic feature such as a valley or other lineament. However, lineaments and river systems can reflect geological features other than faults and the drainage system may owe its geometry to ancient geological history.

4.4.5 Periglacial shears

Another type of fault that can cause considerable problems because of low shear strength is that formed close to the Earth's surface due to periglacial processes. Such shear surfaces can be formed by a number of different mechanisms and can be extensive laterally (Spink, 1991). Numerous failures of slopes and embankments have been attributed to their presence (Early & Skempton, 1972) including Carsington Dam, during construction as described in Chapter 8. Norbury (2010) notes that such surfaces are often difficult to find even when you are fairly certain that they are there. Patient and detailed logging will be required, possibly with trial pits left open for several weeks to allow the shear surfaces to become apparent as the ground dries out and stress relief occurs.

Figure 4.25 Thin fault zone with clay gouge, hammer for scale, Po Selim Landslide, Malaysia.

4.4.6 Joints

Joints are fractures in rock that, by definition, show no discernible displacement relating to their time of origin, which distinguishes them from faults. All fractures are the consequence of overstressing the rock or soil material. The nature of joints – their orientations, roughness and persistence – is controlled by the local stress conditions that caused them to develop together with the strength of the rock, other conditions including temperature and water pressure and subsequent history. Many joints occur as sets of fractures, pervasive through large volumes of rock, and owe their origins to processes such as cooling, burial or orogenic events (e.g. Hancock, 1985; Mandl, 2005). A "set" comprises a roughly parallel series of joints. Sets that are apparently related in terms of origin are called "systematic". Joints can also be non-systematic or random. Joints can be regarded as essentially:

- primary – associated with the geological formation of rock,
- secondary – caused by tectonic and gravitational stress, including the result of uplift and bending or
- tertiary – due to local geomorphological or weathering influences.

Many of these begin as "proto-joints" that develop with time – they begin as general planes of weakness that only become visible traces and later, mechanical discontinuities, on uplift and exposure. Rock masses that have few or no joints (as visible traces or mechanical fractures) include deep-seated, un-weathered igneous plutons (Martin, 1994) as illustrated in Figure 4.26. The water-lain sandstone in Figure 4.27 also has very few visible joints – presumably the rock was not sufficiently over-stressed during burial for the formation of hydraulic joints as discussed later and the later uplift that must have occurred did not involve tension, bending or relaxation to cause differential stresses sufficient to induce fracturing.

Much effort is made to try to characterise joint networks in rock masses in geotechnical engineering – orientation, spacing, persistence and aperture especially. Guidance

Figure 4.26 Massive layered gabbro with no visible joints, Loch Scavaig, Isle of Skye, Scotland.

Figure 4.27 Miocene sandstone with very few visible joint traces, Capela da Sra da Rocha, Nr Porches, Algarve, Portugal.

is given in BS5930 (British Standards Institute, 2020), International Society for Rock Mechanics (1978) and by Priest (1993). The standard guidelines for practitioners however treat joints largely as statistical entities rather than geological features and this is an area where geology has the potential to offer great insight and time-saving in geotechnical engineering; an opportunity that has been rather disregarded to date. This is partly because joint origin is still a difficult, rather poorly understood and highly debated subject (Pollard & Aydin, 1988).

The following discussion refers to the stress conditions that initiate fracturing, relative to the strength of the soil or rock at the time of joint formation. This is explained through reference to Mohr's stress circles in some detail in Chapter 6 in introducing triaxial testing and the reader is recommended to go through that section in order to understand the following discussion. Mohr's circles are also well explained in most soil mechanics textbooks (Craig, 1992), rock mechanics textbooks (Hudson & Harrison, 1997) and structural geology textbooks (Davis & Reynolds, 1996), which demonstrate the importance of these concepts to different scientific disciplines.

Most joints are thought to develop as extensional fractures (in tension), parallel to a compressive major principle stress, σ_1. In a cooling igneous body, the extensional stresses might be due to contraction. Alternatively, the tensile stress σ_3 might be tectonic due to pulling apart of plates as at the Mid-Atlantic Ridge or along the East African Rift Valley or the result of bending and relaxation during uplift or exhumation (Price, 1959; Price & Cosgrove, 1990; Rives et al., 1994). It can also be due to excess pore water pressures in thick sediment piles (Engelder & Peacock, 2001). The condition for formation of extensional joints is illustrated by the left-most Mohr's Circle in the lower Figure 4.28 where the minimum principle stress is tensile and equal to the tensile strength of the soil or rock mass and the maximum principle compressive stress (assumed vertical) is less than 3 times the tensile strength of the intact rock. Under those stress conditions joints would form in the plane of σ_1 and σ_3 with the "dihedral" angle, 2θ equal to 0 (Joint E in the upper diagram). At higher levels of differential stress shear fractures (S in the upper diagram) would form as illustrated by the right-most Mohr's circle in Figure 4.28. The stress circle is tangential to the rock failure envelope with a dihedral angle at 60 degrees (assuming a friction angle of 30 degrees). In between these extremes there is a possibility for "hybrid" fractures to form with dihedral angles between 0 and 60 degrees (sub-vertical), and a potential "joint spectrum" as illustrated in Figure 4.29. Details are given in Hancock (1985) who notes that the regular arrangement of structures such as joints within large areas (>1000 km^2) of weakly deformed rocks gives confidence that they are indeed linked to tectonic processes as per theory. Engelder (1999) however questions the predictive validity of the Mohr-Coulomb approach in detail.

The concept of fracture formation in a large mass of rock governed by a Mohr-Coulomb strength law and under uniform stress conditions rock is helpful in explaining joint formation at a site but generally geological history and local stress conditions and constraints means that the situation is more complex. The soil or rock mass is unlikely to be uniform and will include intrinsic flaws, pre-existing discontinuities and variable hydraulic conductivity (controlling effective stress). Pre-existing bedding, schistosity and faults will have a controlling influence on the way joints develop (e.g. Rawnsley et al., 1992). Where the rock mass has a long and complex geological history there may be several generations of fracturing, each influenced by the former condition (see Rawnsley et al., 1990). Deciphering that history is made more difficult once it is appreciated that all fractures that we now see as obvious visible mechanical discontinuities at the Earth's surface may have only been incipient or integral "proto-joints" at the time of later joint formation and therefore might have had little influence on the formation of the later joints (Hencher et al., 2011).

The proto-joint network provides relatively easy directions for breaking otherwise massive rock such as the rift and grain directions in granite quarrying (Fujii et al., 2007) or as preferential directions for breakage in laboratory testing (Douglas & Voight, 1969). Proto-joints develop as persistent, mechanical fractures later, following the pre-imposed geological blueprint (location, orientation and spacing) through weathering processes and/or stress changes as illustrated for different rocks. The development of each joint is progressive as microfractures merge and extend over geological time (Hencher, 1987; Selby, 1993; Hencher, 2006; Hencher & Knipe, 2007; Hencher, 2013; Rogers & Engelder, 2004). At any particular moment a joint may be made up of open sections, sections where a trace is visible but where there is still considerable tensile strength and sections where the rock is apparently intact (rock bridges). Only microfractures mark the line of the future development of a mechanical fracture. That this is so is evident from the obvious tensile strength of many rock joints even though they are clearly visible as traces (Figures 4.30 & 4.31).

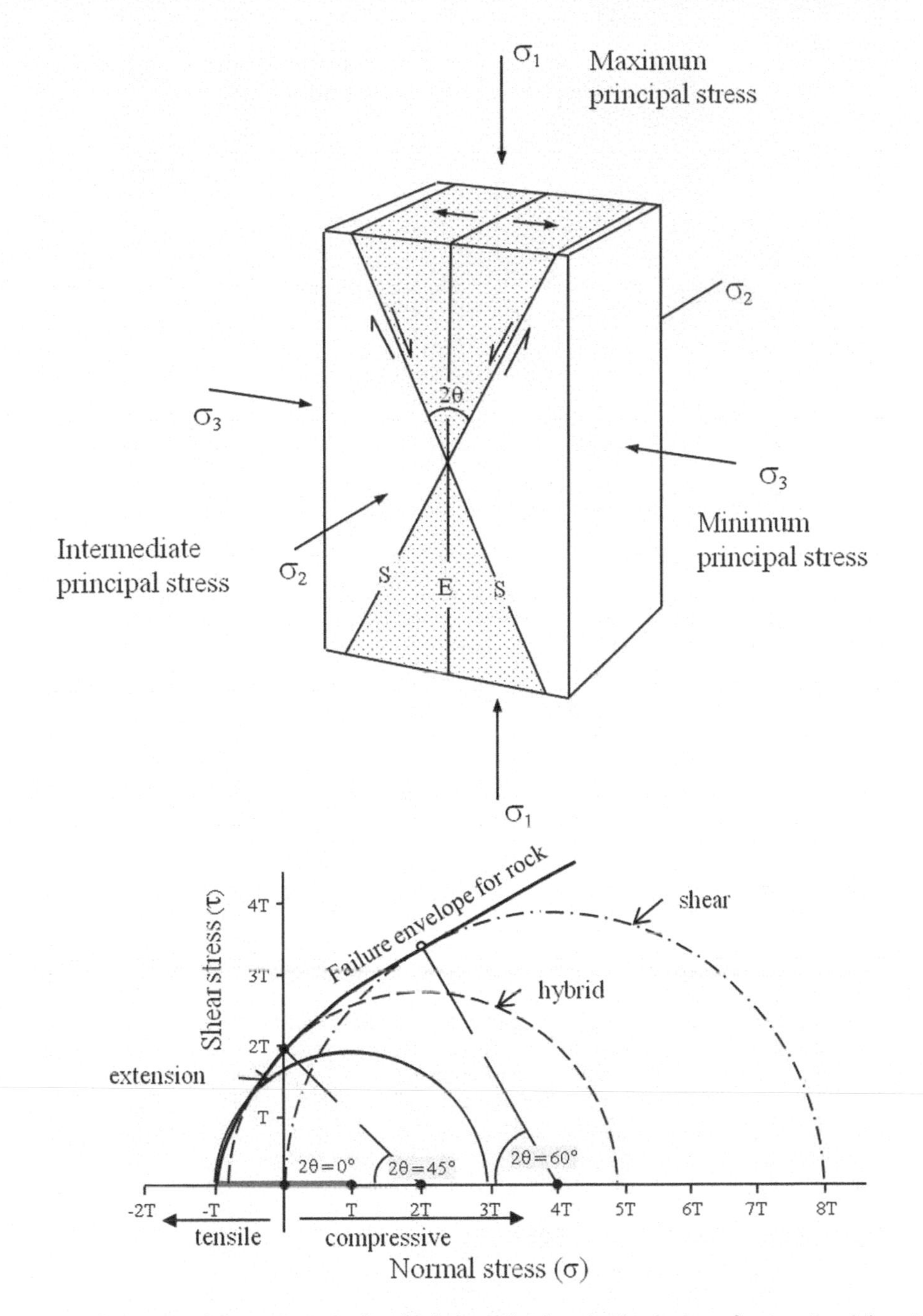

Figure 4.28 Stress conditions expressed as Mohr's circles for the formation of extensional (tension), hybrid and shear joints. The tangential line to the various circles is the overall failure envelope for the rock under different stress conditions.

Source: After Hancock, 1985.

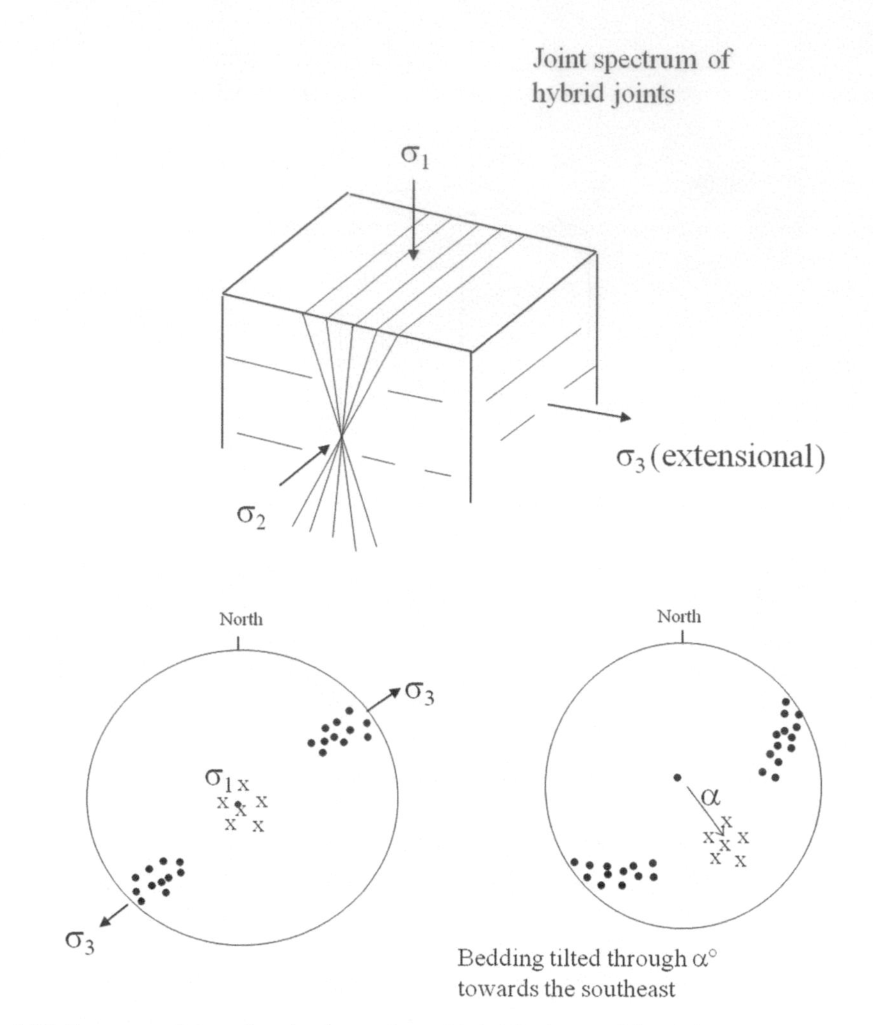

Figure 4.29 Stress conditions for the formation of hybrid joints and how these would appear on a stereographic representation. The centre crosses represent a horizontal set of discontinuities (say bedding in a sediment or flow banding in an igneous rock). The hybrid system plots from the circumference (vertical, extensional) inwards up to dips of about 60 degrees to horizontal. The lower right figure shows how this pattern might appear if the whole rock mass was tilted through α degrees Source: After Hancock, 1991.

4.4.7 Differentiation into sets

Joint sets are generally differentiated for rock mechanics analysis according to their geometries. The basics of stereographic projection representation of rock joints is dealt with in Chapter 4 and explained in detail in Wyllie & Mah (2004). If the poles representing discontinuities plot closely on a stereographic projection then they are considered to comprise the same set. For example, when one defines a set in the numerical simulation programme UDEC from Itasca (Chapter 6), this is done by inputting a mean dip and dip direction with a variation (such as plus or minus 5 degrees).

Figure 4.30 Exposure in granite with exposed and incipient joints, New Territories, Hong Kong. The hammer is for scale, 290 mm in length.

Programs such as DIPS from Rocscience (Chapter 6) can be used to identify sets statistically according to various methods such as the Fisher distribution. This is a useful tool but a number of things must be borne in mind:

1. The original data might be biased or partial. Some joint sets may not be fully developed at the point of observation or might by misrepresented in terms of population because of the geometry of the exposure.
2. Joints of similar geometry might include different sets geologically (in terms of time of formation) and which have characteristics that are quite different despite their parallel orientation.
3. Important but rare geological features, such as a fault, might be overshadowed by the rest of the data and even removed from consideration by statistical manipulation as discussed by Hencher (1985).
4. A better approach is first to try to interpret the distribution and nature of joints in terms of a model for the geological history at the site (Rawnsley et al. 1990; Hencher & Knipe, 2007; Hencher, 2013). Data such as surface textures and mineral coating can also be very helpful for differentiating between joint sets, especially for high level interpretations such as for nuclear waste studies (e.g. Bridges, 1990).

Figure 4.31 Sandstone cliffs near Guoliang, China. Cliffs are more than 100 m high and contain bedding plane traces at, perhaps, 1–2 m spacing and extensive traces of vertical joints. Photograph courtesy of P. Evans.

Some aids for understanding joint origin on the basis of their geometrical expression as seen in stereographic projections are set out below. These are often very helpful for interpreting geological history but as Price and Cosgrove (1990) put it: "be warned – many fractures resist all attempts at interpretation".

4.4.8 Orthogonal systematic

Many joint sets are orthogonal; two sets occur perpendicular to one another and perpendicular to some planar fabric such as bedding, schistosity or flow banding in an igneous pluton. Examples of such joint sets in sandstone and granite are presented

Figure 4.32 Orthogonal fractures in sandstone, Table Mountain, Cape Town, South Africa.

in Figures 4.32 & 4.33. The formation history can be quite complex with one set being formed initially, the second following stress relief due to the development of the first set or perhaps a general stress reversal as discussed by Rives et al. (1994). Interpretation may need detailed study of cross-cutting relationships. For the practicing engineering geologist, the important thing is that this joint arrangement is very common in a variety of rock types and this can aid in interpretation of sets from field data. Figure 4.34 shows the typical distribution of poles that might be seen in a stereographic projection of horizontally bedded strata with two orthogonal joint sets and the geometrical expression if the strata have been tilted. The interpreter should always be looking for angular relationships between sets and linking these relationships back to possible modes of origin. Orthogonal sets may be essentially primary, formed orthogonal to the intermediate and minor principal stress directions during burial and diagenesis of sedimentary rock (hydraulic joints of Engelder, 1985) or during cooling in igneous rock. It has been demonstrated experimentally also that orthogonal sets can also be formed as a secondary phenomenon by flexing layers (Rives et al., 1994) and this is probably a common origin during erosion, uplift and general doming (Price & Cosgrove, 1990).

4.4.9 Non-orthogonal, systematic

These joints are formed where the stresses perpendicular to the maximum principal stress, σ_1, are uniform. The most common joints in this category are those that develop

Figure 4.33 Orthogonal jointing in granite, Mount Butler Quarry, Hong Kong.

in extensive sheet lavas as illustrated in Figures 4.35 to 4.37. They form as tensile, planar zones of microfractures arranged around centres of cooling as primary structures and may develop incrementally during cooling (DeGraff & Aydin, 1987) or even later, once the basalt is cooled and proto-joints develop as mechanical fractures, incrementally. Figure 4.38 shows deformed columnar jointed basalt lava and one explanation is that a vertical columnar framework was established at an early stage of cooling but then part of the lava sheet was "nudged" by additional lava movement, whilst the rock was still plastic, and the full joints developed as fractures later (Hencher et al., 2011). Figure 4.39 shows how columnar joints in a general lava flow would appear on a stereographic projection originally and when tilted. Additional cross joints through the lava probably developed through twisting forces as the non-perfectly, vertically-sided, joints opened up and gravity took on a role.

Non-orthogonal, systematic columnar fractures can occur in rocks other than lavas. Young (2008) describes their occurrence in sandstone. In Chapter 8, the case example is presented of the Po Selim landslide in Malaysia. The pervasive jointing, which played a major role in allowing the failure to develop, had no preferred orientation other than it was at right angles to the planar schistosity (Figure 4.40). These are secondary joints probably formed during a relatively late stage of the regional tectonics responsible for the schistosity.

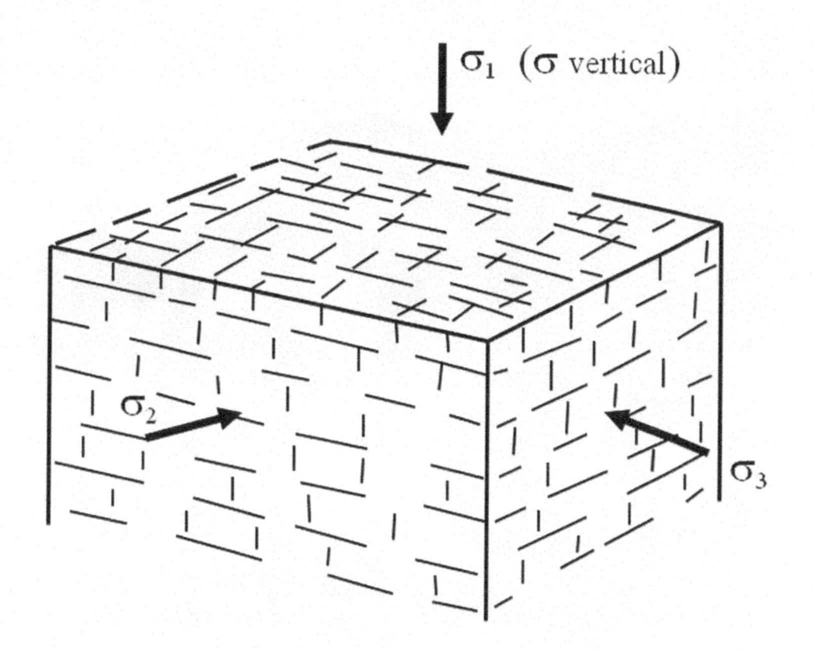

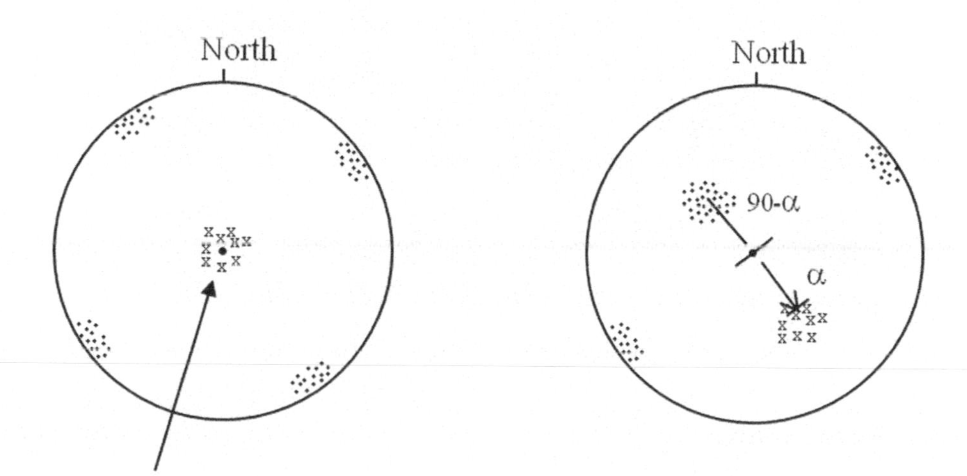

Figure 4.34 Schematic representation of orthogonal fractures and their representation on a stereographic projection assuming one set (e.g. bedding or cooling surface) is horizontal. Lower right diagram shows typical pattern of poles to orthogonal fracture sets if the model were tilted through α degrees by folding.

Figure 4.35 Vertical jointing through andesite lava, Hallasan, Cheju Island, South Korea.

4.4.10 Shear joints

Pollard and Aydin (1988) dismiss the concept of shear joints as "sheer nonsense" but this seems to be a bit tongue-in-cheek (the paradox being that once shear takes place a joint becomes a fault by definition). Fractures certainly do develop in shear directions as they do in triaxial testing (Chapter 6). In the example shown in Figure 4.41, some sections of the shear "joints" show no visible displacement but over other lengths of the same discontinuity, there is displacement. The argument about shear joints is largely academic in that whereas the joints propagate in the shearing direction predicted from Mohr's circle representation, in detail the joint is probably made up of coalesced sections, which are strictly tensile, originating from minor flaws in the

Figure 4.36 Columnar joints in basalt lava flow, Isle of Staffa, Scotland.

Figure 4.37 Columnar joints as seen at right angles to cooling surface (looking down, vertically). There is no directional control on joint formation in the horizontal plane ($\sigma_2 = \sigma_3$). Isle of Staffa, Scotland.

Figure 4.38 Collapsed columnar joints. This has apparently happened after the pattern of columns had been defined in the cooling lava sheet but whilst the lava was still plastic. Am Buachaille (the Herdsman), Isle of Staffa, Scotland.

rock (Kulander & Dean, 1995) and Engelder (1999) extends the discussion to hybrid joints. The appearance of shear joints on a stereonet, before and after tilting, is shown in Figure 4.42.

4.4.11 Complex geometries

As discussed above many joints follow some systematic geometrical pattern relating to the principal stress directions and magnitudes at the time of their formation. In some field exposures however, the fracture network can be very complex and difficult to unravel, especially when a rock mass has been through several structural events with each event resulting in a new episode of fracturing, which will be influenced by any pre-existing fractures (Rawnsley et al., 1990). Examples of major structures influencing the geometrical development of joints are given by Rawnsley et al. (1992) and one of these is illustrated in Figure 4.43. The joint geometries clearly follow stress trajectories that were strongly influenced by the pre-existing major fault.

4.4.12 Sheeting joints

Sheeting joints, which are sometimes also referred to as exfoliation fractures, are unlike other joints in that their geometry is not pre-defined by ancient geological history but

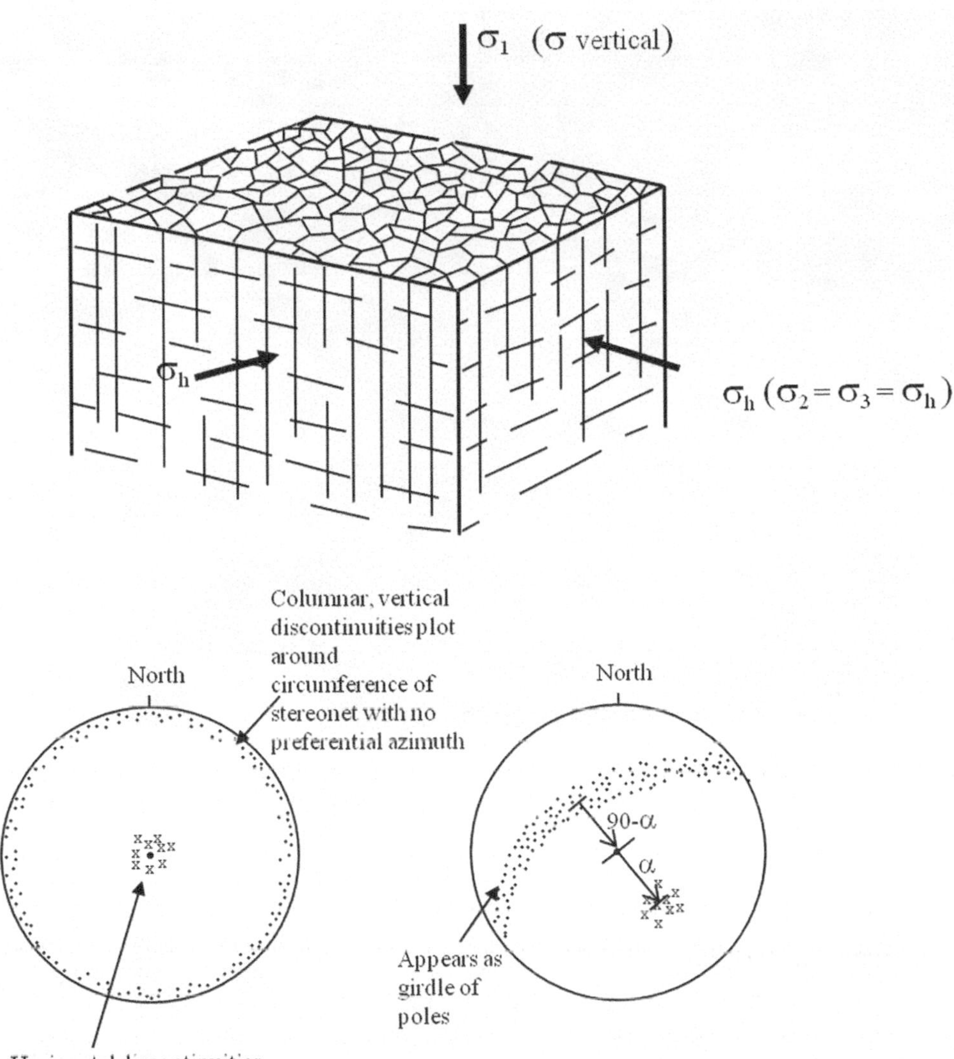

Figure 4.39 Schematic representation of columnar fractures and their representation on a stereographic projection assuming one set (e.g. schistosity or cooling surface) is horizontal and formed in an isotropic stress field in the horizontal direction. Second diagram shows typical pattern of poles if the model were tilted, say through folding. The original pattern of vertical fractures, plotting around the circumference, are expressed as a girdle of poles following a great circle, centred on the originally horizontal set.

Figure 4.40 Joint pattern in schist. All joints are approximately at right angles to the schistosity but otherwise random in orientation. Po Selim Landslide, Malaysia. See Chapter 8 for more details of this case history.

instead they develop in response to near-surface stress conditions reflecting locally prevailing topography (Hencher et al., 2011). These are "tertiary" joints as defined earlier. Sheeting joints are a striking feature of many landscapes and they have been studied for more than two centuries (Twidale, 1973). They run roughly parallel to the ground surface in flat-lying and steeply-inclined terrain and generally occur close to the surface, typically at less than 30 metres depth. They can often be traced laterally for 100's of metres. Most sheeting joints are young geologically and some have been observed to develop explosively and rapidly as tensile fractures in response to unloading (Nichols, 1980). Others are propagated to assist in quarrying using heat or hydraulic pressure (Holzhausen, 1989). Their recent origins and long persistence without rock bridges differentiates them from most other joints and from most bedding, cleavage or schistosity-parallel discontinuities.

Sheeting joints are common in granite and other massive igneous rocks but also develop more rarely in other rock types, including sandstone and conglomerate (Figures 4.44 & 4.45). Some sheeting joints develop at shallow dip angles, for instance, during quarrying, where high horizontal compressive stresses are locked in at shallow depths. In Southern Ontario, Canada, for example, high horizontal stresses locked in following glacial unloading often give rise to quarry floor heave and pop-up structures

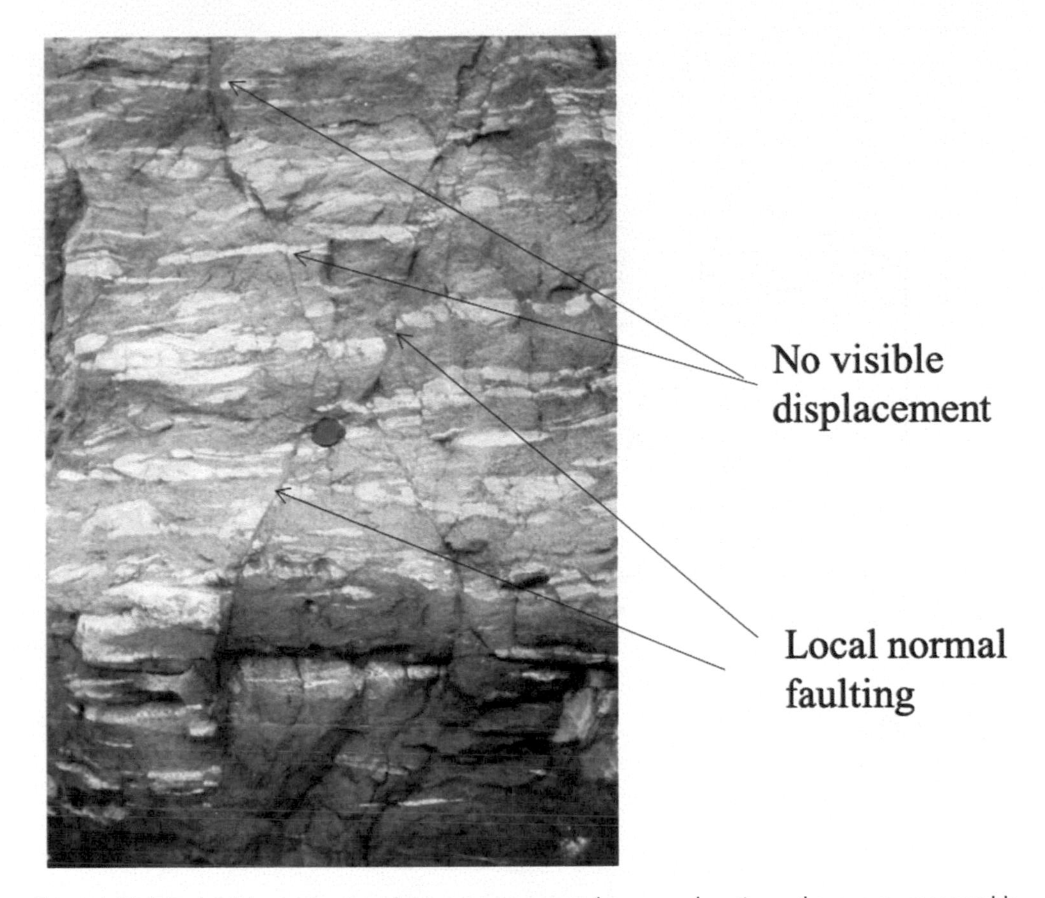

Figure 4.41 Joints with conjugate shear arrangement. At some locations there are measurable displacements so the shear joints grade into small displacement faults. Near Austin, Texas, USA.

accompanied by opening up of pre-existing incipient discontinuities, such as bedding planes and schistose cleavage (Roorda et al., 1982). Where there are no pre-existing weakness directions, new sub-horizontal fractures may develop in otherwise unfractured rock. Holzhausen (1989) describes propagation of new sheeting joints under a horizontal stress of about 17 MPa at a depth of only 4 metres where the vertical confining stress due to self-weight of the rock is only about 100 kPa. The mechanism is similar to a uniaxial compressive strength test where tensile fracture propagates parallel to the maximum principal stress (σ_1). Such exfoliation and tensile development of sheeting joints is analogous to the sometimes-explosive spalling and slabbing often seen in deep mines (Diederichs et al., 2004; Diederichs, 2018; Hoek, 1968).

From a worldwide perspective, however, the joints most commonly recognised as sheeting structures are observed in steep natural slopes. These joints are also thought to develop as tensile fractures where the maximum compressive stress due to gravity is re-oriented to run parallel to the slope as demonstrated by numerical models (Yu

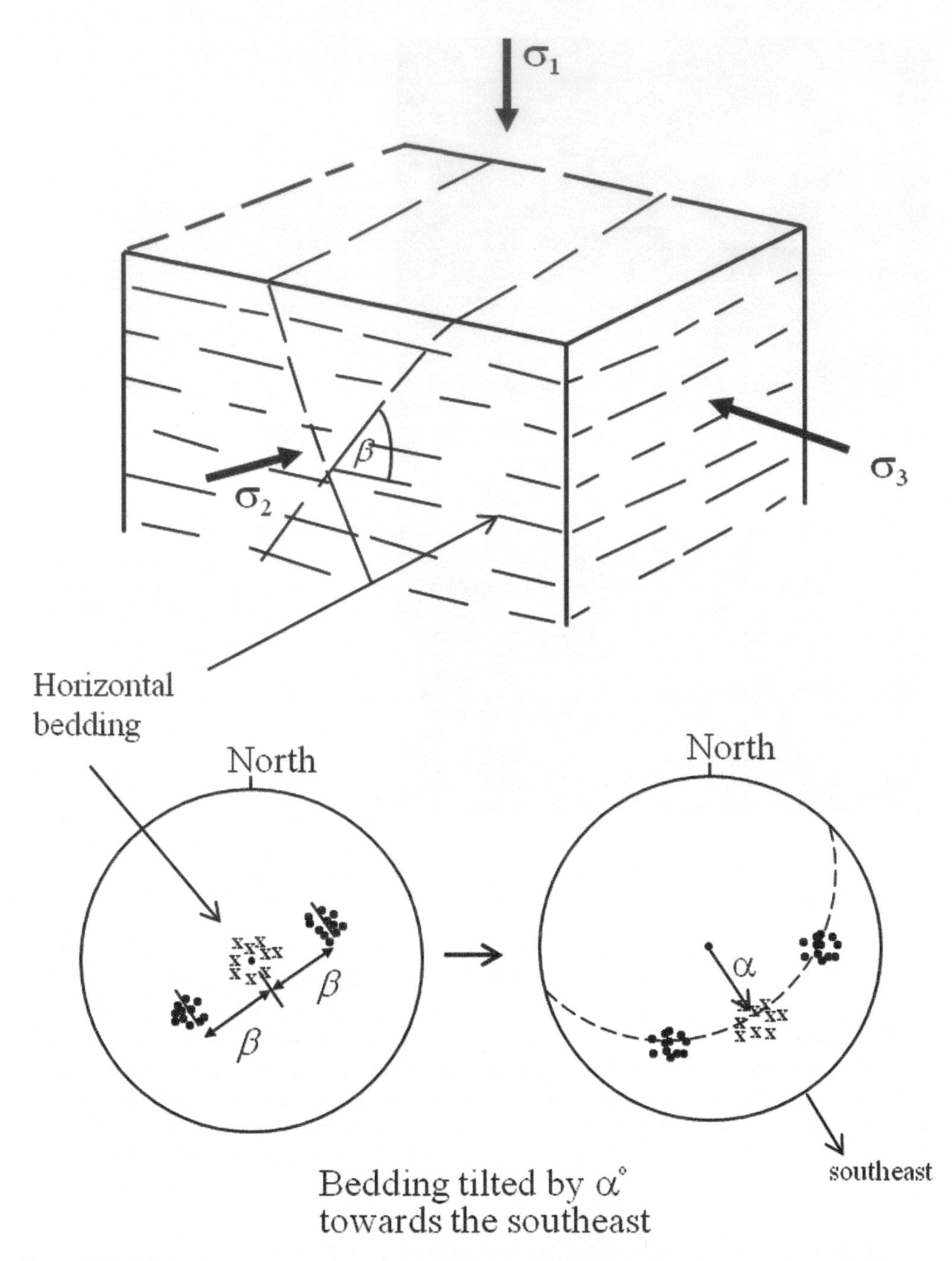

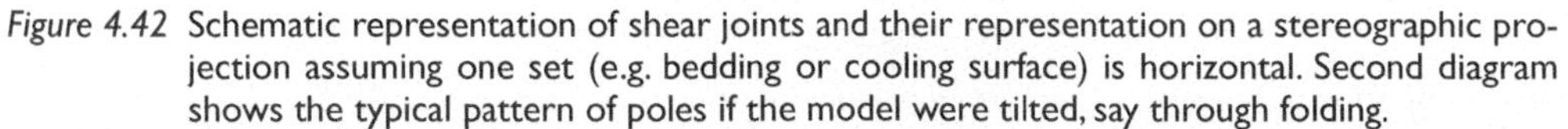

Figure 4.42 Schematic representation of shear joints and their representation on a stereographic projection assuming one set (e.g. bedding or cooling surface) is horizontal. Second diagram shows the typical pattern of poles if the model were tilted, say through folding.

Figure 4.43 Systematic joints whose geometries of formation were clearly influenced by a pre-existing mechanical discontinuity (the Peak Fault). Ravenscar, near Robin Hood's Bay, North Yorkshire, England.

Figure 4.44 Sheeting joint through arkosic sandstone, Uluru (Ayers Rock), Northern Territory, Australia.

Figure 4.45 Sheeting joints through conglomerate. Kata Tjuta (the Olgas), Northern Territory, Australia.

& Coates, 1970; Selby, 1993) and discussed in detail by Bahat et al. (1999). Sheeting joints also develop parallel to the stress trajectories that curve under valleys where there has been rapid glacial unloading or valley down cutting. Failure and erosion are continuing processes, with the formation of new sheeting joints following the failure of sheet-bounded slabs. Wakasa et al. (2006) calculated an average erosion rate of 56 metres in one million years from measurements of exposed sheeting joints in granite in Korea (Figure 4.46), which is significantly higher than erosion rates on other slopes without sheeting joints. Whilst many exposed sheeting joints are evidently very recent, others are much older. Jahns (1943) and Martel (2006) note the apparent dissection of landscapes post-dating sheet joint formation. Antiquity is also indicated by preferential weathering along some sheet joints. Additional evidence for the great age of some sheeting joints is the fact that they can sometimes be observed cutting through otherwise highly fractured rock. Most sheeting joints occur in massive, strong rock and it is argued that if the rock mass had been already highly fractured or weathered then the topographic stresses would be accommodated by movements within the weak mass rather than by initiating a new tensile fracture (Vidal Romani & Twidale, 1999). Therefore, where sheeting joints are found in highly fractured rock masses it is likely that they pre-date the gradual development of the other joints as mechanical fractures during unloading and weathering (Hencher, 2006; Hencher & Knipe, 2007). Figure 4.47 shows the stereographic representation of sheeting joints at a site in Hong Kong together with cross joints, perpendicular to the sheeting joints and at right angles to the azimuth of dip, which indicates their likely tensile origin.

Figure 4.46 Sheeting joints in granite, Mount Bukansan, near Seoul, South Korea. Climbers show the scale.

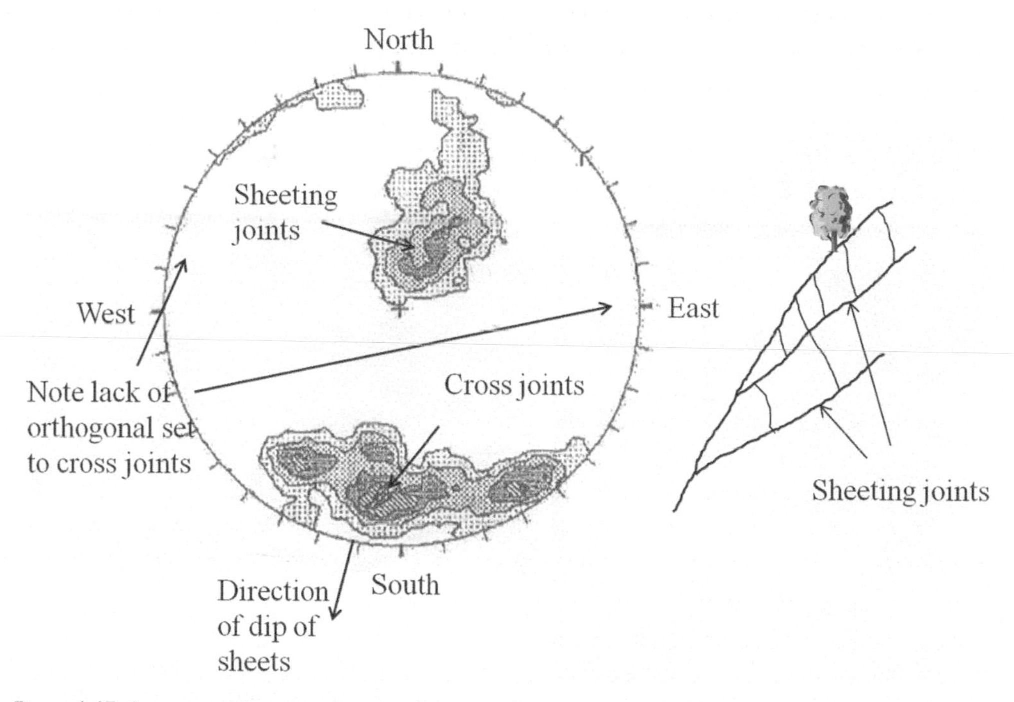

Figure 4.47 Stereonet showing sheeting joints and cross joints at 90 degrees. Tuen Mun Highway, Hong Kong.

Source: After Hencher et al., 2011.

4.4.13 Morphology of discontinuity surfaces

The shape of discontinuity surfaces is important to rock engineering, not least because of its influence on shear strength and this is dealt with in Chapters 6 and 7. As for overall joint set orientation and spacing, discontinuity morphology is related to geological history and mode of propagation, and surface features should be observed and described. The description, analysis and interpretation of discontinuity surface morphologies and the ways these can be linked to interpretation of causative stresses, mechanisms and subsequent evolution is a relatively undeveloped area of scientific study called fractography (Ameen, 1995). It is difficult to be specific, that one type of joint can be expected to have certain morphological characteristics compared to others, but it is generally recognised that tensile fractures are particularly rough and variable. Generally, roughness is characterised by measurement. Waviness is the deviation from mean direction at the scale of metres. Smaller scale roughness (sometimes termed second order) is either measured objectively or estimated with reference to a scale called Joint Roughness Coefficient (JRC). Examples of smooth and rougher joints where there is cross-jointing are given in Figure 4.48. Details of descriptive methods are given in International Society for Rock Mechanics (1978) and Barton (1973). Roughness is described in the field using terms such a stepped or slickensided (Chapter 6 and Appendix C). At the smallest scale, surface texture, together with mineralogy will control basic friction.

4.4.13.1 Sedimentary rocks

Sedimentary bedding planes may show a variety of distinct morphologies as illustrated by Stow (2005). Many are flat and planar but others are rough. Ripple marks as seen

Figure 4.48 Smooth and rougher joints in the same tuffaceous sediments, Hong Kong.

on beaches are commonly preserved in sandstone and in turbidites surfaces scour marks can make them very rough, interlocking with the overlying layer. Bedding surfaces can also be roughened by biological activity such as burrowing. Sometimes bedding surfaces are exposed, weathered and eroded before the overlying rocks are emplaced and again, such features need to be interpreted correctly both in terms of understanding the geological situation and because they will affect geotechnical properties.

4.4.13.2 Tension fractures

The first major work on fracture morphology was by Woodworth (1896). He observed and illustrated what are now considered classic expressions of fractures propagating from a flaw with typical features, including "feather markings" and arrest or hackle marks (Figure 4.49). Most researchers consider that most such markings involve extension (tension) although other features probably involve some shear. They are recognised in all rocks and even drying sediments. They are also commonly observed in drill core and distinguishing natural from induced fractures is an important task for those logging the recovered rock (Kulander et al., 1990). Such features are significant in their own right geotechnically but are also important for the interpretation of the geological history of a site.

4.5 WEATHERING

4.5.1 Weathering processes

Weathering is the process by which rock deteriorates until it eventually breaks down to a soil. It occurs close to the Earth's surface and depends very much on climatic influences – rainfall and temperature. Ollier (1975) and Selby (1993) provide good overviews.

In hot, humid climates the following are the most important processes:

- *Decomposition*: The result of chemical changes on exposure to the atmosphere (H_2O; CO_2; O_2). The original rock minerals, stable at the temperatures and pressures operative at the time of formation, break down at the Earth's surface to sand, clay and silt.
- *Disintegration*: Inter- and intra-grain crack growth and coalescence of cracks to form fissures and propagation of large scale joints.
- *Eluviation*: The soft, disintegrated (or dissolved) material is washed out from the parent rock fabric through open joints or from the porous skeletal structure and deposited elsewhere (illuviation).

Weathering affects not only strong rocks but in weak masses including materials that might be regarded as engineering soils even in their "fresh" state. Processes include softening and chemical change (e.g. Moore & Brunsden, 1996; Picarelli & Di Maio, 2010).

The rock mass in tropical areas is commonly severely weathered to depths of 10's of metres and occasionally over 100 metres. Weathering is manifested by changes from

Figure 4.49 Shallow bedding plane with ripple marks. Near vertical surface with typical arrest and hackle marks associated with tensile fracture propagation. Tsau-Ling Landslide, Tai Wan.

the original rock state (fresh), including mineralogy, colour, degree of fracturing, porosity and thereby, density, strength, compressibility and permeability.

In colder climates chemical decomposition is less active and rocks tend to deteriorate due to frost and ice action (Figure 4.50). Mechanical deterioration is also the dominant process in desert environments sometimes associated with expansive formation

Figure 4.50 Disintegrated granite with corestones, above Lake Tahoe, Nevada, USA.

of salt in pores. The depth of weathering in cold climates is far less than in tropical environments.

In temperate climates, such as the UK, weathering is rarely very significant and certainly far less so than in countries such as Brazil, Malaysia and Singapore where weathering has extremely important consequences for investigation, design and construction (e.g. Shirlaw et al., 2000). In the UK and much of Northern Europe, most weathered rocks were stripped from the landscape by recent glaciations.

4.5.2 Weathering profiles

Weathering reduces the strength of rock material as illustrated in Figure 1.4 and discussed in Chapter 7 in terms of geotechnical parameters. Weathering processes generally operate at upper levels in the saturated zone and in the vadose zone above the water table although it should be noted that water tables change periodically and have done so over the millions of years that it will have taken for the development of some thick weathered profiles. Therefore, current water levels may not be related to depth of weathering at a site.

As a general rule, weathering works in from free surfaces where chemicals in water (including the water itself) can attack the parent rock (Figure 4.51). Eventually it may leave a framework of "corestones" of less weathered rock separated by severely weathered zones marking out the loci of the original joints (Figure 4.52). The process is illustrated in Figure 4.53. The wide varieties of conditions that can be encountered

Figure 4.51 Stained joints, volcanics, Hong Kong.

in weathered terrain are discussed by Ruxton & Berry (1957) and Figure 4.54 is based on their interpretation of one type of weathered profile in Hong Kong. Despite the conceptual usefulness of Ruxton & Berry's profiles, exposures are often encountered that do not conform and such exposures provide challenges to mass weathering classifications such as the current European standard as discussed in Chapter 4 and Appendix C. Furthermore, other rock types often weather without the development of

Figure 4.52 Corestone development in granite, Stubbs Road, Hong Kong.

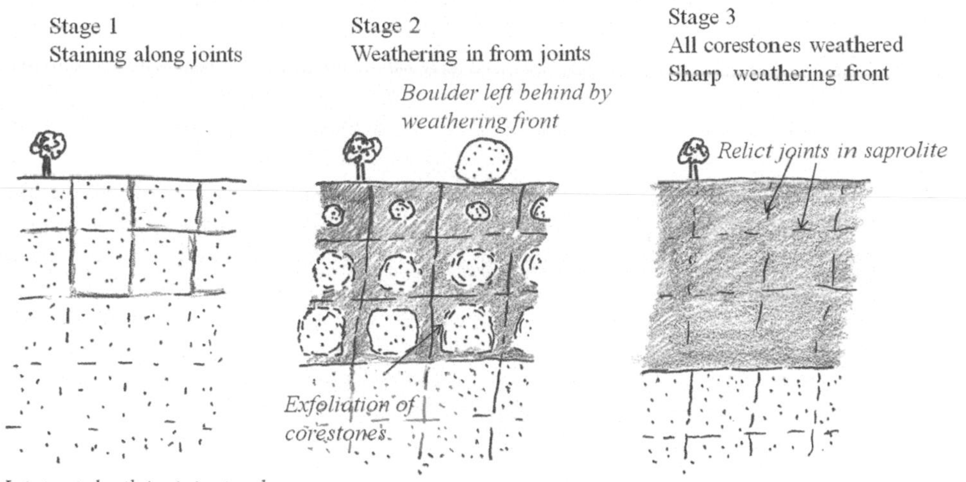

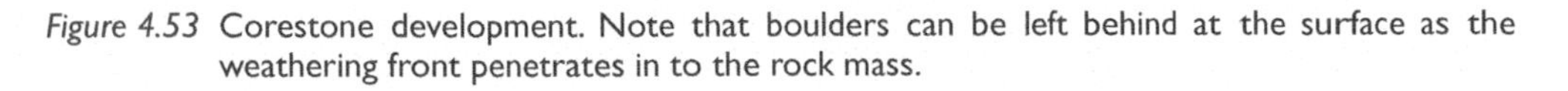

Figure 4.53 Corestone development. Note that boulders can be left behind at the surface as the weathering front penetrates in to the rock mass.

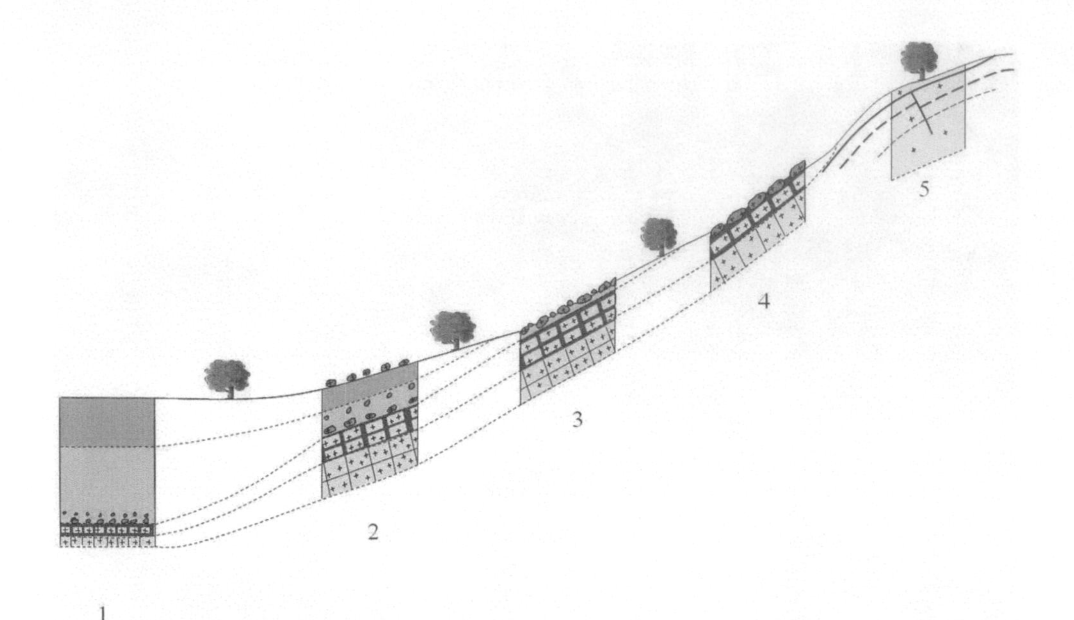

Figure 4.54 Schematic weathering profile with corestone development after Ruxton & Berry (1957) modified by Hencher et al. (2011).

corestone type profiles. Mudstone sequences tend to develop a gradational weathering profile as characterised for the Keuper Marl by Chandler (1969). Limestone is often karstic with large caves and open joints as illustrated in Figure 4.55 and classified by Fookes & Hawkins (1988).

At any particular location the weathering profile is a function of parent geology, groundwater conditions and the geological and geomorphologic history of the site. The profiles may be ancient and bear little relationship to current geomorphologic setting. Given these and other factors, weathering profiles can be rather unpredictable from examination of the current topography. Valleys might be associated with deep weathering along faults (e.g. Shaw & Owen, 2000) but not always.

Chemical weathering rarely occurs to depths of more than 100 metres even in tropical and sub-tropical areas but this should only be used as a general guide to current situations; sometimes weathered profiles are encountered at much greater depths, having their origins in some past landscape and time. For example, Younger & Manning (2010) describe tests on a highly permeable zone in granite at a depth of 410 metres and attribute the high permeability to weathering during the Devonian when the granite was exposed at the surface. Ollier (2010) provides other examples of ancient weathering profiles buried by later sediments.

Figure 4.55 Limestone pavement with proto-joints etched out by dissolution weathering. Above Malham, West Yorkshire.

4.6 GEOLOGICAL HAZARDS

4.6.1 Introduction

Numerous hazards can be regarded as essentially geological, including the potential for subsidence, swelling, clay shrinkage and natural noxious gases as addressed in Chapters 5 and 6, when considering elements to be targeted during site investigation. The most important natural geological hazards in terms of loss of life however are landslides, earthquakes and volcanoes, partly considered in Chapter 2.

Water as an active eroding force, is an additional part of the jigsaw for consideration, and this is dealt with in this chapter, as a hazard for the ground model. It is dealt with in more detail in Chapter 6 in terms of effective stress, as a consideration for throughflow into slopes, tunnels and underground excavations and in terms of investigation. Change of climate, with time, and rising and falling sea-levels are also dealt with, as part of the Second Verbal Equation (of Knill, 1976), in Chapter 5.

4.6.2 Water as a hazard

Water is critically important to many geotechnical projects. Surface water causes erosion and flooding; groundwater controls effective stress and therefore frictional strength and can be a major problem for tunnelling. Water problems for most civil engineering design and construction is dealt with by measurement and monitoring using piezometers backed up by numerical modelling and analysis (Chapters 5 & 7). Groundwater levels do not generally fluctuate too much in response to individual rainstorms other than close to the surface, which is significant for shallow landslides as discussed below, but not for most other engineering works. More significant are groundwater changes brought about by engineering works either deliberately (e.g. dewatering to carry out excavations in the dry) or as an unintended consequence, for example where tunnelling below a site. Lowering of the water table inevitably causes water migration and potentially internal erosion, a loss of buoyancy, increase in effective stress and self-weight compaction of soil and rock. This may result in settlement and damage to adjacent structures; piles can become overloaded by negative skin friction (Chapter 3). Similarly rising water levels can cause difficulties due to buoyancy, which could require holding down anchors or piles. Rising water might also weaken the ground supporting a structure, sometimes due to dissolution of cementing agents.

For tunnels and other underground structures, inflows can be very difficult to predict accurately because it depends so much on the geological situation, which is rarely well understood before the works commence. Given knowledge of water pressure and hydraulic conductivity then inflows can be predicted following standard equations or by numerical simulation but the controlling parameters are difficult to measure or estimate and predictions can be wildly out as found for trials for nuclear waste investigations (e.g. Olsson & Gale, 1995). This is especially so for tunnels passing through variable geology as discussed by Masset & Loew (2010). The pragmatic solution is to probe ahead of the tunnel face periodically and if inflows from the probe holes are high then to improve the ground in front of the tunnel, usually by injecting cement or silica grout to reduce the hydraulic conductivity. Alternatives are dewatering the ground or freezing the ground temporarily. If predictions of groundwater conditions are badly incorrect this can have major consequences for the suitability of tunnelling method or machinery (e.g. the degree of waterproofing of equipment). Where there is a risk of high-water inflow it is normal practice to drive the tunnel uphill to reduce the risk of inundation, danger to workers and damage to machines. Where tunnelling under the sea, lakes or rivers there may be a risk of disastrous inflows as occurred during the construction of the Seikan tunnel in Japan (Matsuo, 1986; Tsuji et al., 1996). Other examples where the severity of groundwater conditions was underestimated with severe consequences include the SSDS tunnels in Hong Kong and Ping Lin tunnel in Taiwan, which are discussed in Chapter 8.

4.6.3 Groundwater response to rainfall

Most landslides are caused by rainfall and in Hong Kong, for example, rises in water level during a storm of more than 10 metres have been recorded (Sweeney & Robertson, 1979). Therefore, there is great interest in trying to predict changes that might occur as these will affect greatly any numerical calculations of slope stability as well as other engineering projects. Geological profiles are generally depicted for groundwater modelling as made up of discrete, homogeneous and often isotropic units of given hydraulic conductivity (Todd, 1980) and most commercially available, hydrogeological software only deal with homogeneous units. To be more realistic, models may need to incorporate local barriers such as fault seals, fracture flow or more variable geological conditions such as local litho-facies (e.g. Fogg et al., 1998).

The wetting band theory first proposed by Lumb (1962) is still used for estimating the likely depth of ground that might be affected by a rainstorm. Given a rainstorm of a certain duration and knowledge of the original saturation of the ground and porosity, the thickness of the surface zone of saturation can be estimated. It is assumed that the saturated layer will then descend until it meets the groundwater table resulting in a rise in groundwater table equal to the thickness of the wetting band. This provides a tool for assessing the design groundwater condition albeit rather crude. Some of the geological conditions that will conspire to make such simple approaches unrealistic are illustrated in Figure 4.56.

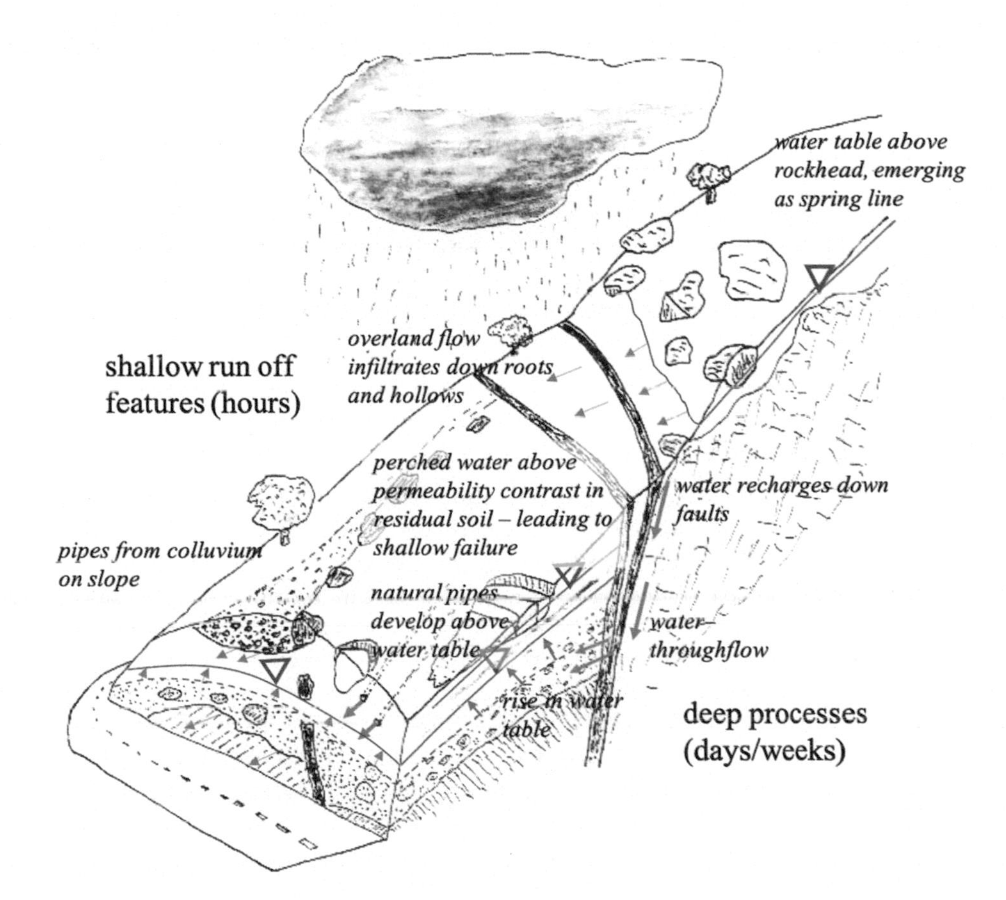

Figure 4.56 Schematic model of water runoff, inflow and throughflow in weathered profile. Note the importance of minor geological Features such as dykes, clay infilled joints or simple permeability contrasts in the profile and the development of natural pipes. In rock, water flow will be controlled by joints and specific channels along those joints. There is often a zone of more highly fractured rock just below rockhead with preferential flow and sometimes upward flow into overlying soil profile.

Source: Hencher, 2010.

More sophisticated attempts have been made to model infiltration and pressure diffusion processes in pressure head response and the triggering of landslides mathematically (see Iverson, 2000). Such methods are useful in visualising mechanisms but again rely on generalised parameters, such as hydraulic diffusivity, which are often difficult to define and only apply to simple geological conditions. More recent modelling methods are described by Blake et al. (2003).

It is often argued that it is more realistic to instrument a site and then to extrapolate rainfall response to a "design" rainstorm (Geotechnical Control Office, 1984a). However, instrumenting slopes to measure critical water pressures is not easy because instruments must be installed (and monitored remotely) at the precise locations where water pressures will develop. Quite often rock and soil profiles are compartmentalised and water flow carried in narrow channels, pipes or through particularly permeable zones as discussed below so the success of any instrumentation programme will depend very much on the geological and hydrogeological insights and skills of the investigating team as well as a degree of good fortune.

4.6.4 Landslides in natural terrain

There are more than 300 fatal landslides each year on average (Petley, 2024). Some of these are in man-made slopes and therefore a matter for engineering design (Chapter 3). Many however are in natural slopes and single incidents can cause many deaths. The debris flow that struck Gansu, China in August 2010 killed more than 1200 people and destroyed at least 300 buildings. 45,000 people had to be evacuated. The risk to sites from natural terrain landslides is therefore an important consideration at many sites.

4.6.4.1 Modes of failure

Natural terrain landslides can be split into those where the detached debris directly impacts a site through gravity and those where the debris becomes channelised and flows down a valley (Figure 4.57). Channelised landslides are relatively easily to deal with conceptually in that the pathway for the debris which often becomes saturated and flows is easily predicted even if the size of the event is not. The best thing to do is to avoid the outlet of any valley but if this is not possible, for example, where building a road or railway, then the size of the "design event" may be predicted by historical studies or from first principles. Hungr et al. (2005a) present a useful review of landslide characteristics that might be considered for design. In assessing existing facilities and structures, sometimes these can be protected by barriers and other engineering devices but occasionally the risks are so high and cost of mitigation too expensive so that relocation is the only real solution.

Open hillside landslides and rockfalls have a much more limited distance of travel. For most landslides in Hong Kong, debris travel distance is less than 100 m, so the area of concern is quite obvious both in terms of source of landslides and structures at risk. It does not make them less dangerous, just that the nature of the hazard and focus for analysis is clear. Hazard assessment can follow standard methods of investigation, analysis and design as outlined in Chapter 5. Risk review can be used to justify cost of mitigation works or taking no action – despite there being a clear and obvious

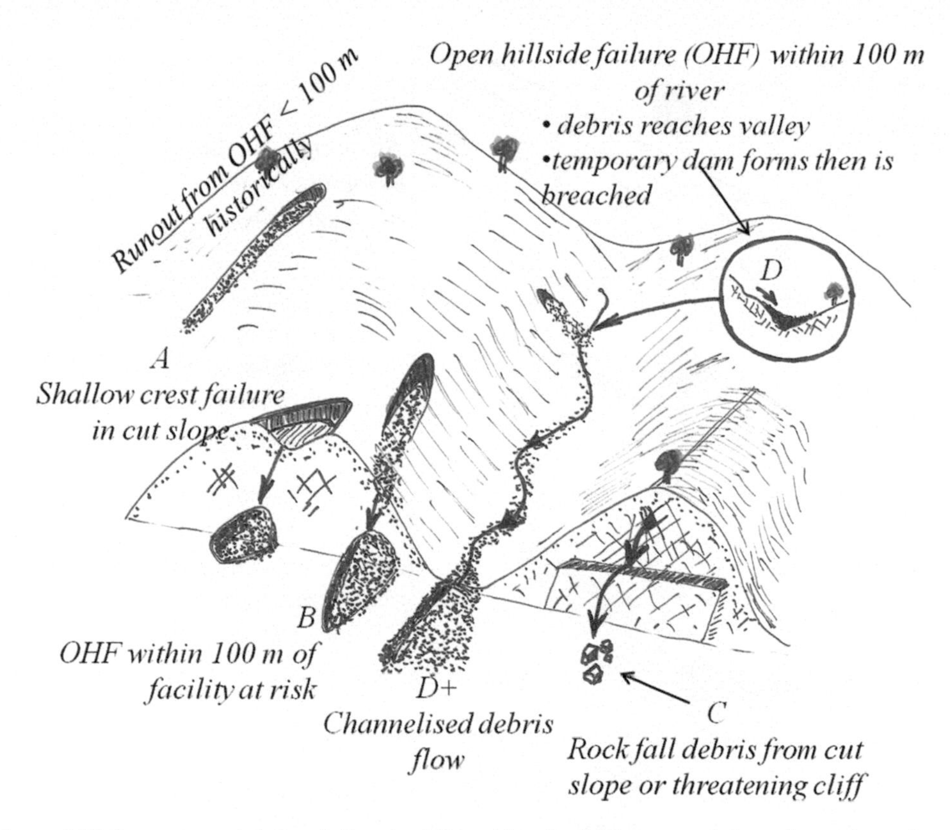

Figure 4.57 Run-out models for shallow landslides. Most landslides are only hazardous where they directly threaten a facility. Remote landslides can however feed into a stream channel where they can be channelised and flow great distances. Things may be made worse where the initial failure produces a landslide dam behind which a lake forms. When the temporary dam is breached a sudden discharge of fluidised debris is released.

danger as, for example, along many roads through mountainous areas. The practical applications of risk concepts are reviewed by many authors in Hungr et al. (2005b). Fell et al. (2005) and Wong (2005) are particularly useful.

4.6.4.2 Slope deterioration and progressive failure

The concept of ripening of slopes prior to failure has been a useful idea for many years but recently evidence for progressive deterioration of slopes prior to detachment has become better documented. This applies to both natural slopes and cut slopes (Malone, 1998; Hencher, 2000; Parry et al., 2000). Factors involved in slope deterioration are illustrated schematically in Figure 4.58 and some of the factors triggering natural terrain failures are illustrated in Figure 4.59.

The gradual deterioration can be represented by a curve in which the Factor of Safety reduces over a period of time, which may be hundreds of years (Figure 4.60). The vertical lines represent temporary reductions in Factor of Safety caused by relatively

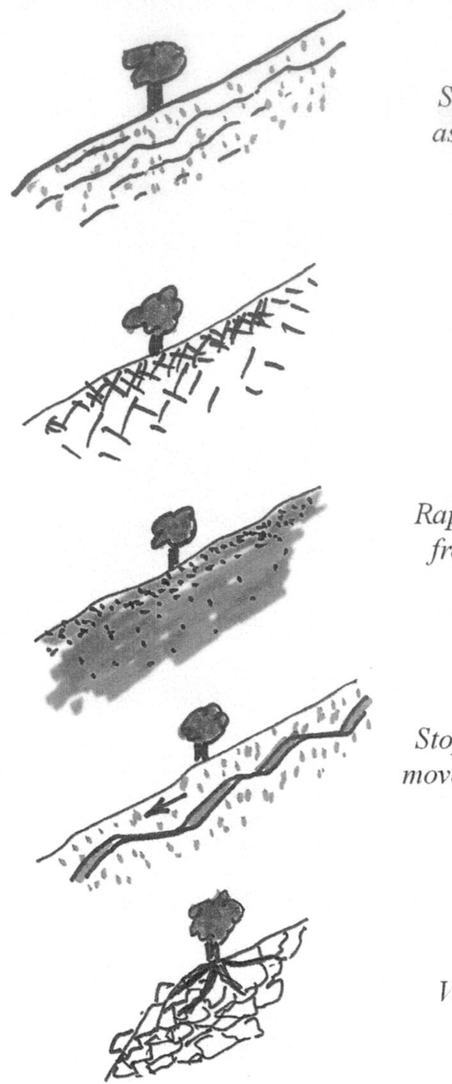

Figure 4.58 Deterioration factors in slopes.

short-term, transient events (days). In the course of time, the slope will deteriorate to the point where it is vulnerable to a transient event – causing a reduction in the Factor of Safety below 1.0. Whether that event results in catastrophic failure or only minor movement and internal deformation depends on many factors, including the severity of the triggering event and how long it lasts. The concept of ripening and progressive failure is discussed in more detail in Hencher (2006). Similar concepts are discussed for claystone slopes by Picarelli & Di Maio (2010).

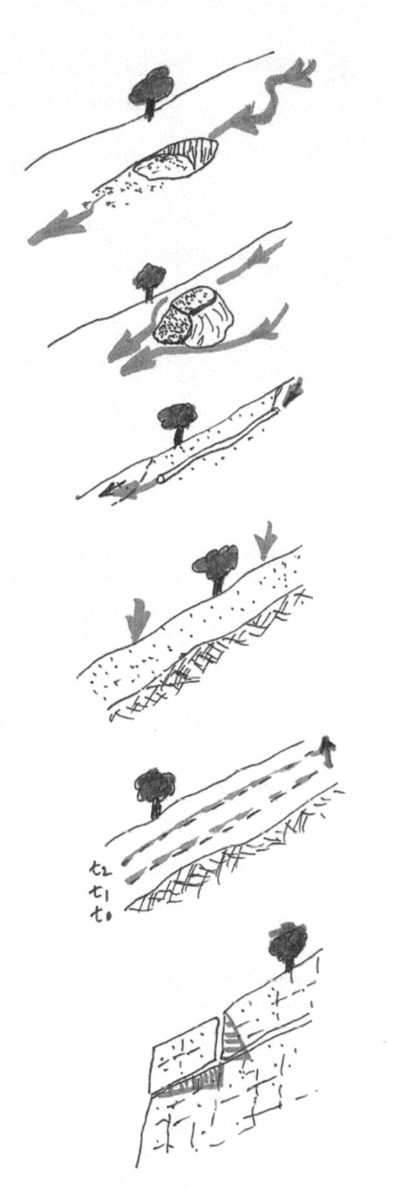

Figure 4.59 Water triggering mechanisms of shallow landslides.

Signs of gradual deterioration or, more likely, the cumulative effect of intermittent triggering events can be seen in many exposures and these can be used during ground investigation to help judge whether a failure is imminent although this may still not be straightforward.

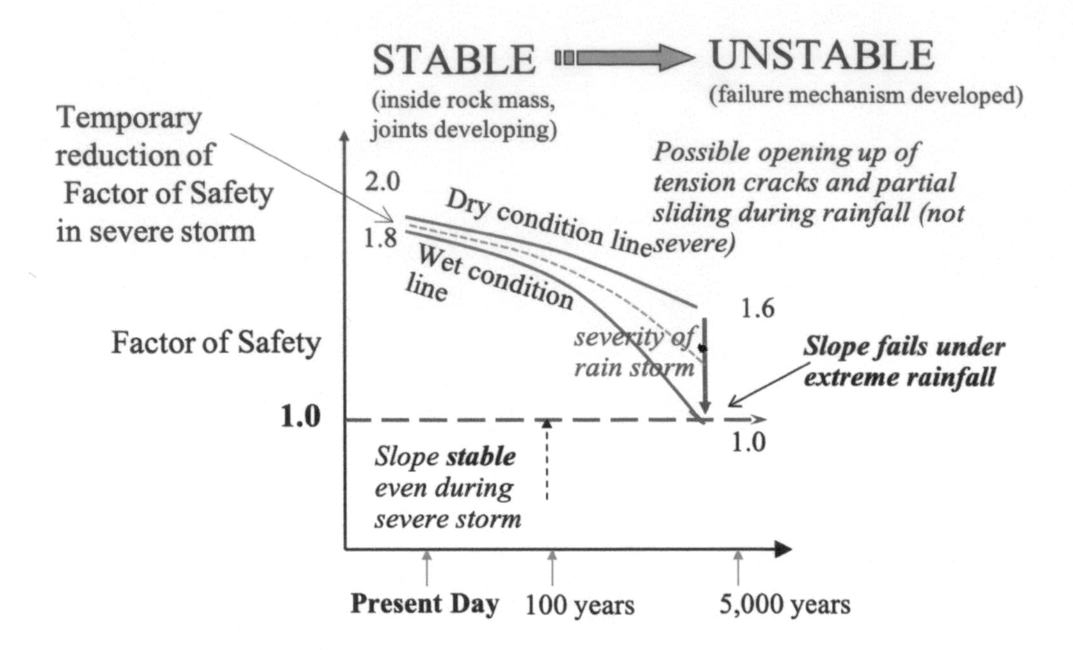

Figure 4.60 The concept of gradual ripening of slopes prior to the development of a full landslide. At some stage the slope will reach the point where it may be moved a little by some transient process such as intense rainfall or an earthquake. Deterioration will then continue and probably accelerate until eventually full detachment occurs. Hillsides can be regarded as having an inventory of different parcels of ground all at different stages of deterioration and therefore susceptibility to a particular triggering event. Depending on the severity of the event, one, two or many landslides will occur.

4.6.5 Earthquakes and volcanoes

Volcanic risk is a clear problem but sources of the hazards are generally well known although surprises do occur as in the case of the mud volcano that erupted disastrously in 2006 (Davies et al., 2011). Clearly if a volcano is active then construction should avoid the potential zone of travel and deposition of very hazardous materials such as lava and ignimbrite. Landslides associated with volcanoes are called lahars, can travel great distances and be hugely damaging. Noxious gases are produced by volcanic activity. A tragic case at Lake Nyos in Cameroon, Africa in 1986 involved the eruption of a bubble of carbon dioxide that suffocated more than 1,700 people and 3,500 livestock in nearby villages Avoidance is again the only real option.

Earthquakes are rather more of a general hazard in that they can occur anywhere in the world, though seismic activity is concentrated along active plate boundaries. The process of assessing earthquake hazard for a site and then design to withstand the potential shaking are dealt with in Chapters 3 and 7.

4.7 GROUND MODELS FOR ENGINEERING PROJECTS

4.7.1 Introduction

Ground modelling is an essential part of engineering design. The ground model for a project will mainly comprise a simplified representation of the site geology that should include all aspects that are likely to affect the project or to be affected by the project. A useful review is given in Geotechnical Engineering Office (2007) and some principles are set out in Box 4.3. Ground models in this context are essentially 3D models of the geological conditions at a site together with environmental influences and hazards. Numerical and physical models can be designed based on the conceptual ground model and might be used in the development of ground models, for example, in simulating the development of *in situ* stresses.

BOX 4.3 PRINCIPLES OF GROUND MODELLING

There is considerable literature considering what a "ground" model should comprise – mostly waxing philosophically about what should or should not be included (e.g. Baynes & Parry, 2022). For what it is worth, I set out below, some principles from my viewpoint...

1. Ground models for a site (for example, geological, geotechnical or hazard models) should be based on *adequate* interpretations of geology and hydrogeological conditions, the parameters which allow mathematical or physical analysis relevant to the works and any hazards, including flooding, changing temperatures and seismicity (Knill, 2003).
2. It can be difficult when logging core to decide on what to include – is RQD necessary to assess rock strength, for example; what about the fossils in the core? Usually one should follow the rote as per the examples in Appendix B but not always. Try to bear in mind the nature of the structure that is to be built and think, "WHAT IS IMPORTANT HERE??"
3. Engineering geologists should consider, early on, whether the site is exceptional and "unforgiving" – and be prepared to shout "STOP" or "HELP"!! This will not happen too often, but beware of such tricky sites.
4. Models should include considerations of time – not just the nominal lifetime of the structure, but for considerably longer-term, up to the eventual collapse or abandonment of an open-cast mine or a tailings dam, for example. The selection and modelling of a nuclear waste repository must allow prediction of changes for **many** thousands of years into the future.
5. Models should be large enough to include consideration of **all** the ground that might be affected by the works. For example, a building will stress the ground significantly to a depth of up to twice the breadth of the foundation footprint. A new dam and reservoir may influence the terrain and environment for a very large area, many kilometres from the actual dam through changes in ground water levels and perhaps induced seismicity.

6. Engineering geological models should consider the **total** geological history at a site (Eggers, 2016; Fookes et al., 2000).
7. Make sure that the models incorporate all the features of the ground important to physical performance as conditions change (e.g. increased or decreased loading by the engineering works, weathering in some circumstances or the application or reduction of fluid pressures).
8. The creation and testing of several simple models exploring the sensitivity of the site to various assumptions will often be more revealing than a single complex model. This is especially true when it comes to using numerical and physical models to model a site (Starfield & Cundall, 1988).
9. Geotechnical engineers and engineering geologists should act as detectives in characterising a site, studying the evidence, hypothesising and testing hypotheses through the collection of additional data including the output from mathematical analyses. Several iterations may be necessary before the models might be considered adequate.

A preliminary geological model based on desk study together with geomorphological interpretation should be used for planning ground investigation, which will allow the model to be checked and refined. Fookes (1997) suggests that a model can simply be a written description or presented as cross sections or block diagrams and plans. It might be focused on some aspect such as groundwater, geomorphology or rock structure but should be targeted at the engineering needs of the project. At a later stage the geological model may be split into units, which can be characterised in terms of engineering properties and anticipated performance. In some locations and for particular projects, rocks that are quite different in origin and age might be lumped together because they can be expected to behave in a similar manner. Complex geology does not always equate with difficult geotechnical behaviour. Conversely an apparently simple geological profile may have subtle variations that affect the success or otherwise of a project.

Once the model includes the range of engineering parameters and ground conditions that need to be considered it becomes a design model (Knill, 2003). It might then be used, for example, to decide what foundation system might be required to carry the load from a building. It is used for making decisions on how to deal with the ground conditions. The full model will include not only geological features but also other site factors including environmental conditions and influences such as groundwater, rainfall, wind and earthquake loading as well as anthropogenic influences such as blasting and traffic vibration.

4.7.2 General procedures for creating a model

The starting point of a model should usually be a three-dimensional representation of the geology of the area and to the depth that will be affected by the project or which the project may affect. The first attempt at a geological model for a site will usually be an interpretation of published maps and the interpretation of aerial photographs and satellite imagery depending upon the location of the project. Unfortunately for some projects that is as far as the geological interpretation goes, sometimes with disastrous results because the maps are either incorrect or at such small scale that they cannot represent the site-specific features that will affect the project. Those working in civil engineering need to appreciate that all published geological maps are professional interpretations of relatively small pieces of reliable data that are then interpolated and extrapolated. Faults may have been interpreted from lineaments and might not exist in reality. Conversely published maps will certainly not show all the major geological discontinuities that may be significant for an engineering project. Most features on geological maps are generally marked as uncertain or inferred but that does not stop the unwary assuming that they are accurate. In all cases, maps and plans that are not site- – or project-specific – should be taken as indicative only and a starting point for detailed investigation as discussed in more detail in Chapter 6. Despite the inevitable limitations of published information, a broad understanding of the geological and geomorphological setting can be used to make predictions of what might be encountered at the site through experience and training. For example, if the site includes a granitic intrusion then one might expect certain joint styles in the granite, a metamorphic aureole around the intrusion where the granite has cooked the country rock and associated minor dykes and hydrothermal alteration as illustrated in earlier sections of this chapter. In limestone country one should expect caves and open fissures, perhaps infilled with secondary sediments even where they have not actually been sampled at the site. The use of "earth science" skills to interpret the available data as a site history is clearly important yet sometimes lacking in civil engineering practice (Brunsden, 2002). The best source of information on what might be anticipated is the geological literature – textbooks on physical and structural geology and sedimentology in particular as background together with local geological reports and memoirs and there are no real short cuts.

4.7.3 Fracture networks

A particular problem with modelling rock masses is defining the fracture network. As addressed earlier in this chapter, consideration of geological origin and an appreciation of the history of development of fractures can be important for creating a realistic model. In reality most fracture models, be they for assessing rock strength or permeability, are generated statistically based on orientation data. Persistence is extremely difficult to judge and most such models start off essentially as "geological guesswork" that can be adjusted and modified as field test data are collected, say in the petroleum industry or from large-scale pump tests associated with water supply or nuclear waste investigations. Particular techniques are used for discrete fracture network (DFN) modelling such as Fracman.

4.7.4 Examples of models

A simple model for a cut slope alongside a road is shown in Figure 4.61. The mass of rock and soil has been split into five units, largely on the basis of strength factors as discussed in later chapters but discrete and possibly important elements such as major adverse discontinuities along which a landslide could occur are identified for special consideration. A simple model of ground conditions for the design of foundations of a building is illustrated in Figure 4.62. The various units will each give some support to the building and the ground-structure interactions need to be assessed if the foundations are to be designed cost-effectively. Models should not be overly complex but must account for all the important features at the site including apparently "minor geological details" that have "major geotechnical importance" such as individual weak discontinuities along which slippage could occur (Terzhagi, 1929; Baecher

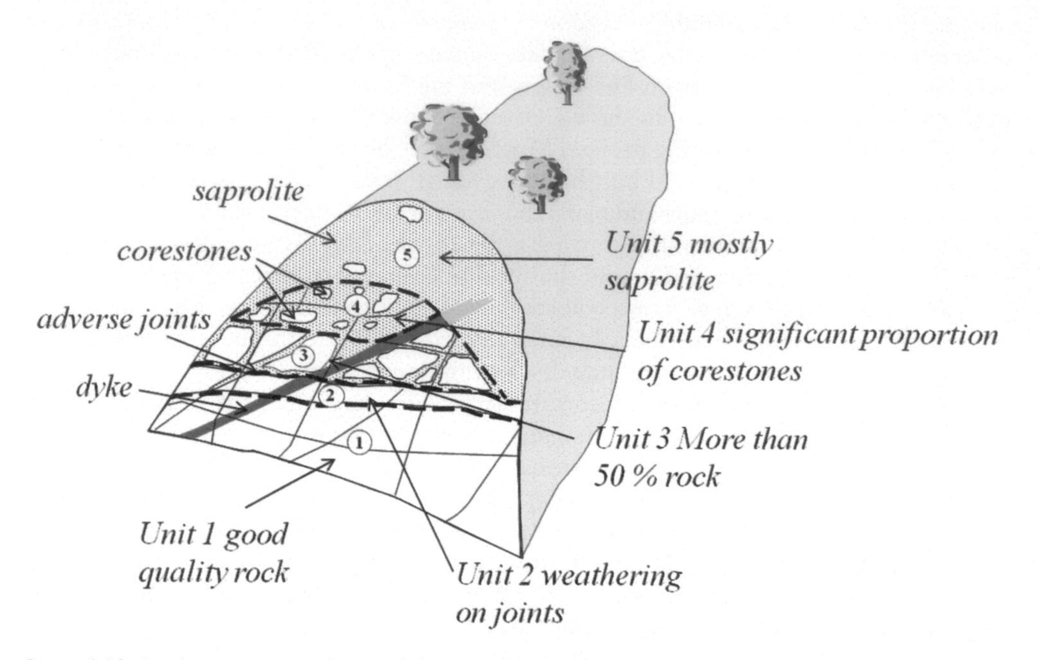

Figure 4.61 A schematic ground model for a weathered rock slope with units defined by degree of weathering, percentage of corestones and degree of fracturing. Particular features such as dykes, faults and adverse master joints need to be included as individual entities.

& Christian, 2003). Many such features can be searched for specifically during investigation provided there is a proper appreciation of the geological and anthropogenic history of the site as discussed in Chapters 5.

The model with assigned geotechnical parameters becomes a design model, which is then used to predict the interaction between the structure and the ground, for example, from a building load, to ensure that failure will not occur (ultimate limit state) and that deformation will be within the tolerance of the structure (serviceability limit state). It must allow failure and deformation mechanisms to be correctly identified – a "mechanism model" – whereby performance is predicted in response to changes brought about by the engineering works or in the long-term from environmental impacts including rainfall, rises in groundwater pressures and vibration shocks. There may be a long sequence of events contributing to the outcome and the geotechnical team need to predict this sequence or to recognise the sequence when carrying out forensic studies of failures. The need for a dynamic approach rather than just static is illustrated by the concept of bore pile design. A pile can be designed to carry load through skin friction as well as through end bearing but the designer needs to take account of how the load is taken up sequentially. In reality, depending on the geological conditions and geometry of piles, the skin friction in the upper part of a long pile will come into play in carrying the carrying load from the structure long before any load will reach the toe of the pile; appreciation of this process will help the designer produce a cost-effective solution.

The design model might be used as the basis for predictive numerical modelling but quite often a conceptual model can be used in its own right to identify the hazards and the best way to proceed at a site. When failures occur, it is often the conceptual engineering

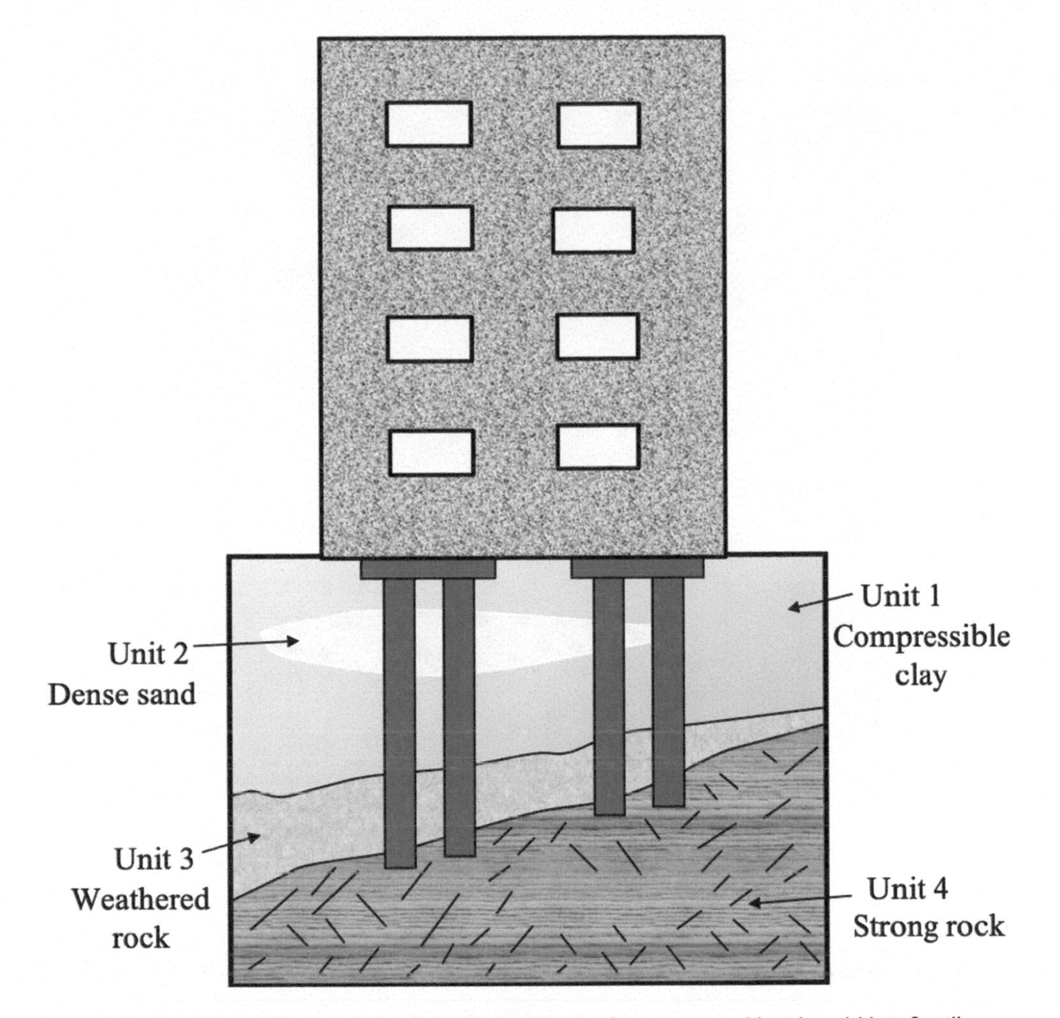

Figure 4.62 Simple ground model to allow design of piles for structure. Unit 1 and Unit 2 will cause drag down on the installed piles adding to the load from the structure. Positive capacity will be provided by skin friction within the weathered rock (Unit 3) and both skin friction and end bearing from the piles founded in the strong rock (Unit 4).

geological model that helps to explain what has happened. A case in point is that of "old terrain" and "new terrain" in natural slopes in Hong Kong, a model prepared by Dr Andy Hansen (1984) and illustrated in Figure 4.63. There are stress concentrations along the interface, where landslides tend to develop. The model was used to explain landslide distribution at changes of slope profile by Devonald et al. (2009).

The parameters are very much secondary – it is rare to be able to be very certain regarding parameters such as strength and water pressures and often a range of possible solutions will fit the facts equally well (Lerouiel & Tavernas, 1981).

An example of an engineering geological model used to explain a major landslide is given in Chapter 8 (the Po Selim Landslide, Malaysia) and discussed in Malone et al. (2008). Experience tells us when a model makes sense in terms of likely strengths and

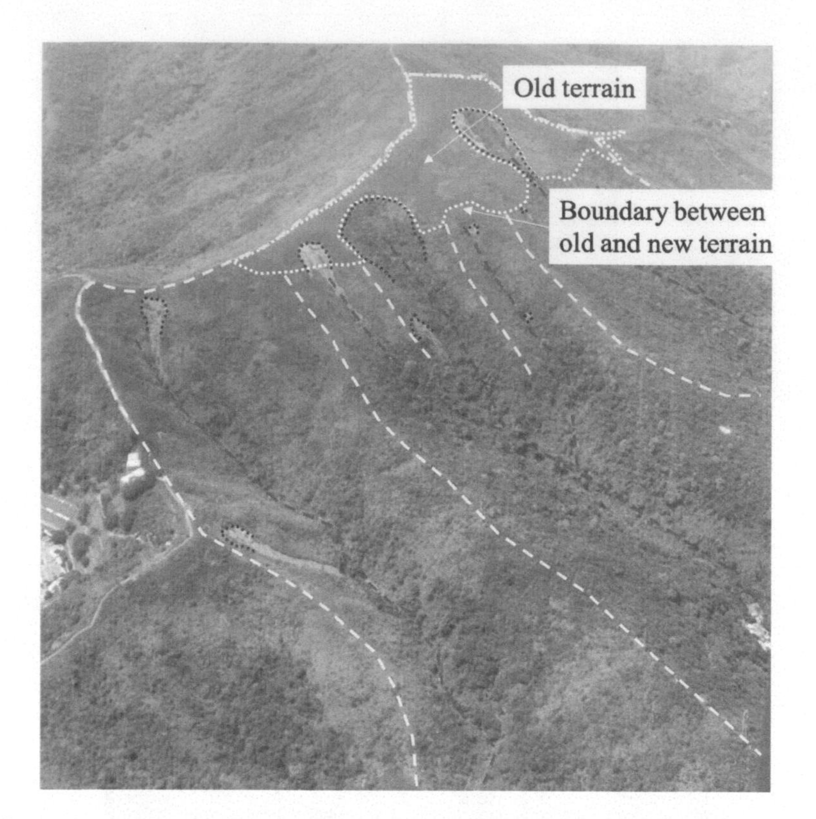

Figure 4.63 Image showing old and new terrain, with landslides delineating the boundary, Hong Kong. (after Devonald et al., 2009).

mechanisms. A complex 3D numerical modelling exercise would often only serve to confirm what we can already tell by judgement.

An example of a simple ground model prepared for the design of a real tunnel project on the basis of very limited ground investigation data is given in Figure 4.64. When designing a tunnel, one needs to predict the ground conditions along the route so that one can decide what tunnelling method needs to be adopted as discussed in detail in Chapter 3. Another requirement of the model is to allow the support requirements to be predicted along the tunnel. For example, in a long drill and blast/hand excavated tunnel, some sections of the ground, in good rock, may need little support, others will need local reinforcement to prevent rock blocks falling. In other sections through weak ground or where there are high water pressures the tunnel might need a thick reinforced concrete liner. A ground model needs to be prepared that includes predictions or rock quality along the route. These predictions are needed so that the team constructing the tunnel know what to expect and where to take special precautions. The predictions are also necessary so that the Contractor can prepare a realistic tender price for the construction and as a reference so that all parties can judge whether conditions were more difficult than might have been anticipated so that additional payment can be made to the Contractor if appropriate as discussed in Chapter 2. The model also needs to allow the influence of the works on other structures to be assessed. In Figure 4.64,

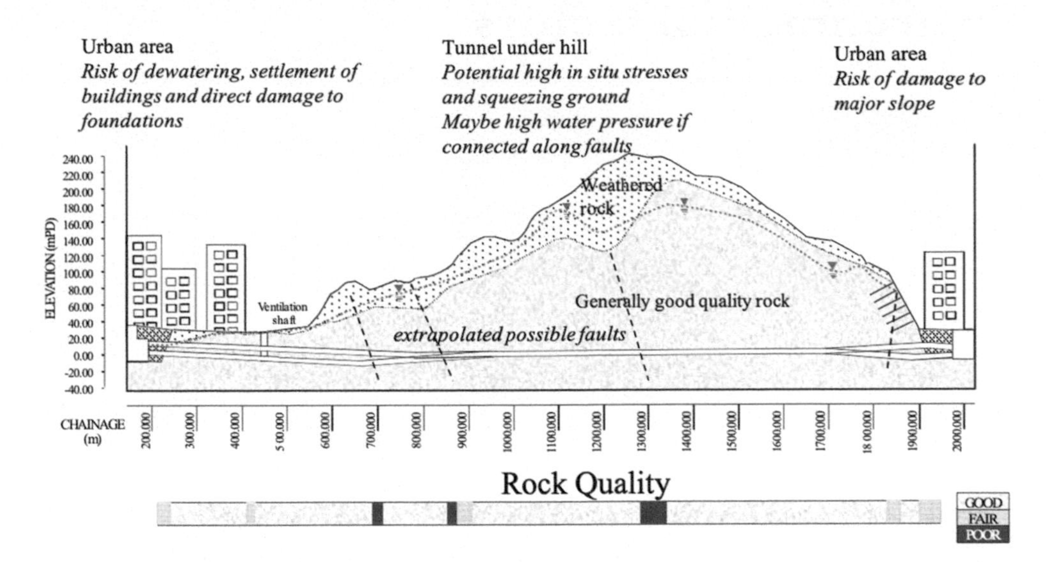

Figure 4.64 Ground model for tunnel. Predicted rock mass characteristics can be used to estimate the amounts of reinforcement such as rock bolts and shotcrete that will be required. It will also be used during construction as part of the risk control, probing ahead as necessary to establish zones of hazardous ground. As illustrated in Chapter 8, tunnelling remains a risky endeavour because very rarely are ground investigations adequate for characterising the ground along their length.

each end of the tunnel terminates in urban areas. The problems of noise, vibrations, dewatering and physical interaction with existing foundations and slopes need to be considered and these are all aspects where a comprehensive ground model is essential. Where the tunnel passes underneath the hill there may be unusually high lateral stresses causing squeezing on the tunnel. The possible faults identified along the route may be associated with particularly poor ground and possibly high inflow of groundwater.

Notes

1 Rocks with even less silica and higher content of Fe and Mg are termed Ultrabasic. Identify using standard geological terminology. Examples are peridotite, pyroxenite and norite and these can have distinctive engineering characteristics (e.g. Dobie, 1987).
2 Rocks types are often grouped together for engineering purposes e.g. all coarse-rock with "free" quartz is called "Granite". This can be an over-simplification as distinctions can be significant. Engineering geologists should use full geological classifications where possible.
3 Micro-granite is sometimes termed fine-grained granite (see Geoguide 3: Geotechnical Control Office, 1988).
4 Diabase in USA practice.
5 Where porphyritic then often called quartz porphyry or feldspar porphyry etc.
6 Clastic means derived from fragments of other rocks. The term detrital is sometimes used essentially synonymously. Some clay is neither of these but newly formed, sometimes from solution.
7 Classification and description of soil including mixed soils is dealt with in Chapter 6 and Appendix B.

Chapter 5

Environmental factors

Don't ever take a fence down until you know the reason it was put up.

G K Chesterton

5.1 THE SECOND EQUATION

Environmental factors can be crucial to project success and make up the second part of the second of three verbal equations that deal with the ultimate project performance:

Geological Factors + Environmental Factors = Engineering Geological Condition

The **Geological Factors** introduced in Chapter 4 include the mineralogy, grain size, type and distribution of rock and soils, and porosity together with the mass features, including joints, fractures and faults, just to name a few. The Geological or Engineering Geological or Geotechnical model is the result of this work. The **Environmental Factors** include all the secondary "influencers" that control the properties of the geological materials such as water as rainfall and as flooding as well as active forces due to earthquakes, and gas pressure as listed in Table 6.4. We also have the time factor, in that a structure will probably need to work successfully for 100 years and in the extreme, for up to 100,000 years in the case of a nuclear waste dump. Just to put that into perspective, 2,000 years ago man was living in stone huts, 20,000 years ago the UK was covered in ice and the sea level was 120 metres lower, so 100,000 years, though easy to say, is almost impossibly long a period to contemplate! That being so, climate change and variation of factors such as sea level over such a period of time, need to be considered.

5.2 HYDROGEOLOGY

5.2.1 Introduction

Water and other fluids in soil and rock masses are important for many reasons:

1. Buoyancy and effective stress.
2. Suction: the tensile force exerted on the meniscus of fluid acting between grains of partially saturated soil.

DOI: 10.1201/9781003348894-5

3. Fundamental geological considerations. Fluids in the rock mass play a major role in hydrothermal alteration, secondary cementation of sediments and mineralisation, hydraulic fracturing, fault propagation and in the development of weathering profiles.
4. Civil and mining engineering projects such as prediction of water flow into tunnels and the design of drainage systems to improve slope stability.
5. Flooding.
6. Water supply from pumped wells.
7. Oil and gas reservoir exploitation.
8. Liquefaction (only occurs in saturated silts/sands).
9. Nuclear waste repository studies.

The principle that flow through the rock mass is driven by a difference in fluid potential (hydraulic head) between two locations is common to all of these applications and to all fluids (oil, gas, water). The following discussion relates primarily to groundwater but the same principles generally apply to oil and gas reservoirs.

5.2.2 Fundamental concepts and definitions

5.2.2.1 Buoyancy and effective stress

"Effective stress" in a saturated soil is the weight of the dry soil minus the buoyancy effect of water as outlined in Chapter 2. This is the calculated stress, averaged through the soil, that is "effective" in causing displacements, relevant to shear strength (as in the process of landslip) and consolidation. Skempton (1960) reviewed the various theories for soil, rock and concrete and deduced that Terzaghi's (1923) original concept, based on experimental data, held true for saturated soil.

Effective stress is calculated from two parameters, total stress (weight per unit area of sample, σ) and pore water pressure (in any direction, u) according to equation (5.1):

$$\sigma' = \sigma - u \tag{5.1}$$

5.2.2.2 Suction

Suction is found in partially saturated soils and weathered rock, where the remnant water clings to grains, with a meniscus on the outside (Figure 5.1).

The tensile force in the meniscus pulls together the soil and rock grains and gives the mass additional strength (apparent cohesion). This is the reason that we can make sand castles at the beach, in partially saturated sand, whereas it is impossible using dry sand. The additional strength is lost once the soil becomes saturated. In the early 80's, in Hong Kong, it was routine to add in an additional 10 kPa cohesion in limit equilibrium calculations to allow for suction in slope stability. Suction was later investigated by Professor Malcolm Anderson of Bristol University, who installed numerous "tensiometers" in the Mid-Levels district and other slopes, which recorded the tensional force, and then monitored it as it disappeared during intense rainstorms.

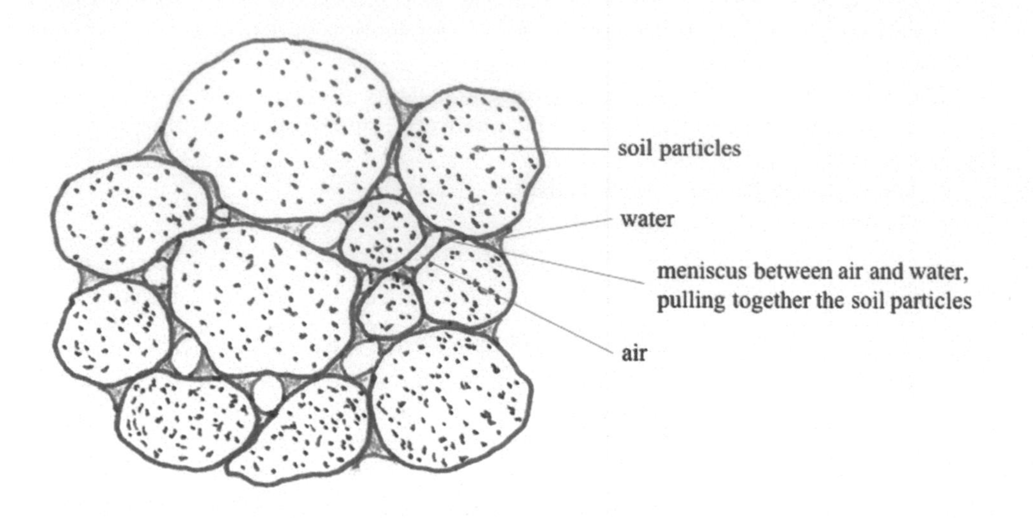

Figure 5.1 Schematic representation of partially saturated soil.

As Stanley Rodin, a geotechnical advisor to the Hong Kong Government said, "what more do we need to know about suction?"

Nevertheless, there are many papers and conferences on partially saturated soils, so there is academic interest and practical application. I recall a landslide at Chung Hum Kok (shown in Box 2.6), where there was a very small physical catchment, which ruled out positive water pressure developing during a heavy rainstorm, and where it was reasoned that the most likely reason why the slope had failed was because of loss of suction. Similarly, the many piles driven at Drax Power Station (see Section 8.7.1) met extremely variable resistance in a thick silty, sandy horizon, resulting in "false sets". In areas underlain by coarser, more permeable sand, the sand dilated as the pile was driven downwards; the soil became undersaturated temporarily, with increased resistance to driving and the pile stopped at a low penetration. Elsewhere, where the horizon contained more silt and clay and was less permeable, the piles advanced rapidly without coming to the required set, until the pile had been driven to the pile carpet. "Oh dear!" said the piling staff… and then, what do you know, when the piling carpet had been dug out, around the pile head, ready to allow the pile to be driven again, the piles were found to be solid and could not be advanced, however many times they were hit with the hammer. I remember Jonathan Gammon, then with WS Atkins, who offered the classic advice: "Effective Stress Rules OK!", and he was right.

5.2.2.3 Porosity

Total porosity is the percentage of a rock mass that comprises voids, which includes discrete voids and fractures – either filled with fluid or gas. The different types of porosity are illustrated in Figure 5.2. In some soils, such as alluvial gravel and sand, the

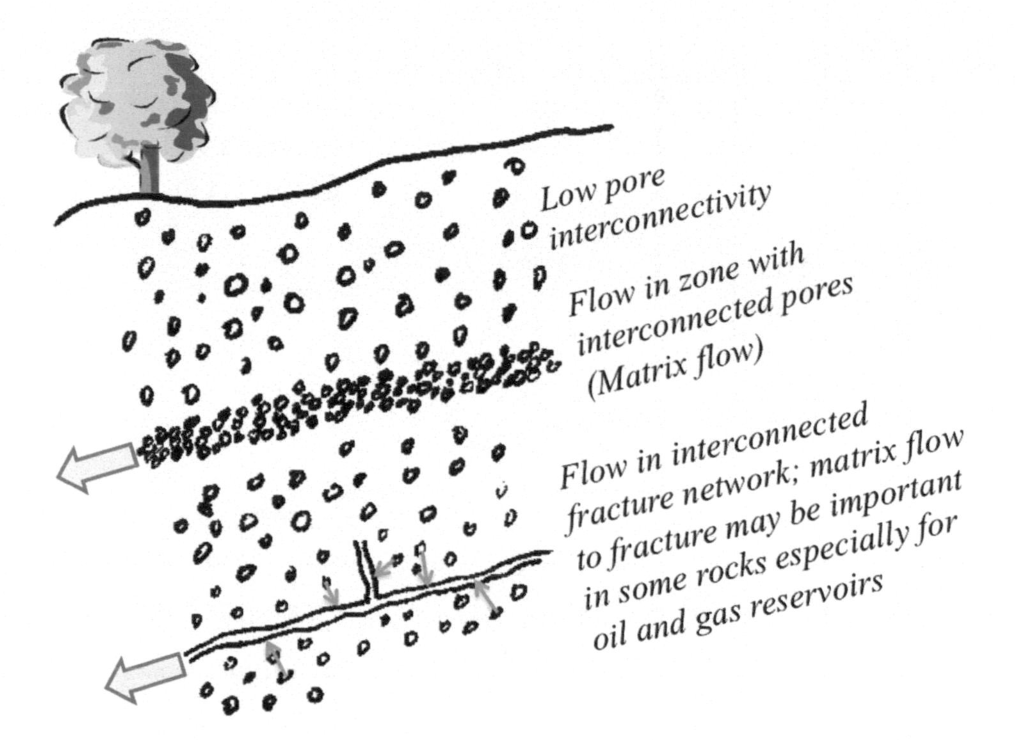

Figure 5.2 Porosity and flow in rock masses.

porosity is dominated by the matrix porosity but in many rocks, the fracture network dominates the interconnected porosity. In some masses (porous sandstone especially) both matrix and fracture porosities are important.

In many rocks only a relatively small proportion of the porosity will be interconnected so that throughflow can occur. An even smaller proportion would drain under gravity. It is a matter of economic importance that, as a consequence of poor connectivity, the percentage of oil and natural gas recovered from reservoirs, without enhanced recovery techniques, is often only 10 to 30%. Enhanced recovery, such as injection of steam and other gases, can improve recovery up to perhaps 50 to 60% but the rest of the oil or gas is left in the ground.

There will be a wide range in distribution of pore sizes in the rock matrix depending on the origin and geological history of the rock. Figure 5.3 shows the change in total porosity with degree of weathering in a series of samples of granite. Range of pore size and total porosity was measured using a mercury porosimeter.

Whereas porosity increases with weathering, the porosity of sediments decreases with burial as illustrated in Figure 1.4. This is partly due to self-weight compaction and partly due to infilling of the pores with cementing agents and finer sediment.

In terms of fracture permeability, the basic model commonly used is of two parallel planes separated by an aperture but that is very simplistic as discussed later. An example of localised channel flow in rock masses is presented in Figure 5.4.

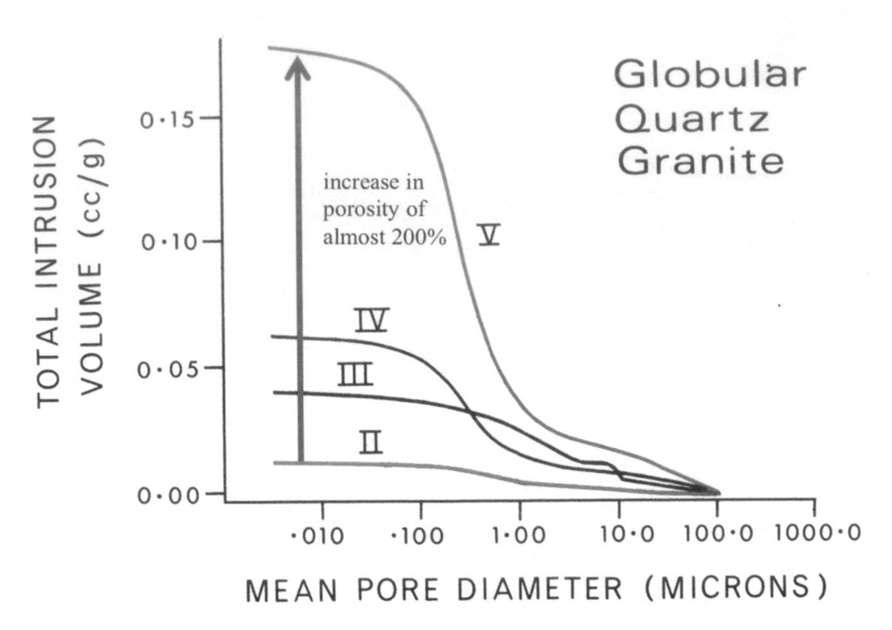

Figure 5.3 Increase in porosity with degree of weathering from Grades II to V granite measured using mercury porosimeter.

Source: After Ebuk et al, 1993.

Figure 5.4 Local water flow from fractures, Drygg Quarry, Horton in Ribblesdale, Yorkshire, UK.

5.2.2.4 *Hydraulic conductivity and permeability*

Fluid movement through the rock mass is driven by fluid potential differential, termed hydraulic head difference. For most situations the head difference is essentially the difference between the levels to which water rises in boreholes or piezometers, although in detail the mechanics and analysis are more complex. Hydraulic conductivity is a measure of the rate at which a fluid moves through a soil or rock mass and is expressed by Darcy's Law:

$$Q = KA\frac{\Delta h}{l}\,\mathrm{m^3s^{-1}} \tag{5.2}$$

where Q is volume of flow per second (m^3/s), A is the cross-sectional area through which the flow passes, Δh is hydraulic head difference between two locations and l is the distance between them along the direction of flow. K is the constant of proportionality otherwise known as hydraulic conductivity (units, e.g. m/s or m/day). The rate of flow clearly will change with the viscosity of the fluid (oil vs. water vs. gas) so there is a need to define a parameter called intrinsic permeability, k, so that:

$$k = \frac{Kv}{g} \tag{5.3}$$

where v is fluid kinematic viscosity (m^2/s) and g is acceleration due to gravity (ms^{-2}). Intrinsic permeability has the dimensions of area (m^2). The oil industry uses a unit of intrinsic permeability called the Darcy, which is approximately equal to 10^{-12} m^2 although values are more commonly quoted in milli-Darcies. In practice, in hydrogeology and geotechnical engineering, the term permeability is used, colloquially (incorrectly), and interchangeably with hydraulic conductivity (with the parameter of ms^{-1} rather than m^2).

5.2.3 Measuring hydraulic conductivity

5.2.3.1 *Difficulties*

Because of the predominance of fracture permeability on large-scale hydraulic conductivity in most rock masses this is one parameter that can certainly not be measured at the lab scale and, even in the field, tests can be extremely difficult to conduct and interpret. There have been major hydrogeological investigation campaigns in association with nuclear waste studies to establish hydraulic characteristics with an aim of predicting migration of radionuclides but results are often disappointing. Fractures identified in boreholes or excavations that are expected to conduct water do not; intersections that are expected to be important to flow paths turn out not to be. This no doubt largely reflects the very low permeability of the rock masses being investigated for nuclear waste studies. In such masses, the flow will inevitably be confined to narrow channels along fracture planes and at their intersections which are very difficult to sample.

Water tests in boreholes can lead to a false impression of the hydrogeological conditions if preferential flow paths are not sampled within the test zones. An example is from the high-quality ground investigations at Sellafield, UK. One of the strata, important to the hydrogeological modelling, was the Brockram, cropping out and found at depth over a large part of the study area. From early tests this stratum was judged to have low hydraulic conductivity and was modelled essentially as an impervious capping layer. Later investigations unexpectedly encountered highly conductive master joints in the Brockram as shown in Chapter 8, and this necessitated a complete change in the way this rock had to be modelled numerically (Hencher, 1996a & 1996b). Baecher & Christian (2003) commenting on the selection of hydraulic conductivity parameters for analysis advise caution over the use of probability distribution function (pdf) representations of the data. The Brockram example illustrates this in that the mass hydraulic conductivity, controlled by very-widely spaced master joints, and measured in later tests, was much higher than the range in the pdf that had been used for numerical modelling up to that time.

5.2.3.2 Water tests in boreholes

In fractured rock masses with higher conductivity, or where the matrix permeability contributes more strongly, testing is easier to conduct. In individual boreholes falling or rising head, "slug" tests are conducted usually using inflated packers to constrain a test interval. Water is added or removed from a borehole and then the time measured incrementally for the hydraulic head to return to its initial ("standing") level. An alternative is to add water continuously to a borehole, maintaining constant hydraulic head, and measuring the water outflow over the test zone. This is known as a constant head test. Methods of interpreting data are provided in standard references such as BS 5930:1999 but these generally assume isotropic matrix permeability. Where dealing with fracture permeability the interpretation is more difficult – flow from or to the borehole may be essentially one-dimensional, along a channel, 2D on a plane or 3D through pervasively and closely fractured rock or masses where the matrix dominates as in porous sandstone or some weathered rocks (Black, 2010).

5.2.3.3 Lugeon testing

A common test used in rock engineering empirically to judge the need for grouting and to establish the efficiency of grouting as a quality-control measure, say in dam foundations, is the Lugeon or packer test. Water is pumped into the ground under constant pressure generally between two inflated packers, then the pressure increased through two more stages. Further testing is carried out at two further stages, reducing the pressure sequentially. The Lugeon value is calculated as volume of water (litres/metre of test interval/minute) multiplied by 1/test pressure (MN/m^2). Assuming isotropic and homogeneous conditions one Lugeon is generally taken as equivalent to hydraulic conductivity of 1.3×10^{-7} m/s (Fell et al, 2005).

Details of the test and its interpretation are given in Houlsby (1976) and Lancaster-Jones (1975). Different ground behaviours can be inferred by the measured hydraulic conductivity K at each stage. For example, if K increases for the higher-pressure stages and then falls again as the pressure is reduced, it can be interpreted that at higher stress, the fractures are jacked open thereby increasing permeability. If the calculated K values increase throughout the five stages then it might be interpreted that infill is being washed out.

5.2.3.4 *Pumping tests*

At a larger scale, pumping tests are carried out, preferably using observation boreholes to measure water drawdown at various distances from the pumping well. Interpretation can be based on the steady state extraction rate and shape of the "drawdown curve" away from the well but there are many options and potential difficulties for interpretation especially where the geology is not uniform and ground sloping (Kruseman & de Ridder, 2000). Where the rock mass is of low conductivity, the drawdown curve will be very steep, and very little water can be extracted from the well; in higher conductivity rocks, the curve will be shallower with water flowing towards the well from greater distance.

For illustration, Figure 5.5 shows an idealised situation with a well fully penetrating an unconfined aquifer. At steady state (once drawdown has stabilised), the pumping volume per second is measured (Q m^3/s) and the water table will have been drawn down to a "cone of depression" around the pumping well. The hydraulic head, far from the well where there is zero drawdown is H metres and the distance from the

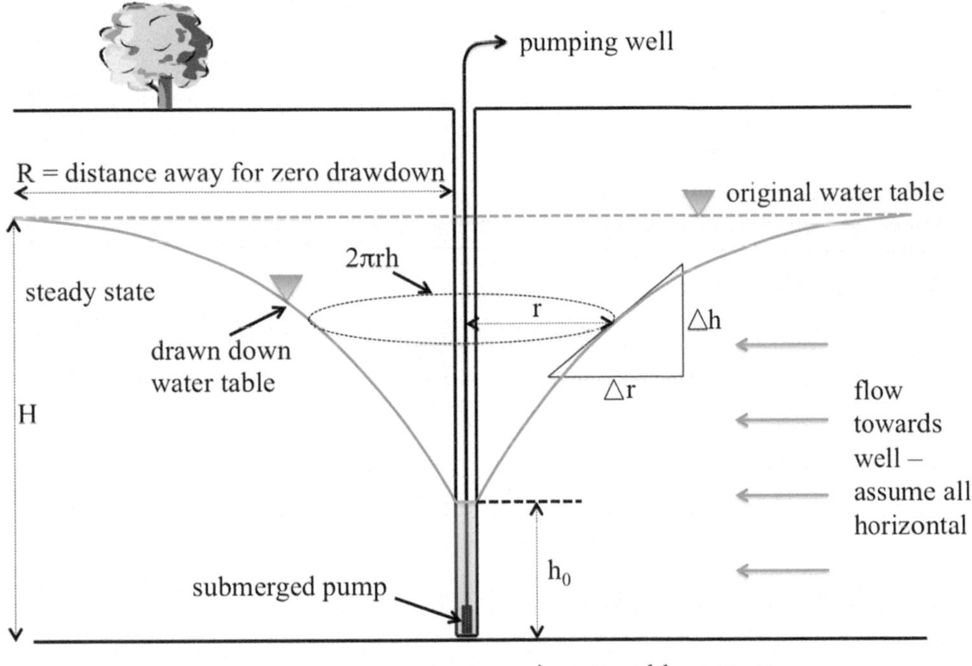

Figure 5.5 Pumping test to measure hydraulic conductivity.

pumping well at which there is zero drawdown is R metres. Then, assuming steady state has been reached in the subsurface (i.e. the cone has ceased to expand) the volume flowing through any cylindrical shape at r metres away from the well must be the same so that

$$Q = Ki\,(2\pi rh) \tag{5.4}$$

where K is the hydraulic conductivity in m/s.

The hydraulic head at various distances from the pumped well is defined as h, so that the hydraulic gradient, i, is the tangent to the drawdown curve:

$$i = \Delta h \,/\, \Delta r \tag{5.5}$$

The radius of the well is a in metres.

Then,

at the well, $r = a$ and $h = h_0$;
at distance, $r = R$ and $h = H$

Integrating between the limits of $r = a$, and $r = R$ yields

$$Q = \pi K\,(H^2 - h_0^2) \,/\, \ln R/a \tag{5.6}$$

which allows K to be found from the pumped well drawdown if an assumption is made about the radius of the cone R (commonly it is assumed that $R = 200a$). There are difficulties with this interpretation, not least that all flow is taken to be horizontal whereas close to the well there will be considerable vertical movement. The interpretation is improved where there are at least two observation wells between the well and the far field. More sophisticated approaches to pumping test interpretation are based on the temporal development of the cone – guidance on both the steady state and transient interpretations is given in Kruseman & de Ridder (2000), Todd (1980) and BS5930:1999 (British Standards Institution, 2020).

Further discussion of the options for measuring hydraulic conductivity specific to rock masses is given by Beale and Read (2014).

In the oil industry pressure tests are carried out in individual wells and these interpreted with respect to the fracture and matrix contributions to flow.

5.2.3.5 Typical parameters

Many rocks have very low permeability as materials (matrix) even where they have high porosity such as in chalk, young mudrocks and volcanic pumice. The main exception to this is sandstone, which may have interconnected high porosity, which allows flow. As noted earlier, flow in rock is often controlled by fractures as is indicated by some of the high value hydraulic conductivity data presented in Table 5.1. These data are indicative only. Other sources give even wider ranges (e.g. Beale & Read, 2014).

Table 5.1 Hydraulic conductivity of rock masses

Sedimentary rock	*Hydraulic conductivity (m/s)*
Sandstone	3×10^{-10} to 6×10^{-6}
Siltstone	1×10^{-11} to 1.4×10^{-8}
Shale	1×10^{-13} to 2×10^{-9}
Limestone/dolomite (non-karstic)	1×10^{-9} to 6×10^{-6}
Karstic limestone	1×10^{-6} to 2×10^{-2}
Salt	1×10^{-12} to 1×10^{-10}
Crystalline rock	
Permeable basalt	4×10^{-7} to 2×10^{-2}
Unfractured basalt	2×10^{-11} to 4.2×10^{-7}
Fractured massive igneous and metamorphic rock	8×10^{-9} to 3×10^{-6}
Unfractured massive igneous and metamorphic rock	3×10^{-14} to 2×10^{-10}
Weathered granite	3.3×10^{-6} to 5.2×10^{-5}; 5×10^{-9} to 4×10^{-3} (according to Anon, 1995)

Source: The main source is Domenico & Schwartz, 1990;Anon, 1995 gives the wider range for weathered granite.

5.2.4 Hydrogeological modelling

5.2.4.1 Modelling geology as isotropic

Geological profiles are generally depicted for groundwater modelling as made up of discrete, homogeneous and often isotropic units of given hydraulic conductivity. Refinements in models might be to incorporate local aquitards and barriers – such as fault seals or more variable geological conditions with local litho-facies represented in some detail (e.g. Fogg et al, 1998). Generally, however, geological profiles are modelled as simple confined or unconfined aquifers (Todd, 1980).

5.2.4.2 Anisotropic flow models

In rock, flow is predominantly through joints and other fractures and the situation can be anticipated to be more complex than for soil. Nevertheless, for practical purposes there is a need to try to compute an equivalent hydraulic conductivity, perhaps by idealising the discontinuity network as regular series of parallel fractures (Hoek & Bray, 1974). Available sophisticated software can model the fracture network fairly realistically. For example, FracMan (Golder Associates) is a discrete fracture simulator (Dershowitz et al, 1996) that allows the geometry of discrete features, including faults, fractures and stratigraphic contacts to be modelled deterministically and/or to meet predefined statistical criteria. The models can then be used to model flow through a unit of the rock mass using the associated software Mafic (Golder Associates). Verification tests (predictive) to predict flow through rock masses have met with mixed success (Olsson & Gale, 1995). Attempts to represent rock masses as uniform units for a site at Sellafield, which is probably the best-investigated, analysed and modelled site in the UK, are discussed in detail by Heathcote et al (1996).

5.2.4.3 Unconfined conditions

An unconfined aquifer is where there is direct connection between the aquifer and the Earth's surface so that water can infiltrate directly from rainfall or from a lake or river (Figure 5.6). The water level in the rock mass can rise and fall without restriction, in response to inflow or drainage. The top of the saturated zone is known as the water table where the pressure is atmospheric. Water pressure below the water table usually increases linearly with depth.

5.2.4.4 Confined conditions

Aquifers are described as confined where there is a low permeable horizon capping the aquifer that prevents the fluid rising to the level matching its pressure within the aquifer. An example is an aquifer of sandstone, overlain by mudstone and where the catchment area (where rainfall infiltrates) is at some distance and at a higher level. If a borehole penetrates the capping layer, fluid will rise in the borehole to match the hydraulic head in the aquifer. Examples of confined aquifers include the London and Paris basins where the chalk is the aquifer, confined by overlying clay beds. Artesian water has also been drawn from deeper boreholes. One such borehole at Slough,

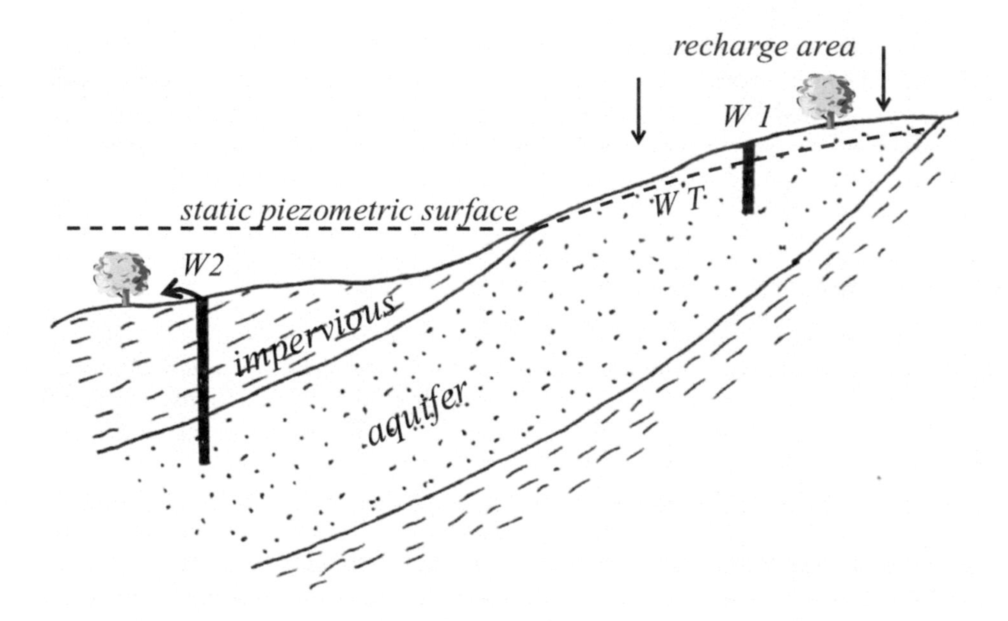

Figure 5.6 Unconfined and confined areas in the same aquifer. Well W1 will encounter water at the water table (WT) and water level will rise and fall with rainfall infiltration. Well W2 is in a confined area and will issue water under artesian conditions reflecting the pressure in the confined aquifer.

Buckinghamshire, UK, yielded almost 500,000 litres per hour from a depth of about 300 metres in the Lower Greensand (Blyth, 1967).

If fluid is taken from a confined reservoir or aquifer, the rock mass will remain saturated, but the fluid pressure will drop. Reduction in pressure leads to an increase in effective stress (weight less water pressure) and therefore compaction due to self-weight as well as a small amount of fluid expansion. The fluid that is released by pumping in a unit volume of the formation, for a unit reduction in pressure, is known as the specific storage and is a function of the density of fluid, porosity and the relative compressibility of the rock mass and the contained fluid. In the case of dual porosity formations both the rock material matrix pore space and the fractures in the rock mass will contribute to the elastic storage; the parameter is used in slope engineering to estimate the quantity of water to be drained to achieve a reduction in water pressure and the rate that this can be achieved (Beale & Read, 2014). Specific yield, also known as the drainable porosity, is a ratio, less than the effective porosity and is an indication of the fraction of the aquifer volume that will drain simply through gravity.

5.2.4.5 Compartmentalisation, aquicludes and aquitards

The heterogeneity of rock masses is reflected in their hydrogeological characteristics, especially close to the Earth's surface in the case of groundwater. Some zones will be readily recharged because of their infiltration characteristics with exposed rock, delayed runoff or simply the presence of surface water such as rivers, lakes or the sea meaning that water can infiltrate and travel under gravity down to the water table. The way that water infiltrates and the rate at which it can do so influences the mode and timing of landsliding in a strong manner (e.g. Iverson, 2000; Hencher, 2010). Some types of shallow and deep landslide mechanisms are illustrated schematically in Figure 5.7.

Low permeability strata such as massive mudstone, igneous dykes and structures such as faults or sheeting joints infilled with clay due to weathering or as sediment infill, will act as barriers to water flow and lead to compartmentalisation of the rock mass. The same is true of hydrocarbon migration at depth within the Earth's crust so that oil and gas accumulates below confining horizons and such traps are the prime targets for geological and geophysical investigation in the oil and gas industries.

In geotechnical engineering the term aquiclude is used to describe an impermeable barrier; an aquitard is a low permeability stratum or horizon that restricts flow. Examples of compartmentalisation and flow concentration by geological variation at a zonal scale are illustrated in Figure 5.8

5.2.5 Flow paths

5.2.5.1 Preferential flow paths through soil

Richards & Reddy (2007) provide a comprehensive review of piping, particularly as related to earth dam construction, which is where piping was first recognised as an

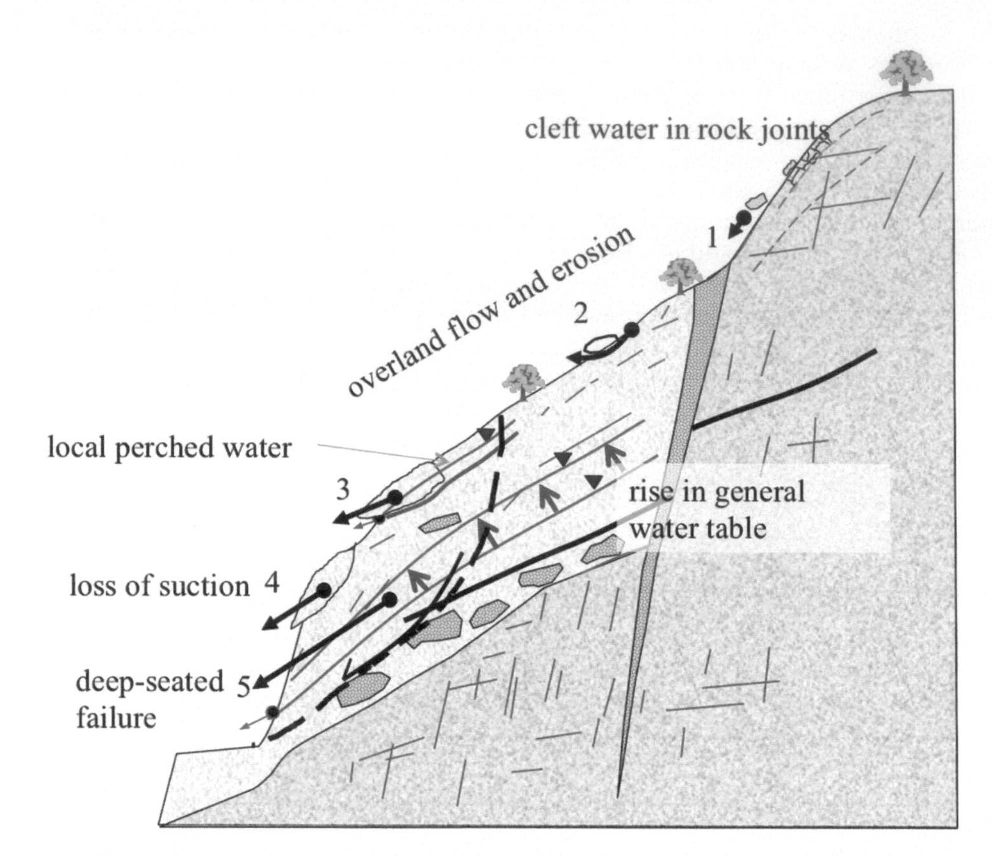

Figure 5.7 Modes of failure in a rocky hill slope, triggered by rainfall and throughflow. Most failures (types 1 to 4) will be relatively small and occur due to cleft water pressures, erosion, minor perching and loss of suction in weathered profiles. Such failures tend to occur during or very shortly after rainstorms. Deeper-seated failures due to rises in deeper groundwater often occur sometime after a storm.

important hazard. They define several types of piping, which are also relevant to natural ground, namely, as follows:

1. Suffosion (or eluviation): the washing out or dissolution of material en-masse leaving a loose framework of granular material, prone to collapse.
2. Dispersion of clay soils by rainwater in the vadose zone.
3. Backwards erosion from a spring. The pipe forms and then material is gradually lost from that opening.
4. Erosion along some pre-existing opening such as a master joint.

The majority of pipes investigated by geomorphologists are confined to the upper few metres in the ground profile (e.g. Jones, 1971; Uchida et al, 2001). They are particularly common in forested areas within shallow soil profiles and are associated with

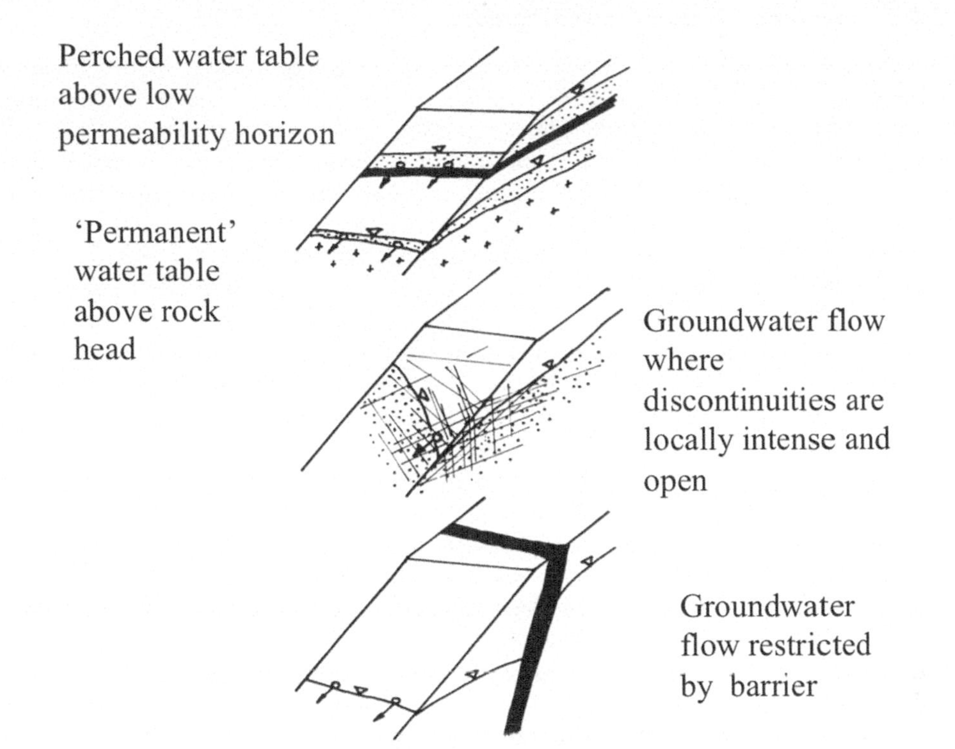

Figure 5.8 Flow paths and restrictions in heterogeneous rock masses.
Source: After Hencher, 1987.

shallow landslides (Pierson, 1983). Anderson et al (2008) used dyes (as have others) to trace pipe networks, exposed by excavation. Not surprisingly the size and connectivity of these shallow features was related to surface catchments. The same is probably not true of deeper features.

Channelised fracture networks from the parent rock often persist through the various stages of weathering. This type of preferential flow needs to be considered in investigation, hydrogeological modelling and design. Such natural pipes probably follow original structural paths (especially master joint or fault intersections), but may also be formed by seepage pressure in weak saprolite or in superficial soils such as colluvium. They also develop at permeability contrasts (e.g. colluvium overlying saprolite). More details are given in Hencher (2010).

Pipes are commonly seen associated with many deep-seated failures in weathered terrain (Hencher, 2006). It is implied that the development of pipes at depth may be linked to early stages of progressive failure as the rock mass dilates and ground water exploits the dilating and deteriorating rock mass. Such deep pipes are probably distinct in origin from pipes found in upper soil horizons. Fletcher (2004) reports that infilled pipes are sometimes encountered up to depths of 80 m below present sea level and these must be associated with ancient times when the sea level was much lower.

5.2.5.2 Flow paths in rock (unweathered)

Fracture flow in rock is poorly understood and extremely difficult to investigate or characterise (Black, 2010). As noted earlier, at the simplest level rock discontinuities are modelled as two parallel plates separated by an aperture (Snow, 1968). The important finding of this analysis is that the flow through a fracture is proportional to its aperture cubed, i.e. flow through rock is highly sensitive to the openness of the fractures.

In reality flow paths are tortuous, localised and extremely difficult to identify. What matters with flow is not only the openness of the fracture network but connectivity along the full transmission path (and head along that path). The author has experience in tunnels at about 150 metres depth below the sea where some zones of highly fractured rock were dry whereas other sections in seemingly intact rock were seeping water. Clearly local observations of fracture state can be poor indicators of total hydraulic conductivity through the rock mass.

Thomas & La Pointe (1995) attempted to discriminate between dry and flowing fractures in the drift at Kiamichi Mine in Japan on the basis of descriptive parameters such as roughness, aperture, orientation as observed in the drift (at the exposed traces of discontinuities in the mine). They used a neural network approach to train the analysis but with limited success. Indeed, it seems clear that connectivity is the most important factor that controls flow through rock rather than locally measurable characteristics such as fracture intensity, aperture or spacing. The fact that the most transmissive and well-connected features might not be adequately sampled during investigation is a major problem for all geotechnical projects (Figure 5.9), not least for potential nuclear waste repositories where it is recognised that large scale conductive features need to be identified and dealt with in a deterministic manner (Black et al, 2007; Nirex, 2007).

Representative Elemental Volume (REV) is the concept of a volume of rock that can be taken to have repeatable hydraulic properties. REVs have been established for numerical models (e.g. Gnirk, 1993) but the idea that the hydrogeology of a rock mass in reality can be represented as an REV is still problematic (e.g. Davy et al, 2006).

If such an REV were to exist, then the rock mass could be modelled as an equivalent continuum. If the REV did not exist, or there were a series of transmissive fractures, then the mass would need to be dealt with as a discontinuous framework and there are software packages that can do that as discussed later. The problem is always a lack of data. Clearly it will never be possible to establish the geometry of the fracture network in any real sense (Black et al, 2007) but one way of approaching the problem is to investigate a particular rock horizon of interest, by researching the degree of rock bridges present in particular weathering zones, by using a method called Forensic Excavation of Rock Masses (FERM) (Shang et al, 2018).

Numerical models as illustrated in Figure 5.10 are put together based on available information and some geological interpretation. In practice, modelled flows seldom match those measured in test and production wells in the oil industry very accurately so models need to be adjusted iteratively, adding or removing fractures and adjusting their properties, to try better to match the model with observed behaviour.

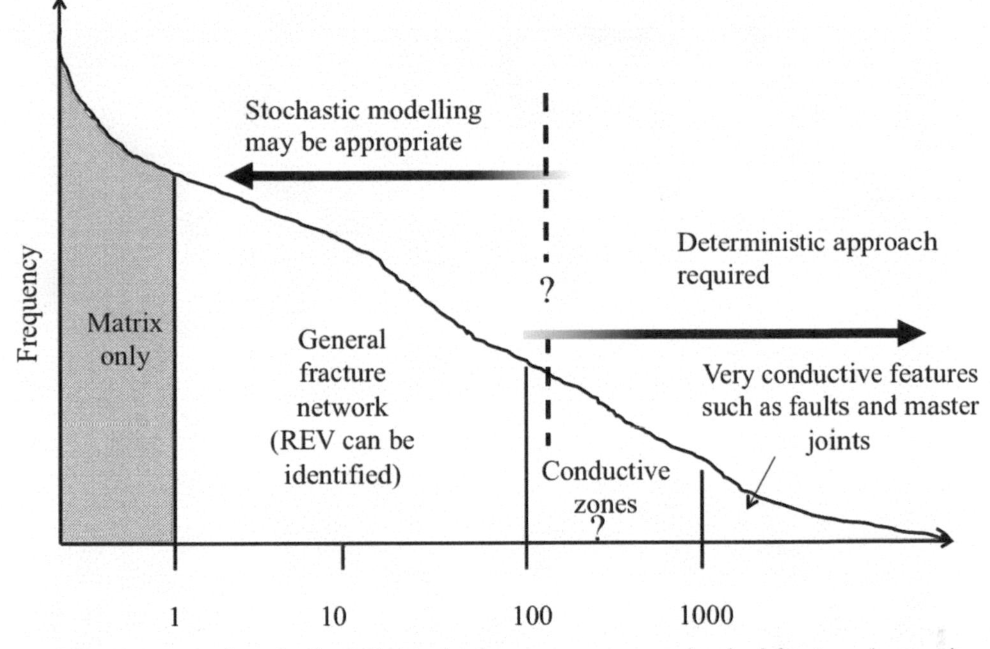

Figure 5.9 Approach for measurement, analysis and numerical modelling of the hydrogeology of rock masses.

Source: After Nirex, 2007.

The importance of single transmissive fractures or sets of fractures, linking over great distances, must be emphasised. Dershowitz & LaPointe (1994) report how new oil wells caused large drops in production to existing wells at distances of several km within 2 days whilst other wells between them were unaffected. Such behaviour could not be predicted without extremely good knowledge of the fracture network and understanding of potential connectivity. There are many similar examples in tunnelling where tunnelling at one location has affected structures at great distance, unexpectedly. Muir-Wood (2000) reports the case of water flow from an adit causing unacceptable settlement to the Zeuzier arch dam in Switzerland, 1.5 km away. This and other cases are discussed by Strozzi et al (2011). Beitnes (2005) discusses major environmental difficulties relating to inflow from above and through the invert of the Romerksporten rail tunnel in Norway. Subsequent grouting of a 2.2 km section of the tunnel cost as much as the construction of the whole 13.8 km tunnel. The spatial intensity of occurrence of the most transmissive features may not be adequately sampled during investigation and this is a major problem for all geotechnical investigations, not least for potential nuclear waste repositories where it is recognised that large-scale conductive features need to be identified and dealt with in a deterministic manner. This is a major, continuing research area in nuclear waste studies.

Figure 5.10 Fracman model of groundmass.

5.2.5.3 Preferential flow paths in weathered rock

In weathered rock profiles, water flow is sometimes concentrated in fractured rock, underlying weathered saprolite or colluvium and this can result in transmission of water (and high water pressures) from one part of a hillside to another or even between catchments where the water then feeds into the overlying weathered mantle and may trigger landslides. During the Mid-Levels Study in Hong Kong it was recognised that confined conditions could occur where material of lower conductivity overlies rock of higher conductivity. Strong upward hydraulic gradients from bedrock to the decomposed rock aquifer were identified in some areas (Geotechnical Control Office, 1984b) and these observations were used in setting up a numerical model of the hydrogeology (Leach & Herbert, 1982). Jiao & Malone (2000) and Jiao et al (2005 & 2006) have extended this concept of a highly transmissive zone at depth to explain several deep-seated landslides and evidence of artesian pressure in Hong Kong. Montgomery et al (2002) report an intensely instrumented site (more than 100 shallow piezometers) in Oregon, USA and similarly noted artesian flow from the underlying bedrock, which they found surprising for such a steep hillside. They also commented that, whilst the seepage from bedrock might effectively determine the specific locations where debris flows might initiate, the distribution and connectivity of the near-surface bedrock fracture system are almost impossible to predict. Similar upward flows from bedrock into the overlying soil mantle are reported for steep granitic terrain by Katsura et al, 2008.

Whilst flow through fresh rock is difficult to investigate and characterise, weathered rock masses are even more complex and consequently their hydrogeological characteristics. Note the very wide range of values in hydraulic conductivity for weathered granite in Table 5.1. In practice and empirically, the mass might be represented by a simple set of parameters as in an REV approach for less weathered rock, but those parameters probably do not actually represent the physical and mechanical processes taking place in anything other than very simple situations.

Channelised fracture networks from the parent rock often persist though the various stages of weathering. This type of preferential flow needs to be considered in investigation, hydrogeological modelling and design. Natural pipe systems develop and probably follow original structural paths (especially master joint or fault intersections), but may also be formed by seepage pressure in weak saprolite or in superficial soils such as colluvium. They also develop at permeability contrasts (e.g. colluvium overlying saprolite).

Pipe systems carry water during and following storms but may become clogged or collapse seasonally (Figure 5.11) so that, as closed-end pipes, they can lead to excess water pressures during the following wet season. Pipe systems can provide a large store of water that can maintain water pressure in a slope far longer than from normally infiltrated water. More details of piping and their link to landsliding processes are given in Hencher (2010).

5.2.5.4 Establishing hydrogeological conditions in weathered rock profiles

To allow prediction of water pressures there is no substitution for realistic hydrogeological models preferably backed up by data from careful instrumentation but this

Figure 5.11 Well-graded alluvial sand in a Mazier sample tube, sandwiched top and bottom by weathered granite at a depth of about 20 metres in Ching Cheung Road cut slope, Hong Kong.

can prove difficult to achieve in practice partly because most ground investigation contracts are conducted in a single campaign, which allows little time for assessment of the hydrogeological situation before the installation of instruments.

Furthermore, in any ground investigation it is very important to establish a preliminary hydrogeological model before installing piezometers. In an investigation of a series of major landslides in Hong Kong, several of which occurred in engineered slopes with ground investigations and instrumentation it was found that in all cases where piezometric data were available these did not reflect peak water pressure at the failure surface as inferred from post-failure observations or from back analysis (Hencher et al, 1985). One problem was the common inability of the installed instrumentation systems to record rapid transient rises and falls in water levels. A further problem is that many piezometers were installed at geological locations where they cannot detect the critical perched water tables, which develop prior to failure.

Obviously identifying hydrogeological situations dominated by natural pipe systems and discrete channels can be particularly difficult and rather "hit or miss". Geomorphological features, such as lines of boulders, vegetation and dampness, may be indicative of shallow sub-surface flow and these can be studied by remote sensing techniques including infrared photography. Trenches can be used to find shallow pipes but deeper systems are much more difficult to identify and are rarely observed in the author's experience. Subsurface investigation of large tracts of land to identify sources of infiltration and throughflow paths is not generally feasible although in a highly populated area subject to high risk, the cost might be justified. Open pipe systems will generally just be seen as poor core recovery and perhaps misinterpreted as core loss. If pipes are suspected, these can be investigated by down-hole investigation techniques, such as TV cameras, borehole periscopes or geophysical tools. Tracer tests using dyes or saline solutions can be used to trace sources of water and get some measure of velocities of throughflow.

Resistivity surveys can be helpful in identifying underground streams and zones of enhanced conductivity as illustrated by the case study at Yee King Road (Hencher et al, 2008). Other geophysical tools such as ground radar proved of little use during that investigation.

5.2.6 Characterisation of hydrogeological conditions for engineering projects

5.2.6.1 Slopes

For rock slope stability, assumptions are generally made as in soil slopes that there is an overall groundwater table with water pressures distributed linearly with depth as illustrated in Figure 5.12. The level of the water table might be measured in boreholes using standpipes or might be inferred from observed dampness in a slope face and the presence of vegetation.

Where fractures are open, allowing rapid infiltration, "cleft" water pressure can develop quite quickly in some situations and to take that pressure fully into account can lead to onerous but rather conservative design assumptions. For a more refined analysis, a better model is needed of the pressures that develop in the fracture network on a site-specific basis. One of the best examples known to the author is the investigation and instrumentation that was carried out by Golder Associates in a densely populated

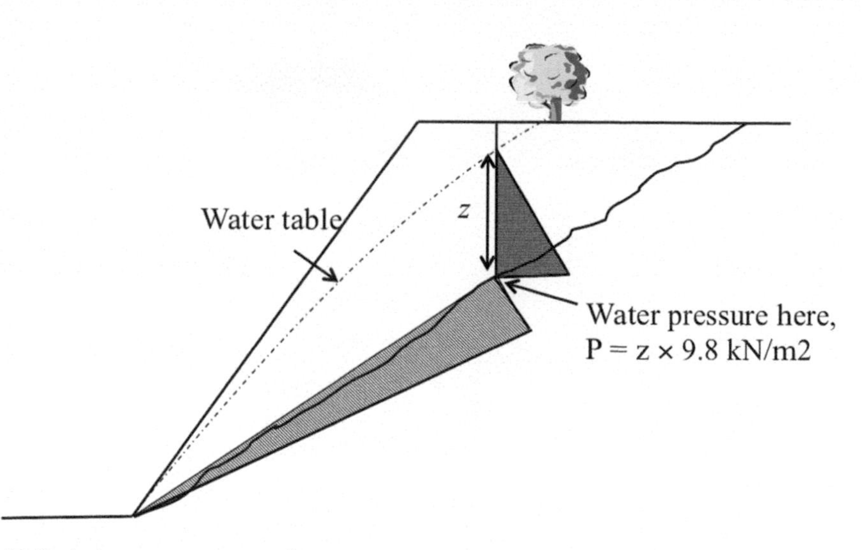

Figure 5.12 Typical assumption regarding water pressure in a rock slope (after Hoek & Bray, 1974). Water exerts a destabilising force in a vertical tension crack; there are often vertical joints in rock slopes. Water pressure also exerts an uplifting force (buoyancy effect) on the potential slip surface.

area of Hong Kong Island (the North Point Study). The area is underlain by granite with many adversely oriented sheeting joints. Some of these are severely weathered. The risk is high but so would be the cost of preventive measures, such as anchors and drains or of re-housing as might be the outcome if a broad, conservative approach was taken to the design parameters, and this justified a more sophisticated examination of the problem.

The design engineers designed a monitoring system with a number of pneumatic piezometers installed specifically on the various sheeting joints identified in boreholes. These instruments allowed water pressure to be monitored against time during storms as illustrated in Figure 5.13. Measurements showed rises in water head above the steady state water table, after major storms, of up to 10 metres. The most important observation however was that different piezometers at different locations responded at different times. They also responded differently from storm to storm. The data indicated pressure pulses travelling down the joints following a storm. These data allowed it to be demonstrated that the conditions for stability were far less onerous than if the peak level in each borehole (adjusted up for even worse storms in the future) were combined in a single piezometric surface for design as often would be done.

5.2.6.2 *Tunnels*

Groundwater flow into a tunnel depends upon a number of factors including the following:

1. Construction below water table with head differential towards opening.
2. Sufficient water in storage within the rock mass to keep flow running for a significant time; in the case of sub-sea or sub-lake tunnels there may be unlimited recharge.

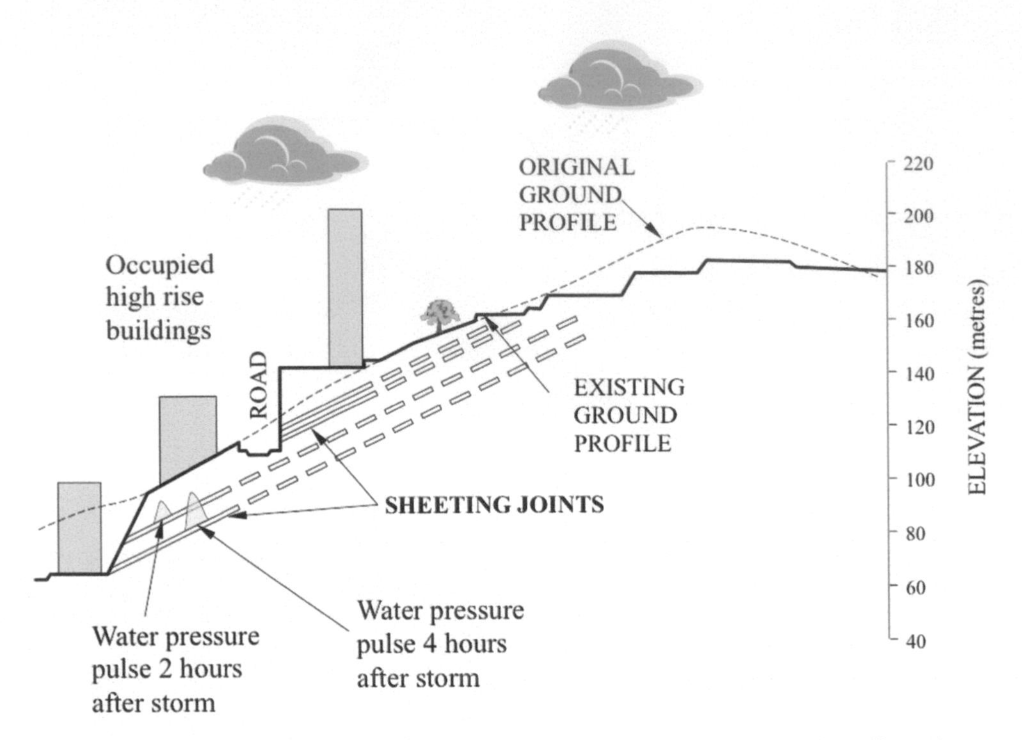

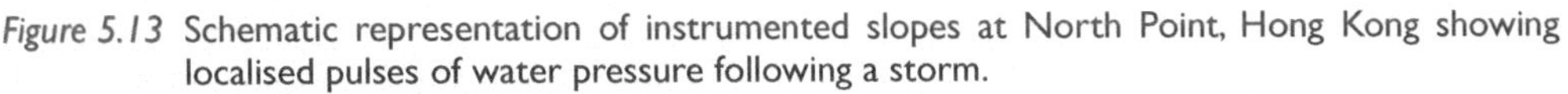

Figure 5.13 Schematic representation of instrumented slopes at North Point, Hong Kong showing localised pulses of water pressure following a storm.

Source: After Richards & Cowland, 1986.

3. A connecting network of fractures, daylighting in the face, each of sufficient hydraulic conductivity to contribute to flow.

It follows that it is not simply sufficient to intersect what appears to be an open fracture – what matters is the connectivity away from the face as illustrated in Figure 5.14. At location A the tunnel intersects a major discontinuity but the connectivity is poor so the flow would soon dry up. At location B there is a smaller fracture intersection but more extensive connectivity. If that series of fractures connected to a major source of water, such as a lake or even a syncline with high conductivity as at the Ping Lin Tunnel in Taiwan, then the flow could be high and continuous.

Odling (1997) describes fracture network models in sandstone, in terms of their connectivity and describes fractures that do not lie on direct pathways through the rock mass as "dead ends"; ignoring the dead ends, we are left with the "backbone", a term she borrowed from percolation theory (Stauffer, 1985). In Figure 5.14 the fractures leading to point B comprise a "backbone".

Most problems with water inflow to tunnels arise where highly conductive and laterally extensive features, such as brittle shear zones and faults infilled with granular materials, are encountered. Where there may be highly permeable zones with high or

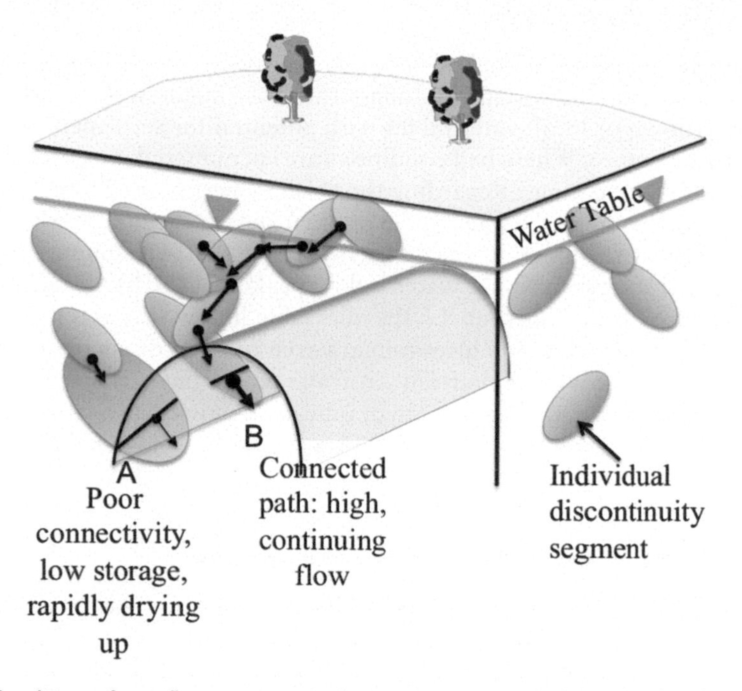

Figure 5.14 Conditions for inflow to tunnels. Representation of discontinuities as ovoid disks or rectangles is often used in numerical models and used here schematically.

unlimited recharge in advance of the tunnel then pre-drilling in advance of the tunnel, with blowout preventers, is highly advisable.

Excessive inflow of groundwater to tunnels and mines can cause delays, collapse and danger to tunnel workers during construction. These major inflows are usually associated with three different geological features:

I. Fractured and crushed shear zones.
II. Solution channels and caves.
III. Super-conducting rock joints.

Anthropogenic works such as old mine workings also can be a major hazard.

Breaching shear zones often leads to a sudden inrush of water. The Seikan railway tunnel in Japan is 53.8 km long and runs between the two most northern islands (Matsuo, 1986). Much of the tunnel is undersea (23.3 km). Despite being a minimum of 100 metres below the sea floor, several inundations occurred. The most serious in May 1976 was due to the tunnel intersecting a zone of fractured tuff that had hydraulic connectivity through to the sea floor. A tunnel boring machine had to be abandoned and the tunnel advanced by drill and blast with extensive grouting necessary in advance, to restrict inflows which led to very slow tunnelling rates and major delays.

5.2.6.3 Setting limits for inflow

At a less catastrophic level, tunnel owners and their designers will generally set a limit for what is acceptable inflow in the long-term. This might well be with a view to avoiding drawdown of local water tables with potential for settlement damage to the overlying infrastructure. Where bad conditions are encountered it may prove very difficult to meet the specifications. Regarding the SSDS project under Hong Kong harbour, described in more detail in Chapter 8, the original specification limit for water ingress for one of the tunnels, tunnel C, was approximately 1,000 litres/minute for the whole 5.3 km tunnel. During tunnelling, water inflow peaked at 10,400 litres/min just for a single discrete fault zone. For tunnel F the allowable inflow was about 200 litres/min/km compared to actual of 1,400 litres/min/km even for the completed tunnel.

Pells (2004) quotes an example from Australia where the permitted inflow was 33 litres/min/km, which equates to a mass hydraulic conductivity of 7×10^{-9} m/s. He notes that this is very optimistic (unachievable) for a tunnel in sandstone and shale, below the water table, even with pervasive grouting.

5.2.6.4 Predicting inflow into an underground opening

Predictions of inflow to tunnels are generally made using analytical solutions such as that of Goodman et al (1965). For a situation with infinite recharge (say below a lake or the sea), the only parameters are the radius of tunnel and depth below water table together with hydraulic conductivity. Such solutions can give reasonable predictions for "general country rock" but of course the key variable is hydraulic conductivity, which is difficult to measure or predict especially for a tunnel at depth where there will be little geotechnical data.

Kong (2011) reviews the various methods for predicting water inflow with reference to measurements taken from directionally drilled boreholes as part of the investigation for the planned drainage tunnels for stabilising the Po Shan area of Hong Kong – an upgrade from the previous scheme of long, inclined drains. Kong uses an equation proposed by Raymer (2001), which is based on that of Goodman et al., but with a correction of one of the parameters and an empirical adjustment from experience of actual flows. The revised equation is as follows:

$$q_s = (2\pi K\,(z + h_1)/\ln(2z/r)) \times 1/8 \qquad (5.7)$$

where

q_s = steady state inflow per unit length of tunnel (m^3/s)
K = equivalent hydraulic conductivity of rock mass (m/s)
z = thickness of ground cover above tunnel (m)
h_1 = depth of standing water above ground surface, if present (m) (e.g. lake or sea)
r = tunnel radius (m).

Kong recommends the use of this analytical approach combined with an empirical method of McFeat-Smith et al (1998) that provides estimation of inflow based on IMS rock class (weathering grade and degree of fracturing) linked to experience of mainly land tunnels in Hong Kong.

5.2.6.5 Experience of inflow

Nilsen (2014) presents a review of inflow case studies to Norwegian sub-sea tunnels. One of his findings is the lack of correlation between large water inflows during construction and pre-tunnel predictions both in terms of hydraulic conductivity and geophysical rock mass quality. In his examples, Lugeon tests were carried out in boreholes at some distance from the tunnels and, he observes that extrapolating from those locations, where there were high Lugeon values, to tunnel level did not correlate with locations of high inflow. There are two main lessons that stem from this observation:

1. Extrapolating observations of open fractures from one location to another, only tens of metres away, is highly dubious without a realistic understanding of geological structure (which might anyway be impossible to achieve), and
2. The Lugeon test only measures water outflow to the rock mass locally – not ingress from a large volume of the rock mass. Locally high measured conductivity is no evidence of extensive connectivity. Conversely, a low local Lugeon value is no guarantee against nearby systems of connective fractures, not sampled over the test length.

Nilsen notes regarding prediction from geophysical tests that, in the Karmsund gas pipe tunnel, inflow occurred at one location with low seismic velocity as might be expected (poor quality ground) but that there were many other features with similar or lower seismic velocity where there was no water ingress. Again, this serves as an illustration of the importance of connectivity.

Holmøy & Nilsen (2014) reviewed the occurrence of inflows to hard rock tunnels and how they related to various "geological parameters" on the basis of six reasonably well-documented case histories. Of the various correlations they attempted, the most interesting findings were the following:

1. Almost all water ingress was from fractures that were sub-parallel to the major principal stress and making an angle with nearby major faults of 45° ±15°. This was explained in that such joints were probably influenced by tectonic stresses of relatively recent geological age (and therefore might be open and not infilled).
2. Water inflow did not decrease with depth of rock cover. This is surprising and contrary to a common assumption that with depth, fracture frequency will be less and aperture tighter and therefore water flow reduced.
3. There is no linear correlation between Q value (rock quality) and water inflow.

The difficulties in understanding or predicting connectivity are well-illustrated from tests for nuclear waste disposal research activities. Even where extensive investigation has been carried out as at the Underground Research Laboratory, Manitoba, prediction of flow is often inaccurate (Martin et al, 1990). One particular fracture through the rock mass was encountered which had characteristics considered ideal for testing methods of instrumentation and analysis. Lang (1988) described the situation as a "unique opportunity". In the event, despite such favourable circumstances, whilst the mechanical response of the fracture was predictable, "predictions of the permeability and hydraulic pressure changes in the fracture, and the water flows into the tunnel, were poor". Similar difficulties and poor results are reported by Rouleau & Raven (1995) for tracer tests in Ontario and by Nirex (1996) at Sellafield UK where high-quality and well-constrained cross hole-testing between two boreholes showed that zones previously identified as flowing showed no observable response in the cross-hole tests. These examples simply demonstrate the complexity of the problem of characterising the rock mass and making predictions about fluid flow and the need to take a cautious, managed risk approach, generally involving due consideration of all potential hazards and ways of dealing with these pre-drilling in advance of the excavation face.

5.2.6.6 Mining

Morton et al (1988) developed a hybrid aquifer-inflow finite element model to predict inflow to a proposed gold mine in South Africa, over a period of 6 years. The model extended over an area of about 100 km^2 and was calibrated using pumping test data. Account was taken of various development stages and due to pumping from existing mine shafts. Comparisons between predicted pumping rates and actual inflows were very good although it was noted that the model would not account for sudden influxes associated with major faults.

Beale & Read (2014) give a detailed and up-to-date review of hydrogeology associated with large open-pit mining including investigation techniques, analysis and mitigation of adverse effects on slope stability.

5.2.6.7 Nuclear waste repositories

The proposed development of repositories and disposal sites for nuclear waste has driven considerable research in rock engineering over the last 40 years or so. Many of the issues for design are essentially the same as for other underground openings but there are particular issues regarding thermal loading, and fluid flow that are especially important to nuclear waste studies. This is because of the extremely long engineering life of such proposed disposal sites – one hundred thousand years compared to perhaps a nominal 100 years for many engineering structures.

Probably the greatest issue is establishing hydrogeological conditions at a site and ensuring through experimentation and analysis that risk targets regarding the escape of polluted fluids to the biosphere will be achieved. Eventually this will need to be demonstrated to a sceptical public. This is no mean task as predictions from water and other fluid experiments to date have been at best, marginally successful (Olsson & Gale, 1995). This no doubt can be put down largely to the geological unknowns in many geological scenarios. The one perhaps most promising observation has been at the Canadian underground research facility at Manitoba where it was found that there are no joints in the massive granite at depth (Martin, 1994). If there are no joints then flow will be extremely slow though a low permeability matrix. It is the lack of joints that also makes salt an attractive potential host rock for repositories.

By comparison the Borrowdale Volcanics near Sellafield, targeted in the UK for nuclear waste disposal, have had a very long geological history with burial, tectonic uplift, joint development and faulting. The prospect of proving a safety case for such rock seems bleak, no matter how intensive the investigations, testing and analysis. Probably the only way forward for such rock masses is through better structural geological characterisation methods (establishing hypotheses and testable rules for fracture network geometries including apertures) although it seems unlikely this will be achieved within many decades without extensive research.

In the meantime, empirical testing and back analysis will allow some advance though it is unlikely to be convincing. Numerical modelling provides a way forward in data-limited situations where there is some tolerance to risk, as in open-pit mining. However, the lack of data, quantifying hydrogeological conditions in a rock masses, means that the results and predictions from numerical predictions will always be questionable. This is considered in more detail in Chapter 8.

5.3 OIL AND GAS

5.3.1 Dual porosity and well testing

Oil and groundwater reservoirs can be considered, conceptually, to range from essentially homogeneous porous media to rock masses in which fractures dominate both storage and flow whilst the intervening intact rock is essentially impermeable. Intermediate to these extreme cases are reservoirs where transport and storage are shared between the matrix blocks and fractures. Such "dual porosity" models are important to oil and gas production. The relative contributions vary with changes in effective stress and distance from the production well during extraction of fluids.

Such fractured reservoirs show somewhat complex behaviour during well tests. The pressure variations are interpreted as reflecting the relative contributions of the fracture network and matrix blocks to the flow system at different times. If such interpretations of the mechanisms controlling flow are correct and can be extrapolated throughout the reservoir then extraction might be optimised.

Warren & Root (1963) provided a solution for the double porosity model, which accounted for pressure changes during production. It is assumed that matrix blocks and fractures are uniformly distributed within their model. It is also assumed that the fractures have a relatively high permeability and a low storage capacity. The fractures carry the fluid to the wellbore. The matrix blocks have a relatively low permeability but a high storage capacity. The role of the matrix is to feed fluid to the fractures. It is also assumed that any reservoir volume contains large numbers of fractures and matrix blocks so that the Representative Elemental Volume (REV) is small.

The solution of the double porosity model by Warren & Root showed that the pressure behaviour within a wellbore is controlled by two parameters, λ and ω.

The flow coefficient, λ, relates to the ease with which fluid can seep from the matrix blocks into the fractures and ω is the ratio of the storativity of the fracture system to that of the rock mass as a whole (fractures plus matrix blocks). These parameters are a function of compressibility, "porosity" and permeability of both matrix and fractures.

These generalised parameters are obviously gross simplifications for real rock masses at the scale of an oil reservoir but provide a starting point for the interpretation of well tests.

5.4 GROUTING

5.4.1 Purpose of grouting

Grouting is used to reduce the permeability of rock masses and therefore throughflows and inflows into tunnels. Examples of its use are as follows:

(a) Beneath dams, to reduce water loss from reservoirs.
(b) In tunnelling, to reduce water flow both during construction and for the permanent works. Grouting might also improve ground conditions through compaction and by providing some cohesion to otherwise open fractures.
(c) To form an isolation barrier.

5.4.2 Options and methods

Usually the need for grouting and results of grouting are assessed beneath dams using the Lugeon test as discussed earlier. Rubber packers are inflated in boreholes to isolate a test section and then water injected using three increments of pressure, ascending and then descending whilst measuring water loss. Once the Lugeon value has been measured then steps can be taken to reduce this by injecting grout into the fracture network from boreholes under pressure. Further Lugeon tests are conducted post-grouting to check the reduction in permeability. In tunnels the need for advance grouting may be judged from inflow measurements from boreholes drilled in advance of the tunnel face (Figure 5.15).

In practice, it often possible to reduce permeability by a factor of one or two orders of magnitude, e.g. from 10^{-7} to 10^{-5} m/s, but it depends upon the nature and connectivity of the fracture network. Grouting is usually done using normal cement, sometimes combined with other material, such as fly ash. If the rock mass is tight but permeability unacceptable, then ultrafine cement might be used, silica gel, other chemical grouts or bitumen.

To inject grout into a tight rock mass requires very high pressure and this needs the rock confining pressure also to be high otherwise grouting might actually open up joints and make matters worse. In dam construction, the grout curtain below the dam is usually completed after the dam is constructed to allow higher pressures to be used. In a tunnel, grouting needs to be done in advance of the tunnel construction. Post-grouting a leaking tunnel is often ineffective (Pells, 2004).

5.5 CLIMATE VARIABILITY

> The start of spring, otherwise known to Al Gore as proof of global warming.
>
> *Bill Clinton*

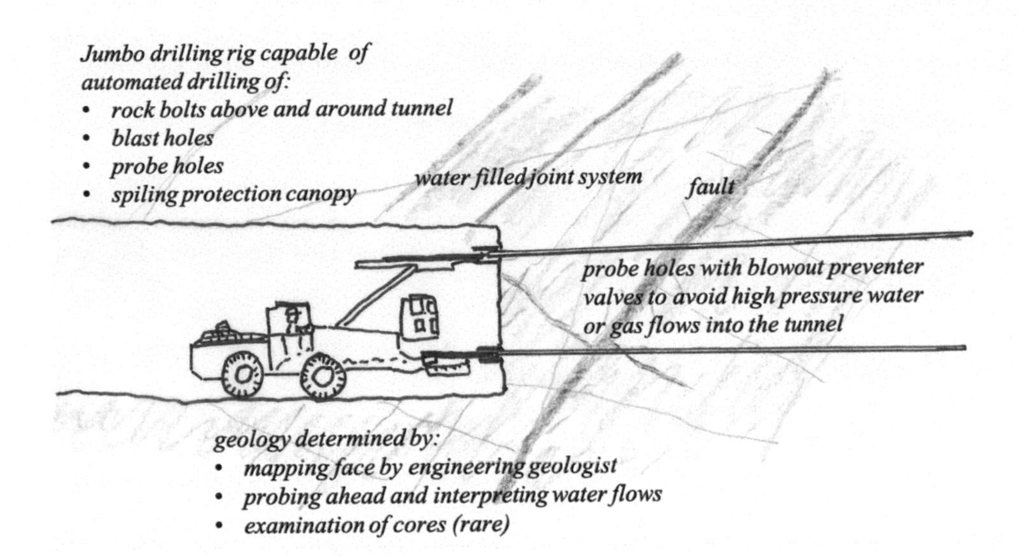
Jumbo drilling rig capable of
automated drilling of:
• rock bolts above and around tunnel
• blast holes
• probe holes
• spiling protection canopy
water filled joint system
fault
probe holes with blowout preventer
valves to avoid high pressure water
or gas flows into the tunnel
geology determined by:
• mapping face by engineering geologist
• probing ahead and interpreting water flows
• examination of cores (rare)

Figure 5.15 Jumbo drilling rig being used to drill ahead to check geological conditions. Rig is equipped with coring drills and special valves to prevent blow-outs if high water pressure is encountered.

5.5.1 Introduction

In a book on engineering geology, it is important to address the changes that might occur with time, to decide on whether the design method should be changed, perhaps in response to anticipated increase in precipitation or rising sea levels. This is even more the case where nuclear waste repositories must remain inaccessible to the population for at least 100,000 years as addressed in more detail in Chapter 8.

There are some extreme views around, such as predicted sea level change according to the Met Office in the UK:

> UK tide gauge records show substantial year-to-year changes in coastal water levels (typically several centimetres).[1]

For London, by 2030, they predict from their numerical models that it will have risen by between 100 mm and 200 mm in 8 years, meaning a sea level rise of between 12.5 mm and 25 mm a^{-1} (UKCP18, Met Office, 2018). For various reasons I believe these figures to be alarmist and irrational as explained later.

> "There are two classes of forecasters: those who don't know and those who don't know they don't know".
>
> *J K Galbraith*

5.5.2 Modelling reliability

Now, firstly, let us ask, hypothetically, which of these might prove more difficult?

(a) Predicting population growth in the UK and its consequences, over a thirty-year time span, or
(b) Predicting the effect of gases (CO_2, CH_4, H_20) escaping into the atmosphere and possibly causing temperatures to rise, which results in sea level change, landslides and hurricane increase, over, again, say thirty years. This is compounded by the need to predict volcanic eruptions.

Both of these rely on numerical models which are combinations of a large number of mathematical equations, using computers, to find an approximate solution to the underlying physical problem.

Well, fortunately we have a conference, which was held to decide on the Optimum Population for Britain, dated December 1969, which can help in deciding between these.

5.5.3 Population predictions for the UK in 1969

> The audience was mainly of professional biologists with their invited visitors and 90% registered their opinion that the optimum population for Britain has already been exceeded.
>
> *Taylor, 1969*

Erlich (1969a) from Stanford University, USA, stated "By any standard you wish to select Britain is grossly overpopulated. People have been reticent to express their opinion on what an optimal population for Britain would be, but I am certain that it would be well under 20 million people" and "based on our calculations at Stanford, upper limits of the world optimum which turns out to be, from almost every point of view, under one billion people".

This number would come as a shock to many people in the UK but, when you think about it, a population of 20 million makes sense with less cars, easier schooling and a less pressurised National Health Service. The current UK population (2023) is estimated to be almost 69 million people; world population is 7.9 billion (newsflash 8 billion). By the way, have you ever heard this debated in detail in Parliament or on the radio? I may have missed it.

Erlich (1969b) continues "If current trends continue, by the year 2000, the United Kingdom will simply be a small group of impoverished islands, inhabited by some 70 million hungry people, of little or no concern to the other 5–7 billion people of a sick world" and

"If I were a gambler, I would take even money that England will not exist in the year 2000, and give 10 to 1 that the life of the average Briton would be of distinctly lower quality than it is today".

Well I do not think that any of his predictions, apart from the one regarding poor quality of life, holds strictly true in 2023, let alone in 2000, so Erlich would have lost his money. So, it can be concluded that his modelling (at Stanford University) must have contained some tricky parameters or dodgy mathematical equations. I will return to considering population growth later.

5.5.4 Climate change

So, what hope for modelling predictions of the effect of CO_2 and other gases for the future climate of the planet – a task far much more difficult than predicting the consequences of population growth? Whether or not the statement about CO_2 concentrations and the effects is true or not, I need here to concentrate on what environmental factors (precipitation, wind etc.) to include for civil engineering design for say the next 100 years.

5.5.5 Climate modelling

Many of the arguments over climate modelling relate to predictions based on various sets of equations. These are brought down to four models with Representative Concentration Pathways "RCP" 8.5, 6.0, 4.5 and 2.5. RCP 8.5 is calculated as producing "radiative forcing" of 8.5 W/m^2 of the planet…

These models, each with complex mathematical equations contained (much more so than used in predicting population and quality of life in the UK in 2000 at 5.5.2 above) have vastly different predictions ranging from 1°C to 6°C by the year 2070 (Hausfather & Peters, 2020).

There are three main sources of errors in climate change modelling (Terando et al, 2020):

1. the natural variability in the climate system,
2. imperfect scientific knowledge about the response of the climate system to changes in gas emissions, and
3. uncertainty about the future of emissions resulting from human actions and policy decisions.

The authors conclude that in the short-term (years to decades), natural climate variability (weather) is the largest source of uncertainty in the climate projections.

5.5.6 Historical data

5.5.6.1 Rainfall

The BBC and press in the UK continually discuss variability of the weather as if it were the same as climate change. Lee (2020) carried out a review of precipitation and temperatures, from instrumental records around the UK. He found that precipitation shows no statistical increase in England and Wales since records began in the 1880's. In Scotland about half the stations show no statistical increase in precipitation.

It is worth bearing in mind the tremendous variety of rainfall that we see around the Earth, that civil engineers and drainage engineers have to deal with.

In Hong Kong, for example, the annual rainfall is about 2.4 metres; in 1997 Hong Kong Observatory recorded an extreme 3.3 metres. The UK annual rainfall is about one third of this ranging from 0.8 to 1.4 metres per year.

Rainfall in Hong Kong is typically associated with tropical cyclones and storms, which can cause huge amounts of rain over a few days. Brand et al (1983) found by studying times that the Fire Services were called out that the triggering rainfall for landslides was when the intensity reached 70 mm per hour. They also found that 24-hour rainfall was a significant data point for landsliding and concluded that if the rainfall was less than 100 mm then it was very unlikely that there would be a major event.[2]

Figure 5.16 shows data from Hong Kong with a relationship that seems to fit quite well (Hencher et al, 2006). It is obviously not accurate because of the variability of slopes along roads and topography, and the remoteness of many of the failures (Hong Kong is mountainous with 84% hilly and unpopulated), and the time of rainfall. Nevertheless, the relationship seems to hold approximately and gels with the impression gained by the author that "all hell had broken loose" on a road in Tsing Yi island where **every** slope (fill, cut slopes in soil and/or rock), showed some damage in August 1982 where the heaviest rainfall had occurred, and led to the idea of there being an "inventory" of slopes in any region, at various points of deterioration, leading to failure as shown diagrammatically in Figure 4.60 (Hencher, 2006).

On 3 October 2020 it was announced that the UK had had its wettest day on record when an average of 31.7 mm fell across the whole of the UK in 24 hours. This would not even make the grade in HK (let alone the need for storm drains etc.[3]).

Elsewhere in the world, the highest 24-hour rainfall measured was at Alishan, Taiwan and was 2,327 mm, which exceeds Hong Kong by some margin.

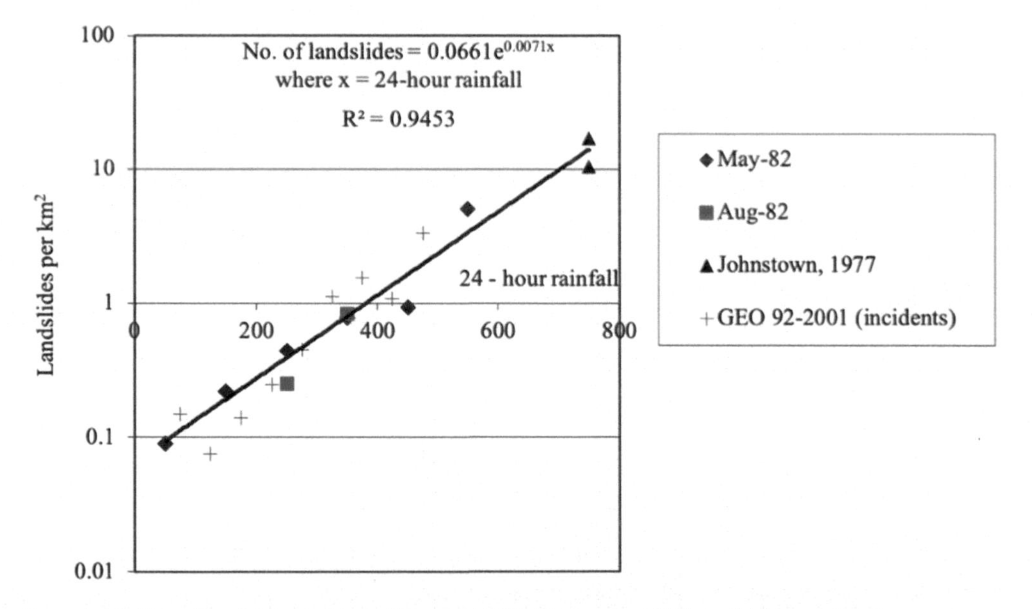

Figure 5.16 No. of landslides in Hong Kong and USA (Pomeroy, 1981) vs. 24 hour rainfall.

I remember when my good friend Professor JJ Hung from Taiwan National University presented a lecture to the Geotechnical Engineering Office (GEO) in 2003, including videos of debris flows. He mentioned that the debris flows were caused by daily rainfall in excess of 1 metre. HN Wong, the head of GEO, commented that such rainfall must be very unusual? JJ Hung said: "Yes, it is very rare to get such rainfall. Perhaps only 3 or 4 times a year".

5.5.6.2 Temperature

Cook (2022) produced a useful table entitled "engineering impacts from climate change in developing countries". He listed these in two columns showing the impact that might be caused – I have added a third column in Table 5.2, showing the factually recorded impact in the UK. It can be seen that in the UK, which has better coverage than most other countries, the effect is minimal. Review of temperature and precipitation showed an average of 1.4°C increase over the last 100 years (Lee, 2020), which seems like a weather warning would be appropriate rather than declaring a climate catastrophe. In detail, Oxford showed a summer average increase of 0.851°C and winter increase of 0.391°C, over 100 years, whereas Eastbourne, 126 miles away, showed a summer increase of 2.917°C and a winter increase of 2.87°C per one hundred years. Closer still, Heathrow airport, 48 miles away from Oxford, showed a "trend" of more than 300% higher than Oxford in maximum temperature and 800% higher in the winter. Various other stations around the Isles showed *no trend* throughout the year (Tiree) or *no winter trend* (North-eastern Scotland and Southern Wales) as shown in Figure 5.17. One might wonder about the interpretation of these statistics when stations so close together geographically show such different data "trends". Presumably the sampling level is insufficient to determine real trends over 100 years or so of measurement. As is well-known, the climate fluctuates, with for example, the medieval period and 1700s being particularly cold, whereas the previous Roman warm period had vineyards flourishing, similar to today.

Geologists, such as those at the Smithsonian Institute in the USA, have reviewed temperatures throughout time through study of indirect clues, such as the chemical signatures of rocks, ocean sediments, fossils, fossilized reefs, tree rings and ice cores. The data for the last million years are presented in Figure 5.18 (with two different horizontal scales). It can be seen that the minimum temperatures have been about 5 to 6 degrees Celsius cooler than current during the various glacial periods and have been 1 to 3 degrees warmer in the interglacial periods (as we are now in). It must be borne in mind that temperature variation is very large around the globe, so Figure 5.18 must be taken with a pinch of salt.

5.5.7 Retreat of glaciers

One of the current concerns regarding climate change is melting of glacial ice. Glacial ice covers more than 700,000 m^2 according to the Randolph Glacial Inventory (RGI, 2017), excluding Greenland and Antarctica. This ice is in decline in the Alps, Himalayas and America, for example, and there is natural concern about water supply and the breaching of natural dams and lakes.

Table 5.2 Climate change impact in first two columns, according to Cook (2022) with actual impact in the UK added as a third column

Potential climate change	*Typical impacts on transport infrastructure*	*Actual changes in the UK and elsewhere*
Increases in very hot days and heat waves.	Deterioration of pavement integrity, such as softening, traffic-related rutting, and migration of liquid asphalt (sustained air temperature over 32°C is identified as a significant threshold). Excessive thermal expansion of bridge expansion joints and concrete paved roads.	Since instrument records began about 150 years ago, the UK temperature has risen on average by about 1.4°C/100 years according to Lee (2020). Ice core records show, however, that warming and cooling events have been approximately +3 to –5°C, over the past 20,000 years – see Figure 5.18.
Increases in very hot days and heat waves and decreased precipitation.	Corrosion of steel reinforcement to concrete structures due to increase in surface salt levels in some locations.	In the UK, the 1976 "heat wave" had a maximum temperature of about 36°C and the temperature was in excess of 32°C for more than 2 weeks (consecutive). The coldest temperature ever recorded in summer in the UK is –5.6 °C recorded on 1 and 3 June 1962 in Norfolk.
Sea level rise and storm surges.	Damage to highways, roads, underground tunnels and bridges due to flooding, inundation in coastal areas and coastal erosion. Damage to infrastructure from land subsidence and landslides. Erosion of bridge supports. Reduced clearance under bridges.	See text regarding sea level change over geological time, which is currently, globally, about 1.4 to 3.5 mm a^{-1} according to NASA. This is not alarming. In the UK, sea level change is estimated at ~1 mm a^{-1}, accounting for geological isostasy (Woodworth, 2018).
Increase in number and intensity of high precipitation events.	Damage to roads, subterranean tunnels, and drainage systems due to flooding. Increase in scouring of roads and earthworks including backfill to bridges and support structures. Damage to road infrastructure due to landslides. Overwhelming of drainage systems. Deterioration of structural integrity of roads, bridges and tunnels due to increase in soil moisture levels. Accelerated chemical weathering.	Annual mean precipitation over England and Wales has not changed since records began in 1766 (Jenkins et al, 2009). Short-term changes occur as at Eskdalemuir, not far north of the Lake District, for example. This station has the highest trend of all stations with 8.77 mm rise a^{-1} in the summer, and in 2011 received the highest rainfall of 2,289 mm over 12 months. Note however the dangers of reading this interesting weather fact and then generalising across the UK (where there is no such trend).

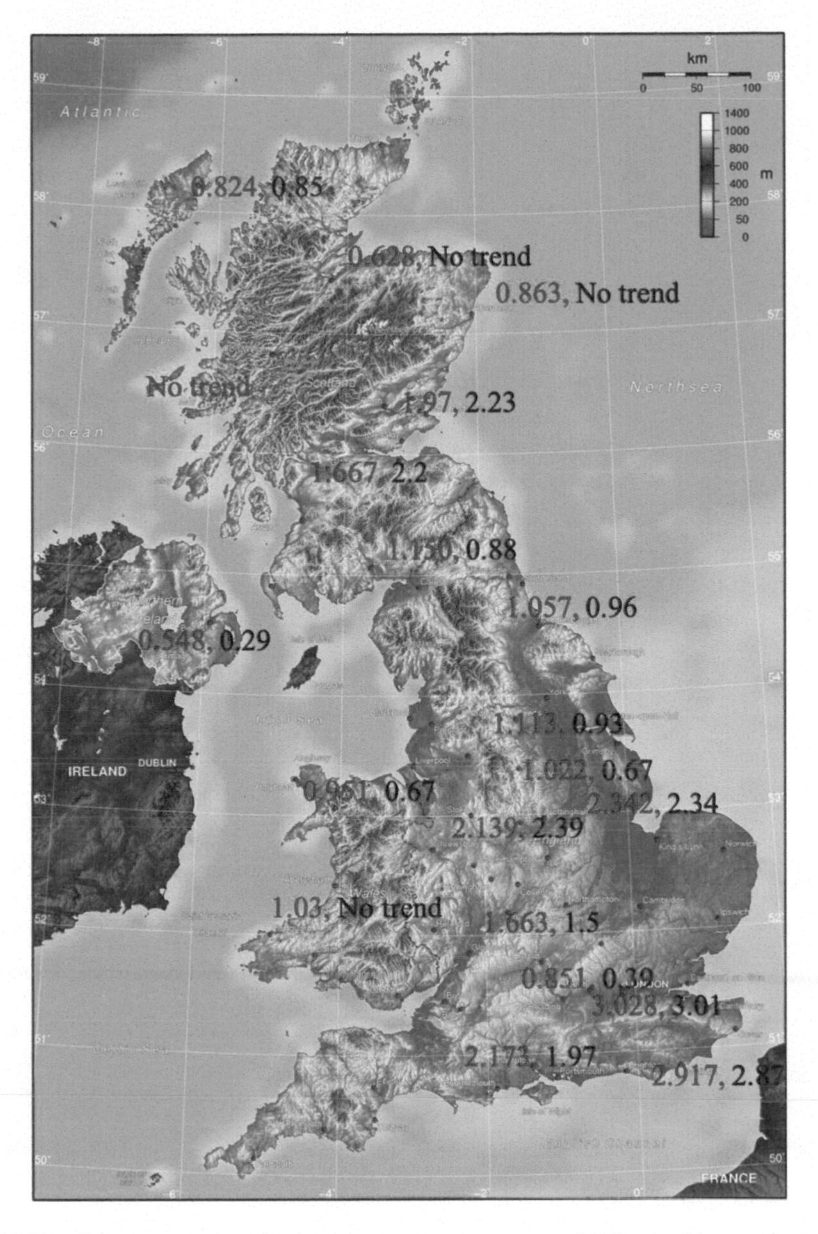

Figure 5.17 Trends in maximum and minimum temperatures recorded by the Meteorological Office, UK, per hundred years as analysed by Lee (2020). Red figures relate to trend in maximum temperature recorded, black figures relate to minimum temperature recorded (per month). Measurements were taken in the shade.

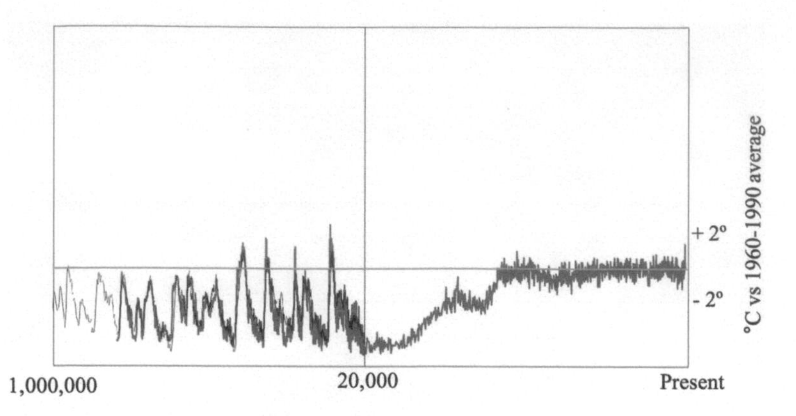

Figure 5.18 Temperatures in the last million years.

Figure 5.19 View of glacial valley (now with no ice), Western Skye, Scotland.

Land ice retreat took over 20,000 years to clear Britain. An example of an ice-free valley in Scotland should cause people to stop and reflect about ice loss (Figure 5.19).

This valley has clearly been carved out by ice and yet the ice has now disappeared.

The location of the ice cap 26 thousand years ago is shown in the left-hand figure in Figure 5.20, and 11 thousand years later to the right. There is a distance from Bristol

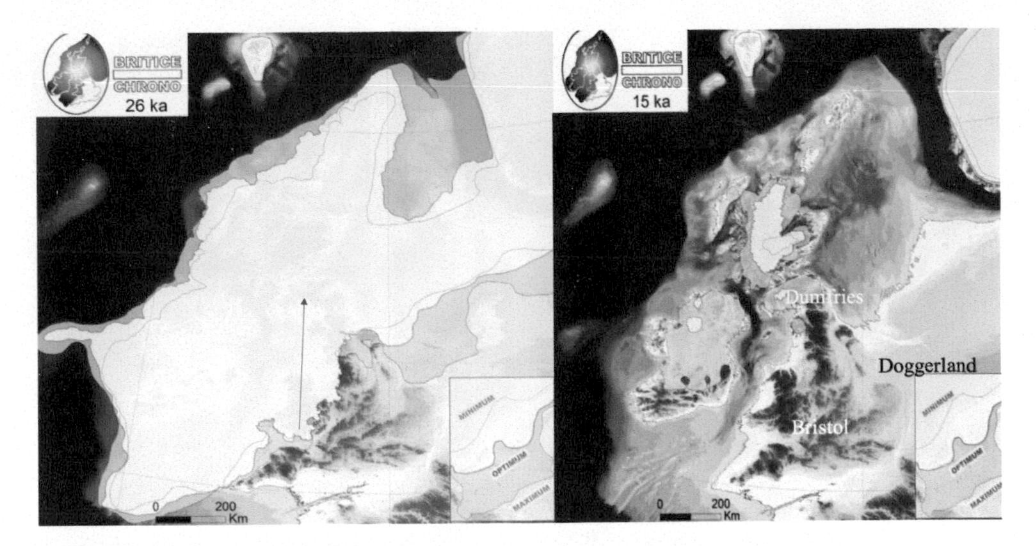

Figure 5.20 Maps showing best determined location of ice over Britain and Ireland from 26 to 15 thousand years ago after Clark et al. (2022).

to Dumfries of 434 km. That means that on average the ice (which at places was more than 2 km thick) retreated at approximately 40 metres per year. If we measured the line between the westerly extent of ice and Dumfries we would measure faster rates. That is a similar rate as glaciers are melting throughout the world, i.e. the process is continuing essentially unchanged. For example, the Mer de Glac at Chamonix shown in Figure 5.21, has retreated at 11.6 metres per year over the period 1934 to 2014 (Daws, 2018). The Pederson Glacier in Alaska has retreated at about 86 metres per year.

5.5.8 Sea ice at the North Pole

Arctic ice melt is an essential part of the ice age cycle in that, prior to a cooling episode, the North Pole needs to be free of ice, to allow evaporation from the sea surface and precipitation of snow at lower latitudes so that the next ice age can develop.

This was pointed out by Ewing and Donn (1956), and repeated by Arthur Holmes in Principles of Physical Geology (1965):

> An open Artic Ocean during the Pleistocene seems to be the only geographic condition which could have produced glacial conditions in northern Canada equivalent to those in Greenland today.

In other words, melting of the Arctic ice cap is an essential and natural precursor of the cycle for the next ice age to progress as shown by Ehlers & Gibbard (2007). The fact that the current melting, with loss of habitats is taking place under the gaze and dutiful observation of human scientists seems to be the one fact that certainly distinguishes

Figure 5.21 Showing Mer de Glac glacier, in France. Note the reduced level of ice of about 100 metres, the current rate of retreat of the glacier is only one third of the rate of ice retreat over Britain over 11 thousand years.

it from the previous interglacial maxima. Another of course is that it might well be happening at a faster speed due to man's polluting influence.

5.5.9 Sea level rise

Regarding sea level change, measurement is prone to numerous errors because of the time the measurement is taken (high or low tides), the barometric pressure, the heating and cooling of the sea and the orogenic movement of the land. Nevertheless, this has been measured at 31 locations from 1859 to the present in the UK, and it turns out that the rate of sea level change is approximately 1 mm per year above what might be expected due to geological changes (Woodworth, 2018). Pirazzoli (1996) lists studies from 21 other groups of tide-gauge records with the vast majority showing between 1 and 1.5 mm per year. The largest study of 517 stations from 1807 to 1986 showed the rise as "indeterminable" (Emery & Aubrey, 1991).

Just to emphasise the difficulties in measurement, the surface of the oceans has a variable roughness with a relief of up to 200 m (Pirazzoli, op cit). Nevertheless, his extensive study confirmed that the most realistic range of recent global sea level rise is 0.9 ±0.3 mm a^{-1} (footnote[4]).

Figure 5.23 shows changes over just the last 20 thousand years, which is a very short time geologically, with a few historical and pre-historical dates added to put it in perspective.

To these figures, I have added variations in global temperature over the last 800,000 years from the European Project for Ice Coring in Antarctica ("EPICA") data set and original paper in Science by Jouzel et al (2007).

It is clear from these data that sea levels are linked to global temperatures (and apparently CO_2 levels) and further, that associated glaciations (ice ages) have

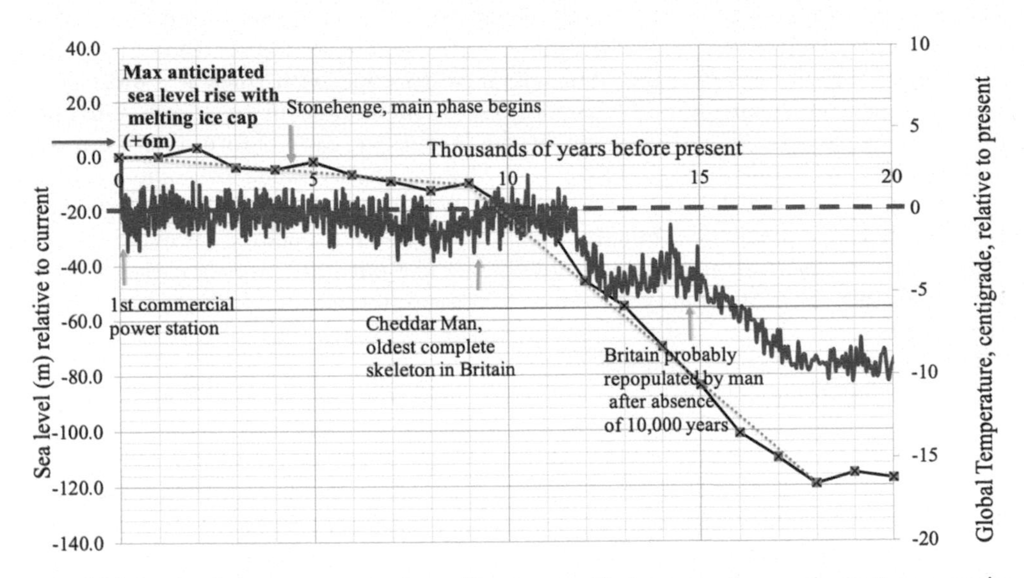

Figure 5.22 Sea level changes over the last million years with data taken from authoritative papers by Miller et al (2005 & 2011). Most of the data were obtained from drill core and analysis of the ratio of O^{16} to O^{18} in foraminifera. A very similar trend is shown in Figure 10.14 from the book by Pugh (1987) based on carbon14 data. The origins of Miller et al's data, errors and assumptions are clearly set out in the referenced papers.

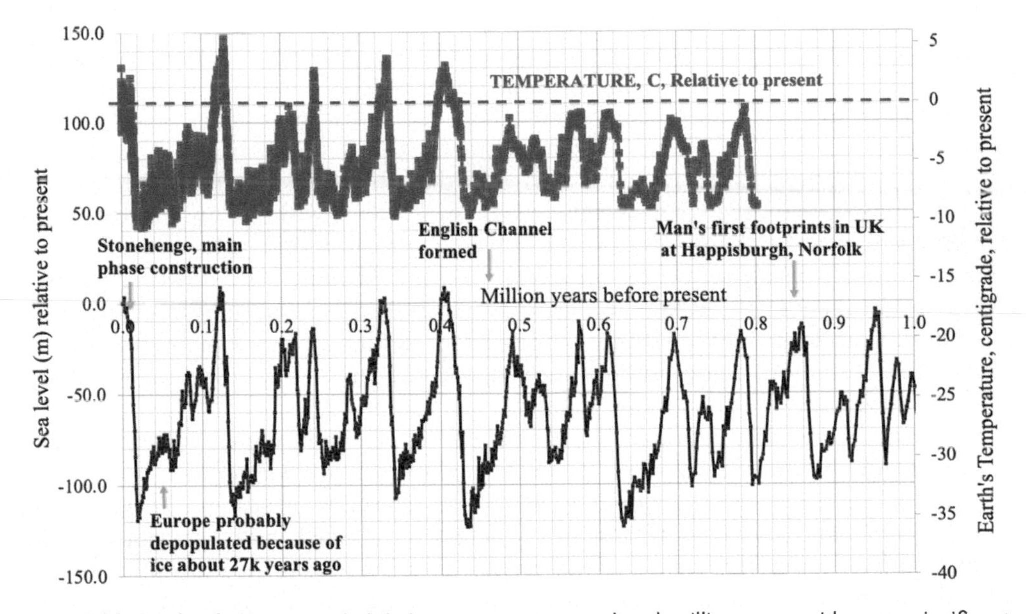

Figure 5.23 Sea level changes and global temperatures over last 1 million years with some significant events.

caused migrations of early man. Only since the last glacial maximum has Europe been consistently occupied. The graphs have a certain regularity that imply that in another few thousand years, man will have to cope with thick ice sheets over much of northern Europe and Canada. The 6 metre rise in sea levels currently predicted is associated with melting of the Arctic ice cap (before a reversal and new cooling cycle takes over). The extent of melting envisaged by some scientists (all of Antarctica and Greenland ice sheets) would cause much higher rises (and the flooding of major cities) but this level of rise has not occurred previously over the Pleistocene period and seems unlikely if the "natural" processes responsible for the ice age cycles (probably related to processes in the Sun – still not fully understood to my knowledge – rather than any trigger on Earth) take the same course as they have over the last million years or so.

Current measurements of rise in sea level range from about 1 to 3 mm per year, which is actually rather slow compared to the rate over the last 20 thousand years following the last glacial maximum as shown in Figure 5.22 when the rise (shown with blue line) was of the order of 12 mm per year between 19,000 and 9,000 year ago. The trend since about 9,000 years ago to present day has been less than 2.5 mm per year. Note that relative sea level rose globally during that time by perhaps 120 metres.

That rise was associated with much land loss due to flooding and erosion (Figure 5.20b shows Doggerland, which stretched from the East coast of the UK across to Holland). Fishermen still find archaeological artefacts from the North Sea. From these data (predating anthropogenic causes of climate change) it seems that further changes are inevitable even though it does seem certain that carbon emissions due to man's activities are contributing to rise in temperature and that curbing such emissions would help postpone matters.

These pre-historic sea level rises and global temperature changes would, of course, have been accompanied by massive changes to weather patterns, desertification and so on as it seems very likely be inevitable in the future.

The current concerns over the sea encroaching volcanic islands such as the Maldives and Kiribati, are not due to some new phenomenon. Charles Darwin (1842), following the Voyage of the Beagle, reasoned correctly that the development of fringe reefs and coral atolls were the result of relative rise in sea level with the coral growth just keeping up with the change (tens of metres over a long period). He thought it was due to gradual subsidence of the underlying volcanic islands. Daly however in 1910 showed that these features "are an inevitable result of the Quaternary oscillations of climate and sea level." (quote from Holmes, 1965). So, however we might fear for the swamping of shallow islands in the oceans due to sea level rises, I am afraid that this is nothing new.

My own view is that it should be that the *consequences* of climate change (which to me seem largely inevitable given the geological record) are where focus is required. These consequences include flooding of course, changes in sea currents, rainfall and drought, and sociological migratory factors (combined with over-population), restricted by national boundaries. It is evident that man was forced to migrate several times from what is now Britain over the last 100,000 years because of cooling temperatures (Stringer, 2006) but would find this difficult in the future politically as the climate cools again as seems likely.

5.5.10 True anthropogenic changes

Anthropogenic changes in the short-term are extremely important and burning carbon definitely causes pollution and increased CO_2 emissions.

In my opinion, these are matters that will take decades to cause major difficulties in terms of temperature and precipitation changes and hundreds of years for sea level rise to cause a real problem. This is given despite the world's population spreading widely into uninhabitable regions (including hotter) and coastal stretches, where the land is sinking for other reasons, such as water extraction in Bangkok and Jakarta. There are plans to re-locate Jakarta inland.

Of more concern is pollution (I grew up in the smog of London and have worked in the smog of Chungking in China) and over-population of countries that cannot sustain life in a reasonable way. Air pollution, water supply and condition of the seas (the dreadful pollution of plastic and dumping of chemicals) and sanitation are key issues and this, in my opinion, is where our attention should be focussed rather than trying to convince people that we can prevent or slow down climate change in some acceptable way.

If the huge scientific resources currently focussed on "geoengineering" climate, were aimed instead or, at least more strongly, at improved water supply, energy supply and air quality for third world countries and for ensuring that such countries can withstand the poor weather that they are and will be increasingly forced to endure then I think the outcomes would be better.

Population change is a great concern as highlighted earlier in Section 5.5.3 where it was argued that the maximum, optimal world population should be about 1 billion people. Figure 5.24 shows the current world population of 7.8 billion people, which it is modelled, based on current trends, will double to 15.5 billion by the year 2100. I know that there are people who will dispute this, arguing that some kind of slowing factor will introduce itself, but simply going on historical records, that is where we are.

This is an extremely difficult problem, but I imagine that provided the difficulties were highlighted to the world ("population alarmism", rather than "climate alarmism"), and small families of one or two children were encouraged, then perhaps this could happen. I am aware of the various "models" of population and the lack of any guilt regarding CO_2 emission by large chunks of the population, but Figure 5.24 (left side) is true and factual.

5.6 EARTHQUAKES

Earthquakes are a serious environmental hazard. On 6 February 2023 a magnitude 7.8 earthquake struck southern and central Turkey and more than 45,000 people died.

There are four major considerations for design:

1. Local ground failure, e.g. because of liquefaction in loose, saturated, cohesionless sand and silt.
2. Rupture because of fault movement which can be significant especially for tunnel design, and

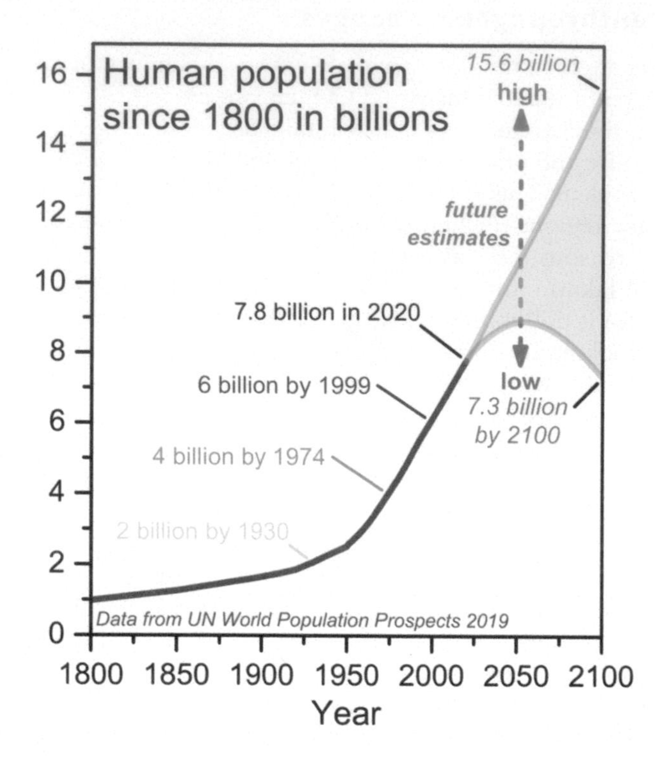

Figure 5.24 World population growth according to UN (2019).

3. Ground shaking causing inertial forces. Buildings and slopes are especially at risk from horizontal shaking.
4. Remote hazards. These will include landslides from adjacent land where debris run-out could impact the site and tsunamis.

5.6.1 Ground motion

Most structures need to be designed to withstand dynamic loading. This includes wind loading (to typhoon levels in countries such as Japan, Korea and Hong Kong), earthquakes and blasting/traffic. The main one of these that requires input from the engineering geologist is earthquake loading. The level of hazard is assessed at the site investigation stage (Chapter 4), and there is often a mandatory design code for a particular country. Alternatively, or as a check, the design team will identify some "design earthquake" or series of such design events with equal probability of occurrence within the lifetime of the structure. For example, statistical analysis of historical earthquake activity might indicate that there is an equal chance of a magnitude 8 (M8) earthquake at 200 km distance as a magnitude 5.2 (M5.2) earthquake at 10 km. These earthquakes would probably result in very different ground shaking at the project site. From study of recorded data using strong motion seismographs, attenuation laws have been derived for different parts of the world. Forces are used for engineering design so acceleration is an important parameter. Equation (5.8) has been shown to fit the available European seismic data reasonably well and can be used for prediction (Ambraseys et al, 1996). Data from North America and elsewhere are not very different.

$$\log(a) = -1.48 + 0.266M_s - 0.922 \log(r) \qquad (5.8)$$

where a is peak horizontal ground acceleration expressed as a fraction of gravitational acceleration, g (9.81 m/s^2). A vertical ground acceleration of 1g would throw an object into the air. M_s is surface wave magnitude and r is essentially the distance between the project site and the earthquake epicentre. Equation (5.8) and Figure 5.25 give median data and can be refined for degree of confidence and for site characteristics (Ambraseys et al, 1996). Unexpectedly high accelerations do occur and this is often the result of local ground conditions or topography that amplify the effect as for the peak accelerations of 1.25g and 1.6g in the abutment of Pacoima Dam, USA during two separate earthquakes – Bell & Davidson (1996). The February 2011 earthquake that caused huge damage in Christchurch, New Zealand involved vertical ground accelerations up to 2.2g and horizontal ground accelerations of up to 1.2g, which are very high for a 6.3 magnitude event and can be largely attributed to the very shallow nature of the earthquake (about 5 km according to the New Zealand Society for Earthquake Engineering).

Peak ground acceleration, although an important starting point is not enough to give an indication of structural performance. What also matters is the time that the strong shaking continues and the frequency spectrum of the waves carrying the energy. The situation is complicated by the way that individual structures respond to repetitive dynamic loading, which is a matter of harmonic resonance. Thus, whilst for the M5.2 design earthquake at 10 km the peak acceleration can be predicted from Equation (5.8)

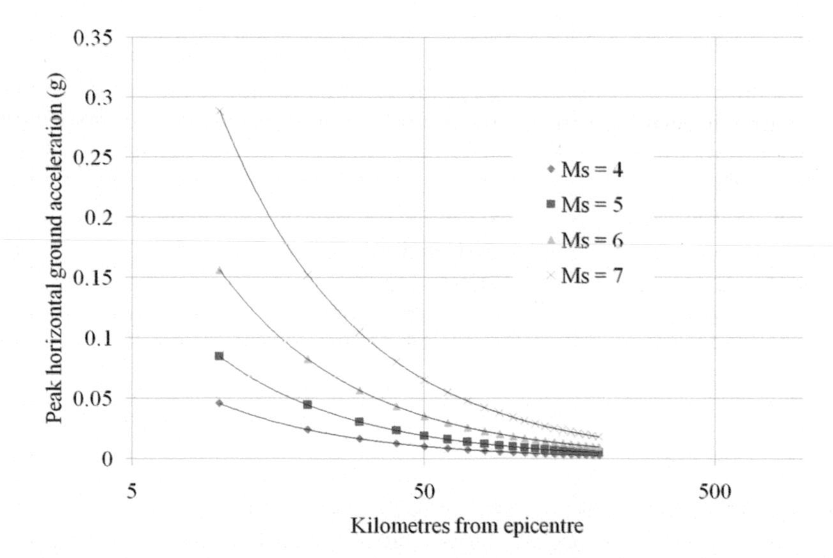

Figure 5.25 Peak acceleration vs. distance for different magnitude earthquakes (European data). Equations given in Ambraseys et al (1995).

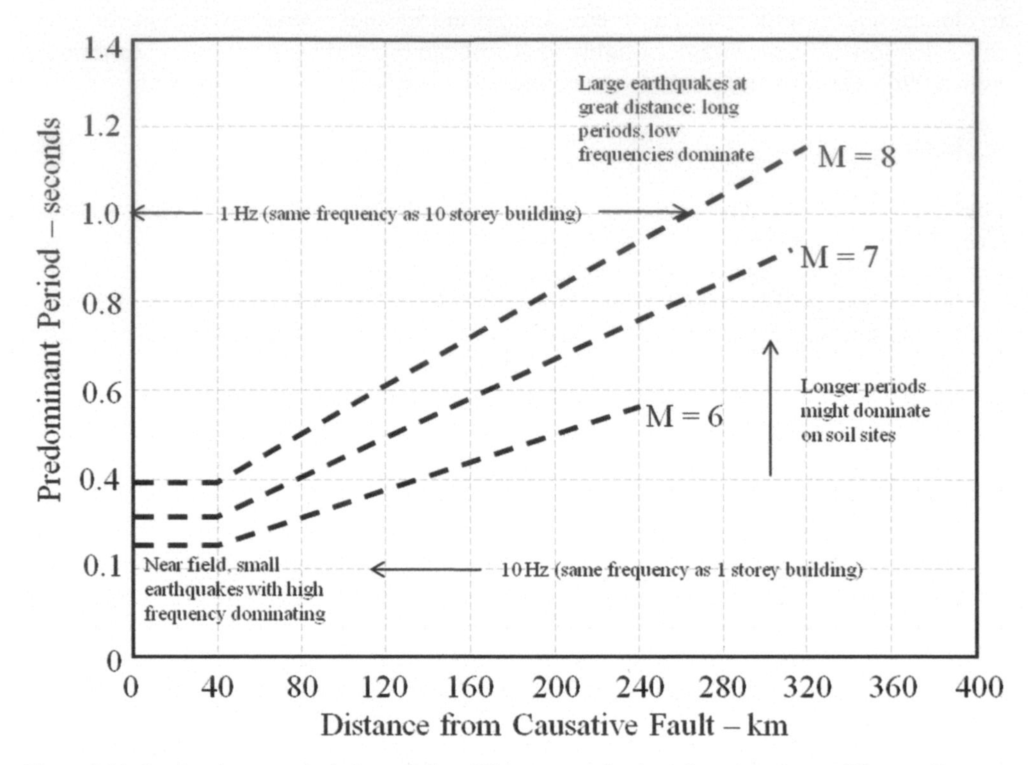

Figure 5.26 Predominant periods in rock for different magnitudes of earthquake at different distances (US data).

Source: After Seed & Idriss (1982).

as 0.12*g* and for the M8 design earthquake at 200 km, 0.04*g*, other characteristics will be very different. Figure 5.26 shows predominant period in ground acceleration records for the western USA (Seed et al, 1968), which indicates for a near field M5.2 quake the predominant period might be less than 0.2 s whereas for the distant M8 quake the predominant period could be more than 0.8 s. Furthermore, the duration of shaking will be significantly longer for the large magnitude earthquake (e.g. Bommer & Martinez-Pereira, 1999). The duration of strong shaking for a M5.2 earthquake might be a few seconds. For the Christchurch February 2011, M6.3 earthquake, the strong shaking lasted about 12 s. For an earthquake of M8, the duration could be over a minute. With longer duration, the potential for amplification will be much greater and fatigue type failure can occur.

5.6.2 Liquefaction

Liquefaction is a common failure mode in natural soils and sometimes in embankment dams during earthquakes. It occurs in loose saturated cohesionless sand and silt, which when disturbed, loses its structure and collapses. Because of its low permeability water cannot escape so natural piping and even general liquefaction occurs as the effective stress and thereby friction reduces to zero. There are many classic examples of whole apartment blocks tilting over and buildings settling. Elsewhere service pipes float to the surface and sea walls collapse as the retained fill flows into the sea. The potential for liquefaction is readily identified during site investigation. The general rules are as follows:

1. It occurs in un-cemented deposits – fill or geologically recent soil.
2. The most susceptible soils are cohesionless (sands and silts) with a liquid limit less than 35 percent and water content greater than 0.9 times the liquid limit (Seed & Idriss, 1982).
3. It generally occurs at depths shallower than about 15 metres.
4. Generally, SPT N value (corrected) less than 30 (Marcuson et al, 1990) or CPT cone resistance less than 15 MPa (Shibata & Taparaska, 1988).

Analysis of the hazard might be refined by considering the liquefaction potential vs. the characteristics of a design earthquake but generally if the area has high seismicity and the granular soil at a site is relatively loose and ground water table high then it is probably wise to carry out preventive measures. These might include compaction, grouting or the installation of stone column drains that will help prevent excess pore water pressure development although they would not prevent settlement. Alternatively, passive mitigation may be the best option – relocate the proposed structures away from the zone of liquefiable soil. If the ground does liquefy then apart from movement of structures in or on the ground, the settled soil might cause drag down (negative skin friction) on any piles installed through that zone.

5.6.3 Design of buildings

For buildings such as one- or two-storey houses there are certain simple rules that if adopted can reduce the risk of failure and would limit injuries, especially in developing nations. These include ensuring that walls are tied together preferably by reinforced ground beams or beams along the tops of walls (Coburn & Spence, 1992).

For larger engineered structures these need to be designed to withstand the repeated force waves. As outlined earlier, given a particular design earthquake one can make estimates about the ground motion characteristics that the structure will have to withstand. These include peak acceleration, predominant frequency and duration of shaking for a given return period earthquake. Typically, the return period used is 1 in 500 to 1,000 years but the choice is rather arbitrary and will depend on the nature and sensitivity of the project and the seismic history. These "bed rock" ground motions may be modified by the local site geology or topography and estimates of the modified shaking characteristics can be made by dynamic analysis using software such as SHAKE or through reference to published ground motion spectra for particular ground profiles. Generally thick, soft soil profiles may lead to relative amplification of longer period waves. The design ground motion then needs to be applied to the structure. Structures have their own dynamic characteristics and if the incoming frequencies match the natural response frequencies of the structure then movements may be magnified (Figure 5.27). Structural engineers will take the incoming design earthquake

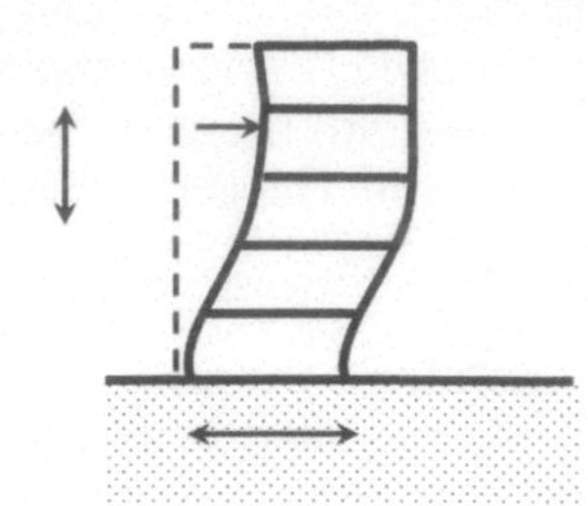

– Displacement of building (or slope) depends on building form and ground motion

– Buildings have natural periods
– T = 0.1 N seconds

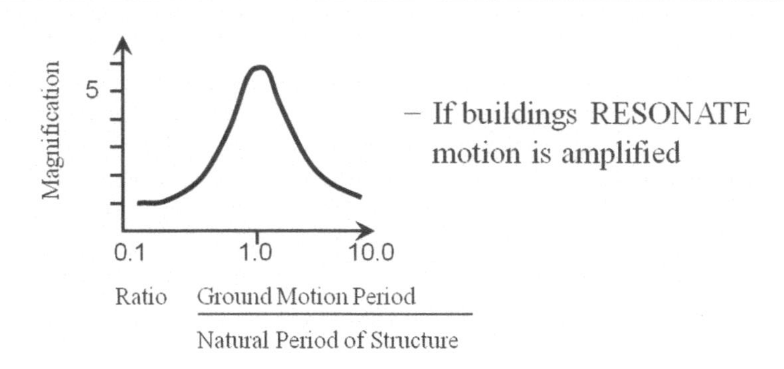

– If buildings RESONATE motion is amplified

Figure 5.27 Responses of buildings to earthquake shaking.

characteristics and calculate the response of the structure. For more frequent, smaller earthquakes, the structural engineer will design the structure as far as possible to behave elastically (no permanent displacement). In the event of an extremely large and less probable event a structure can be designed to be fail-safe. Redundant elements such as additional steel beams can be included that yield under extreme loads but also change the fundamental frequency of the building, damping the response to the shaking. Other options are to put a building on springs of some kind or to include hydraulic actuators or pendulums that again reduce the structural shaking. An example of an innovative aseismic design is the foundations for the Rion-Antirion Bridge constructed in Greece in 2005. The cable stay bridge, with five main spans extending 2.25 km across a fault zone, was designed to withstand horizontal accelerations of 0.5*g* at ground level and up to 2 m offsets between adjacent towers. Underlying each tower is thick soil and the depth of sea is up to 65 m. The towers were founded on 90 metre diameter cellular structures placed on a 3.6 m layer of gravel placed on the natural soil, which was reinforced by up to 200, 2 m diameter tubular piles to depths of 30 m. The foundation structure is not attached to the piles; the gravel acts as a fuse, limiting the transfer of load to the superstructure. The piles in the underlying soil are there to prevent rotational bearing failure. Details of the design of the foundations are given by Combault et al (2000) and further references are given at the web page for the bridge.

Two recent earthquakes however show that even with good design practice earthquakes can cause damage to a level that is not anticipated. As a result of the February 2011 Christchurch, NZ earthquake, many small one- and two-storey buildings were destroyed or badly damaged as one might expect near the epicentre of an earthquake with magnitude exceeding 6.0 where the ground motion might be expected to be dominated by high frequencies. Widespread liquefaction was also a major contributor to the damage of these smaller buildings. However, for this earthquake, because of its shallow nature and possibly other factors that served to concentrate and amplify the ground motion, unexpectedly large accelerations and forces were generated. In the case of the March 2011 earthquake that struck NE Japan (East of Honshu), most engineered buildings on mainland Japan withstood the very strong shaking associated with this 8.9 or even 9.0 earthquake (10,000 times as strong in terms of overall energy release than the Christchurch earthquake) and this is testament to the skill and knowledge of the civil engineer designers. The huge damage and large number of deaths caused by the Japan earthquake resulted from a 10 m high tsunami wave that came ashore and destroyed whole villages. Regarding engineered structures, several nuclear power stations had been constructed along the shore line in the impacted region. The structures apparently performed well in terms of withstanding seismic shaking but severe damage did occur because of failure of cooling systems. The initial shaking caused safe shut down of the reactors as is the required procedure for nuclear power stations impacted by a major earthquake but the loss of electrical power stopped the flow of cooling water required to prevent the fuel rods overheating. Back-up diesel generators kicked in and provided the necessary power for an hour or so but then they failed because of the tsunami. In hindsight, no doubt the secondary power sources could and should have been designed to survive inundation as they are for more modern installations and the risk properly identified using an event tree approach.

5.6.4 Tunnels during earthquakes

Tunnels and mines tend to be safer than surface structures during earthquakes, and this safety increases with increased depth (Power et al, 1998). Except where the tunnel passes through particularly poor ground or intercepts active faults, earthquake-resistant design is generally not a high priority. Of course, where the support in a tunnel is inadequate or marginal under static loading conditions then earthquake shaking might well trigger failure. This is especially true at portals of tunnels; landslides and especially rock falls are very commonly triggered by earthquakes as discussed in the next section. Failures in some tunnels and especially the failure of Daikai subway station during the Kobe earthquake in 1995 have caused a re-think on seismic stability of underground structures. Hashash et al (2001) provide a very useful review and examples of aseismic design. A reinforced concrete lining should have significantly better seismic resistance characteristics than an unreinforced lining. If the tunnel intersects a fault that is suspected of being active then special measures will be required or, preferably, the fault avoided. Key considerations are the estimated magnitude of the displacement and the width of the zone over which displacement is distributed. If large displacements are concentrated in a narrow zone, then the design strategy may be to enlarge the tunnel across and beyond the displacement zone. The tunnel is made wide enough such that the fault displacement will not close the tunnel and traffic can be resumed after repairs have been made. In some cases, an enlarged tunnel is constructed outside the main tunnel and the annulus backfilled with weak cellular concrete or similar. The backfill has low yield strength to minimise lateral loads on the inner tunnel liner, but with adequate strength to resist normal ground pressures and minor seismic loads. If fault movements are predicted to be small and/or distributed over a relatively wide zone it is possible that fault displacement may be accommodated by providing articulation of the tunnel liner using ductile joints. This detail allows the tunnel to distort into an S-shape through the fault zone without rupture, and with repairable damage. This may not be feasible for fault displacements more than 75–100 mm. An alternative approach is to accept that damage will occur and to make contingency plans to control traffic and to carry out repairs as quickly as possible in the event of a damaging earthquake.

5.6.5 Landslides triggered by earthquakes

Landslides are commonly triggered by earthquake shaking, especially in mountainous areas. The Wenchuan earthquake in Sichuan Province, China of 12 May 2008 was very large (M8) and quite shallow (14 km) and the active faults ran through populated valleys surrounded by high slopes. Landslides, including rock fall caused more than 20,000 deaths with one individual landslide killing more than1600 people (Yin et al, 2008). One of the main consequences was the damming of streams, which necessitated emergency engineering works to lower the water levels in the lakes that formed behind the landslide debris before they were overtopped or burst uncontrollably.

5.6.6 Landslide mechanisms

Slopes affected by strong earthquake shaking can be categorised in three classes as set out in Table 5.3. These are as follows:

1. Stable Slopes

These are defined as situations where the shaking is not strong enough to cause permanent displacement in a slope. This may be because the peak forces are insufficient to overcome the strength of the ground or because different parts of the same slope are out of phase so that whilst some parts are being driven towards failure other parts are being accelerated in the opposing direction.

2. Permanently Displaced Slopes

The key aspect of dynamic loading, whether it is from earthquakes or blasting, is its transient nature. The waves pass through the ground and induce inertial forces. In the same way as discussed in Chapter 7 regarding a laboratory experiment, at a critical acceleration (k_c) a slope will start to move. The continued positive acceleration above critical will cause the displacement to increase in velocity. However, after a short time (typically a fraction of a second), the acceleration will decrease and then change direction so that the inertial force is back into the slope. This will stop the movement unless

Table 5.3 Performance of slopes under dynamic loading

Outcome	*Pre-quake condition*	*Consequence*	*Details and examples*
1 Elastic deformation – stable	Earthquake loading is less than critical acceleration to cause permanent displacement	No permanent displacement	
2 Permanent minor displacement	Earthquake acceleration higher than critical acceleration for displacement but: - Residual strength ensures FoS > 1.0 after shaking has finished - Overall geometry is essentially stable	- Small permanent deformation - Post-quake configuration stable - FoS post-quake > 1.0 - May be prone to deterioration from rainfall, weathering because of opening of joints and tension cracks	

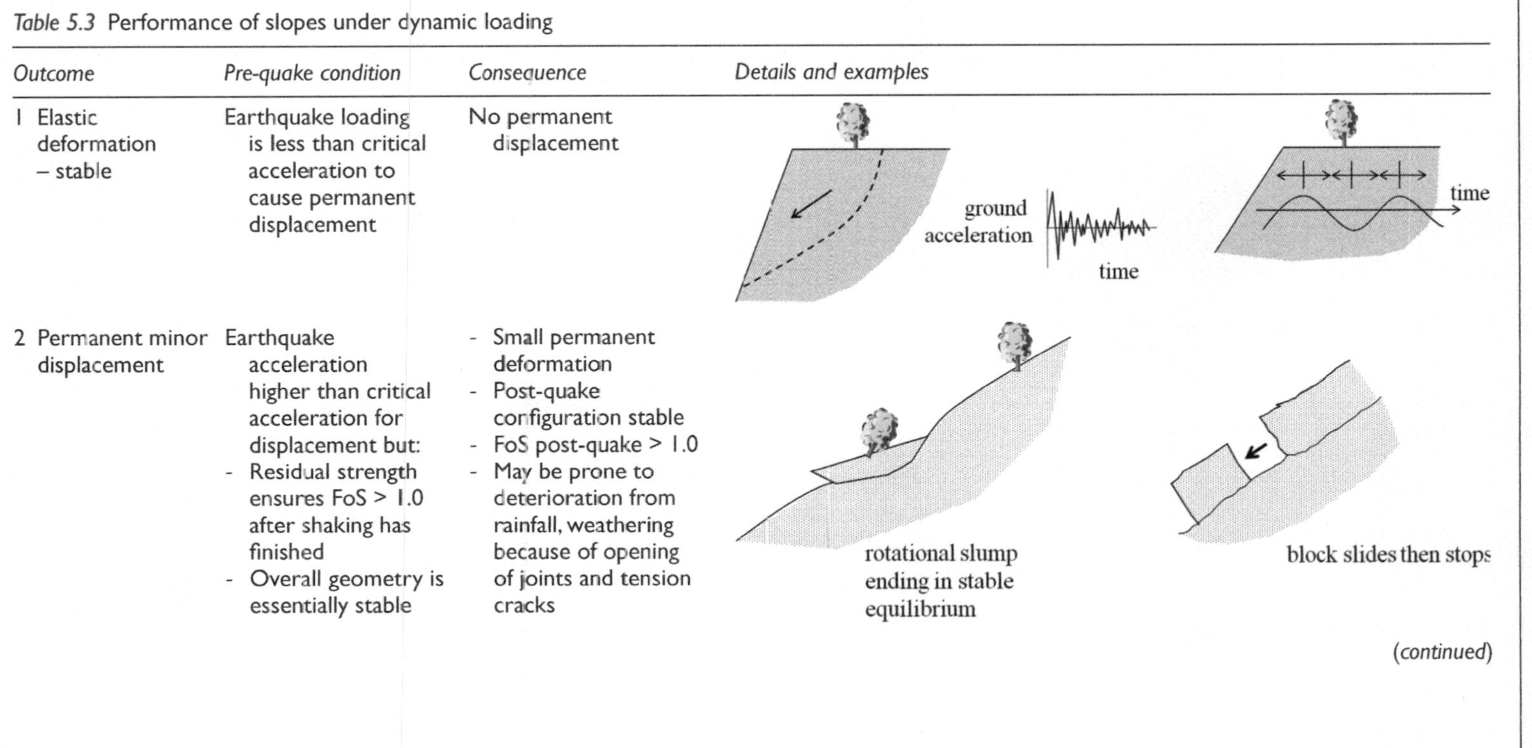

(continued)

Table 5.3 (Cont.)

Outcome	*Pre-quake condition*	*Consequence*	*Details and examples*
3 Failure	Residual strength low so that FoS is less than 1.0 after earthquake shaking has finished	- Possibly large-scale movement with kinetic energy providing momentum	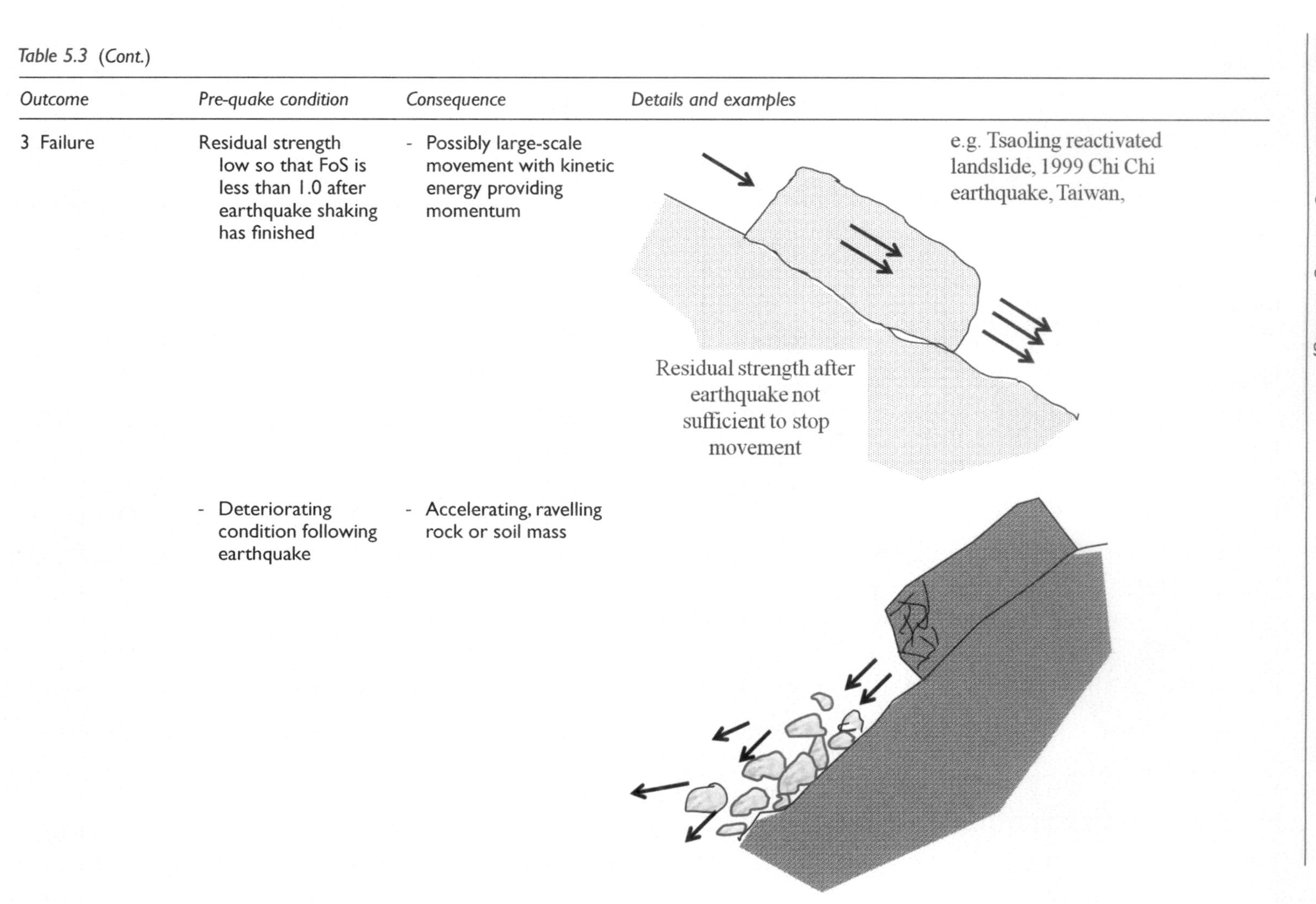
	- Deteriorating condition following earthquake	- Accelerating, ravelling rock or soil mass	

- Geometrically unstable	- Collapse and entrainment of debris	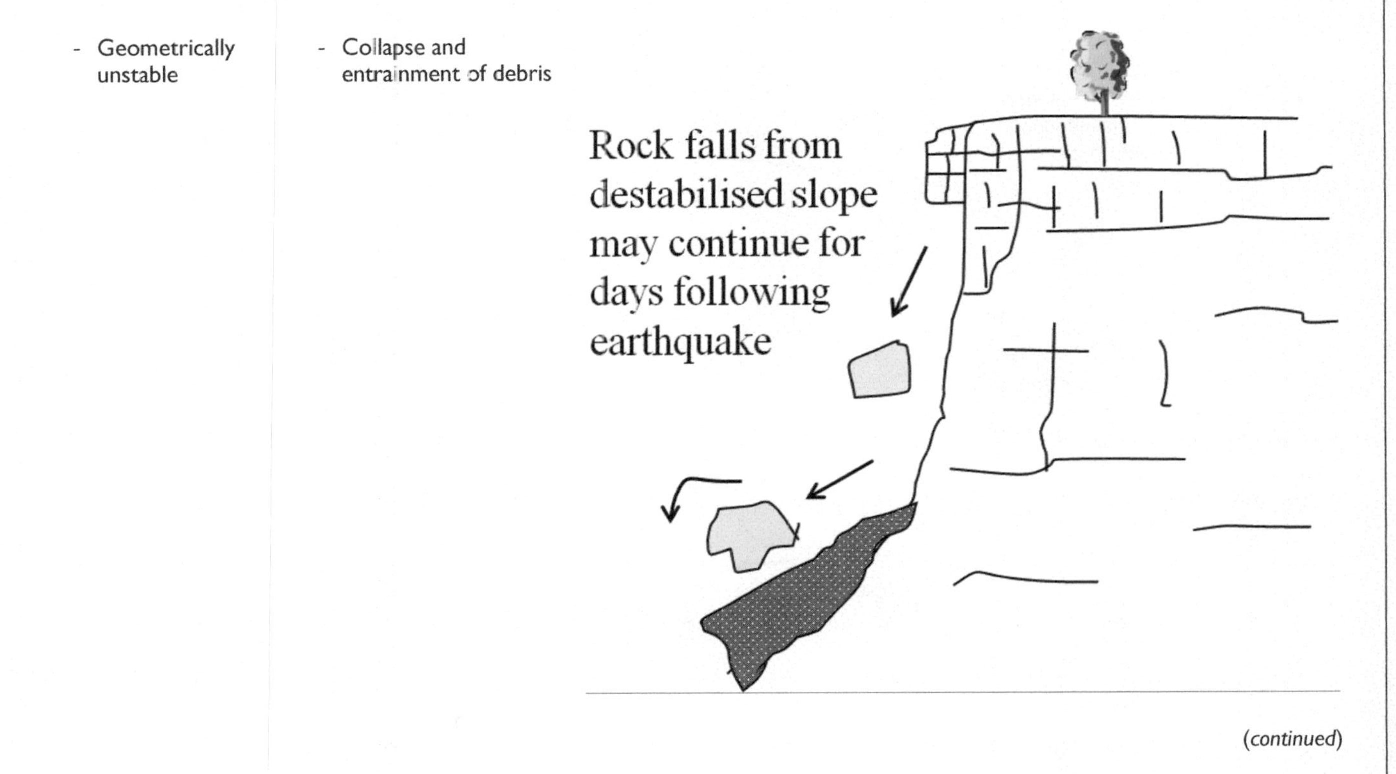

(continued)

Table 5.3 (Cont.)

Outcome	*Pre-quake condition*	*Consequence*	*Details and examples*
	- Rise in water pressure due to collapse of soil structure (in loose saturated sand and silt) or regional changes in hydrogeological conditions	- Liquefaction or water induced failure - Possible debris flow	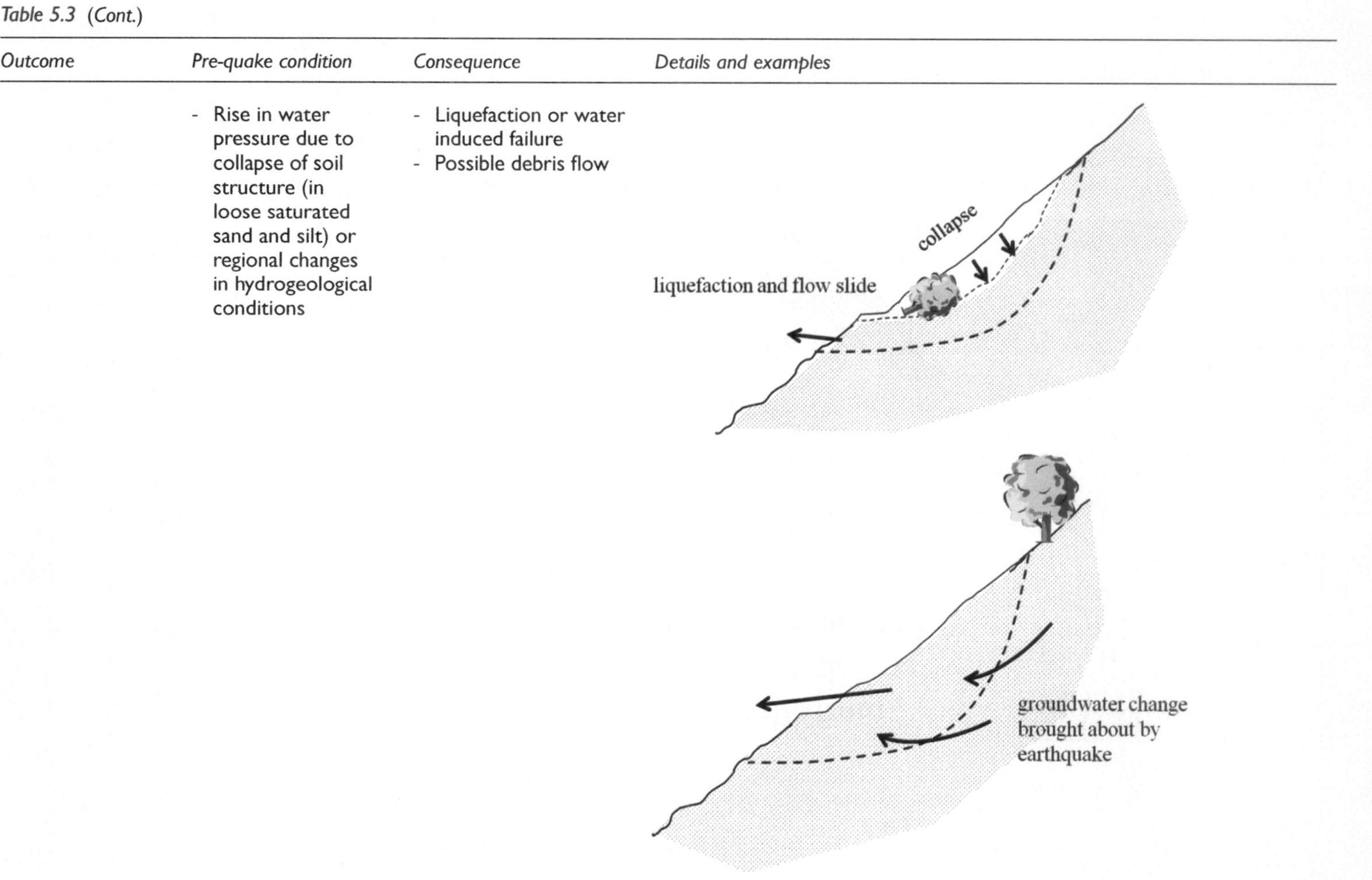

the slope is metastable as discussed later. Sliding friction can be lower than residual (Hencher, 1977; 1991; Crawford & Curran, 1982; Tika et al, 1990) and by employing a pessimistically low shear strength, total displacement can be calculated for a series of acceleration pulses and this can be used as part of a design decision. Generally, even for a very large earthquake, the permanent displacement in a slope directly attributable to inertial loading will be small, of the order of millimetres or centimetres (Newmark, 1965; Ambraseys & Srbutov, 1995). Nevertheless, small permanent displacements will make the slope prone to accelerated weathering and deterioration if not protected or repaired.

3. Failed Slopes

Catastrophic landslides during earthquakes can be the result of four different conditions, viz:

- *Low Residual Strength*. The inertial displacement during the earthquake reduces shear strength to a residual value so that even after the earthquake shaking the slope continues to move. Examples of large-scale failures involving sliding on bedding planes with reducing strength are described for the Chi-Chi earthquake (1999) by Chen et al (2003) and Chigira et al (2003) and for the Niigata earthquake (2004) by Chigira et al (2006).
- *Deteriorated State*. The structure of rock or soil mass is disturbed so that it collapses and a flow can develop.
- *Geometrically Unstable Equilibrium*. The initial displacement caused by the earthquake shaking results in unstable equilibrium. A typical example is rock fall from exposed rock cliffs. Once displaced the rock will fall, sometimes as a progressive failure several days after the earthquake. Rock falls may become entrained and develop into debris avalanches.
- *Water-induced Failure*. Firstly loose, saturated soil can collapse and liquefy down to depths of about 15 m on slopes inclined at only a few degrees. The collapsed material can spread and flow. As a second mechanism the general groundwater flow paths can be affected by earthquake loading and this can trigger slope failures.

5.6.7 Empirical relationships

Keefer (1984, 2002) identifies 14 individual types of earthquake-induced landslide. The three main categories are as follows:

1. Disrupted Slides and Falls

These include highly disrupted landslides that move down slope by falling, bouncing or rolling or by translational sliding or by complex mechanisms involving both sliding and fluid-like flow. They typically originate on steep slopes, travel fast and can transport material far beyond the slope in which they originate. Other than large rock avalanches, failures in this category are thin with initial failure depths less than 3 m.

2. **Coherent Landslides**

These include translational slides and rotational slides. Such failures are typically relatively deep seated (greater than 3 m) slow moving and displace material less than 100 m^3.

3. **Lateral Spreads and Flows**

Fluid flow is the dominant mechanism and this mode of failure is typical of liquefied soils.

The most common failures according to Keefer are rock falls, rock slides and disrupted soil slides. This follows from the analysis of Table 5.3 where it can be seen that significant landslides will only occur where there are pre-disposing factors such as a topographic setting that is in unstable equilibrium or strain softening (due to collapsing structure or low residual strength (for example, through the loss of rock bridge-cohesion during the earthquake shaking). Keefer (op cit) compiled data from many earthquakes and plotted the area affected by earthquakes vs. magnitude of the earthquake. The upper bound is rather well-defined. For a magnitude M5, the affected area might be about 100 km^2, 1,000 km^2 for M6, and 10,000 km^2 for M7. Keefer also presents data on the maximum distance of landslides triggered by earthquakes of given magnitude. He provides separate "upper bound" curves for disrupted, coherent and flow type failures. Disrupted landslides such as rock falls, which are the most common type of earthquake-triggered landslides, are also shown as the most likely to occur at far distances from the epicentre. Rodriguez (2001) has carried out a further review of data, including more recent data from Japan and his data demonstrate the considerable scatter that can be expected and therefore difficulties in prediction on a site-specific basis. For example, some M7 earthquakes only cause landslides within an epicentral distance of 10 km whereas others of the same magnitude cause landslides 200 km away. This might be attributed in part to resonance affects associated with ground frequency spectra and duration as for buildings (Hencher & Acar, 1995).

5.6.8 Slope design to resist earthquakes

Traditionally and in most software packages there are two main approaches to slope design to withstand dynamic loads (mostly earthquakes). The options for landslide prevention are essentially the same as for the static condition (change geometry, reinforce, reduce water pressure, protect the site below or move the facility at risk).

5.6.8.1 Pseudo-static load analysis

One approach is simply to include a horizontal inertial load into the analysis (some authors argue for an inclined force but it really makes no difference considering the inexact nature of the method) and to determine whether or not the FoS reduces below 1.0. The problem with this approach is that if one includes the peak predicted particle acceleration [say from equation (5.8)] then very often the slope will be shown to fail, whereas in reality the permanent displacement would be negligible because of the extremely short time that acceleration would be acting. As confirmation many vertical slopes in quarries are acted on by accelerations approaching or exceeding 1g during production blasting but landslides due to blasting are very rare. Engineers therefore often choose to use some arbitrarily reduced acceleration such as a nominal 0.1g as a pseudo-static force in the stability analysis to check that the slope (or dam) has some degree of resistance to horizontal loading but this is clearly rather unsatisfactory.

5.6.8.2 Displacement analysis

As discussed earlier, given a predicted acceleration against time record it is straightforward to calculate the likely displacement that might be caused in a slope during an earthquake and there are options to do so in software such as SLOPEW. Those displacements will always be small however, no matter how large the earthquake, and what matters more is the residual state of the slope after the earthquake – is there a situation where the ground is strain softening or is it in unstable equilibrium? These are considerations for the engineer who must decide whether additional reinforcement might be necessary or other protective measures such as nets and barriers. Other software such as FLAC and UDEC (Itasca) can be used to study the seismic susceptibility of slopes. These being time-stepping software, the mode of failure can be identified, expressed visually and perhaps as a movie. It might, for example, be possible to test the potential failure mechanism of soil nails during an earthquake, each nail modelled specifically. That said, as ever, the models can only be as good as the input data and results will only be indicative.

5.7 CONSTRUCTION VIBRATIONS

5.7.1 Blasting

Blasting causes noise, ground vibrations, air overpressure and fly rock. All of these can be controlled – generally by using less or different types of explosive and limiting the number of charged drill holes that are detonated at the same time. In particular, using millisecond delays between lines of drillholes will reduce the vibration level considerably. Details are given in Dowding (1985) and many other publications. Safety is a major issue and an engineering geologist working in a situation where blasting is being conducted may well be involved in blast monitoring, checking fragmentation and reviewing the overall suitability of the blast design given the changing geological situation as the rock is excavated.

5.7.2 Piling vibrations

The other major source of potentially damaging vibrations in civil engineering is from driven piles. Potentially damaging levels are generally limited to about 10 metres distance although this depends on the sensitivity and state of repair of the structure. Predictions can be made using empirical formulae into which the main inputs are hammer energy and distance (Head & Jardine, 1992) but these are rarely very accurate.

5.8 VOLCANIC HAZARDS

Like earthquakes, there is little one can do to prevent or even to monitor the hazards from volcanoes, but the hazard is very clear.

Mount St Helens volcano, which erupted in 1980, killed 57 people and destroyed 200 homes, 47 bridges, 24 km of railways and 298 km of highways.

Over a lightly longer period, the Laki volcano in Iceland of 1783 affected the whole of Europe and resulted in many deaths. It also was accompanied by a hot year (+3° C) followed by a cold year (–2° C) even though these were not apparently the direct result of the eruption according to Zambri et al. (2019). Nevertheless, the disruption would have been a magnitude greater than the closing down of air services of the eruption of 2010 for a month, with huge economic and sociological consequences.

5.9 OTHER HAZARDS

There are many other factors that need to be considered under the auspices of Equation (5.2), according to the site and the project as illustrated in Chapters 4 and 8.

These include wind speed, flooding, tsunamis, noxious gas and gas pressure, all of which can cause problems for construction and projects.

Wind speed and flooding are assessed according to historical records and generally a 1 in 100 or 1 in 1000 criteria are used for design, rather similar to statistical assessment to earthquake risk (also done by deterministic approach). Flooding, of course, is rather a strange one in that river valleys and flood terraces are often a clue (I am being facetious), and buying or building a house at a location called "Fish Lake" might be treated with caution.

Notes

1 This is actually incorrect (apart from the obvious changes between high and low tide).
2 Note however that this short-term rainfall does not predict the occurrence of large deep-seated landslides such as those on Ching Cheung road that occurred a week or more after heavy rainfall had ceased (see Chapter 8).
3 The actual record 24 hour rainfall for the UK is 341 mm per 24 hours at Honister Pass in the Lake District in 2015.
4 Compare these data with those of the Met Office of 12.5 to 25 mm a^{-1}, which are 1,000% higher!

Chapter 6

Site investigation

> If you do not know what you should be looking for in a site investigation you are not likely to find much of value.
>
> *Glossop, 1968*

This much-quoted quote is worth repeating because it sums up the philosophy of site investigation very well. Critical features need to be anticipated and looked for. Without care, the important details might be hidden within a pile of essentially irrelevant information. The difficulty and skill, of course, is in recognising what is critical and more on this later.

6.1 SAFETY FIRST

In the first edition of this book there were several pictures showing workers without high-visibility clothing, steel capped boots, glasses or gloves, all measures which are very familiar for workers in the UK. Several of these figures were from Portugal, Malaysia and other countries where, I am afraid, that safety measures are still lacking. No apologies, the pictures told their own stories, nevertheless, safety first should be the rule.

I recall one of our engineering geologists coming back into Halcrow's office in Hong Kong, reporting that one of the cables on a ground investigation rig had snapped and taken off the operator's arm. I similarly remember walking (with helmet on and my feet in steel-toed wellington boots), across the construction site at Drax power station in 1979, when suddenly a 7 metre deep pre-bore hole, that had had a thin soil cover, opened up beneath me. I was just saved by putting out my arms to catch on the sides of the hole, so that a quick-witted piling inspector could help me up again. After that occasion, an instruction was sent out to the contractor stating that after the arisings from the pre-bores had been scraped across, as normal using a bulldozer, a metal cap should be placed over the top of each pre-bore pile location as a safety precaution. You learn some things and improve, but we must remember that construction sites remain risky places.

BS: 5930:2020 states that a site safety plan should be prepared for any ground investigation works, identifying any hazards that exist and procedures that should be followed if an accident occurs. As a start, following an electronic search for active cables and gas and water pipes, a pit should be hand dug at each proposed location of drilling or probing, to ensure that services are not damaged by sub-surface works. Any

DOI: 10.1201/9781003348894-6

trial pit should be constructed safely and shored-up, like any temporary works, before anyone is allowed down to examine the soil or rock. It needs to be emphasised that the site safety plan is a different document to the risk register as addressed in Appendix D.

6.2 NATURE OF SITE INVESTIGATION

At any site, the ground conditions need to be assessed to enable safe and cost-effective design, construction and operation of civil engineering projects. This will generally include sub-surface ground investigation (GI), which needs to be focused on the particular project's needs and unknowns. The requirements for GI will be very different for a tunnel compared to the design of foundations for a high-rise building or for stability assessment of a cut slope. There needs to be a preliminary review of the nature of the project, the constraints for construction and the uncertainties about the engineering geological conditions at the site. The British Code of Practice for Site Investigation, BS 5930 (British Standards Insitution, 1999; British Standards Intitution, 2020), sets out the objectives broadly as follows:

1. *Suitability*: to assess the general suitability of a site and its environs for the proposed works.
2. *Design*: to enable an adequate and economic design, including for temporary works.
3. *Construction*: to plan the best method of construction and, for some projects, to identify sources of suitable materials, such as concrete aggregate and fill and to locate sites for disposal of waste.
4. *Effect of changes*: to consider ground and environmental changes on the works (e.g. intense rainfall and earthquakes) and to assess the impact of the works on adjacent properties and on the environment.
5. *Choice of site*: where appropriate, to identify alternative sites or to allow optimal planning of the works.

6.3 SCOPE AND EXTENT OF GROUND INVESTIGATION

6.3.1 Scope and programme of investigation

The scope of site investigation is set out in Box 6.1. This should include everything relevant to use of the site, including site history and long-term environmental hazards and not just geology. All authorities (e.g. Association of Geotechnical & Geoenvironmental Specialists, 2022) agree that site investigation should, ideally, be carried out in stages, each building on the information gained at the previous stage, as outlined in Box 6.2. A preliminary engineering geological model should be developed for the site from desk study and field reconnaissance. That model should then be used to consider the project constraints and optimisation (e.g. the likely need for deep foundations or the best location for a dam) and for designing the first phase of GI. For a large project, this first phase is usually carried out during the "conceptual phase". Further GI campaigns might be carried out for "basic design", for "detailed design" and often additional works during construction. Engineering geologists should readily appreciate that all sites do not require the same level of ground investigation. Some have simple ground conditions, others more complex. At some locations, existing exposures will allow the broad geology to be assessed and reduce the need for GI. Projects may be situated in areas where the geology and ground conditions are already well understood. For example, if designing piles in London Clay, because of the wealth of published data and industry experience, GI requirements should be fairly routine[1] – little should be needed in the way of testing to determine parameters for design.

BOX 6.1 OVERALL SCOPE OF SITE INVESTIGATION

1. Hazards and constraints during construction and in the longer term	- Previous site use – obstructions, contamination - Any history of mining or other underlying or adjacent projects (e.g. tunnels or pipelines) - "Sensitive receivers" – such as neighbours that might be affected by noise, dust, vibration and changes in water levels - Regulatory restrictions - Natural hazards, including flooding, wind, earthquakes, subsidence and landslides
2. Assess and record site characteristics	- Access constraints for investigation and construction - Need for traffic control, access for plant and waste disposal - Access to services - Site condition survey (partly as a record for any future dispute)
3. Geological profile at site	- Distribution and nature of soil and rock underlying the site, to an adequate degree, to allow safe and cost-effective design - Usually this will require a sub-surface ground investigation
4. Physical properties of soil and rock units and design parameters	Key parameters: - mass strength (to avoid failure) - deformability (to ensure movements are tolerable) - permeability (flow to and from site, response to rainfall and loading/unloading) Other factors: - chemical stability (e.g. reactivity in concrete, potential for dissolution) - potential for piping and collapse - abrasivity (sometimes a major consideration for construction)
5. Changes with time	- install instruments to check physical nature of the site – e.g. groundwater response to rainfall - install instruments to monitor settlement and effect on adjacent structures during construction - consider the potential for deterioration and need for maintenance

BOX 6.2 STAGES IN A SITE INVESTIGATION

Stage 1: Desk study at project conception stage
- Identification of key geological and environmental hazards and benefits at option sites, based on broad desk study and site walk-overs.
- Assess balance between site constraints, engineering considerations and economic factors.

Stage 2: Detailed desk study and reconnaissance survey
- Collect and review all documents relevant to the selected site, including topographic and geological maps, aerial and terrestrial photographs and any previous investigation reports. Review site history – previous building works, mining, etc. Look for hazards such as landslides.
- Site visits and surface mapping, possibly with advance contract for safe access, vegetation clearance and trial pits or trenches.

THE PRELIMINARY GROUND MODEL

Develop a preliminary geological and geotechnical working ground model that can be used as a reference for the rest of the ground investigation.

This preliminary model should be used by all the team, including those who are going to be logging any samples obtained by sub-surface drilling. The loggers need to know what to expect and be able to identify anything that will necessitate revisions to the ground model.

Site-specific ground investigation should be aimed at verifying the model, answering any unknowns and allowing design parameters to be derived.

Stage 3: Preliminary ground investigation linked to basic engineering design
- Consider use of geophysical techniques to investigate large areas and volumes.
- Preliminary boreholes designed to prove geological model (rather than design parameters).
- Instrumentation as appropriate (e.g. to establish groundwater conditions and seismicity).

Stage 4: Detailed ground investigation
- Further investigation to prepare detailed ground model and allow detailed design.
- *In situ* and laboratory testing to establish parameters.
- Detailed instrumentation and monitoring.

Stage 5: Construction
- Review of ground models during construction.
- Testing to confirm design parameters.
- Revision to design as necessary.

Stage 6: Maintenance
- Ongoing review – e.g. of settlement, slope distortion, groundwater changes and other environmental impacts, possibly linked to a risk management system.

6.3.2 Forgiving and unforgiving sites

Taking this further, experience shows that the majority of sites world-wide do not have any particularly inherently hazardous conditions and might be categorised as "forgiving". Even with no, or no competent investigation, the project is often completed without geotechnical difficulty. Such sites need little investigation – enough to establish that there are no particularly adverse hazards. In a review of the scope of ground investigations for foundation projects in the UK, Egan (2008) found that GI was either not conducted or were lacking borehole plans for 30% out of 221 projects, but it is intriguing that he spotted no adverse consequences. In other words, the engineers took a risk, perhaps on the basis of previous experience in an area, and apparently got away with it, although, as Egan points out, a ground investigation might have allowed more cost-effective solutions.

Unfortunately, the world also has relatively rare *unforgiving* sites with inherently difficult geotechnical conditions that need careful and insightful investigation if problems are to be avoided. Examples include the major slope near Brisbane in Australia where the initial ground investigation proved hopelessly inadequate and it took 4 years and more than 5 km of boreholes to understand the geology, and the landslide at Po Selim, Malaysia where the failure occurred at right angles to the main discontinuities, both of which are described in Chapter 8. Further examples could be the Heathrow Express tunnel collapse, also described in Chapter 8, where the contributing factors are still not fully understood in my opinion. The big problem is identifying whether any particular site is unforgiving and in what way.

It is one task of the engineering geologist, through his knowledge of geological and geomorphological processes, to anticipate hazardous geological conditions at a site, and to make sure that a GI is properly focussed. A "brainstorming" checklist approach to hazard prediction is strongly advocated below, to avoid missing something significant.

6.3.3 Fast tracking

Typically, the cost of a site investigation is only a small part of the overall project cost (less than a few percent), yet clients often require some persuasion that the money will be well spent and might be especially reluctant to allow a staged approach because of the impact on programme. He might be unwilling to allow thinking and planning time as the ground investigation data are received and especially unenthusiastic to pay for a revised design as the ground models are developed and refined. Sometimes the Engineer might adopt a "fast-track approach", whereby GI, design and construction are carried out in parallel although this approach carries the risk that information gained later might impact on earlier parts of the design and even on constructed parts of the works. The programming can sometimes go awry, as on a site in Algeria where the author was trying to set out locations for drilling rigs in the same area as a contractor was preparing to construct foundations, which obviously did not make sense. It turned out that the design engineers had made assumptions about the ground conditions without waiting for the GI, thinking that surface footings would be adequate. This proved incorrect. The design needed to be revised completely involving the installation of sheet piles around the foundations, to contain the ground.

6.3.4 The observational method and the "what if" approach to design

6.3.4.1 Observational method

In a similar manner to fast tracking, an observational method is often adopted, especially for tunnelling, whereby ground conditions are predicted, often on rather sparse data, and provisions made for change if and when ground conditions turn out to be different from those anticipated (Powderham, 1994). The observational method generally relies upon monitoring of ground movements, measured loads in structural members, or water levels, whereby performance is checked against predictions.

Nicholson (1994) summarises Peck's (1969) concept of the observational method as firstly, to develop an initial design based on the most probable conditions, together with predictions of behaviour. Calculations are made and these are used to identify contingency plans and trigger values for the monitoring system. If the monitoring records exceed the predicted behaviour, then the pre-defined contingency plans are triggered. The response time for monitoring and implementation of the contingency plan must be appropriate to control the work. There are nowadays, numerous instrumentation companies, which can install and monitor ground behaviour throughout the project, allowing the observational method to be applied almost routinely.

This can go seriously wrong where the ground behaves outside predictions – perhaps because the geological model is fundamentally incorrect or because instrument systems fail or are not reacted to quickly enough. Examples of where instruments were not reacted to early enough include the Heathrow Express tunnel (Muir Wood, 2000) and the Nichol Highway collapse in Singapore (Hight, 2009); these are described in some detail in Chapter 8.

6.3.4.2 "What if" approach

An alternative to the observational approach is the "what if" approach, by which I mean that given the absence of data, one can "imagine" from experience and research, what ground conditions might be encountered so that the project can be designed to deal with the worst-case scenarios. deFreitas (1993) summarised the process as follows:

> It is always worth considering a conceptual model at an early stage in ground investigation planning. It can be thought of basically under 3 headings: (a) what we know, (b) what we need to know and (c) what we do not know.

As an example, a tunnel boring machine was designed to drive a tunnel through the Himalayas so that it could withstand the most challenging, anticipated conditions of bad ground conditions, including squeezing, faults and water pressure using a variety of techniques (Hencher, 2015, page 291). This was based on a rather patchy ground model deduced from surface mapping, together with a few remote or targeted boreholes, just by using pure engineering geological skill. The tunnel was formed with many stops but no serious delay – a prime example of Howard Harding's proposal that "a tunneller should be prepared to be surprised but never astonished".

An "observational" or "what if" approach should also generally be adopted for rock slope construction, although it is seldom referred to as such. It is very difficult to characterise the complete rock fracture network and to estimate the connectivity and presence of rock bridges from a few boreholes and therefore it is extremely important to check any design assumptions during construction by observation, and to be prepared to come up with different solutions for stabilisation as the rock is exposed and structures are identified and mapped (see Box 1.1).

6.3.5 Extent of ground investigation

A large part of any site investigation budget will generally be taken up in sub-surface investigation and characterisation of the ground conditions (Items 3 and 4 in Box 6.1). Important questions are, "how much ground investigation is required and how should it be done?" There are no hard and fast rules, even though some authors try to provide guidance on the basis of site area or volume for particular types of operation or on hypothetical considerations (e.g. Jaksa et al., 2005). In reality, it depends upon the complexity of the geology at the site, how much is already known about the area, the nature of the project and cost. For sites with simple geology, the plan might be for boreholes at 10 to 30 m spacing, for discrete structures like a building (BS 5939: 1999). For a linear structure like a road or railway project, the spacing might be anywhere between 30 and 300 m spacing, depending on perceived variability (Clayton et al., 1995). West et al. (1981) consider the particular difficulties in planning investigations for tunnels. So much depends upon the depth of tunnel, the topography and variability of geology. Often, considerable reliance is made on aerial photography interpretation, geological mapping, a few widely spaced preliminary boreholes and other boreholes targeted at particular perceived hazards such as faults that might be associated with poor-quality rock and high-water inflows. For example, Figure 6.1

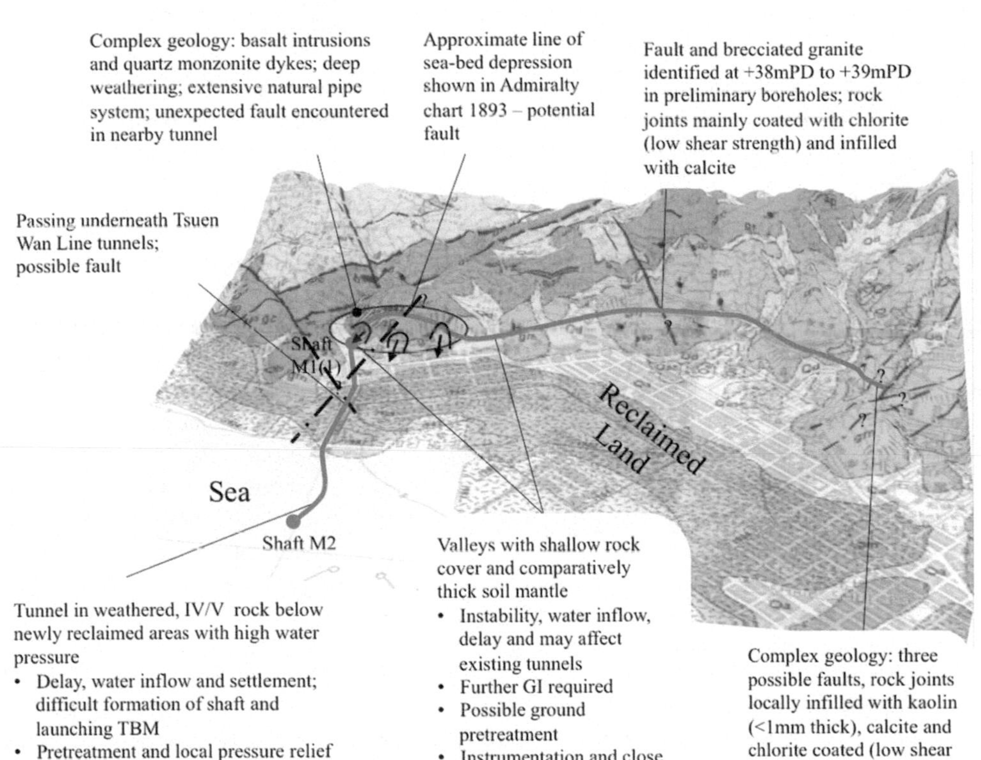

Figure 6.1 Preliminary assessment of ground investigation requirements for a new tunnel, Hong Kong.

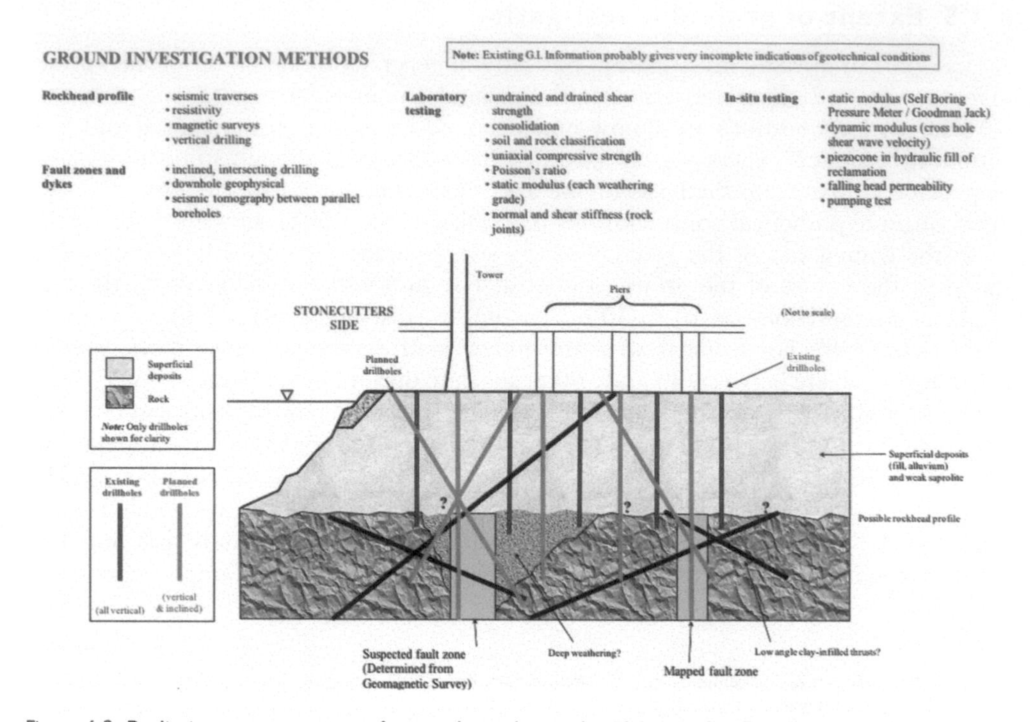

Figure 6.2 Preliminary assessment of ground conditions by Halcrow for East Tower of Stonecutters Bridge, Hong Kong, and need for inclined boreholes to investigate major fault structures.

shows the route of a planned tunnel in Hong Kong, close to the coast and with one shaft (M2) off shore with potential hazards identified, together with a rationale for their mitigation and additional GI. Where steeply dipping geological structures such as faults are anticipated, inclined boreholes may be required. Figure 6.2 shows an assessment of possible conditions under the Eastern Tower of Stonecutters Bridge in Hong Kong at tender stage, based on desk study together with a proposed borehole investigation targeted at likely faults and zones of deep weathering. Broad details of what was actually found are given in Fletcher (2004) and consequences by Tapley et al. (2006).

Requirements and practice for ground investigation vary around the world. In Hong Kong, for example, it is now required practice to put down a borehole at the location of every bored pile (called a "pre-drill"). Elsewhere, a pattern of perhaps three, four or five boreholes might be adopted below each pile cap for a major structure. For example, for the 2nd Incheon Bridge in South Korea, opened in 2009, for each of the main cable stay bridge towers (Figure 6.3) there were four boreholes per pile cap, each of which was about 70 m by 25 m in plan and supported by 24 large-diameter bored piles. For the Busan-Geoje fixed link crossing completed in 2010, also in South Korea, there were two cable-stayed bridge sections, one with two towers and main span of 475 m, the other with three towers. The towers were founded on gravity caissons sitting on excavated rock (Chapter 3) and with plan dimensions of up to 40 m × 20

Figure 6.3 Incheon Bridge, South Korea, shortly before opening.

m. For each of these foundations, there were usually about six boreholes, typically one put down at the conceptual stage, three for the basic design and two for detailed design. For most of the other viaduct piers, with plan caisson dimensions of 17 m ×17 m, there were from one to five boreholes – less where the geology was better known, close to shore.

Obviously, where the site reconnaissance, together with desk study or findings from preliminary boreholes, indicate potentially complex and hazardous conditions, it may prove necessary to put down far more boreholes. For the design of the new South West Transport Corridor near Brisbane, Australia, the preliminary investigation over a critical section comprised four or five widely spaced boreholes and a few trial pits, mostly along the centre line of the road. As the earthworks were approaching completion, minor landslides occurred at road level, together with some indications of deeper-seated movements (Figure 6.4). Over the next few months, an additional 70+ deep boreholes were put down, 56 trial pits and 54 inclinometers installed, despite almost

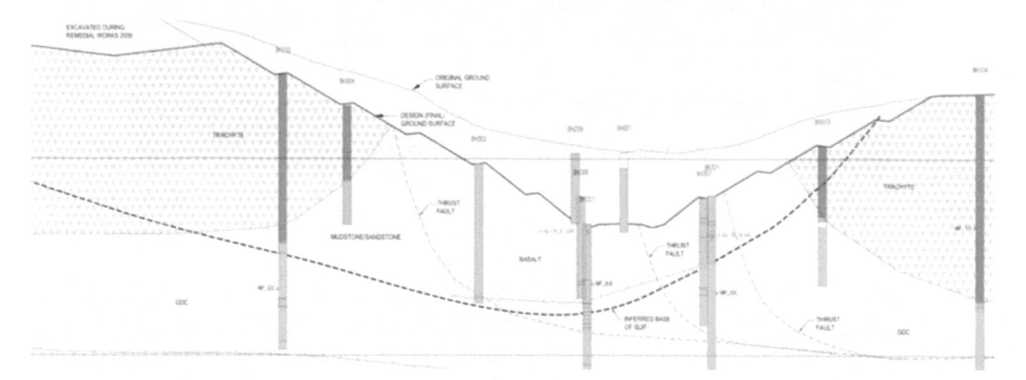

Figure 6.4 Complex geology in slopes near Brisbane, Australia, which required numerous deep drillholes, surface mapping and age-dating to understand.

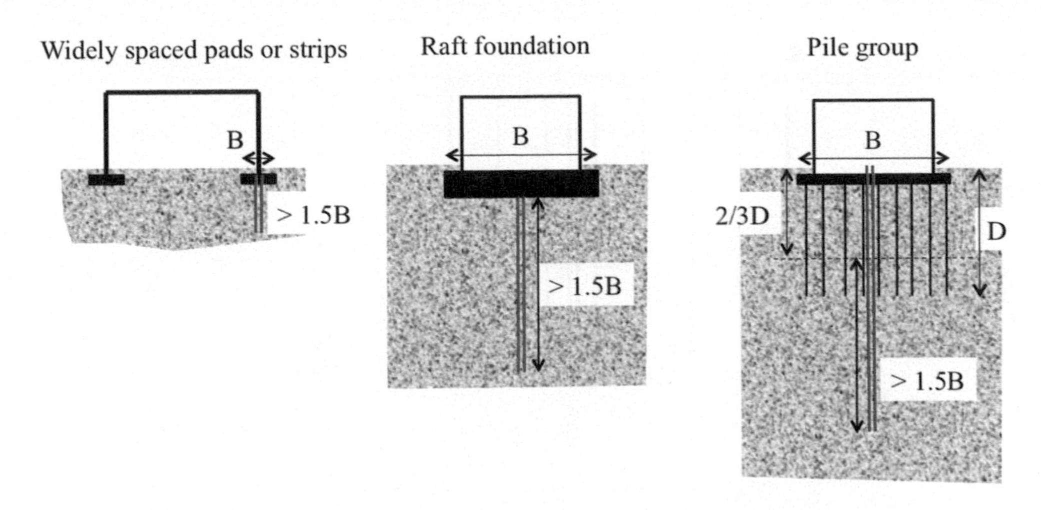

Figure 6.5 Criteria usually adopted for investigating the ground for foundations. Where geology is or may be complex, ground conditions might need to be proved to greater depth and several boreholes might be required. Similarly, these criteria do not apply or limit the need to consider particular site hazards, such as slope stability above or below the site.

100% rock exposure in the cuttings (which were carefully examined and mapped). This intensive investigation allowed the landslide mechanisms to be identified in this very complex, "unforgiving" site and remedial works to be implemented, which permitted this important road to be completed on time (Starr et al., 2010). In hindsight, the preliminary boreholes, which would have been more than adequate for a normal stretch of road, gave no indication of the degree of difficulty and complexity at this site, which only became clear following intensive work involving a wide range of experts. In a similar manner, the landslide at Pos Selim, Malaysia, described in Chapter 8, could not have been anticipated from a few boreholes. The mechanism was at a very large scale and involved too many components to have been unravelled and predicted before the major displacements occurred.

As a general rule, at any site, at least one borehole should be put down to prove ground conditions to a depth far greater than the depth of ground to be stressed significantly by the works. Generally, for foundations, at least one borehole should be taken to at least 1.5 times the breadth (B) of the foundation (Figure 6.5). For pile groups, it is generally assumed that there is an equivalent raft at a depth of $2/3D$ where D is the length of piles and the ground should be proved to at least $1.5B$ below that level. This is only a general guideline – if there is any reason to suspect more variable conditions and, where the geology is non-uniform, one borehole will probably not be enough (Figure 6.6). Poulos (2005) discusses the consequences of "geological imperfections" on pile design and performance. Boreholes are often terminated once rock has been proved to at least 5 m, but this may be inadequate in weathered terrain to prove that "bedrock" is not large boulders. Hencher & McNicholl (1985) describe a case where the boreholes were indeed stopped prematurely, following HK Government guidelines,

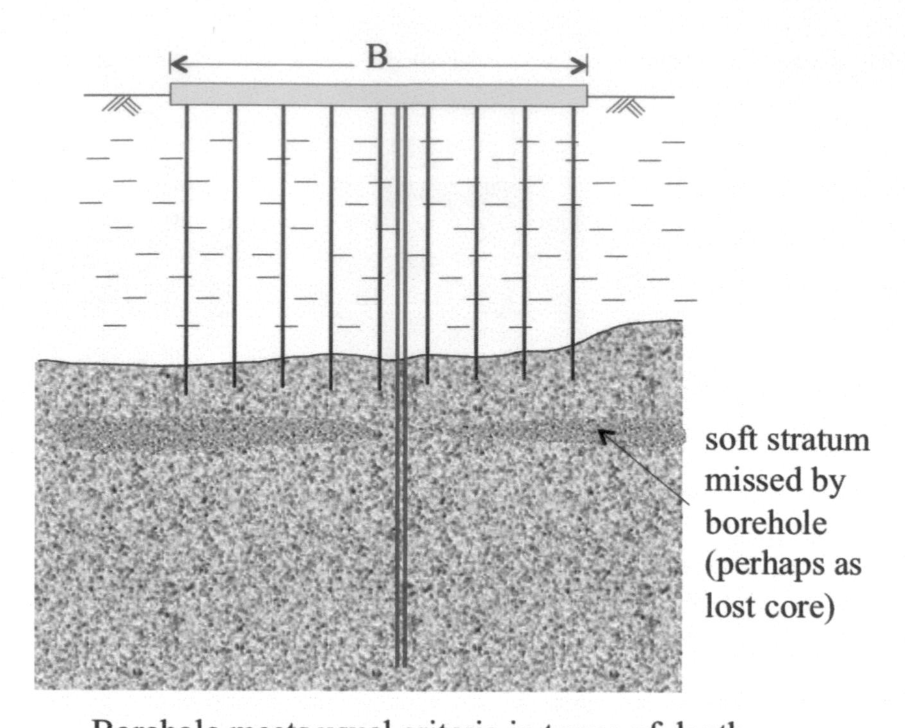

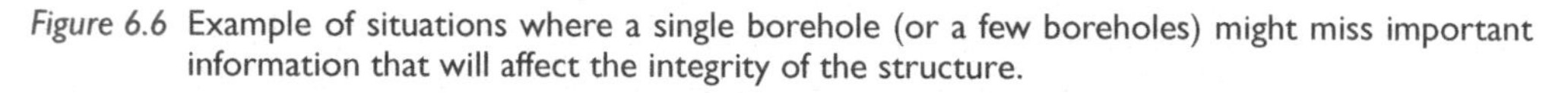

Figure 6.6 Example of situations where a single borehole (or a few boreholes) might miss important information that will affect the integrity of the structure.

and the slope failed at depth in the saprolite, which surrounded the boulders. Whether or not one has reached *in situ* bedrock might be established by geological interpretation of consistent rock fabric or structure across a site, but elsewhere it may be more difficult, in which case it is best to take one or more boreholes even deeper if important to the design.

6.4 PROCEDURES FOR SITE INVESTIGATION

6.4.1 General

Guidance on procedures and methodologies for site investigation is given for the UK by Clayton et al. (1995) and for the USA and more broadly by Hunt (2005). The British Code of Practice for Site Investigations, BS 5930 (British Standards Institution, 1999; British Standards Institution, 2020), provides comprehensive advice on procedures and techniques and for soil and rock description for the UK. Other codes exist for different countries (e.g. Australia, China and New Zealand). Generally, there is consistent advice over the overall approach to site investigation, although terminology and recommended techniques differ. All agree, however, that the first step should be a comprehensive review of all available maps and documents pertaining to a site – this is called a "desk study".

6.4.2 Desk study

6.4.2.1 *Sources of information*

For any site, it is important to conduct a thorough document search. This should include topographic and geological maps. Hazard maps are sometimes available. These include broad seismic zoning maps for countries linked to seismic design codes. In some countries, there are also local seismic micro-zoning maps showing locations of active faults and hazards such as liquefaction susceptibility. Sources of information for the UK are given in BS 5930 (British Standards Institution, 2020) and Clayton et al., 1995. The Association of Geotechnical and Geoenvironmental Specialists (AGS), whose contact details are given in Appendix A, also give useful advice and sources of reference. Records of historical mining activity and previous land use are especially important. In the UK, the British Geological Survey (BGS) has made available a digital atlas of hazards, including mining (but not coal), collapsible materials, swelling and compressible soils, landslides and noxious gas. Landslide hazard maps are published in the USA for southwest California and in Hong Kong, as discussed later.

6.4.2.2 *InSAR and LiDar*

Interferometric synthetic aperture radar (InSAR) and light detection and ranging (LiDar) are relatively recent methods using reflected light beams for measuring topographic relief remotely (Jaboyedoff et al., 2012). The method, which in 2012 was accurate to ±1.5 cm over 1,000 metres, relies on various mathematical algorithms to determine a shape from three or more points of reflection (i.e. on a plane). InSAR is usually carried out from the ground by scanning a site, or from satellites, whereas LiDar can be used on the ground or from flying aircraft or drones over a site, taking radar images of the ground. Dense vegetation and trees cause a problem but otherwise the images can be interpreted as a set of photographs with respect to past activity.

6.4.2.3 *Air photograph interpretation*

Air photographs can be extremely useful for examining sites. Pairs of overlapping photographs can be examined in 3D using stereographic viewers, and skilled operators can provide many insights into the geology and geomorphological conditions (Allum, 1966; Dumbleton & West, 1970). Historical sets of photographs help to reveal the site development and to assess the risk from natural hazards such as landslides. In Hong Kong, it is normal practice to set out the site history for any new project through air photo interpretation (API) of sets of photos dating back to the 1920s (Ho et al., 2004). The role of API in helping to assess the ground conditions at a site is illustrated in Box 6.3.

BOX 6.3 ROLE OF AIR PHOTOGRAPH INTERPRETATION (API)

Air photos used to interpret the nature of terrain, landslides and site history (Figure B6.3.1).

Overlapping air photos allows a skilled earth scientist to examine the site topography in three dimensions. According to Styles (personal communication), in order to do it well you must put yourself on the ground mentally and walk across the terrain looking around in oblique perspective. Topographic expression and other features such as the presence of boulders, hummocky ground, arcuate steps and vegetation, can be interpreted in terms of terrain components and geomorphological development: landslide morphology, degree of weathering and distribution of superficial deposits such as colluvium and alluvium. Broad geological structure such as major joint systems, faults and folds, may be observed, interpreted and measured in a way that would be more difficult working only be mapping exposures on the ground (Figure B6.3.2).

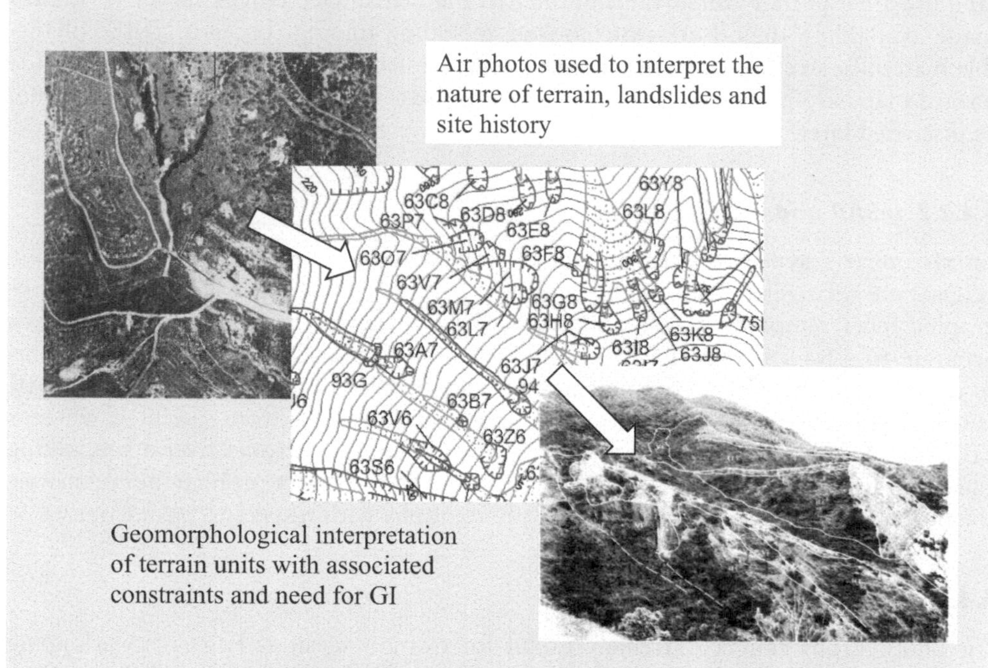

Figure B6.3.1 Process of API. Pairs of overlapping photographs can be examined stereographically to give a 3D image. Major terrain features can be identified and if historical series of photographs are available, then land development and site history can be ascertained, in this example, in terms of landslide history. In the second image above, interpreted landslides have been mapped (with date of the photo in which the landslide is first seen). These interpretations can then be checked in the field (Devonald et al., 2009). In addition, terrain can be split into units on the basis of surface expression, underlying geology, "activity" and vegetation, as described by Burnett et al.. (1985). Third photo and overlay provided by Kevin Styles (Chairman of the Hong Kong Regional Group of the Geological Society of London).

Figure B6.3.2 Major structural lineations visible in aerial photograph and controlling river development of Zambezi River above Victoria Falls between Zimbabwe and Zambia.

Where landslides are identified on photographs, debris run-out can be measured, which may help in assessing the degree of risk for existing and future developments. River channels can be traced and catchments measured. Where a series of historical photographs is available, an inventory of landslide events can be compiled and related to historical rainfall records. Anthropogenic development and use of sites can be documented.

It is important that API is checked by examination in the field and this is known as "ground truthing", which is an integral part of site reconnaissance and field mapping. Similarly, interpreted site history should be checked and correlated against other documentary evidence such as old maps and photographs. The preliminary ground model developed from API and field studies can then be investigated further by trial pits and boreholes, as necessary. Conversely, a ground investigation in an area of variable topography, without prior API, reconnaissance and desk study, may be ineffective and poorly focussed. An introduction to the use of air photographs, with particular consideration of landslide investigations, is given by Ho et al.. (2006).

Even with little training, the importance of air photographs can be immediately clear, as in Figure 6.7, which is an air photograph from 1949 on to which has been marked the route of the Ching Cheung Road in Hong Kong, constructed in 1963. Various ground hazards are evident in the photo (landslides and deep gullying) and it

Figure 6.7 Aerial photograph of Hong Kong in 1949 with route of Ching Cheung road superimposed – failures occurred in cut slopes at C in 1982 on a dry day, 5 days after rainfall and at E in 1972, 10 days after rainfall.

is no surprise that these led to later problems with the road, as addressed in Chapter 8 and discussed by Hudson & Hencher (1984).

Systematic interpretation of air photographs for determining geotechnical hazards has been carried out in several countries. For example, the whole of Hong Kong was mapped, in terms of perceived geotechnical hazard, from air photographs in the 1980s at a 1:20,000 scale and locally at 1:2,500 and, whilst never intended for site-specific interpretation, these were very useful for urban planning (Burnett et al., 1985; Styles & Hansen, 1989). Air photos can be used for detailed measurement by those trained to do so. Topographic surveys can also be carried out using terrestrial or airborne LiDar surveys and these can be repeated to monitor ongoing movements in landslides or in volcanic eruptions (e.g. Jones, 2006). In some situations, especially for remote sites lacking good air photo coverage, satellite images may be helpful, although often the scale is not large enough to provide the detailed interpretation required and stereo imagery is impossible – unlike for purpose-flown aerial photograph sequences. Use of false spectral images such as infra-red can help interpretation, for example, of vegetation and seepage.

6.4.3 Planning a ground investigation

BS 5930 and most textbooks on site investigation provide good information on techniques and procedures but little advice on how to plan a ground investigation or on how to separate and characterise geotechnical units within a geological model. They also say little about how to anticipate hazards, which is a key task for the engineering geologist. It is important to take a holistic view of the geological and hydrogeological setting – the "total geological model" approach of Fookes et al. (2000) – but the geological data need to be prioritised to identify what is really important to the project and to obtain the relevant parameters for safe design.

The problem is that there are so many things that might potentially go wrong at sites and with alternatives for cost-effective design that it is sometimes difficult to know where to start in collecting information. One might hope that simply by following a code of practice, that would be enough, but, in practice, the critical detail may be overshadowed by relatively irrelevant information collected following routine drilling and logging methodology.

One approach that can be useful for planning and reviewing data from a ground investigation, and focusing on critical information, is to consider the different aspects of the site and how they might affect the project in a checklist manner (Knill, 1976, 2003; Hencher, 2007). The various components and aspects of the project and how different site conditions might affect its success are considered one by one and in an integrated way. This is similar to the rock engineering systems methodology of Hudson (1992), in which the various parameters of a project are set out and their influence judged and measured in a relative way (Hudson & Harrison, 1992). This is also akin to the concept of a risk register for a civil engineering project at the design and construction stages, whereby each potential hazard and its consequence is identified and plans made for how those risks might be mitigated and managed.

The three verbal equations of Knill (1976) are set out in Table 6.1. The first equation addresses purely geological factors: material and mass strengths and other properties.

Table 6.1 Engineering geology expressed as three verbal equations

Equation 1 GEOLOGY
MATERIAL PROPERTIES + MASS FABRIC = MASS PROPERTIES
The first equation includes the geology of the site and concerns the physical, chemical and engineering properties of the ground at small and large scales. It essentially constitutes the soil and rock ground conditions.
Equation 2 + ENVIRONMENT
MASS PROPERTIES + ENVIRONMENT = ENGINEERING GEOLOGICAL SITUATION
The second equation relates to the geological setting within the environment. Environmental factors include climatic influences, groundwater, stress, time and natural hazards.
Equation 3 + CONSTRUCTION
ENGINEERING GEOLOGICAL SITUATION + INFLUENCE OF ENGINEERING WORKS = ENGINEERING BEHAVIOUR OF GROUND.
The third equation relates to changes caused by the engineering works such as loading, unloading and changes to the groundwater levels. It is the job of the engineer to ensure that the changes are within acceptable limits

Source: After Knill, 1976.

Table 6.2 Examples of material-scale factors that should be considered for a project

Factor	*Considerations*	*Examples of rock types/situations*
mineral hardness	abrasivity, damage to drilling equipment	silica-rich rocks and soils (e.g. quartzite, flint, chert)
mineral chemistry	reaction in concrete	olivine, high temperature quartz, etc.
	Oxidation – acids	pyrites
	swelling, squeezing and dissolution	swelling clays and mudrocks, salts, limestone
	low friction	clay-infilled discontinuities, chlorite coating
loose, open texture	collapse on disturbance or overloading, liquefaction, piping, low shear strength	poorly cemented sandstone, completely weathered rocks (V); loess; quick clays

The second equation is to assess the influence of environmental factors, such as *in situ* stress, water and earthquakes on the geological/geotechnical model. The final consideration is how these factors affect, and are affected by the construction works.

6.4.3.1 Equation 1: geological factors

The first equation encourages the investigator to consider the ground profile (geology) and its properties at both the material and mass scales.

Material scale

The material scale is that of the intact soil and rock making up the site. It is also the scale of laboratory testing, which is usually the source of engineering parameters for design. Typical factors to review are given in Table 6.2. They include the chemistry, density and strength of the various geological materials and contained fluids making up the geological profile. Hazards might include adverse chemical attack on foundations or ground anchors, liquefaction during an earthquake, swelling or low shear strength due to the presence of smectite clays, abrasivity or potential for piping failure. Inherent site hazards associated with geology include harmful minerals, such as asbestos and erionite. Granitic areas, phosphates, shale and old mine tailings are sometimes linked to relatively high levels of radon gas, which is estimated to cause up to 2,000 deaths each year in the UK (Health Protection Agency). Talbot et al. (1997) describe investigations for radon during tunnelling. Gas hazards are especially important considerations for tunnelling and mining but are also an issue for completed structures, as illustrated by the Abbeystead disaster of 1984 when methane that migrated from coal-bearing strata accumulated in a valve house and exploded killing 16 people (Health and Safety Executive, 1985). These are all material-scale factors linked to the geological nature of the rocks at a site.

Locating sources for aggregate, armourstone and other building materials is often a task for an engineering geologist. Other than the obvious considerations of ensuring adequate reserves and cost, one must consider durability and reactivity, and this will involve geological characterisation and probably testing. Two examples in Chapter 8 relate how adverse material properties of sourced fill and aggregate material led to severe consequences. At Carsington Dam, UK, a chemical reaction was set up between the various rocks used to construct the dam, which resulted in acid pollution of river courses and the production of hazardous gas, with the death of two workers; at

Pracana Dam, Portugal, the use of reactive aggregate led to rapid deterioration of the concrete. The latter phenomenon has been reported from many locations around the world and is associated with a variety of minerals, including cryptocrystalline silica (some types of flint), high-temperature quartz, opal and rock types ranging from greywacke to andesite. Details of how to investigate whether aggregate may be reactive and actions to take are given in RILEM (2003).

Mass scale

Mass-scale factors include the distribution of different materials in different weathering zones or structural regimes, as successive strata or as intrusions. It includes structural geological features such as folds, faults, unconformities and joints (Table 6.3). Discontinuities very commonly control the mechanical behaviour of rock masses and some soils. They strongly influence strength, deformability and hydraulic conductivity.

One of the main geological hazards to engineering projects at the mass scale is faults. Faults can be associated with zones of fractured and weathered material, high permeability and earthquakes. Alternatively, faults can be tight, cemented and actually act as barriers to flow, and natural dams rather than zones of high permeability. Faults should always be looked for and their influence considered. There are many cases of unwary constructers building on or across faults, with severe consequences, sometimes leading to delays to projects or a need for re-design. Consequence is sometimes difficult to predict but should be considered and investigated. Other examples of mass factors that would significantly affect projects include boulders in otherwise weak soil, which might preclude the use of driven piles or would comprise a hazard in a steep slope.

An example of where a formal review of the potential for large-scale structural control might have helped is provided by the investigation for a potential nuclear

Table 6.3 Examples of mass-scale factors that should be considered for a project

Factor	*Considerations*	*Examples of rock types/situations*
lithological heterogeneity	difficulty in establishing engineering properties, construction problems (plant and methodology)	colluvium, un-engineered fill, interbedded strong and weak strata, soft ground with hard corestones
joints/natural fractures	sliding or toppling of blocks, deformation, water inflows/collapse, leakage/migration of radioactive fluids	slopes, foundations, tunnels and reservoirs, nuclear repository
faults	as joints, sudden changes in conditions, displacement, dynamic loads	tunnels, foundations, seismically active areas
structural boundaries, folds, intrusions	heterogeneity, local stress concentrations, changes in permeability – water inflows	all rocks/soils
weathering (mass scale)	mass weakening; heterogeneity (hard in soft matrix), local water inflow, unloading fractures	all rocks and soils close to Earth's surface, especially in tropical zones; ravelling in disintegrated rock masses
hydrothermal alteration	as weathering, minerals low strength	generally igneous rocks

waste repository at Sellafield in the UK, as explained in Box 6.4. It appears that early boreholes and tests did not sample relatively widely spaced master joints within the stratum and, therefore, an incomplete picture was formed of the factors controlling mass permeability. In hindsight, the true nature of the rock might have been anticipated by desk study and field reconnaissance of exposures.

BOX 6.4 ANTICIPATING MASS CHARACTERISTICS

THE BROCKRAM AND THE SELLAFIELD INVESTIGATIONS

The UK government specification for acceptable risk from any nuclear waste repository was set to be extremely onerous and necessitated intensive investigation combined with intensive modelling. Ground investigation has been conducted at Sellafield, Cumbria, since 1989, aimed at determining whether or not the site is suitable as a repository for radioactive waste. The target host rock is the Borrowdale Volcanics at a depth of more than 500 m. Part of the modelling has involved trying to predict groundwater flow and the movement of radio-nuclides. For this, a good ground model was necessary with estimates of permeability for the full rock sequence. Several high-quality boreholes have been put down at the site and logged very carefully. A general model has been developed, as illustrated in Figure B6.4.1 (ENE is to the right). The geological model has the Borrowdale Volcanics, which contain saline water, separated from the overlying sandstones, containing fresh water, by a bed called the Brockram, which is typically 25–100 m in thickness and cut by faults. The Brockram and associated evaporites and shale further west evidently play a very important potential role as a barrier to flow of groundwater (flow into the repository) and, hence, radio-nuclides migrating away for the repository.

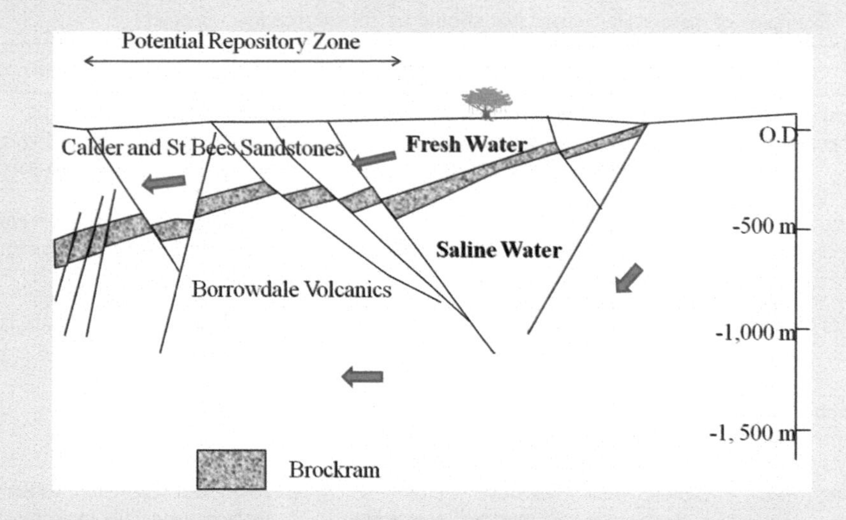

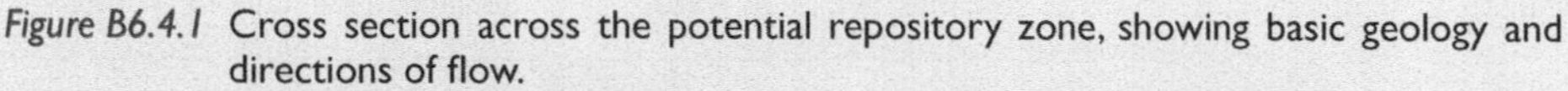

Figure B6.4.1 Cross section across the potential repository zone, showing basic geology and directions of flow.

Source: Modified from Chaplow, 1996.

Figure B6.4.2 Brockram at Hoff's Quarry in the Vale of Eden, UK. The rock is a cross-bedded, limestone-rich, strongly cemented breccia. It contains fossiliferous blocks of Carboniferous Limestone and rather rarer rocks such as Whin Sill dolerite. It has the appearance of a wadi-type deposit – poorly sorted, probably rapidly deposited by flash floods. From its field appearance, it has a low permeability as a material.

EARLY MODELLING

For most early numerical simulations, the Brockram was modelled with very low conductivity (2×10^{-10} to 1×10^{-9} m/s), based largely on borehole tests and "expert elucidation" (Heathcote et al., 1996). These values are similar to those measured for the Borrowdale Volcanics – 50% measured over 50 m lengths, with permeability 1×10^{-10} m/s, according to Chaplow (1996).

LATER TESTS

At a later stage, field tests were carried out that found "significant flows" in the Brockram, and the earlier modelling had to be revised. Michie (1996) reports hydraulic conductivity measurements within the Brockram with a maximum of 1×10^{-5} m/s, i.e. four orders of magnitude higher than adopted for the early models.

WAS THIS A SURPRISE?

The changed perception for this important stratum might be considered just part of what is to be expected in any progressive ground investigation. However, the potential for locally high permeability associated with extremely widely spaced and persistent joints, at spacing such that they will be rarely sampled in boreholes, could have been anticipated, partly because such joints can be observed elsewhere at exposures in the Lake District. At Hoff's Quarry to the east of the Lake District, the rock can be examined, and at a material scale a low permeability would be anticipated (Figure B6.4.2).

At a larger scale, as shown in Figure B6.4.3, the rock at Hoff can be seen cut by near-vertical master joints, which would affect the mass permeability in a dramatic way (as

Figure B6.4.3 More distant view of Brockram at Hoff's Quarry. Note the fully persistent, near-vertical master joints about 40 m apart. Note that vertical joints are unlikely to be intersected by vertical boreholes.

evidenced from the Sellafield test). There were also indications from the literature that the Brockram might be permeable at a scale of hundreds of metres. As an additional clue, Trotter et al. (1937) commented on the possibility of pathways through the Brockram, with reference to the distribution of haematite mines within the Carboniferous Limestone underlying the Brockram.

Lessons: it is very important not simply to rely on site-specific data when elucidating parameters for design. There is a need to consider the geological setting, origins and history – with all that entails. Furthermore, when looking at data from boreholes, especially ones with a strong directional bias, one should consider all the field evidence (as at Hoff's quarry) that might offer some clues as to the validity of the expert elucidation process.

6.4.3.2 Equation 2: Environmental factors

Environmental factors, some of which are listed in Table 6.4, including hydrogeological conditions, should be considered part of the ground model for a site, but are best reviewed separately from the basic geology, although the two are closely interrelated. In this edition, Chapter 5 now highlights how environmental factors are to be assessed and measured. The environmental factors to be accounted for depend largely on the nature, sensitivity and design life of structures and the consequence of failure. It is usual practice to design structures to some "return period" criterion such as a 1 in 100 year storm or 1 in 1,000 year earthquake, the parameters for which are determined statistically through historical review. In some cases, engineers will also want to know the largest magnitude event that might occur, given the location of the site and the geological situation. Then some thought can be given as to whether or not it is possible to make some provision for that maximum credible event. For earthquakes, for example, a structure might be designed to behave elastically (without permanent damage) for a 1 in 1,000 year event but for a, very unlikely, maximum credible event, some degree of damage would be accepted.

The factors to review at this stage include natural hazards such as earthquake loading, strong winds, heavy rain and high groundwater pressures or flooding. Anthropogenic factors to consider include industrial contamination and proximity of other structures and any constraints that they may impose.

6.4.3.3 Equation 3: Construction-related factors

The third verbal equation of Knill (1976) addresses the interaction between the geological and environmental conditions at a site and the construction and operation constraints (Hencher & Daughton, 2000). Excavation will always give rise to changes in stresses and the ground may need to be supported. Excavations may also result in changes in groundwater and the consequences need to be addressed and mitigated if potentially harmful. Similarly, loading from structures has to be thought through, not only because of deformations but also because of potentially raising water pressures, albeit temporarily.

Table 6.4 Examples of environmental factors that should be considered for a project

Factor	*Considerations*	*Examples of rock types/situations*
stresses	high stress: squeezing, overstressing, rock bursts low stress: open fractures, high inflows, roof collapse in tunnels	mountain slopes and at depth, shield areas, seismically active areas, extensional tectonic zones, unloaded zones, hillside ridges
natural gases	methane, radon	coal measures, granite, black shales
seismicity	design loading, liquefaction, landslides	seismically active zones, high consequence situation in low seismic zones
influenced by man	unexpectedly weak rocks, collapse structures	undermined areas
	gases and leachate	landfills, industrial areas
groundwater chemistry	chemical attack on anchors/nails foundations/ materials	acidic groundwater, salt water
groundwater pressure	effective stress, head driving inflow, settlement if drawn down	all soils and rocks
ice	ground heave, special problems in permafrost/ tundra areas, freeze-thaw jacking and disintegration	anywhere out of tropics
biogenic factors	physical weathering by vegetation,	near-surface slopes
	rotted roots leading to piping,	weathered rocks
	insect attack	causing tree collapse

There will also be hazards associated specifically with the way the project is to be carried out. For example, a drill and blast tunnel is very different to one excavated by a tunnel boring machine and will have specific ground hazards associated with its construction. Similarly, the construction constraints are very different for bored piling compared to driven piles (Table 6.5). The systematic review and investigation of site geology and environmental factors, discussed earlier, needs to be conducted with specific reference to the project at hand. This will hopefully allow the key hazards to be identified and design to be robust yet cost-effective. Nevertheless, models are always simplifications, and the engineer must adopt a cautious and robust approach when designing, especially where the geological conditions are potentially variable and where that variability might cause problems.

6.4.3.4 Discussion

It is evident that site investigation cannot provide a fully detailed picture of the ground conditions to be faced. This is particularly true for tunnelling, because of the length of ground to be traversed, the volume of rock to be excavated and often the nature of the terrain, which prevents boreholes being put down to tunnel level or makes their cost

Table 6.5 Examples of the influence of engineering works

Factor	*Considerations*
loading or unloading the ground – statically or dynamically	settlement, failure; opening up of rock joints; increased permeability in cut slopes; vibrations due to blasting or driven piles
	loss of support and chimneying to the surface from tunnels; roof collapse if inadequate support; blowing out of debris from excess pressure used in earth balance and slurry TBMs
change in water table	increased or decreased pressure head, change in effective stress; drawdown drainage to tunnels leading to settlement of structures at the ground surface; induced seismicity from reservoir loading
denudation or land clearance	increased infiltration, erosion, landsliding

unjustifiable. Instead, reliance must be placed on engineering geological interpretation of available information, prediction on the basis of known geological relationships and careful interpolation and extrapolation of data by experienced practitioners. Factors crucial to the success of the operation, need to be judged and consideration given to the question: *what if*? It is generally too late to introduce major changes to the methods of working, support measures, etc. at the construction stage, without serious cost implications.

Site investigation must be targeted at establishing those factors that are important to the project and not to waste money and time investigating and testing aspects that can be readily estimated to an acceptable level or aspects that are simply irrelevant. This requires a careful review of geotechnical hazards, as advocated above. Even then, one must remain wary of the unknowns and consider ways in which residual risks can be investigated further and mitigated, perhaps during construction.

There is a somewhat unhealthy belief that standardisation (for example, by using British Standards, Eurocodes, Geoguides and ISRM Standard Methods) will provide protection against ground condition hazards. Whilst most standards certainly encompass and encourage good practice, they often do so in a generic way that may not always be appropriate to the project at hand and they may not provide specific advice for coping with a particular situation. Ground investigations are often designed on the basis of some kind of norm – a "one size fits all" approach to ground investigation. It is imagined that a certain number of boreholes and tests will suffice for a particular project, essentially irrespective of the actual ground conditions at the site. This ignores the fact that ground investigations of average scope are probably unnecessary for many sites but will fail to identify the actual ground condition hazards at rare, but less forgiving sites. Similarly, an averaging-type approach will mean that many irrelevant and unnecessary samples are taken and tested whilst the most important aspects of a site are perhaps missed or poorly appreciated. This is, unfortunately, commonplace.

If the hazards are considered in a systematic way, as discussed earlier, then the risks can be thought through fully and this will help the ground investigation to be better focussed. The process is illustrated in Box 6.5 for a hydroelectric scheme involving the construction of a dam, reservoir, power station and associated infrastructure.

BOX 6.5 PLANNING A SITE INVESTIGATION FOR A NEW HYDROELECTRIC SCHEME (SEE FIGURE B6.5.1)

Project Concept: High arch dam with high-pressure penstock tunnels (120 metre hydraulic head) leading to underground power house; tailrace tunnels and surge chamber. Structures to be considered include reservoir, ancillary buildings, roads, power lines and diversion tunnel. Sources of concrete aggregate need to be identified as well as locations for disposing construction waste.

General Setting: Valley with narrowing point suitable for arch dam (high stresses). Topography and hydrology adequate for reservoir capacity. Steep slopes above reservoir.

Geology from Preliminary Desk Study: Major fault along valley, maybe more. Right abutment (looking downstream) in granitic rock, sometimes deeply weathered. Left side ancient schist, greywacke, mudstone and some limestone. Folded and faulted with many joints. Alluvial sediments along valley.

KEY ISSUES FOR INVESTIGATION (SEE TABLE B6.5.1):

Dam: Stability of foundations and abutments, settlement, leakage, overtopping from landslide into reservoir, silting up.

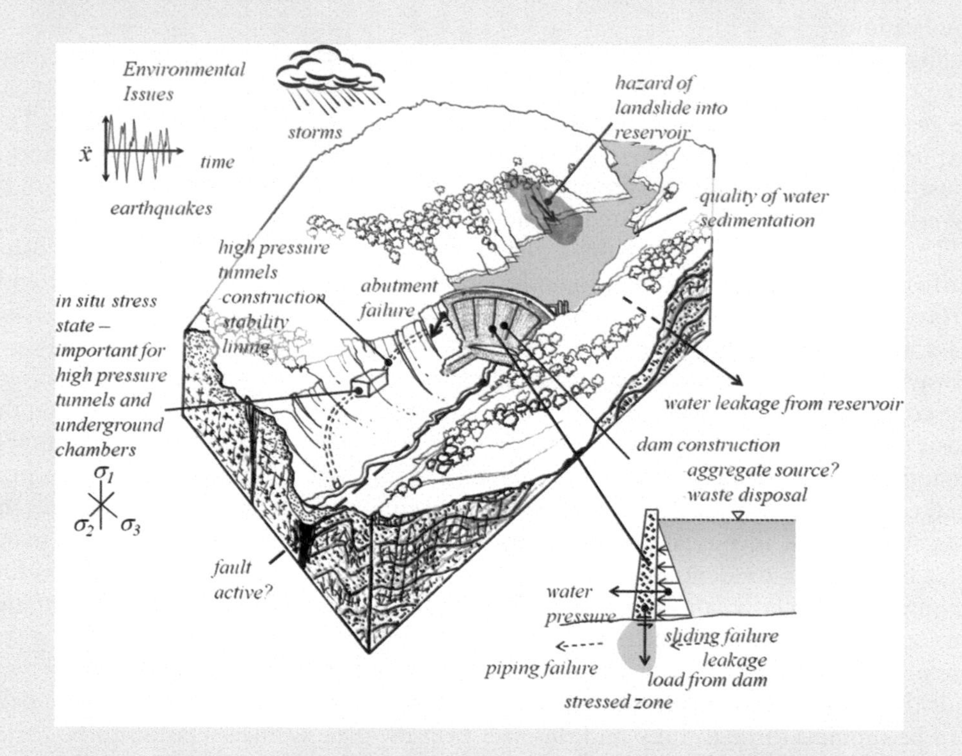

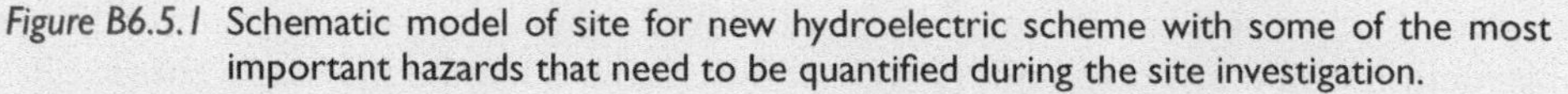
Figure B6.5.1 Schematic model of site for new hydroelectric scheme with some of the most important hazards that need to be quantified during the site investigation.

Tunnels and Powerhouse: Rock quality, *in situ* stress state, construction method, stability, lining and support requirements.

Reservoir: Leakage, siltation, water quality.

Construction: Source of aggregate, waste disposal, access, river diversion.

Table B6.5.1 Main geotechnical considerations when conducting site investigation for dam and surrounding area

Issues	*Equation 1*		*Equation 2*	*Equation 3*
	Geology		*Environment*	*Construction*
	Material	*Mass*		
Arch Dam Stability and Construction	Strength, deformability and durability of foundation materials including highly stressed abutments	Geological profile; depth to bedrock Presence of discontinuities allowing failure in abutments or sliding failure below dam Fault reactivation	Seismicity. Water pressure in foundations and abutments. Check history of mining	Adequate source of non-reactive aggregate Waste disposal locations (fill embankments)
Leakage below dam and from reservoir	Permeability (need for grouting/ cut offs) Potential for piping	Leakage on main fault and other faults/ weathered zones Limestone might be karstic	Groundwater profile in surrounding terrain. Existing throughflow paths.	Options for grouting and/ or cut off structures
Landslides into reservoir	Material strength	Adverse discontinuities; aquitards allowing perched water pressure Landslide history	Response of groundwater to storms and to lowering of water in reservoir Seismic loading	Need for stabilisation such as drainage or option to remove hazardous ground
Powerhouse and high pressure tunnels	Rock strength Abrasivity for tunnel equipment	Fracturing (rock mass classification allows judgement of stabilisation required) Weathered zones	*in situ* stress state (potential squeezing or leakage and need for steel liners) Groundwater pressure and permeability (inflows or water loss for operating tunnels)	Method of excavation Ground movement due to excavation Blasting vibration Groundwater changes

Figure 6.8 (a) Two MSc Engineering Geology students in Portugal discussing the nature of fault with Dr Chris Fletcher (ex BGS and GEO structural geologist). (b) Three junior engineering geologists from Mott Macdonald, mapping complex geological conditions in Turkey, again under the guidance of Dr Fletcher.

6.5 FIELD RECONNAISSANCE AND MAPPING

6.5.1 General

At many sites, geologists can get a great deal from examining the landscape, mapping and interpolating information from exposures, and this is one of the most important aspects of geological education and training (Figure 6.8). This, together with desk study information, should allow preliminary ground models to be developed, which can then be used to form the basis for planning any necessary ground investigation. The preliminary model should allow an initial layout of the components of the project and, for buildings, some insight into the types of foundation that might be required. For tunnels, decisions can be made on locations for portals and access shafts. The degree to which walk-over studies and field mapping can be cost-effective is often overlooked, as illustrated by a case example in Box 6.6.

BOX 6.6 CASE EXAMPLE

COST-EFFECTIVENESS OF SITE RECONNAISSANCE–BRIDGE ABUTMENT, LAKE DISTRICT, UK

The first ground investigation that the author was involved with was for a bridge abutment in the Lake District, UK. Figure B6.6.1 is a view of the rock cliff that was to form the abutment, and halfway down the cliff is a platform. Figure B6.6.2 is a side view of the platform. The man in the middle of the photograph is logging a borehole, using a periscope that has been inserted into a hole, inclined at about 45 degrees, drilled into the rock from the

Figure B6.6.1 Drilling platform on cliff.

Figure B6.6.2 Borehole periscope in use.

same platform. In the foreground, rock can be seen with a fabric dipping roughly parallel to the cliff. For reasons that are unimportant now, a question arose regarding the geological structure being logged by the periscope.

The site engineer was asked for his geological map of the rock along the river (including the 100% exposed cliff). He replied, "what map?"

There we were, perched on a precarious and extremely expensive platform. A drilling rig had been brought in and lowered down the cliff to drill an inclined borehole of perhaps 73 mm diameter, at great cost, and we had been brought to the site from London to log the hole using a periscope. Meanwhile, the full rock exposure was available to be mapped and interpreted at very little cost, which would have allowed a much better and more reliable interpretation of the geological structure than was possible from a single borehole.

Lesson: Use the freely available information first (desk study and walk-over/mapping) before deciding on what ground investigation is necessary at a site.

BOX 6.7 THE SLAKE TEST

DEFINITION 1

The "slake" test as used throughout this book, is a very simple index test to see whether a sample of rock disintegrates when placed in water. It is a quick and effective test that can be carried out when logging core or as a test in the field.

In Hong Kong, Singapore, Australia and other countries, the slake test is used to help classify the degree of weathering. The test was originally defined by Moye (1955) to differentiate between completely decomposed granite (slakes) and highly decomposed granite (doesn't slake).

A Plate from the CHASE study guidance document (1982) for description and classification. is presented as Figure B6.7.1

There are 4 categories for behaviour when water is added, after 2 minutes, and then when the sample is gently agitated:

1. Does not slake
2. Slakes on agitation to fragments
3. Slakes on agitation to a slurry
4. Slakes completely on adding water

It can be seen that sample 1 (top left) and sample 2 (bottom left) did not disaggregate when placed in water, but sample 2 did on agitation. Both of these samples were classified as Grade IV (highly decomposed). Sample 3 (upper right) clearly had rock texture before adding water, and after 45 seconds had "slaked" completely. It was classified as Grade V (completely decomposed). Sample 4 (bottom right) also slaked completely but, because it had no remnant rock texture at the start, was classified as Grade VI (residual soil). The test is extremely useful for splitting Grades IV and V (although not definitive) and describers should use a "basketful" of tests to define weathering grades as recommended by Hencher (1986).

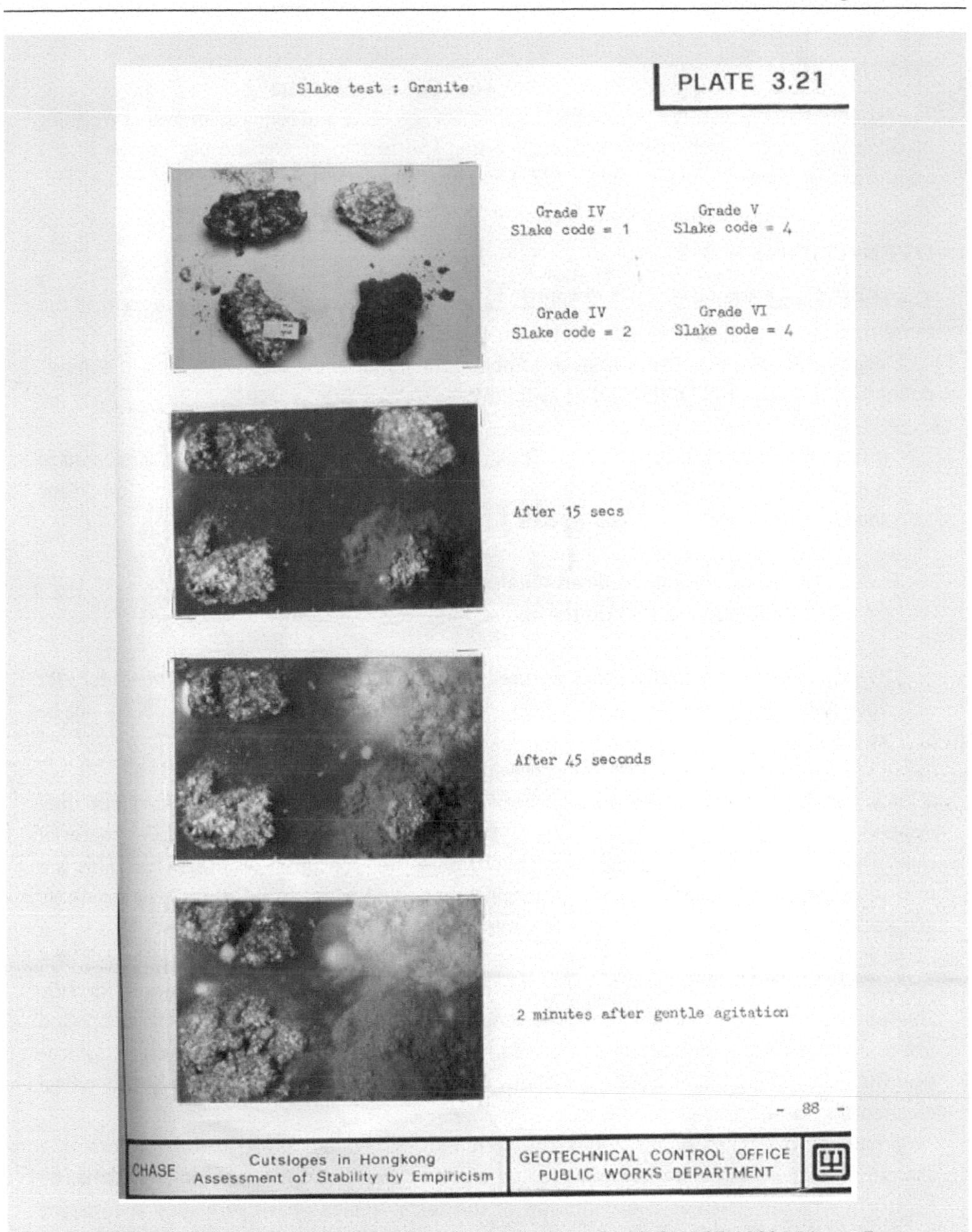

Figure B6.7.1 Slake test on samples of granite, figure taken from CHASE, 1982, Volume F.

DEFINITION 2

The "slake and dispersion" test is used specifically for soil classification for agricultural purposes (Emerson, 1987). This is not the same as the test defined above and the two should not be confused.

DEFINITION 3

The "slake-durability test" is a test on the durability of aggregate and requires an oven, a weighing machine and a net which revolves through water. The test measures the percentage of the aggregate that is worn down (Franklin, 1972) and was later "standardised" by ISRM and ASTM.

DEFINITIONS 4 & 5

The slake test as referred to in BS5930: 2022, in the weathering grade table therein, is the same test as referred to above at Definition 1. i.e. Definition 4 = Definition 1.

Despite this "correct" use, whoever compiled the Eurocode 7 decided to add in a "new" definition of slaking (DEFINITION 5) as follows:

> It is a requirement of the EN that "*The degradation of rock material when it is exposed to a new water or atmospheric environment should be assessed where the relevant conditions shall be determined*".

Three new terms were introduced - Stable, Fairly stable and Unstable
Hencher (2008) commented on the above:

> "*This requirement is rather poorly phrased and it is unclear whether this is mandatory for each encountered material - one hopes not - it would take a long time and probably not be very productive.*"

It is not known where the definitions of these new terms have come from but they depend on submersion of samples in water **for 24 hours**. "Fairly stable" includes material that "slakes" as distinguished from "Unstable" where the "specimen disintegrates". This is a misuse of the term "slakes" which is generally taken itself to mean disintegration or disaggregation of a rock - and actually in a few minutes rather than 24 hours.

It is of some importance to tunnelling in weathered rocks (see for example the discussions by Shirlaw, and Knill 2003). Incidentally and in passing, these discussions by Shirlaw and Knill would no longer make sense if one was trained to follow the proposed EN documents. If the authors of this document wished to introduce some new classification then they should have done so without redefining terms already used in geotechnical engineering to mean something else.

Whatever my criticism (in 2008), the terminology and envisaged tests (immersing a sample of rock in water for 24 hours to see whether the "**specimen surface crumbles highly**"), which is given as the definition of the term "slakes", have remained a constant from the original BS/EN of whatever date, through BS5930: 2015, through to BS5930:2022 Table 28 (May the dear Lord protect us....).

Mapping can be done in the traditional geological manner, using base maps and plans, or onto air photographs, which may need to be rectified for scale. Observations such as spring lines (Figure 6.9) are not only important in delineating probable geological boundaries but also in their own right for hydrogeological modelling. Observation

Figure 6.9 Spring line revealed following heavy rain at base of Carboniferous Limestone, north of Kilnsey Crag, West Yorkshire, UK.

points can be marked in the field, to be picked up accurately later by surveyors (Figure 6.10). Alternatively, locations can be recorded by GPS and input directly into a computer, as illustrated in Figure 6.11. The success of preliminary mapping can be enhanced by letting an early contract to clear vegetation, allow safe access and to put down trial pits and trenches on the instruction of the mapping geologist (Figure 6.12).

Soils and rock can be examined, described and characterised in natural exposures and in trial pits and trenches, and full descriptions should be provided, as discussed later. Samples can be cut by hand for transfer to the laboratory, with relatively slight disturbance (Figures 6.13 and 6.14).

Access can be facilitated by using hydraulic platforms or by temporary scaffolding (Figure 6.15). Trial pits and trenches should not be entered unless properly supported, and care must be taken in examining any steep exposure; as a general rule, for safety reasons, field work should be conducted by teams of at least two people with safety plans as necessary, and access to mobile phones.

Apart from the general benefits to be gained from mapping freely available or cheaply created surface exposures to determine local geology, they are particularly important for characterising aspects of rock structure, such as roughness and persistence of discontinuities, which cannot be determined in boreholes. As for all measurements,

Figure 6.10 Mapping complex geology. The numbers mark boundaries between basalt dykes and the country rock in the foundations of the Queens Valley reservoir in Jersey, UK, Surveyors can follow up by mapping the complexity accurately.

Figure 6.11 Hand-held computer with ortho-corrected air photographs and terrain maps, used to navigate to natural terrain landslide. GPS used to get accurate locations of identified features.

Figure 6.12 Local labourers employed to dig some trial pits during preliminary field mapping. Tlemcen University and hospital site, Algeria.

however, extrapolation should only be made with caution and with awareness that structure and rock quality may change rapidly from location to location (Piteau, 1973). Exposed soil may be desiccated and stronger than soil at depth; exposed rock will often be more weathered with closer and more persistent fractures than rock only a few metres in from the exposed surface.

Figure 6.13 Hand trimming a sample to size in the field, for transportation to laboratory and triaxial testing.

Information gained from desk study and site reconnaissance can be analysed and draped over 3D digital models using GIS, which greatly assists visualisation, interpretation and planning of GI, including access.

6.5.2 Describing field exposures

The task of describing a large field exposure, say in a cut slope, can be daunting, and the following procedure is recommended. The exposure (natural or man-made)

Figure 6.14 Block sample trimmed by hand to fit into a "Leeds" direct shear box.

Figure 6.15 Cherry picker platform used to examine recently failed rock slope to allow remedial action to be determined, Hong Kong.

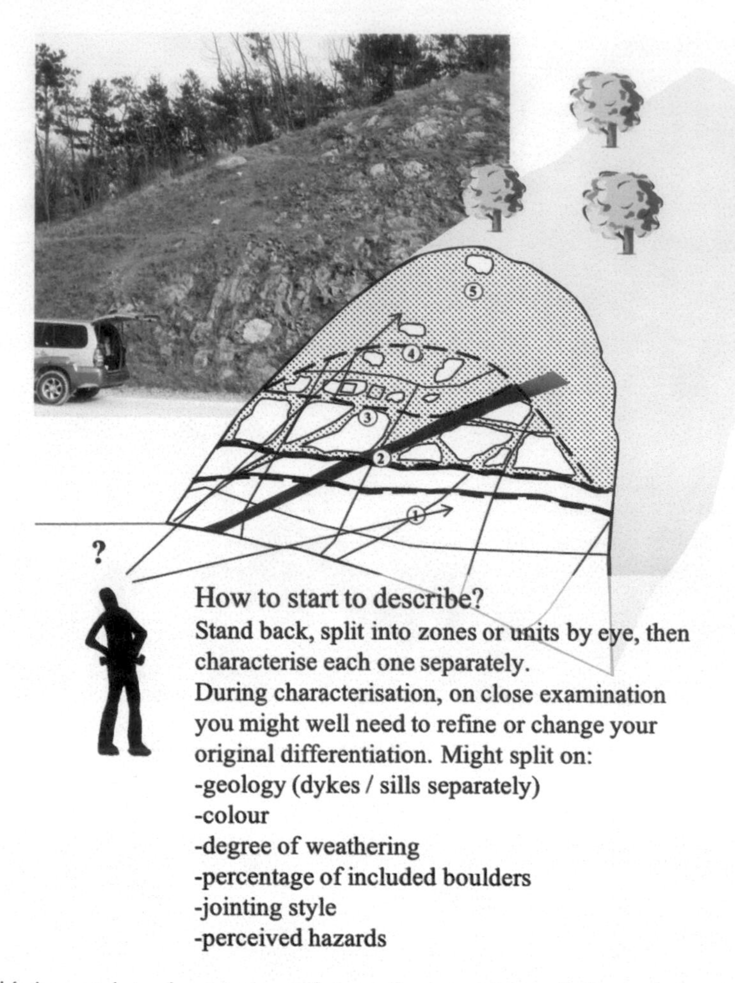

?
How to start to describe?
Stand back, split into zones or units by eye, then
characterise each one separately.
During characterisation, on close examination
you might well need to refine or change your
original differentiation. Might split on:
-geology (dykes / sills separately)
-colour
-degree of weathering
-percentage of included boulders
-jointing style
-perceived hazards

Figure 6.16 Approach to characterise rock mass. First stage is to split into units by eye. Units/zones will be used in later analysis and design.

should be split initially into zones, layers or units, by eye. The primary division will often be geological, i.e. rock and soil units of different age, but then differentiated by rock or soil mass quality such as degree of weathering or closeness of fracturing. Differentiation on strength can be made quickly by simple index tests such as hitting or pushing in a hammer. The split might be on structural regime, i.e. style and orientation of discontinuities. The process is illustrated in Figure 6.16. Once the broad units or zone boundaries have been identified, then each needs to be characterised by systematic description and measurement, as shown schematically in Figure 6.17. Evidence of seepage should be noted; lush vegetation can be indicative of groundwater. The distinction between engineering geological mapping and normal geological practice is the emphasis on characterising units in terms of strength, deformability and permeability, rather than just age (Dearman & Fookes, 1974).

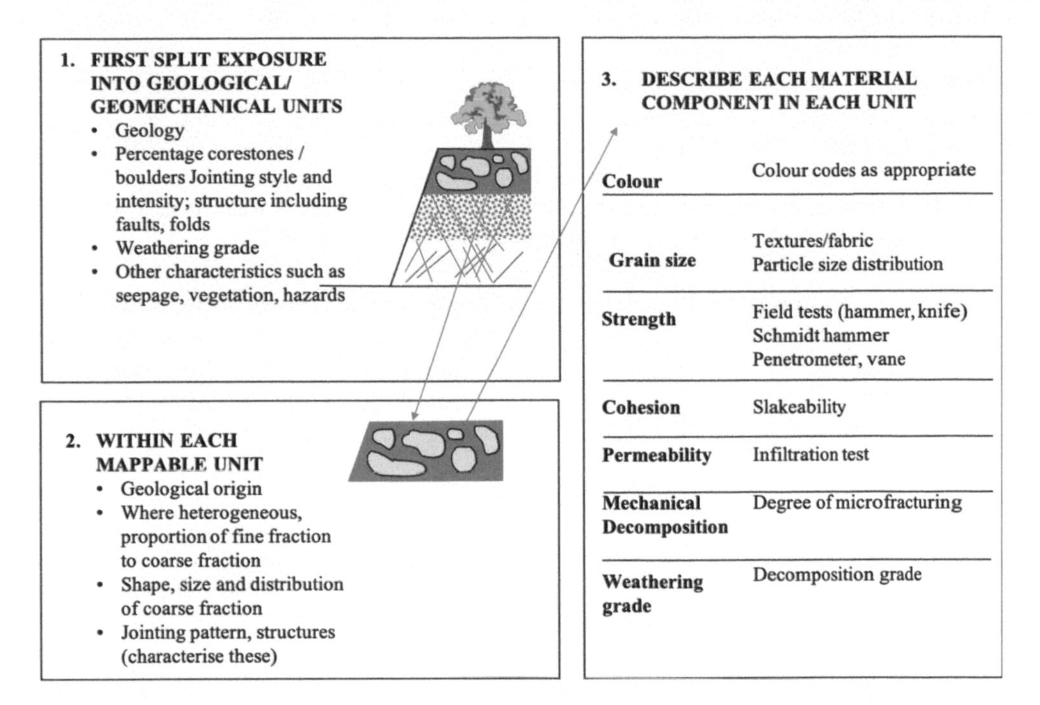

1. FIRST SPLIT EXPOSURE INTO GEOLOGICAL/ GEOMECHANICAL UNITS
• Geology
• Percentage corestones / boulders Jointing style and intensity; structure including faults, folds
• Weathering grade
• Other characteristics such as seepage, vegetation, hazards
2. WITHIN EACH MAPPABLE UNIT
• Geological origin
• Where heterogeneous, proportion of fine fraction to coarse fraction
• Shape, size and distribution of coarse fraction
• Jointing pattern, structures (characterise these)
3. DESCRIBE EACH MATERIAL COMPONENT IN EACH UNIT
Colour
Colour codes as appropriate
Grain size
Textures/fabric
Particle size distribution
Strength
Field tests (hammer, knife)
Schmidt hammer
Penetrometer, vane
Cohesion
Slakeability
Permeability
Infiltration test
Mechanical Decomposition
Degree of microfracturing
Weathering grade
Decomposition grade

Figure 6.17 Once the broad units/layers have been identified, each needs to be characterised.

Some of the equipment that might be used in field characterisation of exposures includes safety harness, tape measures, hammer, knife, hand penetrometer, Schmidt hammers (type N and L), compass/clinometer and hand lens. Water and a container are useful for conducting index tests such as slake tests and for making estimates of soil plasticity and grading. Where appropriate, strength can be measured using such tools as a hand vane, and point load testing, which can be carried out on irregular lumps of rock. Whatever measurements are taken at exposures, the end user needs to be aware that it may be inappropriate to extrapolate properties because of the effects of drying out or softening from seepage and possibly the effects of weathering.

Guidance on geological mapping and description is given in a five-volume, well-illustrated handbook series by the Geological Society of London, which deals with Basic Mapping, the Field Description of Igneous, Sedimentary and Metamorphic Rocks and Mapping of Geological Structure, each with more than 100 pages (www.geolsoc.org.uk). Much of the detail that could be recorded by a geologist, however, might prove irrelevant to an engineering project, but what is or is not important might not be immediately obvious. Nevertheless, it is worth bearing in mind the observations of Burland (2007):

> It is vital to understand the geological processes and man-made activities that formed the ground profile; i.e. its genesis. I am convinced that nine times out of

> ten, the major design decisions can be made on the basis of a good ground profile. Similarly, nine failures out of ten result from a lack of knowledge about the ground profile.

Despite this observation, current standards codes and textbooks dealing with ground investigation tend to take a very simplified, prescriptive, formulaic approach in their recommendations for the description of geological materials and structure. The reason dates back to the 1960s when Deere (1968) noted the following:

> Workers in rock mechanics have often found such a classification system [geological] to be inadequate or at least disappointing, in that rocks of the same lithology may exhibit an extremely large range in mechanical properties. The suggestion has even been made that such geologic names be abandoned and that a new classification system be adopted in which only mechanical properties are used.

Deere went on to introduce classifications based on compressive strength and elastic modulus and the Rock Quality Designation (RQD), and these or similar classifications are now used almost exclusively for logging rock core, with geological detail rarely recorded.

Deere at the same time noted, however, "the importance to consider the distribution of the different geologic elements which occur at the site". This sentiment would have been echoed by Terzhagi (1929), some of whose insightful observations on the importance of geological detail are extracted and highlighted by Goodman (2002, 2003). Restricting geological description to a few coded classifications, as in industry standards, is over-simplistic but it is a fine balance between providing too much geological information and too little.

Generally, GI loggers tend to provide minimal summary descriptions, as per the examples given in BS5930 and other standards, and avoid commenting on unusual features, although it varies from company to company and, of course, the knowledge and insight of the logger. Some guidance on standard logging is given in Appendix B and examples of borehole logs are provided in Appendix C and discussed later. Fletcher (2004) provides many examples of the kind of geological information that can be obtained from logging of cores for engineering projects, most of which would be missed if following standard guidelines for engineering description and classification.

There is much to be said for the Engineer informing the GI Contractor of his preliminary ideas regarding the ground model, based on desk study and reconnaissance, so that the Contractor knows what to look for and can update the model as information is gained.

Rock exposures are particularly important for characterising fracture networks. Orientations are usually measured using a compass clinometer, as illustrated in Figure 6.18, with different diameter plates used to help characterise the variable roughness at different scales (Fecker & Rengers, 1971). Electronic compass/clinometers are under development, which will avoid the need to level the instrument, which can be difficult, especially in underground mapping in tunnels.

Figure 6.18 Joint survey underway using a Clar compass clinometer attached to aluminium plates. Investigation for Glensanda Super Quarry, Scotland.

Data are usually collected by systematic scan-line or window surveys but these are tedious to carry out, seem to be routine to the unknowledgeable, and therefore sometimes delegated to junior staff who may be unable or reticent to exercise independent judgement on what is or is not significant. Such surveys can give a false impression of rigorous measurement, whilst the important element of geological interpretation, best

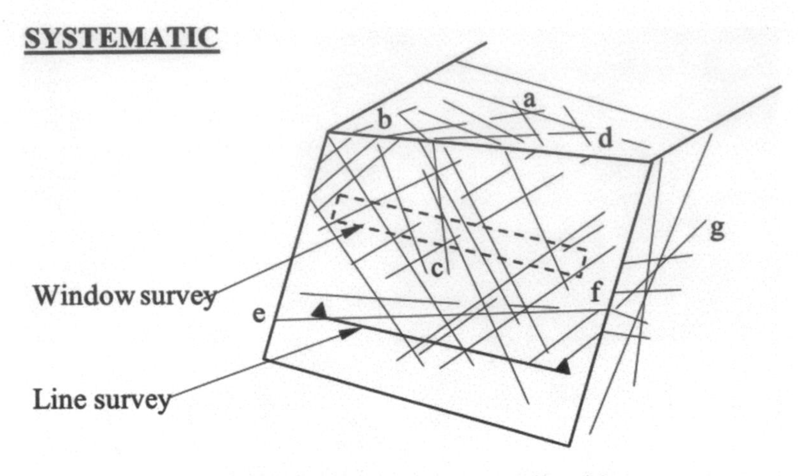

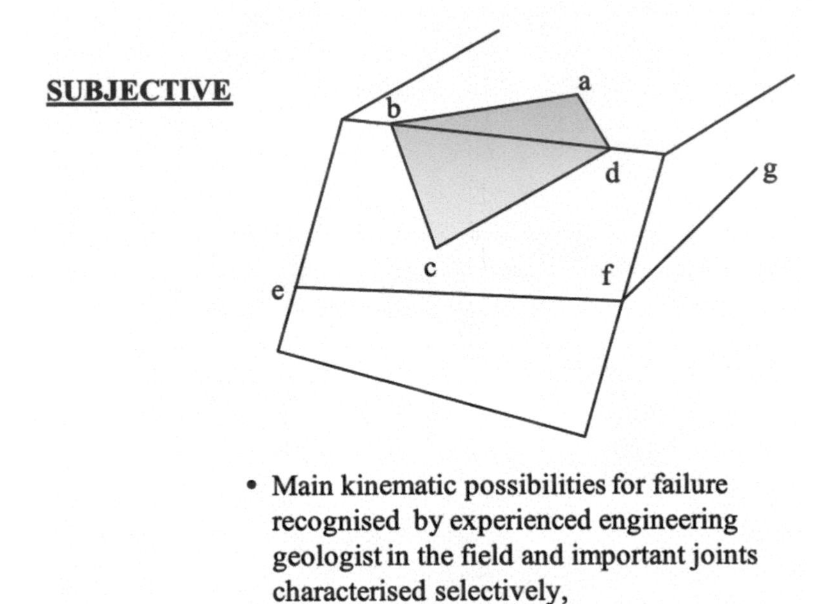

SUBJECTIVE
a
b
d
g
c
f
e
• Main kinematic possibilities for failure recognised by experienced engineering geologist in the field and important joints characterised selectively,

Figure 6.19 Distinction between objective surveys (line/window) and subjective surveys.
Source: From Hencher, 1987.

done in the field, is lacking. Experienced engineering geologists with training in structural geology should be able to assess the rock conditions by eye, both with respect to the geological conditions and potential for instability in a slope and therefore can carry out a subjective survey (Figure 6.19). The recommended approach for collection and interpretation of discontinuity data from rock exposures is set out in Box 6.8.

BOX 6.8 COLLECTION OF DISCONTINUITY DATA IN EXPOSURES (MODIFIED FROM HENCHER & KNIPE, 2007)

1. First take a broad view of the exposed rock. Examine it from different directions.
2. Develop a preliminary geological model and split it into structural and weathering zones/elements. Sketch the model.
3. Broadly identify those joint sets that are present, where they occur, how they relate to geological variation and what their main characteristics are, including spacing, openness as mechanical fractures (or otherwise), roughness, infill and cross cutting or terminations in intact rock or against other discontinuities. Surface roughness characteristics such as hackle marks should be noted as these are indicative of origin and help differentiate between sets.
4. With respect to incipient joints, try to characterise these in terms of strength, relative to the adjacent rock – high (close to intact rock strength), moderate or low [could be broken easily with a hammer (see Figures B6.8.1, B6.8.2 and B6.8.3)]
5. Measure sufficient data to characterise each set geologically and geotechnically. Record locations on plans and on photographs. This might be done using line and window surveys but quite often these are time consuming and not very productive. It is generally best to decide what to measure and then measure it, rather than hope that the answer will drop out of a statistical sample.

Figure B6.8.1 Bedding planes shallowly dipping and cleavage, near vertical. To the left of the exposure the strength is generally high, with occasional cleavage parallel joints at about 0.5 m spacing. To the right the cleavage is more weathered at about 25 mm spacing. Description of this exposure needs to be placed in context of its state of weathering relative to the less-weathered rock a hundred metres away.

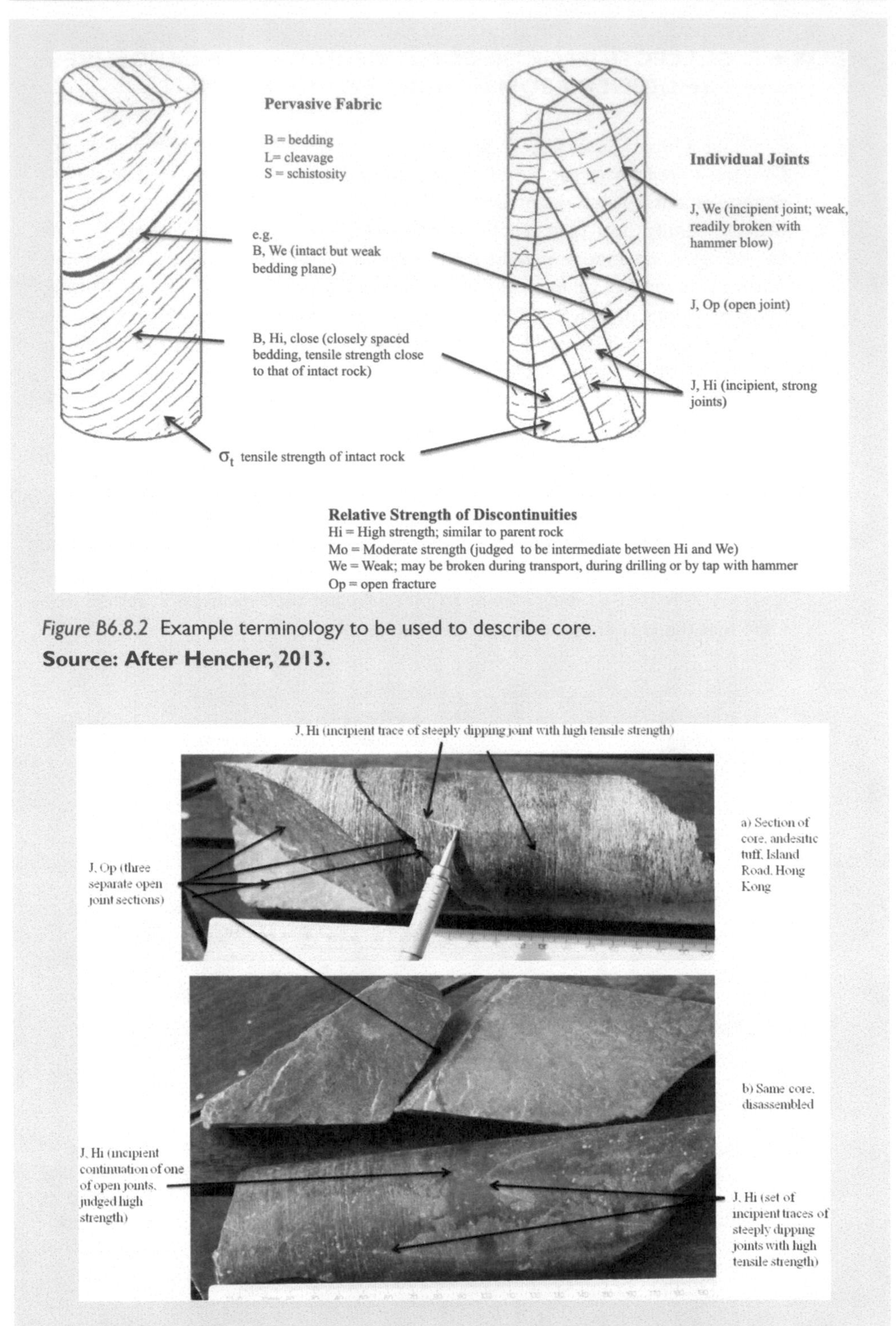

Figure B6.8.2 Example terminology to be used to describe core.

Source: After Hencher, 2013.

Figure B6.8.3 Example of use of terminology to describe fractures in core after Hencher (2013).

6. Plot data and look at geometrical relationships. Consider how the various sets relate to one another and to geological history as evidenced from faults, folds and intrusions (Chapter 6).
7. Search for missing sets that might have been expected given the geological setting.
8. Analyse and re-assess whether additional data are required to characterise those joints that are most significant to the engineering problem.

Figure 6.20 Ground-based radar being used to generate a digital image of cut slopes near Seoul, Korea. Point clouds can be used to remotely measure discontinuity geometry.

Remote measurement of fracture networks is becoming more reliable using photogrammetry (Haneberg, 2008) or ground-based and airborne radar (Figure 6.20) and research is progressing into the automatic interpretation of laser-scanned data into the orientation and spacing of rock sets (Slob, 2010). Currently, this approach, however, lacks any link to an interpretation of origin of the discontinuities and their geological inter-relationships, which would make it much more valuable. In the author's opinion, probably the best use for laser scanning at the moment is as an aid to the field team, in particular for measuring data in areas of an exposure with difficult access or for measuring displacements over a period of time remotely in order to predict failure of rock slopes (see Kromer et al., 2015).

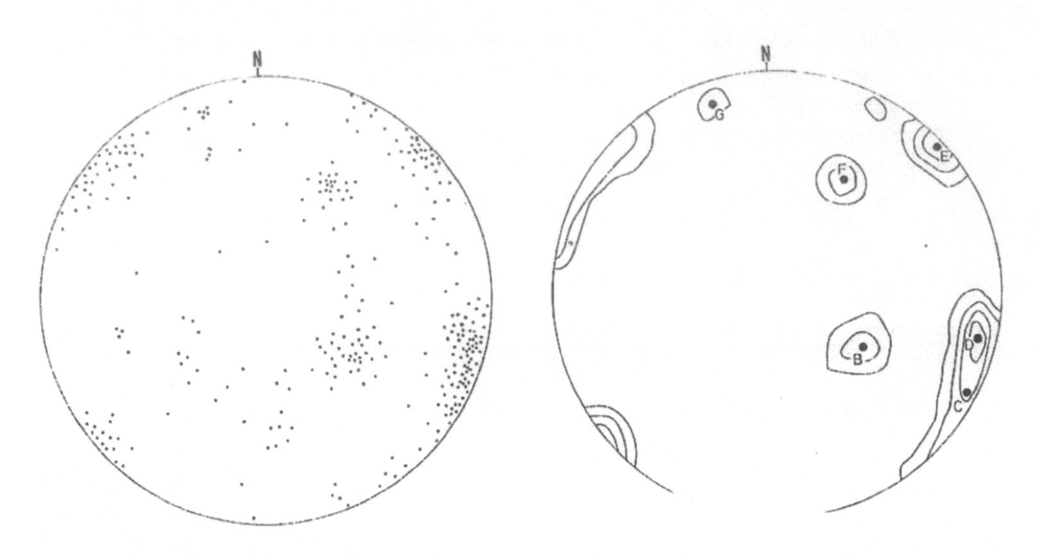

Figure 6.21 Representing discontinuity data on stereographic projections.
Source: After Hencher, 1985.

Rock joint data are generally represented on stereographic projections as explained clearly for geological purposes by Lisle & Leyshon (2004) and illustrated in Figure 6.21. The technique allows sophisticated analysis of geological discontinuity data (Phillips, 1973), but its most common use in engineering geology is for determining the potential for specific rock discontinuities to cause a failure in a cut slope or in an underground opening (Hoek & Bray, 1974). Plotting of data, statistical grouping and comparison to slope geometry is now easily done using software such as Dips (Rocscience), but care should be taken in interpretation and especially against masking important but relatively rare data (Hencher 1985). Bridges (1990) demonstrates the importance of differentiating sets on the basis of geological characteristics rather than just geometry.

6.6 GEOPHYSICS

Geophysical techniques are used to identify the disposition of soil and rock units, based on differences in physical properties, such as strength, density, deformability, electrical resistance and magnetism. They can sometimes be used successfully to identify cavities such as mine workings or solution hollows and for identifying saturated ground. Geophysics really comes into its own for offshore investigations where drilling is very expensive. Geophysics can provide considerable information on geological structure and rock and soil mass quality, which is relevant to engineering design, although such techniques are rarely used by themselves but as part of a wider investigation involving boreholes. Many engineering geologists and geotechnical engineers have both good and bad experience of engineering geophysics. Darracott & McCann (1986) argue that poor results can be often be attributed to poor planning and the use of an inappropriate technique for the geological situation. More specifically, key constraints are as follows:

- penetration achievable;
- resolution;
- signal-to-noise ratio; and
- lack of contrast in physical properties.

When geophysics works well, the results can be extremely useful and the method cost-effective. The main options and constraints are set out in BS 5930 and Clayton (1995).

6.6.1 Seismic methods

Seismic refraction techniques, using an energy source ranging from a sledgehammer to explosives, can be useful on land and in shallow water for finding depth to bedrock, for example, to identify buried channels that could otherwise only be proved by numerous boreholes or probes. Large areas can be investigated quite cheaply and quickly. The method works best where there is a strong contrast in seismic velocity between the overlying and underlying strata and some knowledge of the geological profile, preferably from boreholes. Otherwise, results will be ambiguous. Where weak (low-velocity) strata underlie stronger materials, these may not be identified by seismic survey. Wave velocity (compressive and shear) can be interpreted directly in terms of rock mass quality, deformation modulus and ease of excavation, as reviewed comprehensively by Simons et al. (2001). Seismic reflection is a key technique in offshore investigations.

6.6.2 Resistivity

Resistivity is another cheap and rapid method that can prove very effective, particularly in identifying groundwater (low resistance) and voids (high resistance). The technique has been used successfully in the investigation of landslide profiles, in particular for identifying water bearing strata at depth. Figure 6.22 shows the results of a resistivity survey in Hong Kong to identify underground stream channels as zones of high resistance (voids), which it did extremely well (Hencher et al., 2008).

6.6.3 Other techniques

There are a host of other techniques reported in the literature, with various success rates. Ground-based radar can be useful for finding shallow hidden pipes, etc. Other techniques such as magnetic and micro-gravity rely on particular physical properties of the rock or feature being searched for. Both have been used for locating old mine shafts – because the brick lining might have a magnetic signature and the void is low gravity. Generally, such techniques are used as a first pass across a site to identify any anomalies, which are then investigated more fully using trial pits, trenches and boreholes. For such investigations, percussive holes, as used for forming holes for quarry blasting (no coring), can be very quick and relatively cheap – the presence of holes is indicated by lack of resistance to drilling and loss of flushing medium. The holes can later be examined using TV cameras, periscopes or sonic devices to try to quantify the sizes of voids. For many reasons, such surveys are not always successful

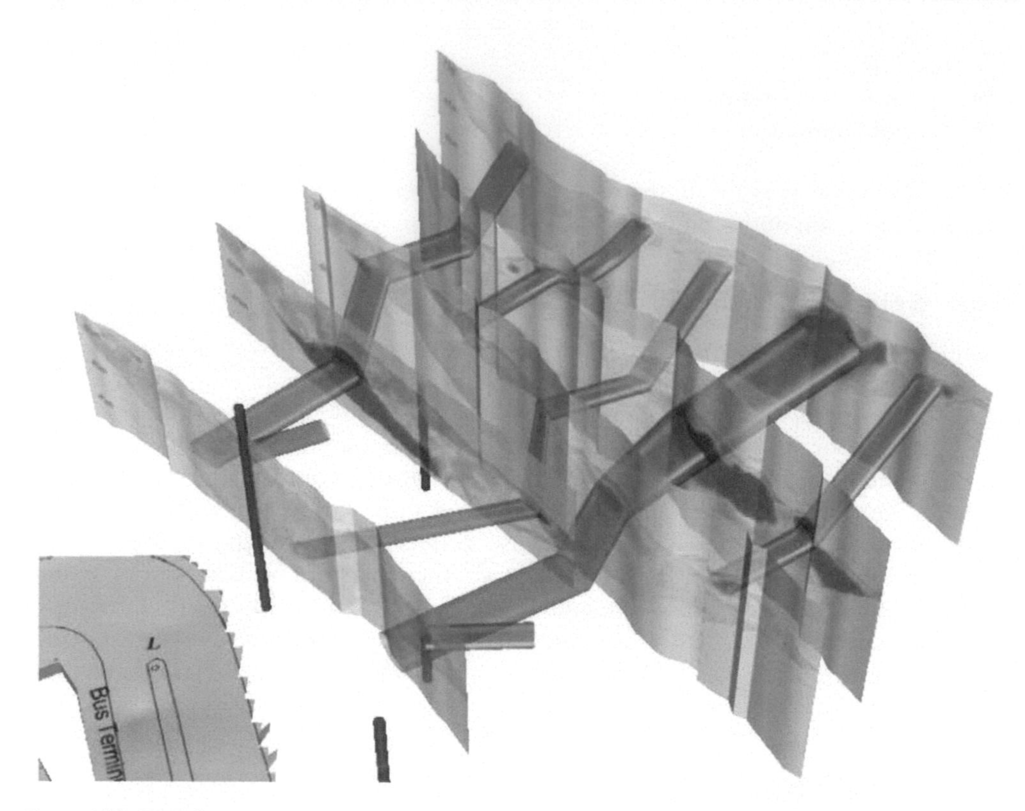

Figure 6.22 Digital image interpretation of resistivity surveys across hillside above Yee King Road, Hong Kong. Tubular features of low resistivity are interpreted as underground streams.

Source: Hencher et al., 2008.

and therefore are not to be relied upon to give a definitive answer (Clayton et al., 1995). Sewell et al. (2000) demonstrate the usefulness of marine magnetic and gravity surveys for identifying geological structures.

6.6.4 Down-hole geophysics

As with seismic reflection, down-hole geophysics is used routinely in oil and gas exploration, in mining and in sophisticated GI linked to nuclear waste disposal studies. Tools can be used to determine minor stratigraphic contrasts and rock properties. These tools are less used for engineering, with the exception of rock joint orientation (using cameras and geophysical tools) and sometimes for identifying clay-rich layers. These tools are discussed below, together with logging and description.

6.7 SUB-SURFACE INVESTIGATION

Methods and techniques for sub-surface investigation are dealt with in many publications, including BS 5930 (British Standards Institution, 1999), Clayton et al. (1995), Geotechnical Control Office (1987), Hunt (2005) and Mayne et al. (2001).

6.7.1 Sampling strategy

There are usually four main objectives in sub-surface investigation, to

1. establish the geological profile;
2. determine engineering properties for the various units within the eventual ground model;
3. establish the hydrogeological conditions; and
4. monitor future changes in ground conditions through instrumentation.

At many sites, it is best to use preliminary boreholes in an attempt to establish the geological profile accurately. This will require sampling over the full depth and with sufficient boreholes to establish lateral and vertical variability. If recovery is low, then boreholes may need to be repeated; it is often the pieces of core that are not recovered that are the most important, because they are also the weakest. It is wise to include a clause in specifications for the GI Contractor, setting out a minimum acceptable recovery, to encourage diligent work. A good driller can generally achieve good recovery in almost any ground, providing he has the right equipment and adjusts his method of working to suit the ground conditions. If he does not have suitable equipment (or flushing medium), then that might be the fault of the Engineer who specified the investigation, rather than the Contractor, and this may need rectification by issuing a variation order to the Contract.

Once the preliminary geological model has been established adequately at a site, then additional boreholes can be put down as necessary to take samples for testing or to carry out *in situ* testing and to install instruments for monitoring changes such as response of water table to rainfall. The same approach (sample first to prove the geological model and to identify any geological hazards, followed by a second phase for testing and instrumentation) should be used for any investigation where geological features may be important. This can only be judged by a competent engineering geologist aware of both the local geological conditions and the factors that will control the success or otherwise of the particular civil engineering project.

In practice, boreholes are often put down using a strategy of intermittent sampling and *in situ* testing within a single borehole, which means that the full ground profile is not seen. This can be cost-effective for design when the site is underlain by relatively uniform deposits and where the ground profile is already well-established from previous investigations. The danger is that site-specific geological features might be missed yet prove important for the project.

6.7.2 Boreholes in soil

6.7.2.1 There are many different tools that can be used to investigate soils and many of these are described by Clayton et al. (1995). In the UK, the most commonly used machine for investigating soils is the "shell and auger", otherwise known as the "cable-percussive rig", as illustrated in Figure 6.23. Such rigs are very manoeuvrable and can be towed behind a field vehicle or winched to the point where the hole is to be put down. They can cope with a wide range of soils, which makes for their popularity in the UK, where mixed glacial soils are common. The

Figure 6.23 Shell and auger rig in action, Leicester, UK in the 1970s. Nowadays in the UK, helmets and other protective clothing would be worn. Casing, used to support the hole, is standing out of ground and a "shell" is being dropped down hole to excavate further. In the foreground is a U100 sampling tube attached to a down-hole hammer, ready for placing down the hole and taking a sample once the hole has been advanced to the required depth. Leaning against the wheel is one of the drillers and also a trip hammer for SPT testing – also awaiting use at appropriate depths and changes in strata.

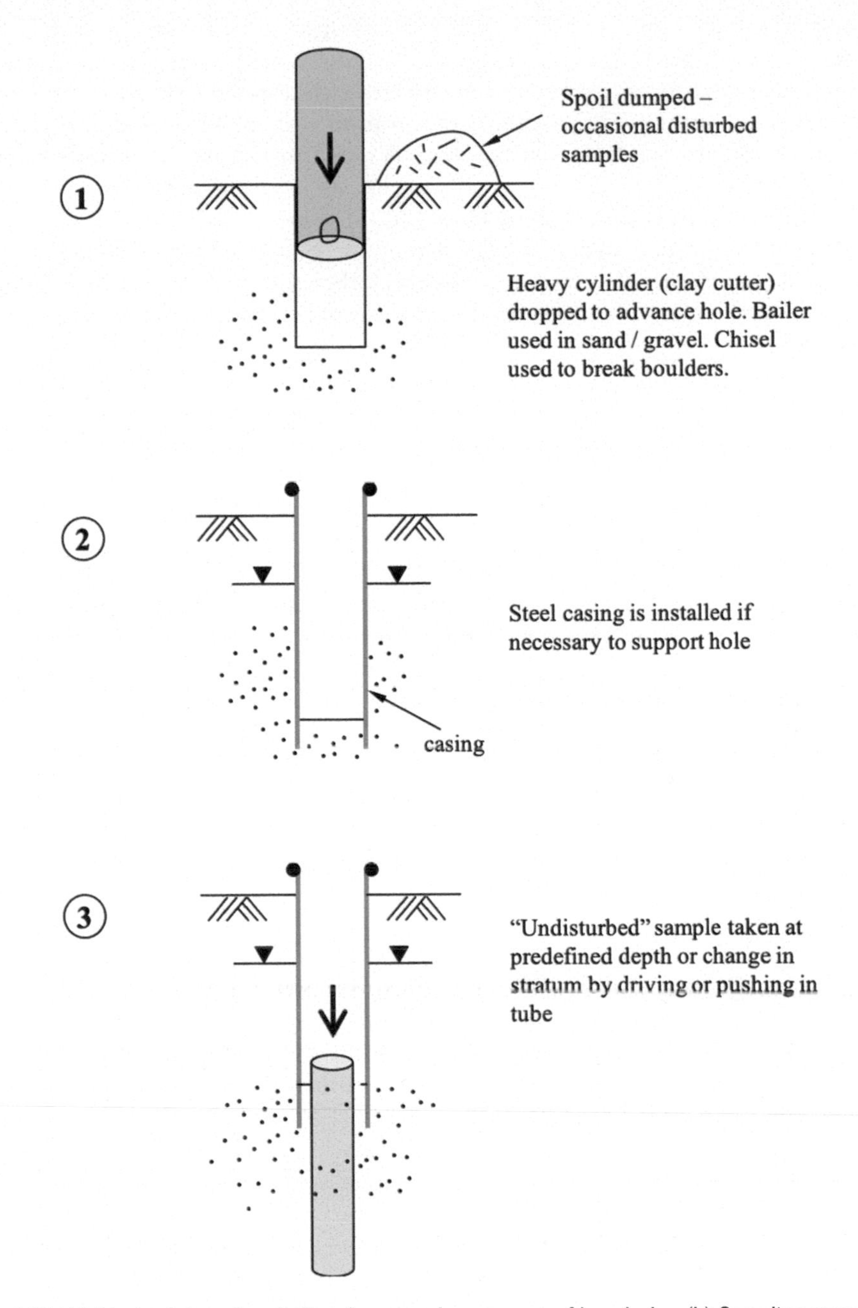
1
Spoil dumped –
occasional disturbed
samples
Heavy cylinder (clay cutter)
dropped to advance hole. Bailer
used in sand / gravel. Chisel
used to break boulders.
2
Steel casing is installed if
necessary to support hole
casing
3
"Undisturbed" sample taken at
predefined depth or change in
stratum by driving or pushing in
tube

Figure 6.24 (a) Methodology for shell and auger advancement of boreholes. (b) Sampling strategy.

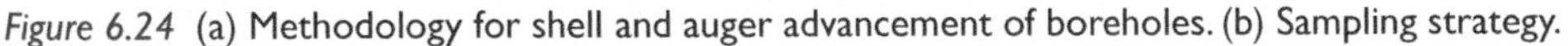

hole is advanced by dropping a heavy "shell" (Figure 6.24). Material between sampling points is usually discarded, although it should be examined and recorded by the drilling contractor and disturbed "bulk" samples are taken in bags, if specified for the Contract. All samples, of course, should be sealed and labelled. If boulders are encountered in the soil profile, these are broken up with a heavy chisel dropped down the hole. Engineers usually specify alternate "undisturbed" samples for laboratory testing and *in situ* strength tests at perhaps 1.5 m intervals or changes in strata. The standard penetration test (SPT) is commonly used to measure strength, as discussed below under *in situ* testing. Vane tests might be carried out rather than SPTs, especially in clay soils. USA practice for investigating and sampling soils is described by Hunt (2005). One cheap and quick way of sampling/testing is to use wash boring, whereby the hole is advanced by water jetting as rods are rotated. SPT tests, and possibly other samples, are taken at intervals. None of these methods gives continuous sampling, so geological detail may be missed.

"Undisturbed" samples are usually taken using a relatively thin walled sampler of diameter 100 mm (U100), and much of the published empirical relationships that are relied upon by designers are based on tests on samples achieved in this way. This sampling method does not, however, meet the more stringent requirements of Eurocode 7 for "Class 1" sampling and testing, because of fears over disturbance. This is rather naïve in that it implies that thinner sampling tubes can take an undisturbed sample, which is not the case. Any sample taken from depth, squeezed into a tube and then extruded at the laboratory, will inevitably be disturbed to some degree. Further disturbance occurs during preparation of samples for laboratory testing and initial loading and saturation, as expressed schematically in Figure 6.25 and investigated by Davis & Poulos (1967). The engineering geologist and geotechnical engineer need to be aware of the likely disturbance to any tested samples and take due care in interpretation. Furthermore, the scaling up of results from laboratory to project scale requires careful consideration because it must include the effect of mass fabric and structure, including fractures and discontinuities.

6.7.3 Rotary drilling

Rotary drilling is used in all rocks but can also be used to obtain good samples in weaker materials, including colluvium (mixed rock and soil), weathered rock and soil. In weaker ground, a similar investigation strategy is often adopted as for soils, whereby sections are cored followed by SPT tests, although as for soils there is the risk that important geological features may be missed.

A drilling rig rotates a string of drilling rods whilst hydraulic cylinders apply a downward force. At the lower end of the drilling string there is a hollow annulus "bit", usually coated with diamonds or tungsten carbide. As the bit is rotated, a stick of core enters into a core barrel at the bottom of the drilling string. The retained core is prevented from falling out as the barrel is brought back to the surface, by some form of core-catching device. Air, water, mud or foam is used to cool the bit and carry rock cuttings back to the surface (Figure 6.26). Where cored samples are not required over a particular length of hole, it can be advanced more quickly using rock roller bits, down-the-hole hammers and water jets, as used in much oil and gas drilling.

At the most basic level, a single-barrel well-boring rig can be used to take core samples but these are often highly disturbed (Figure 6.27). Most drilling is carried out

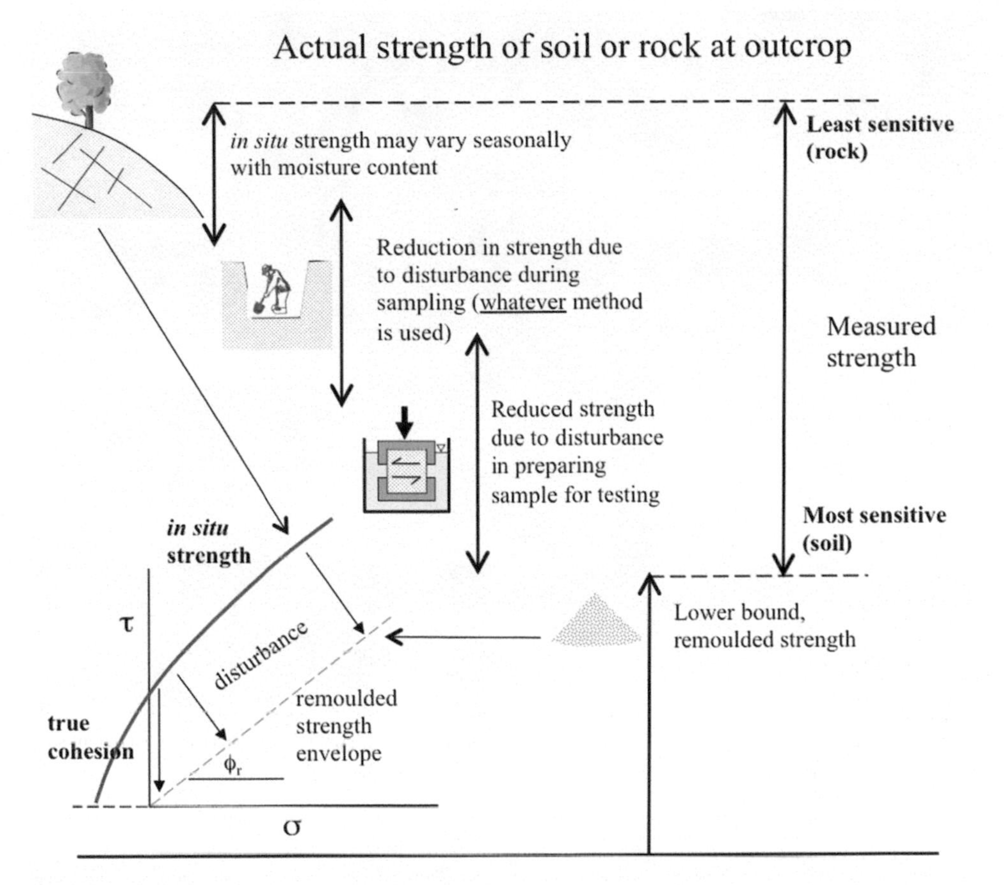

Figure 6.25 Potential sources of sampling disturbance leading to much lower strengths being measured in the laboratory compared to those *in situ*.

using double-barrel systems in which the outer barrel rotates around an inner barrel that takes in the core. A problem with the double-tube system is that the flushing medium flows between the core and the core barrel and can wash away some of the cored material, but it is still used internationally because it is relatively inexpensive and can be mass produced. The problems can be reduced by using a triple-tube system. In this system, the core enters a split inner tube, which does not rotate; the flushing medium flows between the inner tube and an outer tube without touching the core. Such equipment has low manufacturing tolerances so must be bought off the shelf, and the bits are very expensive and only last perhaps 8 to 12 m of coring before they need to be replaced, which precludes its use on many projects.

Usually, the larger the diameter of the core barrel, the better the recovery and quality of sample, and it is prudent to start using a large diameter and reduce diameter as necessary with depth. The wide range of casing, core barrel and drill rod sizes are listed in American Society for Testing and Materials (1999), which also discusses good practice. When there is good-quality rock overlying soil material, retrieving the softer

Figure 6.26 Rotary drilling above landslide at Yee King Road, Hong Kong. Polymer foam (white) is being used as the drilling flush to try to improve recovery.

material can be a problem. As for soil boring, the hole may need to be cased temporarily during drilling to prevent it collapsing. Drillers generally try to recover about 1.5 m of core per "run" before pulling all the drill string back to the ground surface and dismantling it all. If recovery is low, then the driller might try to reduce the core run to 1 m or even less, but this does not always produce better results. Other parameters such as thrust, torque and flushing medium may have more influence on recovery, and much depends on the experience, knowledge and attitude of the drilling crew.

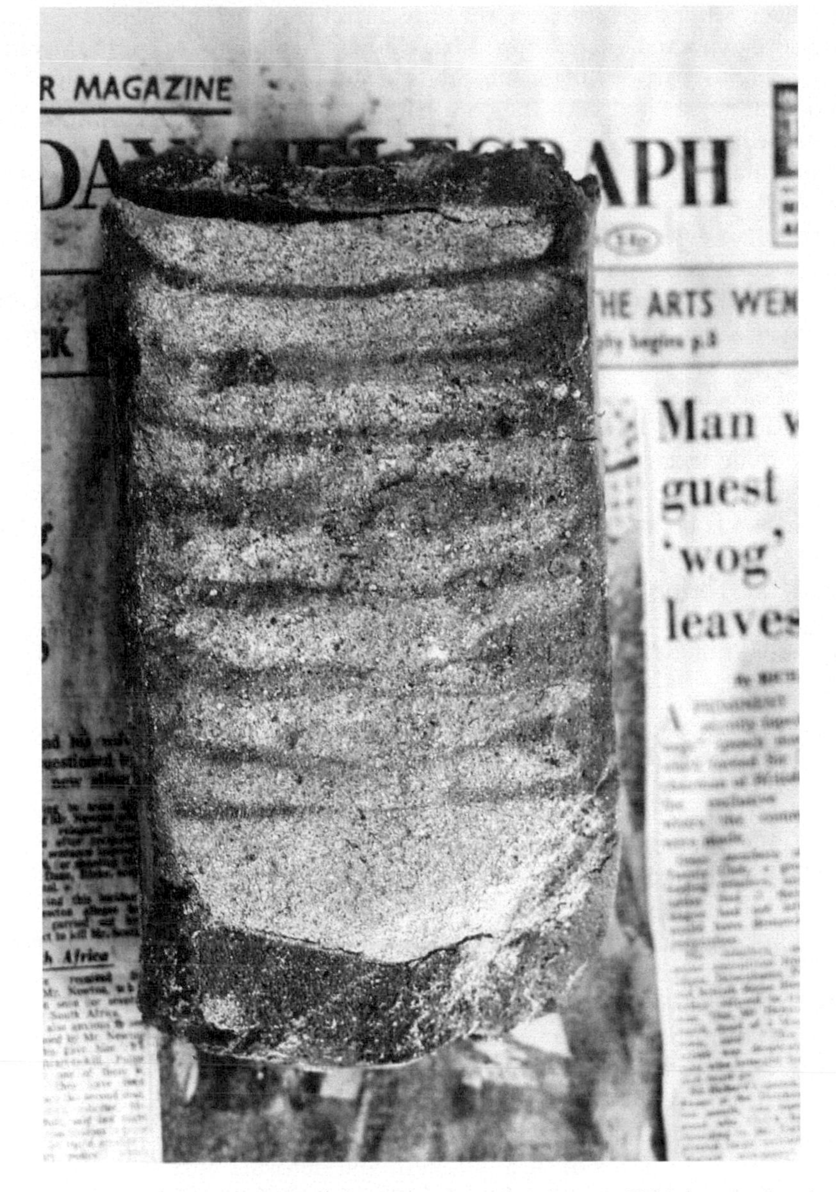

Figure 6.27 Sample obtained from single-barrel Russian well drilling rig, El Hadjar steelworks, Annaba, Algeria (see preface). Previous logging of similar samples had interpreted the layering as some kind of varved sequence of silt and sand. Actually, the horizon *in situ* is fairly uniform weathered (Grade IV) gneiss (the pale material). The dark-brown silt horizons represent occasions when the Algerian driller, bored with the slow drilling progress from his worn-out bits, raised the drilling string and then dropped it again with some force down the hole, letting in a layer of the silty drilling mud, which then became baked by the heat from the drilling process... The thickness of the pale layers are an indication of the driller's boredom threshold – generally pretty consistent.

Wire line drilling employs large-diameter rods, which effectively support the hole as it advances. After each core run, the core barrel is pulled up the centre of the drill rods, the core extracted, then dropped back down the hole to lock into the bottom of the hole, ready to start drilling again. The cutting bit stays at the bottom of the drill rods and is not extracted with the core barrel. To change the cutting bit, however, the whole drill string has to be removed. A system that is very commonly used in Hong Kong and elsewhere for sampling weathered rock and mixed rock and soil is a Mazier core barrel. This has a soil cutting shoe, which is spring loaded and extrudes in advance of an outer rock cutting bit when cutting through relatively weak soil-like material (Figures 6.28 and 6.29). As conditions get harder, the soil cutter is pushed back and the outer coring bit takes over. This system, especially where combined with polymer foam flush, has been shown to produce good recovery of material in weathered and mixed materials (Phillipson & Chipp, 1982). The sample is taken in a plastic tube, which is later cut open so that the sample can be examined, described and tested (Figure 6.30). Drilling contractors will not open Mazier samples without instruction to do so, and, in practice, geotechnical engineers sometimes order Mazier samples (from the office) but then never get around to opening and examining the samples, which is poor practice. The author was recently involved in an arbitration where 20 boreholes had been put down with alternate Mazier sampling in soft clays and then SPTs. The project was then designed on the basis of the SPT data alone and went badly wrong, ending in arbitration. The Mazier samples had not been opened up for examination or testing. A similar system to the Mazier, used in the USA, is the Dennison sampler (Hunt, 2005).

6.8 *IN SITU* TESTING

Many parameters are obtained for design by laboratory testing, as discussed in Chapter 7, but the potential for disturbance are obvious, as discussed earlier, especially for granular soil that disaggregates when not confined. There are therefore many reasons for attempting to test soil and to a lesser extent rock *in situ*. Most tests are conducted in boreholes, but some are conducted by pushing the tools from the ground surface or from the base of a borehole to zones where the soil is relatively undisturbed. A self-boring pressuremeter, suitable for clay and sand, drills itself into the ground with minimal disturbance before carrying out a compression test at the required level.

The SPT is probably the most commonly used *in situ* test, whereby the number of blows to hammer a sample tube into the ground is recorded. Soft soils are penetrated easily, hard soils and weak rocks with more difficulty. The SPT data can be interpreted in terms of shear strength and deformability (Chapter 7) and for making predictions of settlement directly (Chapter 3). The split spoon sampler used for the SPT is a steel tube with a tapered cutting shoe. It is lowered down the borehole, attached to connecting rods, and then driven into the ground by a standard weight, which drops a standard height, as illustrated in Figure 6.31 and shown in action from a rotary drilling rig in Figure 6.32. The number of blows for each penetration of 75 mm is recorded; blows for the first 150 mm are recorded but essentially ignored (considered disturbed); the blows for the final 300 mm are added together as the "*N*" value. Care must be taken in soil that the external water table is balanced,

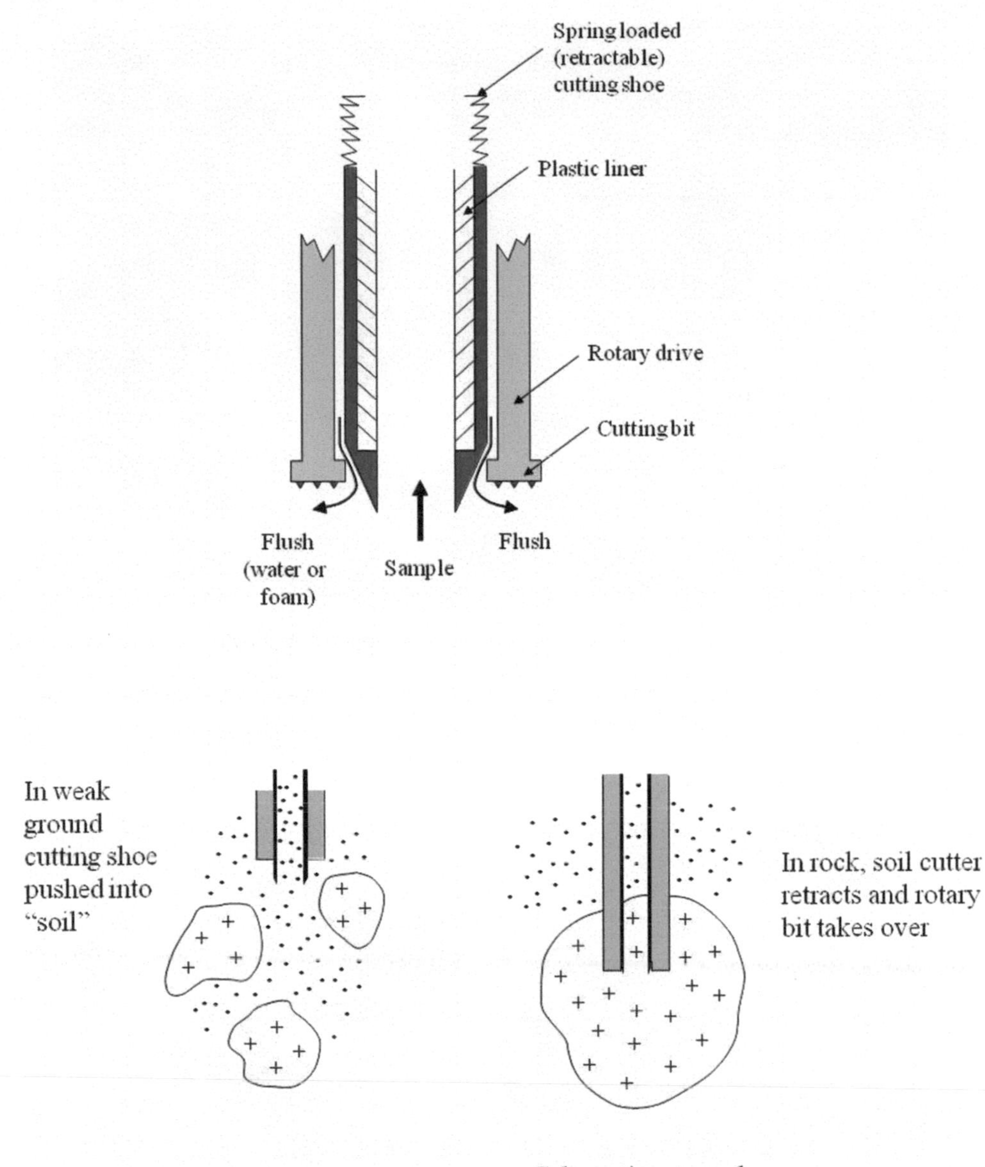

Figure 6.28 Principles of Mazier sampling.

otherwise water may flow in from the bottom of the hole, causing softening and too low an N value. There are various corrections suggested for tests conducted in silty sand and for depth of overburden. Details are given in Clayton (1995). The SPT test is much maligned for associated errors but nevertheless is still the most common basis for design in many foundation projects, mainly because no-one has come up with anything better. It is also actually quite a useful sampling tool, as illustrated in

Figure 6.29 Mazier sampler with nicely recovered weathered granite – the right side third is stained with iron oxides. Spring-loaded cutting shoe is seen extending from the rock cutting bit outside. When the material strength becomes too high for the cutting shoe (exceeds spring stiffness), the outer bit takes over the cutting.

Figure 6.30 Mazier sample plastic tube being cut for examination.

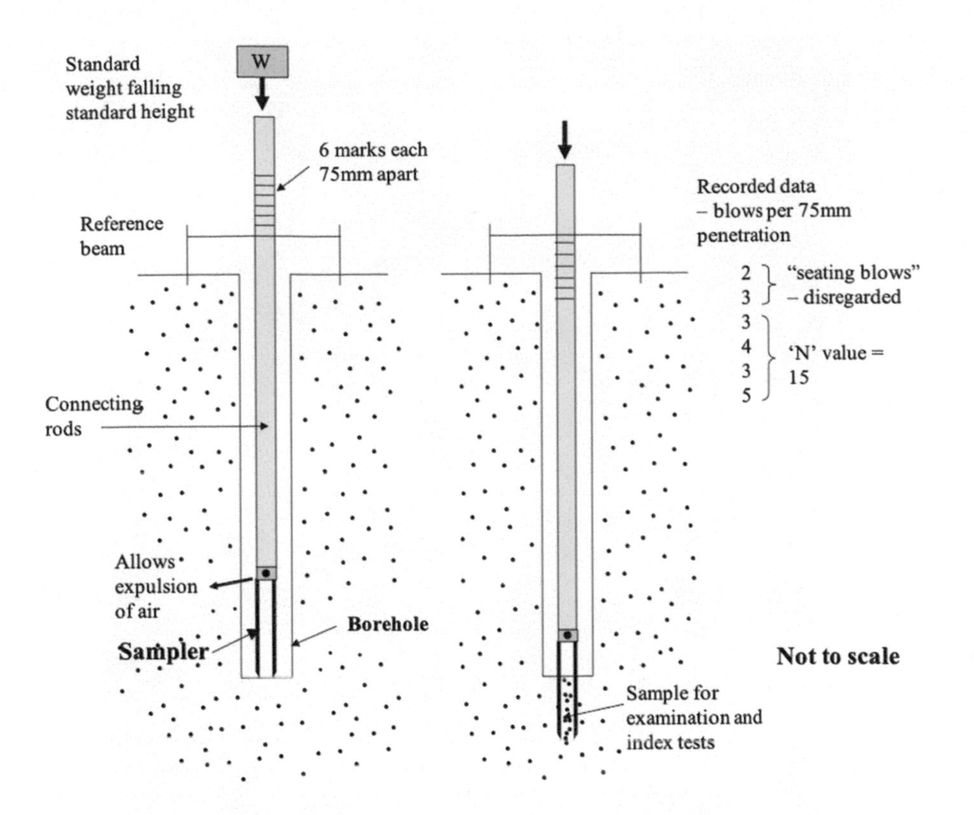

Factors that will influence results and may require corrections to be made

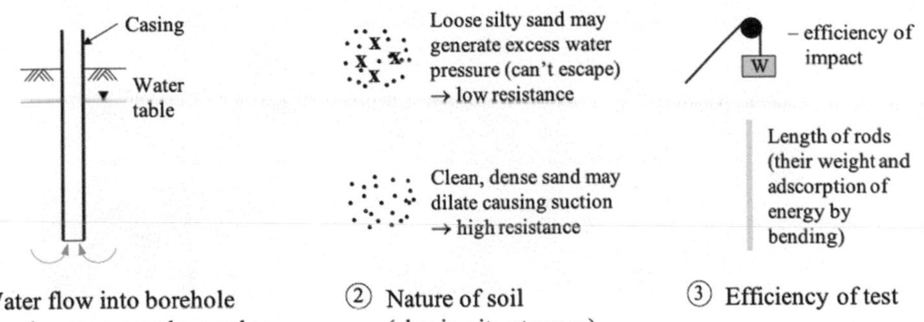

① Water flow into borehole may loosen ground – need to add water to balance (but may be too late to prevent softening)

② Nature of soil (also in-situ stresses)

③ Efficiency of test

Figure 6.31 Principles and details of the SPT test.

Figure 6.32 SPT test underway in Hong Kong (1980s). Nowadays, a helmet and other protective clothing such as boots and gloves would be worn.

Figure 6.33. In the UK, it is normal to stop a test when 50 blows fail to advance the split spoon the full 300 mm and instead to record the penetration achieved for the 50 blows. Depending on the ground conditions and sample retrieved, it might be valid to extrapolate the blow count to an equivalent "*N*" value *pro rata*. Overseas, it is common practice to continue the test for 200 blows or more in weathered rock (see discussion in Box 2.2), and designs are often based almost solely on such data, which is rather questionable practice in a profile that might comprise a heterogeneous mix

Figure 6.33 Split spoon sample of completely weathered granite. Note presence of relict joints and lack of visible disturbance.

of harder and softer materials. Tests carried out in this way may damage equipment and are tedious for the drilling contractor, who might well be tempted to cut corners if no-one is supervising.

The vane test involves rotation of a cruciform steel tool at a slow rate within the soil (Figure 6.34). The test is especially suitable for soft clay where SPTs are inappropriate

Figure 6.34 Field vane used for measuring strength of clay down borehole or sometimes pushed from the ground surface. Once at test location, the vane is rotated to measure shear strength of the cylinder of soil defined by the vane geometry. To the left is a sleeve used to protect the vane during installation.

because of the indeterminate nature of pore pressure changes brought about by rapid loading. The vane test is assumed to give a direct measure of undrained shear strength for the shape sheared by the rotating tool but interpretation can be difficult, especially in bedded soils.

The static cone penetrometer is a conical tool that is pushed rather than driven into the ground, usually from a heavy lorry (Figures 6.35 and 6.36). The end force on the cone tip, and drag on the sides of the tool, are measured independently and can be interpreted in terms of strength and deformability. Clay, being cohesive, grips the side proportionally more than sand or gravel, so the ratio between end resistance and the side "friction" can be used to interpret the type of soil as well as strength. A further refinement (piezocone) allows water pressures to be monitored as the cone is pushed in, which again can help in interpreting the soil profile.

Large-scale direct shear tests are sometimes carried out in the field in the hope that scale and disturbance effects might be reduced. In reality, lack of control in the testing process, as well as questions over representation of samples, however large, often outweighs any advantages. The derived data are generally less reliable than those from a series of laboratory tests, which themselves would need very careful interpretation before use at the mass scale.

Figure 6.35 Electric static cone penetrometer with piezometric ring. Forces on the cone tip are measured independently from the force on the shaft section above. A combination of all three measurements (including water pressure) gives a good indication of soil type as well as strength characteristics.

Figure 6.36 Heavy lorry being used to conduct static cone penetrometer tests.

Figure 6.37 Core being examined by engineering geologist at Izmit bridge investigation, Turkey.

Small-scale deformability tests down boreholes include the use of inflated rubber packers in soil (pressuremeter) or the Goodman jack in rock where two sides of the borehole are jacked apart. All such tests are very small relative to the mass under consideration and need to be interpreted with due care as to their representativeness. Deformation at project scale is better predicted from loading tests involving large volumes. The inclusion of very high capacity Osterberg jacking cells set within large diameter, well-instrumented bored piles gives the prospect of deriving much more representative parameters (e.g. Seol & Jeong, 2009). In practice, most rock mass parameters tend to be estimated from empirical relationships derived from years of project experience together with numerical modelling, rather than small-scale tests.

Field tests are really the only option for measuring hydraulic conductivity (also for oil and gas). Simple tests include falling or rising head tests in individual boreholes, whereby water is either added to or pumped out of a hole and then the time taken for water to come back to equilibrium measured. For realistic indications of behaviour at field scale, however, larger-scale pumping tests are required. Even then, water flow is often localised and channelled so tests may not always be readily interpreted.

6.9 LOGGING BOREHOLE SAMPLES

Data from ground investigations are generally presented in a report comprising factual data as well as an interpretation of conditions (if the GI Contractor is requested to do so). One of the important jobs for an engineering geologist is to examine and record the nature of samples retrieved from boreholes (Figure 6.37). The data from individual boreholes is usually presented in a borehole log, which provides a record not only of the ground profile but many details of how the borehole was carried out. In the oil industry, where the hole is advanced by a rock-roller bit or similar destructive method, logging is done by examining small chips of rock carried in the flushing mud (well logging); in civil engineering, we generally have rather better samples to examine.

Logging is generally conducted using a checklist approach and employing standard terminology to allow good communication, for example, on the apparent strength of a sample. Such standardisation can, however, result in over-simplification and lack of attention to geological detail. The task might be delegated to junior staff who might not have the experience and training to fully understand what they are examining. In addition, GI contractors will not routinely describe all features of samples recovered, partly because they want to avoid disturbing the samples before the Client/design engineer has made a decision on which samples he wishes to select for laboratory testing. Several examples of borehole and trial pit logs are provided in Appendix D. The examples prepared by GI contractors in the UK and Hong Kong demonstrate good practice, whereby the whole process of drilling a hole, testing down the hole and sampling are recorded. The materials encountered are described following standard codes and normal practice. Given the limitations discussed above, designers and investigators may need to examine samples and core boxes themselves and not rely on those produced by the Contractor. In Appendix D, examples are given of logs prepared by engineers who have the responsibility for the overall site investigation. These are supplementary to the logs produced by the GI contractors. The Australian example is from an intensive investigation of a failing slope that was threatening a road. There is considerable attention to detail, especially regarding the nature of discontinuities and far more so than in the Contractor's logs. In practice, even this level of logging may be inadequate to interpret the correct ground model, and selected samples and sections of core will need to be described in even more detail by specialists, perhaps employing techniques such as thin-section microscopy, radiometric dating and chemical analysis. In the example given in Appendix D of a trial pit description from a landslide investigation, the individual geological features have been examined, measured and sketched to provide a very detailed record of what was seen. In all cases and at all levels, logs should be accompanied by high-quality photographs with scales included, together with standard colour charts.

As discussed in Appendix C, guidance on standardised terminology is given in BS 5930:1999 and 2015, in the GEO guide on rock and soil description (Geotechnical Control Office, 1988) and the ISRM guidance on rock mass description (Brown, 1981; Ulusay & Hudson, 2007). There are many different standards and codes of practice in use world-wide – USA practice is far removed from that in that UK, as is that for Australia, China, New Zealand, Japan and Korea, which leads to confusion, particularly as similar terminology is often used to mean different things. A consequence of this fuzzy standardisation is that when projects go wrong geotechnically, as they sometimes do, then legal arguments often hinge on incorrect or misinterpretation of terminology. The engineering geologist needs to do his homework before practising in any region.

Another criticism made earlier regarding field mapping, but equally applicable to logging, is that standard guides and codes to rock and soil description tend to comprise a series of limited classifications that one has sometimes to force on an unwilling rock mass. For example, rock masses, as exposed in quarries, can seldom be simply described as "widely" or "closely" jointed, but loggers are required to apply such classifications to core samples. In the author's opinion, it is far better to concentrate on recording factual data, which can then be interpreted later as the overall ground model becomes clearer. An example of over-simplified rock classification terminology is given in Box 6.9 with reference to the term "aperture". The problem is that by using such terms it is implied that the feature has properly been characterised, which is not the case. de Freitas (2009) discusses the same point and also notes that many terms and indeed measured values such as porosity are "lumped" parameters and therefore rather insensitive and uninformative.

BOX 6.9 DEFINING APERTURE

AN EXAMPLE OF POOR PRACTICE BY GEOTECHNICAL CODING COMMITTEES

This example is used to illustrate the inadequacy of current geotechnical standards for soil and rock description to convey an accurate or realistic representation of the true nature of the geological situation.

Mechanical aperture is the gap between two rock discontinuity walls (three-dimensional) and a very important characteristic with respect to fluid flow and grouting. It is expressed in most codes and standards as a one-dimensional scale of measurement, in the same way as joint spacing. The various attempts at revising description of aperture over 25 years (leading to the current BS/Eurocode 7 requirements discussed later) have simply re-invented the measurement scales and terminology but have failed to address or inform users about the fundamental difficulties in measuring and characterising this property.

WHAT IS APERTURE?

It is the mechanical gap between two walls of a rock discontinuity such as a joint or a fault. An example of a small section of joint with a gaping aperture (because the block has moved down slope and dilated over roughness features) is shown in Figure B6.9.1, which is a photograph of a section of sheeting joint in granite from Hong Kong.

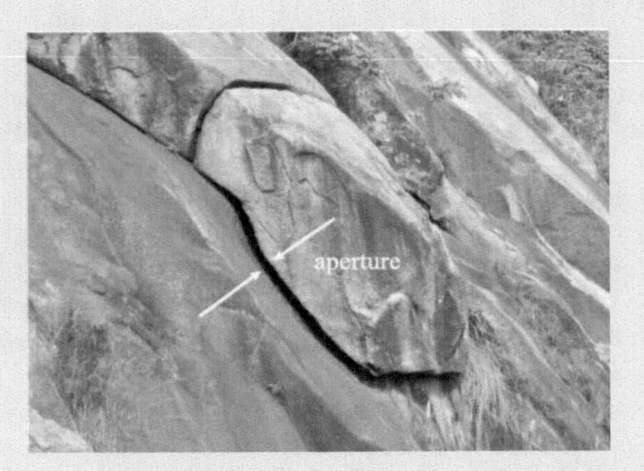

Figure B6.9.1 Part of sheeting joint with gaping aperture where seen. Evidently, away from the exposure the aperture is tight and the rock walls are in contact. Example is near Sau Mau Ping, Hong Kong.

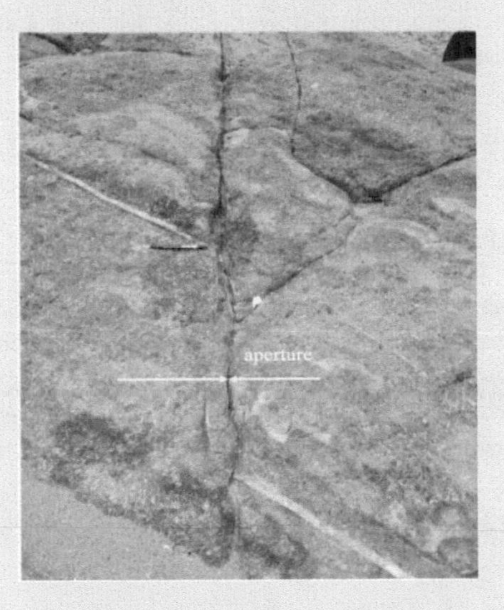

Figure B6.9.2 Minor fault exposed on beach, Peng Chau Island, Hong Kong.

In the second example, of a fault exposed at a beach, also in Hong Kong (Figure B6.9.2), it is not quite so easy; there is a groove along the feature but the astute geologist might interpret this as preferential erosion. Some authors advise measuring aperture using feeler gauges. Others have attempted to characterise aperture volumetrically by injecting resin or liquid metals.

DOES IT MATTER?

It is an extremely important property of the rock mass, controlling fluid flow and also related to shear strength. The problem is, it is a very complex and unpredictable characteristic, as is the associated fluid flow. A single joint can be locally tight and impermeable, whilst elsewhere can be open allowing huge volumes of water to flow, as discussed by Kikuchi & Mito (1993). Investigation and characterisation can be a nightmare – if a borehole hits a conductive section, then high permeabilities will be measured and an installed instrument will be responsive to changes in water pressure, but this is literally a "hit or miss" business, as evidenced by many examples in investigations associated with nuclear waste (e.g. Thomas & La Pointe, 1995). The author has the experience of working in a deep tunnel 150 m below the sea, where, in one section, the rock was highly jointed but dry, but elsewhere, at the same level, there was a steady inflow through what was apparently intact rock. Clearly, it is not just local aperture that matters, but the characteristics of the full fracture network and its connectivity leading to the point of observation. It is an important area for research and for observation linked to geochemical and structural studies together with an appreciation of coupled mechanisms (e.g. Olsson & Barton, 2001; Sausse & Genter, 2005). Without getting to grips with the concept of channelised flow on rock joints and through joint networks, it may be impossible to ever make a safety case for nuclear waste disposal, with all the corollaries, i.e. no nuclear power, global warming and the end of civilisation. Well, perhaps slightly overstated, but not that much.

APART FROM THE NATURAL VARIABILITY OF FRACTURE NETWORKS, ARE THERE ANY OTHER CONSIDERATIONS?

Yes. Most rock joints are sampled in boreholes where aperture simply cannot be measured. Furthermore, it is very unlikely that any borehole sample would be representative of the discontinuity at any great distance. Down-hole examination with cameras and periscopes can be used to examine borehole walls, but again there is a problem with sampling and representativeness. In exposures such as quarries or tunnels, exposure is better but there is a question of disturbance – blasting, stress relief and block movement and whether observations at one location are relevant to the rock mass as a whole.

SO WHAT ADVICE IS GIVEN IN RECOMMENDED METHODS AND STANDARDS?

1978: ISRM. The discussion on aperture is very useful. Its importance is recognised and many of the difficulties in measurement and interpretation are highlighted.

For description purpose and where appropriate, apertures are split into closed, gapped and open features, each sub-divided into three. It is advised that

a. modal (most common) apertures should be recorded for each discontinuity set;
b. individual discontinuities having apertures noticeably wider or larger than the modal value should be carefully described, together with location and orientation data; and

Table B6.9.1 Terms for the description of aperture

Aperture size term	*ISRM 1978*[1]	*BS5930 1999*	*ISO 14689-1:2003 BS5930 2015*
<0.1 mm	Very tight	Very tight	Very tight
0.1–0.25 mm	Tight	Tight	Tight
0.25–0.5 mm	Partly open		Partly open
0.5–2.5 mm	Open	Moderately open	Open
2.5–10 mm	Moderately wide	Open	Moderately wide
10–100 mm	Very wide	Very open	Wide
100–1000 mm	Extremely wide		Very wide
>1,000 mm	Cavernous		Extremely wide

Note: [1]In detail, there is further confusion in that ISRM also defines a term *wide* for *gapped* features >10 mm; the other terms above, also for apertures >10 mm are for *open* features but the difference is not fully obvious.

c. photographs of extremely wide (10–100 cm) or cavernous (>1 m) apertures should be appended.

1999 UK BS5930 (BSI, 1999). Says little about aperture other than noting that it cannot be described in core. Five classes are introduced, which use some of the same terms as ISRM but with different definitions.

2003 INTERNATIONAL STANDARD ISO 14689-1 (BSI, 2003) (for Eurocode 7 users). Provides a new mandatory terminology for one-dimensional measurement that differs from that of BS5930: 1999 and ISRM (1978), as illustrated in Table B6.9.1. The 2015 version of BS5930 reverts to the same classification…

HYDRAULIC APERTURE VS. MECHANICAL APERTURE

For completeness, it is worth emphasising here that even if we could measure mechanical aperture meaningfully, the actual associated flow characteristics of the rock mass (hydraulic aperture) would be very difficult to estimate or predict. It clearly makes sense to observe and characterise rock masses as best we can, with respect to openness of the fracture network, but hydraulic conductivity can only be measured realistically using field tests, as discussed elsewhere, and even these are often open to different interpretations (e.g. Black, 2010).

CONCLUSIONS

After 25 years to digest the ISRM discussion and intensive international experience on research in measuring gaps in discontinuities and associated fluid flow, especially with respect to nuclear waste disposal investigations, the requirement for site investigation in Europe is a new set of linear measurements, which are inconsistent with previous ones. No mention is made of the difficulty of characterising aperture in this way. Meanwhile, the New

Zealand Geotechnical Society (2005) has produced yet another classification for aperture, which uses a selection of the same terms as in the above table but defined differently (e.g. wide = 60 mm to 200 mm) and introduces a new set of classes for the middle range: *very narrow, narrow, moderately narrow*.

APOLOGIES

Apologies for being so critical, but it seems to this author that many codes and classifications over-simplify geological description and constrain/stifle good practice. This is especially so where it is mandated that some particular but fundamentally inadequate terminology shall be used. Unfortunately, inexperienced geotechnical engineers and engineering geologists are led to believe that such codification adequately deals with description and characterisation of the feature, which is not the case.

6.10 DOWN-HOLE LOGGING

Down-hole logging technology has largely come from the oil industry and partly from mining. At the simplest level, a TV camera or borehole periscope is lowered down an uncased borehole and used to identify defects or to examine discontinuities. A borehole can be pumped dry of water and observations made of locations of water inflow, although this might need to be inferred from temperature or chemical measurements (Chaplow, 1996). Borehole impression packers were introduced in the 1970s and can be used to measure the orientation of discontinuities. Using an inflatable rubber packer, paraffin wax paper is pressed against the walls of the borehole and when retrieved, the traces of indented joints are clearly visible (Figure 6.38). Dip of the joints is easily determined from the geometry of the borehole but measuring direction relies upon whatever device is used to orientate the packer and, from experience, this can be a major source of error. It is good practice when using the impression packer to specify overlapping sections of measurement down the hole (by perhaps 0.5 m) so that consistency can be checked. In one borehole we found a 70 degree difference between consecutive sections, resulting from the packer being deflated before the compass had set in position so the Contractor was asked to redo the work. A more modern tool is the Borehole Image Processing System (BIPS), which gives a continual visual record of the borehole wall (Kamewada et al., 1990). The tool is lowered down the borehole and a video camera takes a 360 degree image millimetre by millimetre down the hole through a conical mirror (Figure 6.39). Despite modern instrumentation for this tool, whereby azimuth can be measured by magnetic flux gates or gyroscopes, a recent study revealed errors of up to 20 degrees in this measurement (Döse et al., 2008). Care must also be taken in interpretation of discontinuities logged in boreholes, especially if boreholes are all vertical. There will be obvious bias to the measurements – steep joints will be under-sampled in vertical boreholes. As an example, during the Ching Cheung Road landslide investigation (Halcrow Asia Partnership, 1998a), BIPS measurements were taken in vertical boreholes and gave a completely different style of jointing to those measured in exposed faces (essentially along horizontal scan lines). The data are

Figure 6.38 Impression packer. Paraffin wax paper has been pushed against the walls of the borehole by a rubber inflatable packer. A series of pale-grey traces can be seen, which represent a set of fairly planar joints dipping at about 70 degrees. Direction is obtained from a compass set in glue at the base of the packer. Other options for orienting devices now include flux gate magnetometers and gyroscopes.

Figure 6.39 Output from BIPS down-hole discontinuity orientation device, being used during logging of rock core, Taejon Station, South Korea.

presented in Figure 6.40 and it can be seen that the borehole data essentially measured a girdle of joints at 90 degrees to the main pole concentration, as measured from the horizontal scan line data. Both sets of data were required to provide the correct geological picture.

Other down-hole tools include resistivity and gamma ray intensity (even in cased holes), which, whilst often useful for oil exploration and coal mining, generally have rather limited application to civil engineering, other than possibly as an indication of clay-rich horizons.

6.11 INSTRUMENTATION

Instrumentation is used to establish baseline ground conditions at a site, most commonly in terms of natural groundwater fluctuations. It is also used to monitor changes

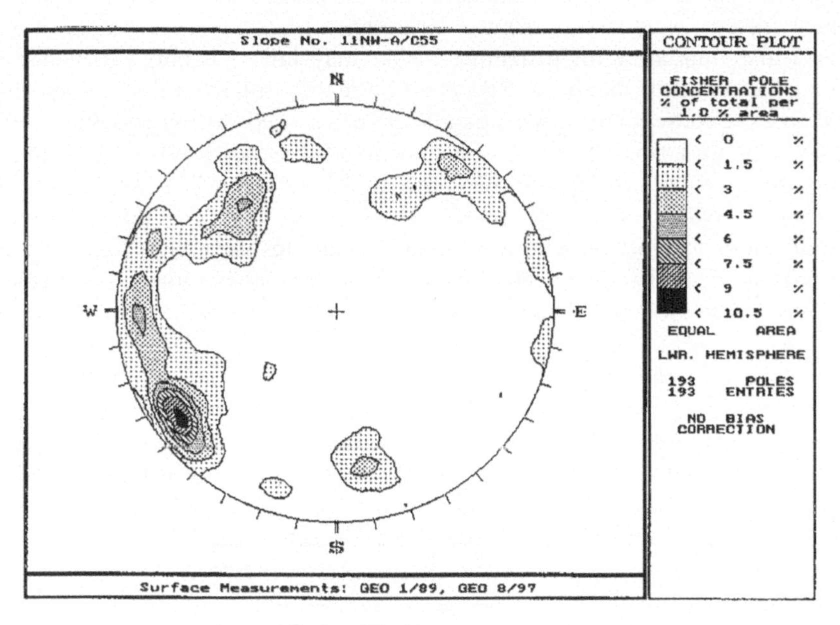

a) Surface Mapping

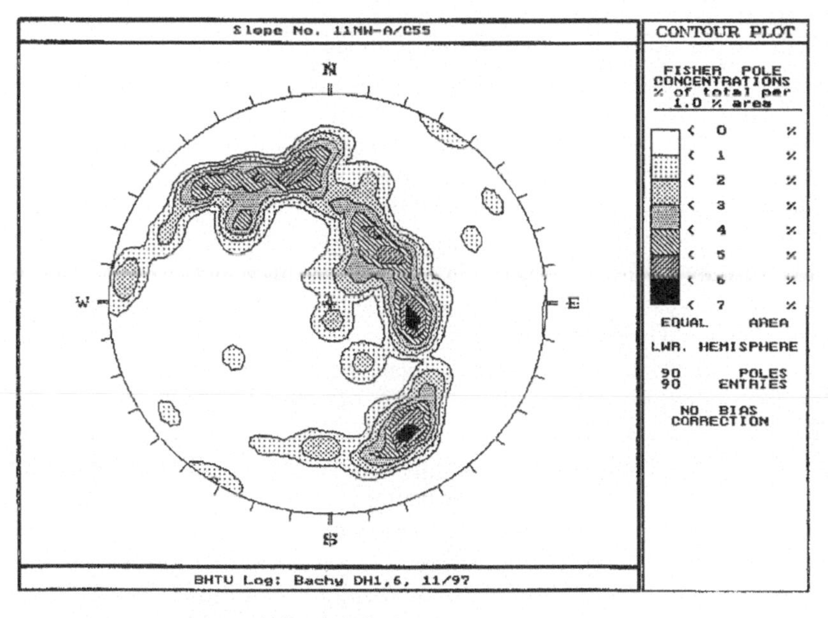

b) Borehole Data

Figure 6.40 Comparison between discontinuity data recorded by BIPS (vertical drillhole) and from surface mapping (horizontal scan lines).

Source: After HAP (1998a).

at a site brought about by construction activities, such as excavation or blasting. Instrument systems need to be designed carefully so that they are reliable; there needs to be built-in redundancy for instruments that may fail or become damaged by site works or by vandalism. Incoming data must be readily interpretable if some action is to be taken as a consequence. Instruments are often used during the works to check performance against predictions. Displacements and water levels can be monitored and compared to those anticipated. First (ALERT) and second (ALARM) level trigger conditions can be defined with prescribed action plans. Data can be sent remotely to mobile phones or by email to engineers who have the responsibility for safety and the power to act, such as closing a road or evacuating a site. Other instruments that might be employed during a large construction project include sound and vibration meters, especially where blasting is to be carried out.

Piezometers are commonly installed as part of ground investigations to measure water pressures. Detailed information on these and other instruments are given by Dunnicliff (1993). The simplest device is an open-tube standpipe with a porous tip, installed in a sand pocket within the borehole, as shown in Figure 6.41. There are also push-in versions available. The water level in the standpipe is "dipped", perhaps on a weekly or monthly basis, using a mechanical or an electronic device lowered down the hole; for an electric dipmeter, the water closes a circuit to activate a buzzer. To measure high rises in water level between visits by monitoring personnel, Halcrow plastic "buckets" can be installed on a fishing line with a weight at the bottom of the string, at perhaps 0.5 m intervals down standpipes. The buckets are pulled out of the hole when the site is visited – the highest one that is filled with water indicates the maximum level of water (Figure 6.42). At a more sophisticated level, standpipes can be set up so that readings are taken automatically at regular intervals using pressure transducers ("divers") or through an air bubbler system (Pope et al., 1982). Data can be recorded on data loggers that can be set up to transmit information by telemetric systems. Other instruments include pneumatic or vibrating wire piezometers that respond very quickly to changes in pressure (Figure 6.43). Because they require almost no water flow to record change of pressure (unlike a standpipe), they can be grouted in place in the borehole and several instruments can be installed in the same hole, which can save cost (Vaughan, 1969; Mikkelsen & Green, 2003).

Instruments that are used to measure displacement include strain gauges, tilt meters, inclinometers and extensometers. They can be mechanical or electrical, for example, using vibrating wire technology. Figure 6.44 shows the end of an extensometer anchored deep behind the working face of a large copper mine in Spain and fitted with lights and a claxon horn to give warning if the anchored point moves towards the mining area. Other instruments used to monitor performance at that site included deep inclinometers and a Leica "total station", whereby numerous targets on the slope surfaces were surveyed remotely and automatically on an hourly basis, with the data sent to the site office (Hencher et al., 1996). An inclinometer is a tubular "torpedo" (with wheels), which is lowered down a grooved tube set into a borehole or built into embankment fill. Figure 6.45 shows a section of inclinometer casing with the two sets of orthogonal grooves for the wheels. The torpedo (Figure 6.46) is first lowered down aligned by the first set of grooves, then removed and lowered down the second set of grooves. The section on the figure also has magnetic "spiders" with magnets, through

Figure 6.41 Standpipe piezometer tip about to be placed in borehole. Another has already been installed at a deeper level. It is not very good practice to install more than one piezometer in a borehole, because of potential leakage between the different horizons being monitored, but can work providing great care is taken in installation. Portsmouth dry dock, UK.

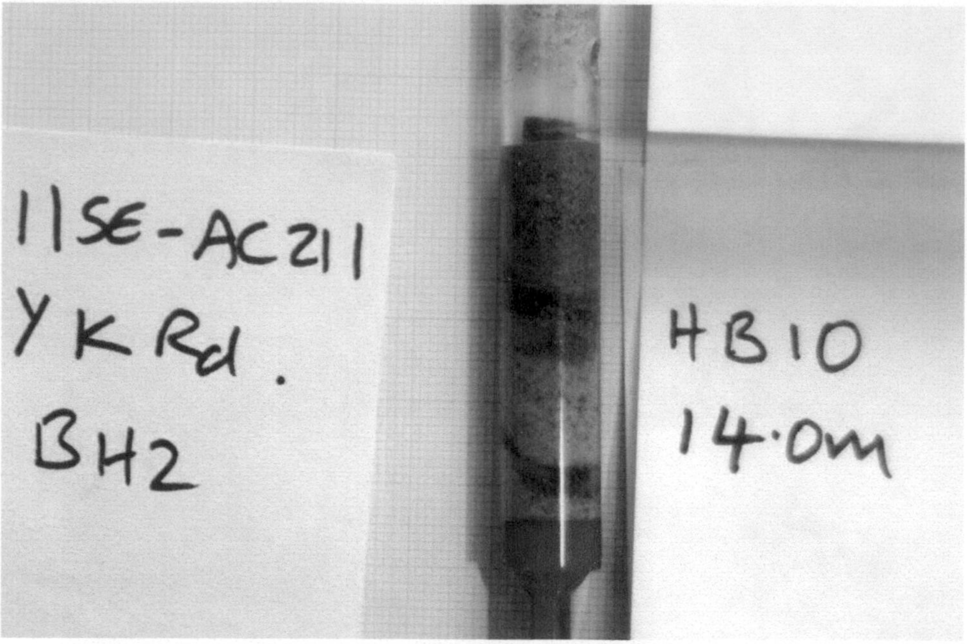

Figure 6.42 Halcrow bucket retrieved at Yee King Road landslide investigation (Hencher et al., 2008). This is unusual in that it contains sediments (from turbulent flows down the borehole). Normally, they would just contain water (or not), indicating the highest level that the water has risen in the borehole between inspections.

which the tube can slide and can therefore be used to monitor vertical settlement where the tube is installed in fill. Strain gauges within the torpedo measure tilt, which is recorded against depth. The orthogonal measurements can be resolved to give the true direction and amount of displacement.

6.12 ENVIRONMENTAL HAZARDS

6.12.1 General

Site investigation needs to include a review of the potential environmental hazards as well as the immediate ground conditions as addressed in Chapter 5. There may be risk from natural landslides and rock fall threatening the project, potential for natural subsidence or collapse (say in areas underlain by salt deposits, old mine workings or karst), coastal erosion, wind, rain or earthquakes (Bell, 1999). As noted earlier, for some locations there are published hazard maps, but such maps cannot usually be relied upon on a site-specific scale. It is up to the site investigation team to identify the potential hazards for the project throughout its life (maybe 50 to 100 years) and to quantify these. In some cases, such an assessment might lead to

Figure 6.43 Pneumatic piezometer being used to take measurements of rapidly changing water pressures during pile driving. Only small volume changes are necessary to measure pressure changes, so readings could be taken every ten seconds or so. Water pressures measured went off scale at about three times overburden pressure (Hencher & Mallard, 1989). Drax Power Station, Yorkshire, UK.

a decision not to proceed with a project, as in the construction of a house at a site where there are serious hazards, as illustrated in Figure 6.47. Elsewhere, the hazard can be dealt with by careful design, and the main example of so doing is the hazard of earthquakes.

Figure 6.44 Extensometer with claxon and flashing lights used to warn workers at Aznalcollar mine, Spain, of danger from moving slope.

Figure 6.45 Exhumed inclinometer tubing. Four grooves inside (ridges outside) are guides for the wheels on the inclinometer instrument. The device with arms is a "spider", which becomes fixed in position against the walls of a borehole whilst the tube can pass up or down inside. It is magnetic and a probe down hole can locate it and measurements can be made of settlement (as well as inclination).

Figure 6.46 Inclinometer torpedo about to be lowered down a grooved tube. Tuen Mun Highway, Hong Kong.

Figure 6.47 Houses constructed in active area of rock fall hazard, South Africa.

6.12.2 Natural terrain landslides

Landslides from natural terrain (rather than man-made slopes) are a hazard in most mountainous regions and can range from minor rock and boulder falls to massive landslides which involve >20 million m^3 of rock and occur on average every 3 or 4 years worldwide (Eberhardt et al., 2004). Landslides like the one that destroyed Yungay, Peru, in May 1970, and killed about 20,000 people, are very difficult to predict and impossible to engineer. All one can do is identify the landform, the degree of risk and perhaps monitor displacements or micro-seismicity, with a plan to evacuate people and close roads if necessary.

Smaller and more common natural terrain landslides can be predicted and mitigated to some degree by engineering works. The starting point is generally historical records of previous landslides, such as incidents on active roads through mountainous regions. These may allow areas of greatest hazard to be identified and some prioritisation of works. It should be noted, however, that small rock falls at one location can be indicative of much larger and deep-seated landslides, and minor incidents should be reviewed in this light. Where there is good historical air photograph coverage, sources of landslides can be identified and these correlated to susceptibility maps prepared using geographical information systems (e.g. Devonald et al., 2009). Typical factors that might be linked to probability of landslide occurrence include geology, thickness of soil, vegetation cover, slope angle, proximity to drainage line and catchment area. Once a best fit has been made linking landslide occurrence to contributing factors, maps can be used in a quantitative, predictive way. Consequence of a landslide depends on location relative to the facility at risk (e.g. road, building), volume, debris run-out, possibility of damming a watercourse and eventually impact velocity. From studies in Hong Kong (Moore et al., 2001; Wong, 2005), it is apparent that the greatest risk is generally from channelised debris flows (outlets of streams and rivers) and to facilities within about 100 m of hazardous slopes (the typical limit of debris run-out in Hong Kong). A broader discussion is given by Fell et al. (2005). A decision can be made on the resources that are justified to mitigate the hazard, once one has determined the level of risk (which can be quantified in terms of risk to life). There are many options, including barriers and debris "brakes" in stream courses and catch nets, especially for rock fall and boulder hazards. In some cases, a decision might be made to stabilise the threatening natural terrain using drainage, surface protection, netting and anchors, as for man-made slopes, dealt with in Chapter 3.

6.12.3 Coastal recession

Coastal recession is a common problem and rates can be very rapid. For example, parts of the Yorkshire coast are retreating at up to 2 m per year (Quinn et al., 2009). Many studies have been carried out on mechanisms, but the harsh fact is that many properties and land near the coast are at risk and many houses have to be abandoned. Coastal protection measures can be designed successfully (much of Holland is below sea level) but these often fail in a relatively short time and constructing works at one location can have consequences for others along the coast, as suspected for the damage to the village of Hallsands in Devon, which had to be largely abandoned (Tanner & Walsh, 1984).

6.12.4 Subsidence and settlement

Excellent reviews on ground subsidence – natural and due to mining, are given by Waltham (2002). Ground subsidence occurs naturally due to lowering of the water table from water extraction, oil and gas extraction, shrinkage of clay, and dissolution of salt deposits, limestone and other soluble rocks (e.g. Cooper & Waltham, 1999). Sub-surface piping can occur associated with landslides in any rocks, including granite (Hencher et al., 2008). The results can be dramatic, with sudden collapses of roads or even loss of buildings. Care must therefore be taken to consider these possible hazards during site investigation.

Underground mining dates back thousands of years in some areas (e.g. flints from chalk) and on a major scale for hundreds of years. Consequently, there are very incomplete records. In desk study, the first approach will always be to consult existing records and documents, but wherever there is some resource, such as coal, that might have been mined, the engineering geologist needs to consider that possibility. Investigations can be put down on a pattern, specifically targeted at the suspected way that mining might have been carried out (pillar and stall or bell pit, for example). Air photograph interpretation will often be useful and geochemical analysis of soil can give some indication of past mining activities.

6.12.5 Contaminated land

Many sites around the world are severely contaminated, often because of man's activities. This means that if the site is to be used for some new purpose, it may need to be cleaned up to be made habitable. Similarly, when constructing near or through possibly contaminated land, this needs to be investigated and the contamination mitigated, possibly by removing the contaminated soil to a treatment area. Barla & Jarre (1993) describe precautions for tunnelling beneath a landfill site. Guidance on investigation is given in British Standards Institution (2001), Construction Industry Research and Infomration Association (1995) and many other sources of information are given by the AGS (Appendix A). Sometimes the contamination is dealt with at site. Desk study can often identify projects where there are severe risks because of previous or current land use. Industrial sites such as old gas works, tanneries, chemical works and many mines are particularly problematical. Severe precautions need to be taken when dealing with such sites and works will probably be controlled by legislation.

6.12.6 Seismicity

(a) Principles

Design against earthquake loading is an issue that needs to be considered in many parts of the world, depending upon the importance of the project and risks from any potential damage. In some locations, because of inherently low historical seismicity (UK) and/or severity of other design issues (e.g. typhoon wind loading in Hong Kong), seismicity might be largely ignored for design other than for high-risk structures like nuclear power plants. Elsewhere, seismicity needs to be formally assessed for all structures and considered for design.

(b) Design codes

Many countries have design codes for aseismic design and these are generally mandatory. Nevertheless, it is often prudent to carry out an independent check and in particular to consider any particular aspects of the site that could affect the impact of an earthquake. For example, the local soil conditions might have the potential to liquefy. These issues are considered in more detail in Chapter 5.

Design codes, where well written and implemented, reduce the earthquake risks considerably. The USA, for example, has a high seismic hazard in some areas but fatalities are few and this can be attributed to good design practice and building control. China also has a high seismic hazard in some areas, but earthquakes commonly result in comparably large loss of life, which might be attributed to poor design and quality of building. Structures can be designed to withstand earthquake shaking, and even minor improvements in construction methods and standards of building control (quality of concrete, walls tied together, steel reinforcement, etc.) can prevent collapse and considerably reduce the likely loss of life (Coburn & Spence, 1992).

(c) Collecting data

The first stage is to consider historical data on earthquakes, which are available from many sources, including the International Seismological Centre, Berkshire, and the US Geological Survey. These historical data can be processed statistically using appropriate empirical relationships to give probabilistic site data – for example, of peak ground acceleration over a 100 or 1000 year period. This can be done by considering distance from site of each of the historical earthquake data or linked to some source structure (such as possible active faults). Dowrick (2009) addresses the process well, and some guidance is presented in Chapter 5. In some cases, estimates are made of the largest earthquake that might occur within the regional tectonic regime and similar regimes around the world, to derive a "maximum credible event". This postulated worst case could be used by responsible authorities for emergency planning and is also used for some structures – a "safe-shutdown" event for a nuclear power station design.

6.13 LABORATORY TESTING

Generally, a series of laboratory tests are specified for samples recovered from boreholes, trial pits and exposures, often employing the same GI contractor who carried out the boring/drilling. Geotechnical parameters and how to measure or estimate them are addressed in Chapters 4 and 7.

6.14 REPORTING

The results of site investigation are usually presented as factual documents by the GI contractor – one for borehole logs, a second for the results of any laboratory testing. In addition, specialist reports might be provided on geophysics and other particular investigations. These reports may include some interpretation, perhaps with some cross sections if the contractor has been asked to do so, but such interpretation may be rather general and unreliable, not least because the GI contractor will not be aware of the full details of the planned project.

Generally, it is up to the design engineer to produce a full interpretation of the ground model in the light of his desk study, including air photo interpretations and the factual GI (that he has specified). This might be done supported by hand-drawn cross sections and block diagrams – which should ensure that the data are considered carefully and should enable any anomalies and errors to be spotted. There is a tendency now to rely upon computer-generated images, with properties defined statistically to define units (e.g. Culshaw, 2005; Turner, 2006), which might reduce the chance that key features of the model are properly recognised by a professional.

Note

1 It does not follow that London Clay is without hazards for construction projects, for example, the Heathrow Express Tunnel collapsed during construction, as discussed in Chapter 8. de Freitas (2009) also provides a warning over geological variation through the London Clay stratum and argues that data banks of geotechnical properties need to be used with care from one area to another.

Chapter 7

Geotechnical parameters

Putting numbers to geology.

Hoek (1999)

7.1 PHYSICAL PROPERTIES OF ROCKS AND SOILS

Civil engineering design involves the use of empiricism, applied mechanics calculations and sometimes physical and numerical modelling; here "empiricism" means the use of design rules based on experience, for example, design of underground support using rock mass classification systems rather than parameters directly (Appendix C). Generally, however, for civil engineering design it is necessary to assign physical properties to each unit of soil or rock within a "ground model". These include readily measurable or estimated attributes such as unit weight, density and porosity. Other parameters that are often needed are strength, deformability and permeability. In the case of aggregates (rock used in construction for making concrete or perhaps as armourstone), important attributes are durability and chemical stability.

Observations and experiments over the last 100 years have shown that the behaviour of soil and rock, at the materials and mass scales, can only be satisfactorily explained in terms of effective stress – the stresses caused by the bulk weight of the soil or rock, modified by water and air pressure (Skempton, 1960). In addition of course there may be other stresses in the ground such as tectonic stress (Hoek & Brown, 1980).

7.2 MATERIAL VS. MASS SCALES

Geotechnical parameters are dealt with under the first of the three "verbal equations", which are considered important, for thinking through the geological, environmental and construction level hazards and risks, *prior* to attempts at geotechnical modelling (Chapters 1 and 2; Knill, 1976; Baynes et al., 2020).

For example, if you were considering driving piles to rock underlain by soil then you would primarily want to know the soil strength variation with depth to rockhead, possibly found by CPT or SPT tests. However, you might also consider the grain size distribution, which might be variable and could affect driving efficiency. The potential for false sets, might need to be reviewed plus future drag-down by negative skin friction. You might also want to know the mineralogy as it might cause problems with the piles in the long-term due to chemical reaction. The underlying rockhead would

DOI: 10.1201/9781003348894-7

need to be investigated; if it was karstic or weathered, this could affect the ultimate capacity of the piles.

Geological factors in the first equation, are first considered at the small scale of intact geological *materials*. At the larger scale, the geological *mass* includes faults and other discontinuities, structures and geological setting, and landslides as an example. Most geotechnical tests are on small-scale samples but these must be scaled up to suit the scale of the project.

Mass strength, deformability and permeability of rock masses are controlled largely by the fracture network (part of the "mass"), rather than intact rock properties.

7.3 ORIGINS OF PROPERTIES

7.3.1 Fundamentals

Geotechnically sensitive geological properties of soils and rocks, such as abrasivity, porosity and mineralogy are highlighted in Chapter 6, and the reader designing a site investigation is advised to think through these carefully to avoid making mistakes by ordering "the standard" format of investigation and tests whilst ignoring the piece of the jigsaw that really makes a difference. Examples where site investigations failed to identify key aspects of the ground model are presented in Chapter 8.

Strength is a fundamental property of both soils and rock, and we need to take some time on this. The strength (tensile, compressive and shear) of soils and rocks comes from friction between individual grains and surfaces. It is a function of the closeness of packing of the mineral grains, even in clay soil (Roscoe et al., 1958). Densely packed soil will be forced to dilate (open up) during shear at relatively low confining stresses as the grains override one another and deform, and the work done against dilation provides additional strength (Rowe, 1962; Bolton, 1986). Strength is also derived from cohesion from inter-granular bonds such as those formed by pressure solution (Tada & Siever, 1989) and from natural cementation filling pore spaces with calcite or other salts (Clough et al., 1981; Al Sanad et al., 1990).

The huge range of property values in soil and rock and how these (especially cohesion) evolve with time, is illustrated for a *single* sample in Figure 7.1. The left-hand picture shows graded sediments. The sand horizons fine upwards to clay, as is typical of sediments deposited from a river into a lake. At the top of the sample, there is a second sand horizon that has been deposited onto the underling sediment. The loading has resulted in the soft, underlying mud penetrating the overlying sand, which shows how soft the soil was at the time of formation. Contrast this with the rear of the same sample in the right-hand figure, which shows conchoidal fractures, in what is actually now, an extremely strong rock. The conversion from soft mud to rock has occurred over a very long time but has occurred naturally. In practice we encounter and need to deal with the full range of materials, transitional between these end members – soft mud to extremely strong rock.

7.3.2 Friction between minerals

Shear strength at actual contact points between grains of soil or between rock surfaces is largely derived from electrochemical bonds over the true area of contact, which is

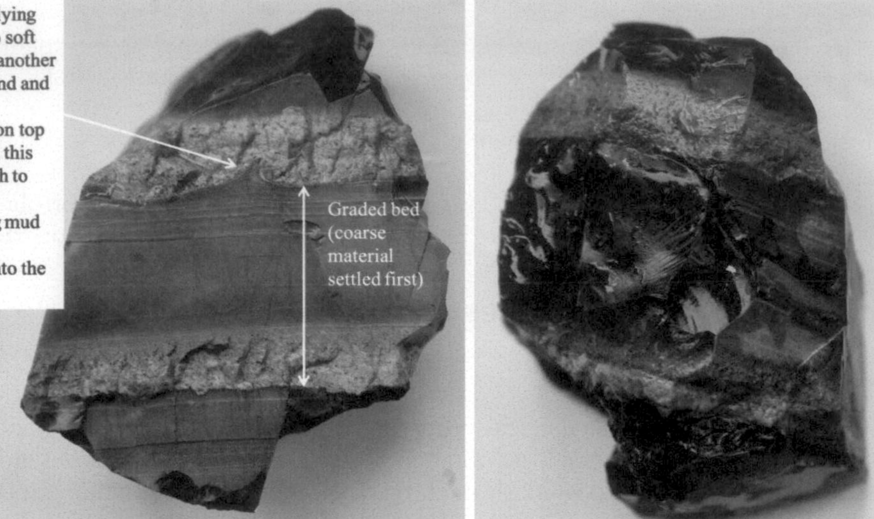
The underlying soil was so soft that when another layer of sand and mud was deposited on top (in a lake), this was enough to cause the underlying mud to deform upwards into the sand
Graded bed (coarse material settled first)
Soft sediment deformation at time of deposition.
Rear of sample showing conchoidal fractures (like glass or obsidian)

Figure 7.1 Sample, illustrating the development of strength with time (Hong Kong).

only a very small proportion of the apparent cross-sectional area of a sample. At each contact between grains, elastic deformation, plastic flow and dissolution may take place, spreading the contact point so that the actual contact area is directly proportional to normal load. The attractive force over the true area of contact gives rise to frictional behaviour.

The same principles apply to rough rock joints or fractured rock masses. Different minerals also have fundamentally different properties – some are more chemically reactive and may form strong chemical bonds in the short-term, some are readily crushed or scratched, whilst others are highly resistant to damage or chemical attack. Minerals such as talc and chlorite, are decidedly slippery and if present on rock joints can result in instability.

Friction angles for dry samples of quartz and calcite are reported as low as 6 degrees but higher when wet (Horn & Deere, 1962; Terzhagi, 1925). The opposite behaviour was reported for mica and other sheet minerals. Some authors question whether these data are valid and argue that there is a contamination effect for the polished samples (of quartz especially) and certainly the same behaviour is not recognised for soil and rock at large scales where mineral grains are rough; wet quartz sand is generally reported to have the same frictional resistance as dry sand (Lambe & Whitman, 1979). Nevertheless, Horn & Deere's data are still held to have some validity by some authors, as discussed by de Freitas (2009) who offers some physical explanations. Furthermore, it is interesting that the mineral species that reportedly give higher friction values when wet are the same minerals that commonly form strong bonds during burial diagenesis through dissolution and authigenic cementation (Trurnit, 1968). It seems possible that, for quartz and calcite, water might aid the formation of larger plastic asperity contacts during experiments in the wet, leading to higher "frictional" resistance (Hencher, 1977). The fact that mica does not show increased frictional resistance when wet might be associated with the reported fact that the presence of mica, chlorite and clay minerals inhibit pressure solution and cementation of quartz (Heald & Larese, 1974) and that the same minerals are rarely associated with pressure solution bonding. In summary, there is evidence that the lower bound frictional resistance between mineral grains, and in particular quartz, may be of the order of 6–10 degrees.

7.3.3 Friction of natural soil and rock

Now, there is a dilemma, in that friction angles, even for planar rock joints and soils, are often greater than 30 degrees (without any contribution from dilation), yet the additional resistance (above the lower bound) is *still* directly proportional to normal load. This additional frictional component varies with surface finish of planar rock joints and can be reduced by polishing (Coulson, 1971) or by reducing the angularity of sand (e.g. Santamarina & Cho, 2004). The fact that friction rules apply to rock (and soil) is illustrated by Figure 7.2, which shows results from two series of direct shear tests on saw-cut surfaces of granite, ground flat with carborundum powder to remove any slight ridges from the saw-cutting process, although at a microscopic scale apparently flat surfaces are still rough (Figure 7.3).

Each data point in Figure 7.2 comes from a separate test with the sample reground beforehand. The upper line (inclined at 38 degrees) is the friction angle measured for moderately weathered (grade III) rock; the lower line inclined at 32.6 degrees is for slightly weathered (grade II) rock. The paradoxical reason for the higher strength

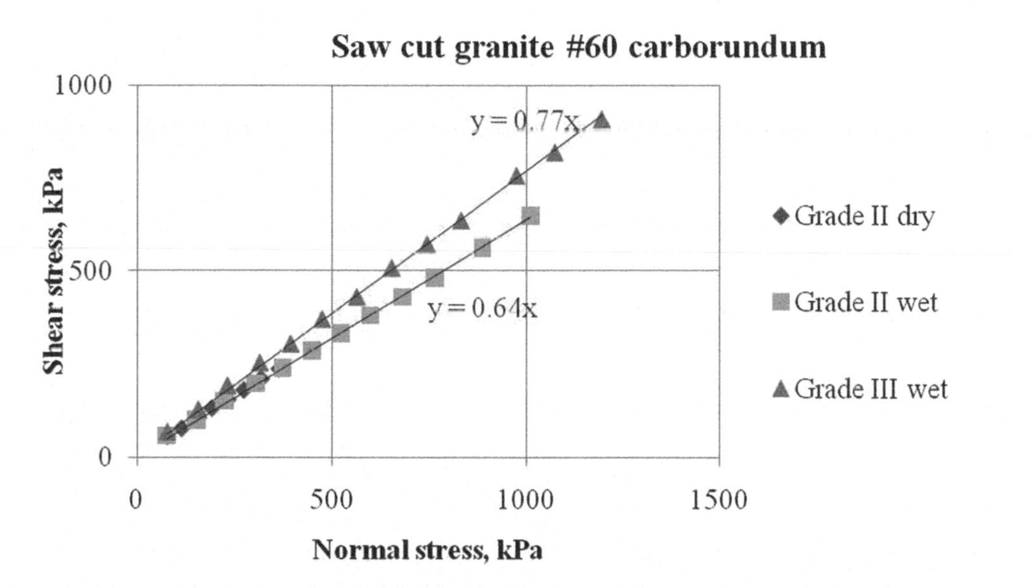

Figure 7.2 Figure demonstrating that two sets of samples with different surface finishes can have different friction angles, yet still obey Amonton's first law.

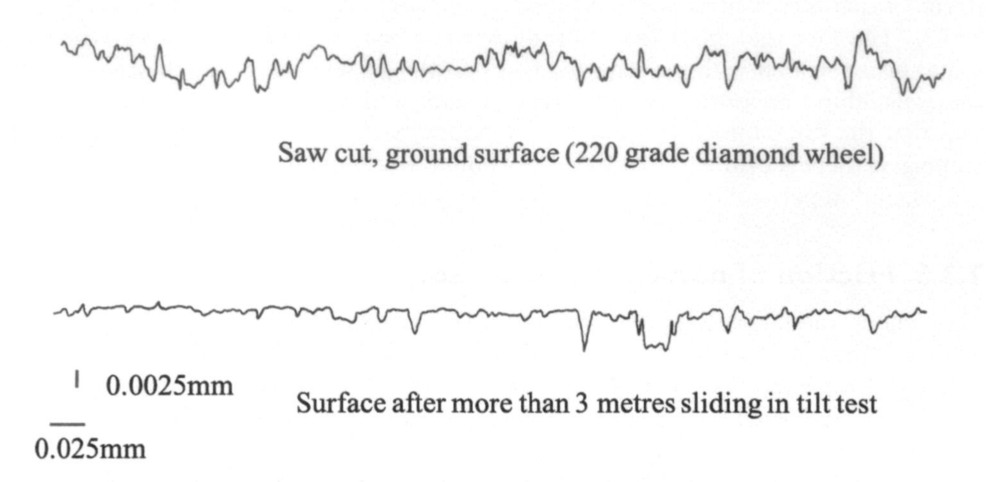

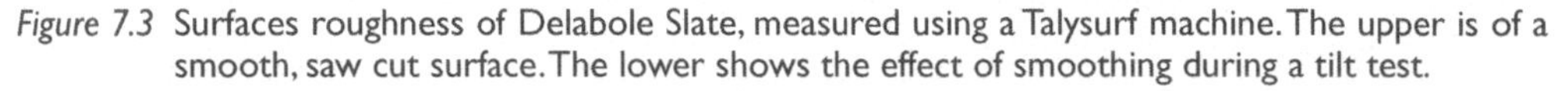

Figure 7.3 Surfaces roughness of Delabole Slate, measured using a Talysurf machine. The upper is of a smooth, saw cut surface. The lower shows the effect of smoothing during a tilt test.

Source: Hencher, 1977.

for the more weathered surface is because the surface finish is slightly rougher, the weathered feldspars being preferentially plucked from the surface during grinding. The *key* observation, however, is the precision of the frictional relationships – an increase in strength that is directly proportional to the level of normal load.

The third contact phenomenon is dilation. Additional work is done against the confining normal load during shear as soil moves from a dense to a less dense state or as a rock joint lifts over a roughness feature. If the raw strength data from a test are plotted against normal stress, then the peak strength envelope may show an intercept on the shear strength axis ("apparent cohesion"), albeit that the peak strength "envelope" may be very irregular, depending upon the variability of the samples tested. If measurements are made and corrections made for the dilational work during the test, in many cases the corrected strength envelope will show only friction: the strength envelope passes through the origin. At higher stresses, all dilation will be constrained and the soil or rock asperities will be sheared through without volume change.

7.3.4 True cohesion

Rocks and natural soil may also exhibit true cohesion, which is due to cementation and chemical bonding of grains. For a rock joint, it is derived from intact rock "bridges" that need to be sheared through. This additional strength, evident as resistance to tension, is essentially independent of normal stress and proportional to sample size.

7.3.5 Geological factors

In Chapter 1 (Figure 1.7), the concept of a rock cycle was introduced whereby fresh rock deteriorates to soil through weathering and then sedimented soil is transformed

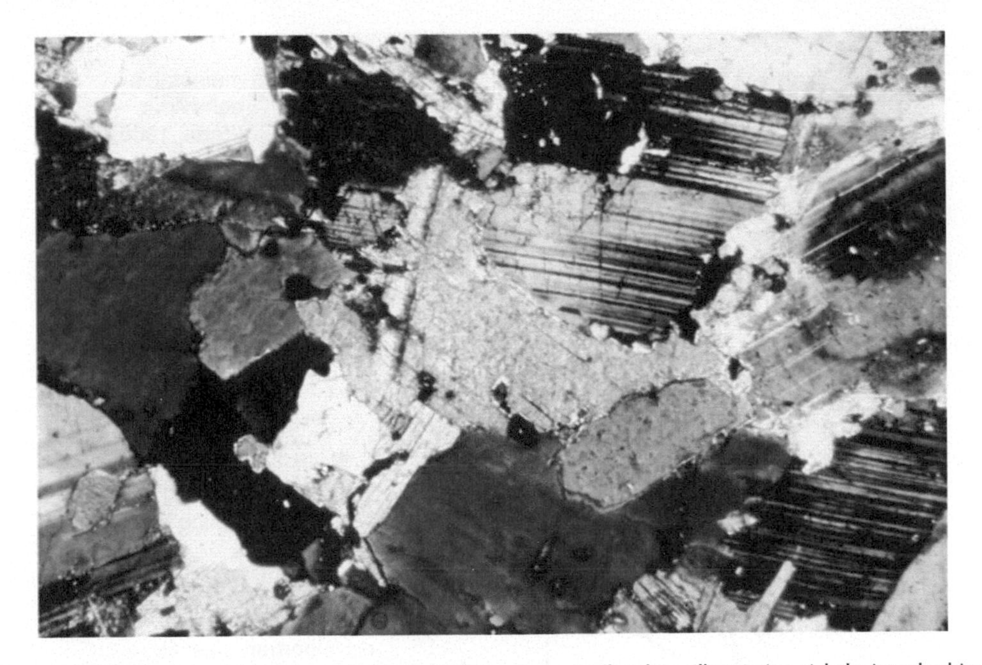

Figure 7.4 This section through granitic rock through crossed-polars, illustrating tightly interlocking fabric. Width of view approximately 20 mm.

again into rock through burial, compaction and cementation. Clearly, at each different stage in this cycle the "geomaterials" will have distinct properties and modes of behaviour.

(a) Weathering

In fresh igneous and metamorphic rocks, the interlocking mineral grains are linked by strong chemical bonds. As illustrated in Figure 7.4, there is almost no void space.

As weathering takes place close to the Earth's surface and fluids pass through the rock, it develops more voids as minerals decompose chemically and weathering products such as clay are washed out. The bonds between and within individual grains are weakened. Figure 7.5 illustrates how rock that starts off with a dry density of about 2.7 Mg/m^3 (typical of granite) becomes more and more porous so that by the "completely decomposed" stage, the dry density may be reduced by more than 70% if weathering products have been washed out. The final stage is collapse to residual soil and an increase in density. Weathering is discussed in detail in Chapter 6.

Strength and deformability at the material scale are linked quite closely to density empirically and, therefore, degree of deterioration from the rock's fresh state. Fresh granite might have a uniaxial compressive strength of perhaps 200 MPa but by the time the rock is highly decomposed the strength is reduced to 10 to15 MPa and when completely decomposed perhaps to 10 to 15 kPa. Where the rock is relatively strong then properties and behaviour of the mass will be dominated by the fractures; for most

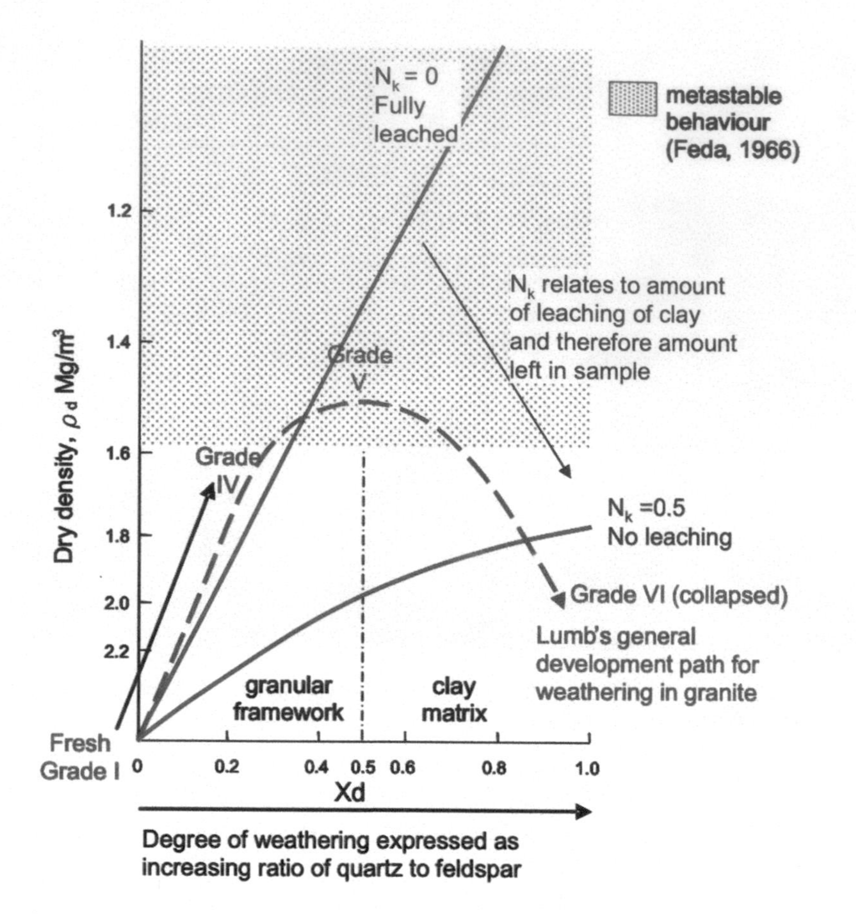

Figure 7.5 Change of dry density in weathered granite. At its fresh state it has a dry density of 2.7. As weathering takes place the voids grow up to Grade V, by which stage the density can be as low as 1.2. At that stage, the material is prone to collapse to a denser, re-worked, Grade VI state.

projects, the point at which material strength begins to dominate design decisions is where the rock can be broken by hand. An example of Grade V granite (note the remnant texture) is shown in Figure 7.6.

At the mass scale in weathered profiles, strength and deformation might be affected by the presence of strong corestones of less weathered rock in a weakened matrix, and the problem of characterisation is similar to that of mixed soils and rock such as boulder clay or boulder landslide colluvium, as discussed later.

Permeability in fractured rock or in weathered profiles can be extremely variable and difficult to predict, with localised channel flow providing high permeability. Elsewhere, accumulations of clay or general heterogeneity in the profile can prevent and divert water flow. The complexities of flow through weathered rock profiles and difficulties in defining permeability are discussed in Chapter 5.

Figure 7.6 Completely decomposed granite (Grade V) with point of hammer pushed in.

(b) Diagenesis and lithification (formation of rock from soil)

As discussed in Chapter 4, soil is transported by water, wind or gravity from the parent rock. During the process of transportation, the sediment is sorted in size. Some soils such as glacial moraine and colluvium remain relatively unsorted. Sediments tend to be continually deposited over a very long period of time, for example, in river estuaries, and each layer of sediment overlies and buries the earlier sediment. The underlying sediment is compacted and water squeezed out. This is termed "burial consolidation" and is a very important process governing the strength and deformability of sediments. Grains become better packed, deformed and may form strong chemical bonds with interpenetration and sutured margins. Voids may be infilled with cement precipitated from soluble grains in the sediment (authigenic cement) or from solutions passing through the sediment pile, as illustrated in Figures 7.7 and 7.8.

Many clay oozes initially have a very high percentage of voids, with the mineral grains arranged like a house of cards. With time, overburden stress and chemical changes cause the flaky minerals to align and the porosity (or void ratio) to decrease markedly, as illustrated in Figure 7.9.

The rate at which the voids are reduced with burial depth in "normal consolidation" has been expressed as a normalised equation by Burland (1990), although there are often departures from this behaviour in natural sediment piles, due largely to cementation (Skempton, 1970; Hoshino, 1993). The changes in property (especially strength and deformability) that ensue from burial, compaction and consolidation are discussed in Section 7.7. At some locations, the upper part of the sediment pile is considerably

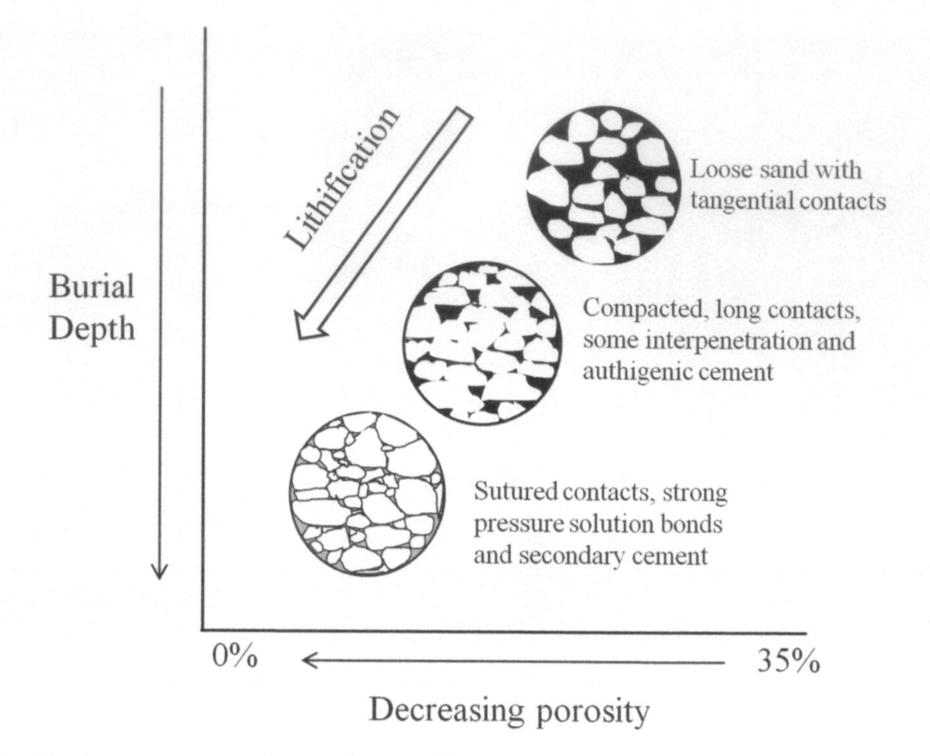

Figure 7.7 Compaction and cementation of granular soils with burial leading to increased strength, reduced deformability and lower permeability.

stronger than might be anticipated from its shallow burial level because it has become desiccated on temporary exposure above water level. Where soils are uplifted and upper levels eroded, or, otherwise loaded and later that load is removed (e.g. by the formation and melting of a glacier), then the strength and stiffness will be relatively high and the soil is termed "overconsolidated". In the case of sand, the history of burial compaction can be expressed by an extremely dense arrangement of the sand particles that cannot be replicated in the laboratory. Such "locked sands", with grains exhibiting some interpenetration and authigenic overgrowths, not surprisingly have high frictional resistance and dilate strongly under shear (Dusseault & Morgenstern, 1979).

(c) Discontinuities and other fracture networks

Natural fractures occur in most rocks close to the Earth's surface and in many soils once they begin to go through the processes of burial and lithification. Figure 7.10 shows rock faces where discontinuities dominate mass geotechnical parameters such as deformability and permeability. They need special consideration and characterisation, as addressed in Chapters 3 and 4 and discussed later.

Fractures, mostly tensile, are similarly developed in relatively recent till deposits at Robin Hoods Bay in North Yorkshire and will also change the permeability and strength characteristics of the deposit (change from FLAC continuum to UDEC block models, with different parameters required).

Figure 7.8 Thin section of aeolian sandstone with rounded grains of quartz, interpenetration of grains and flattened surfaces where in contact, with some pressure solution, plus authigenic cementation of grains by silica and iron oxides. As a result of these diagenetic processes, the material has been turned from loose sand into a strong rock. Triassic Sandstone, UK. Large grains about 7 mm in diameter.

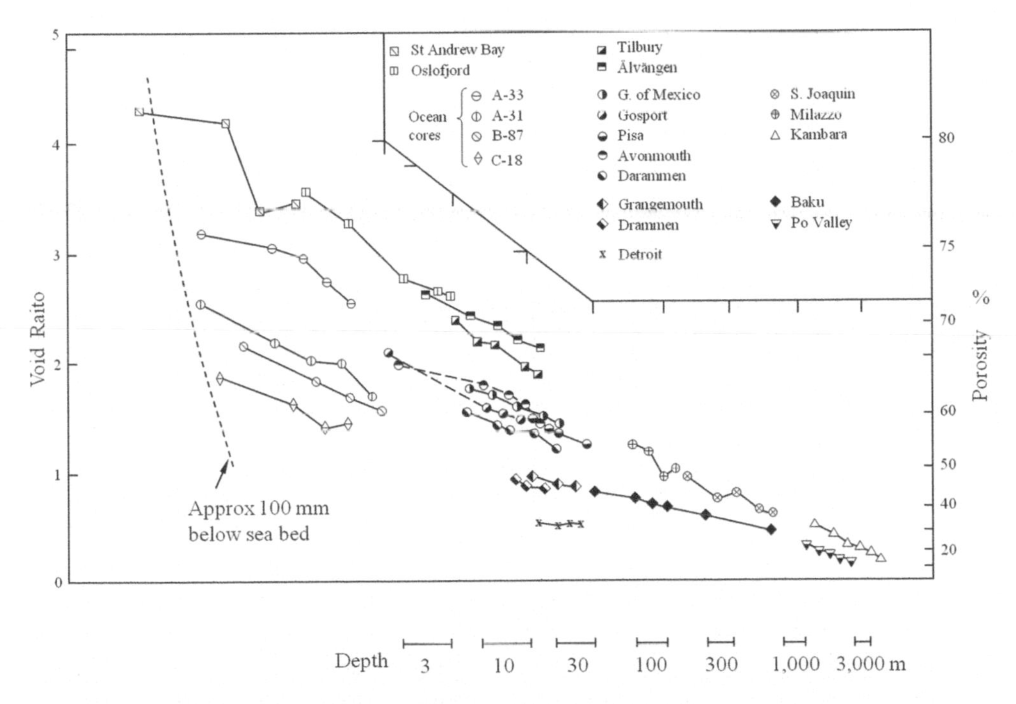

Figure 7.9 Compression curves for naturally consolidated and partially cemented clay.

Source: Modified from Skempton, 1970.

Figure 7.10 Rock masses where discontinuities dominate behaviour. Left photo, view of Bakusan, South Korea; right photo, McGuigan's tree in columnar basalt lava, Isle of Mull, Scotland (courtesy of BGS).

(d) Soil and rock mixtures

Many soils such as glacial boulder clay and colluvium comprise a mixture of finer soil and large clasts of rock, and these need special consideration in terms of their properties. Weathered rocks can similarly comprise mixes of weak and hard materials but there is also the added complication of relict rock fabric and structure (Figure 7.11).

The overall nature of the mass will strongly affect the options for engineering assessment, as illustrated for slopes in Figure 7.12. Geotechnical parameter determination for such mixed deposits is considered in Section 7.8.

7.4 MEASUREMENT METHODS

Testing methods, including preparation of samples, are mostly specified in standards such as BS: 1377 in the UK for soil (British Standards Institution, 1990), BS: 7930 for several field tests (British Standards Institution, 1999) and ASTM, more generally in the USA. Other methods of testing are recommended by bodies such as the International Society for Rock Mechanics (Ulusay & Hudson, 2007). There are some conflicts in recommendations between different standards, for example, over sample dimensions and testing rate, so care has to be taken in practice that an appropriate method is being adopted and referenced. Different tests are used in an attempt to measure the same parameters but inevitably with different outcomes. For example, small strain, dynamic and large-scale loading tests may give very different answers for soil stiffness, each of which might be appropriate to some aspect of numerical analysis and design within a single project (Clayton, 2011). It should also be remembered that, however much they are standardised, all tests on soil and rock are really experiments. There will be many variables, not least the geological nature and moisture content of the sample to be tested, so interpretation and judgement is always required. Further judgement is required before attempting to apply laboratory test results to the field,

Figure 7.11 Corestone development above Devils Causeway, Northern Ireland, UK.

where large-scale fabric, structural factors and representativeness must be accounted for. In this respect, it is often reported that the larger the rock sample tested, the lower the strength and greater the deformability, mainly because of the influence of included flaws and fractures (see review by Cunha, 1990).

7.4.1 Compressive strength

Intact rock, clay ("cohesive" soil that exhibits tensile strength) and concrete are generally described in shorthand by their unconfined compressive strength (UCS). In terms of principal stresses at failure, the axial stress causing failure UCS, or σ_c, is σ_1 and the

Option	Schematic diagram	Approach for defining parameters and analysis
1. Treat as uniform (continuum)	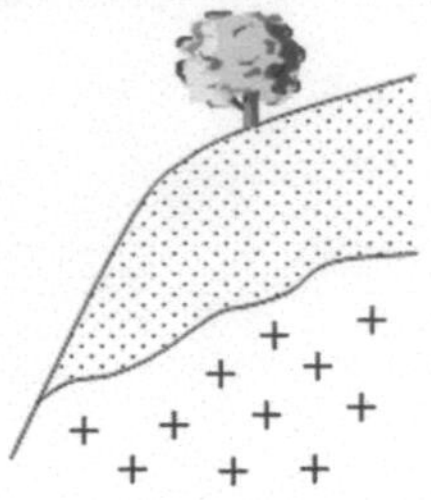	• parameters from laboratory or *in situ* tests taken to be representative of zone
2. Treat as uniform but weakened by discontinuities (continuum)	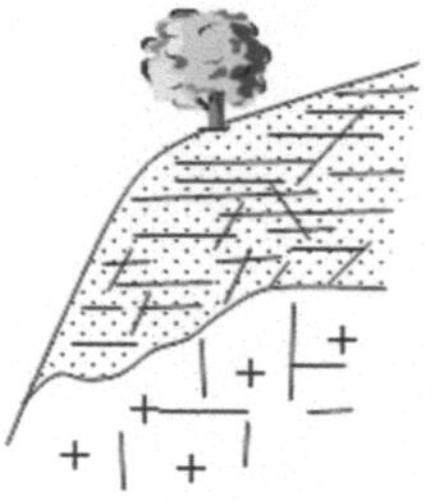	• allowance made for influence (but not control) of discontinuities on mass properties (e.g. Hoek-Brown)
3. Treat as heterogeneous (continuum)	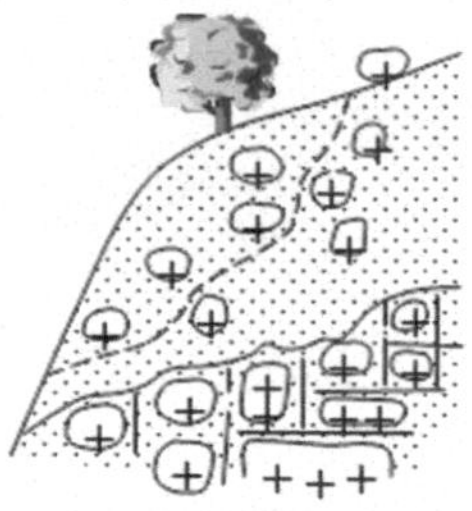	• consideration given to influence of strong inclusions with deviated failure paths
4. Treat as discontinuous due to structural control	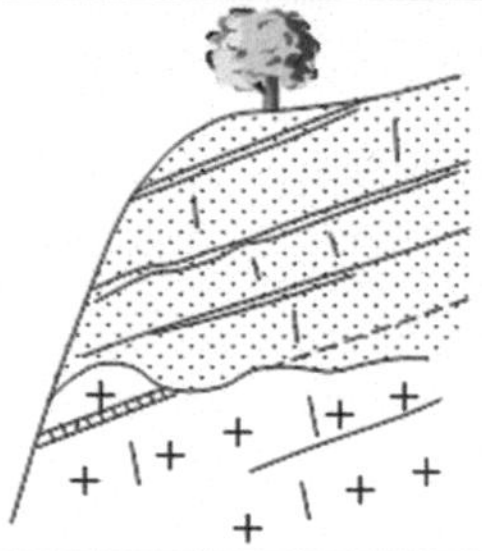	• discontinuity controlled

Figure 7.12 Options for slope stability analysis. After Hencher & McNicholl (1985).

confining stresses (σ_2 and σ_3) are zero. For concrete, unconfined compressive strength is used as a quality assurance test on construction sites. Samples of concrete being poured as part of a structure are collected in cubic or cylindrical moulds and then tested after a time, as specified for the project. Concrete increases rapidly in strength over 24 hours

Table 7.1 Indicative unconfined compressive strengths for some rock, soil and concrete

Material	*Uniaxial Compressive Strength, UCS MPa*	
Natural rock and soil		
Fine-grained, fresh igneous rock such as dolerite, basalt or welded tuff, crystalline limestone	>300	Rings when hit with geological hammer
Grade I to II, fresh to slightly weathered granite	100–200	Difficult to break with hammer
Cemented sandstone (such as Millstone Grit)	40–70	Broken with hammer
Grade III, moderately weathered granite	20–40	
Chalk and Grade IV highly weathered granite	7–30	Readily broken with geological hammer. Weaker material broken by hand
Overconsolidated clay	0.6–1.0	Difficult to excavate with hand pick
Clay-rich soil	0.08–0.17	Indented with finger nail
Concrete		
High-strength concrete (e.g. Channel Tunnel liner)	70–100	
Typical structural concrete	30–70	
Shotcrete in tunnel	20–40	

and then more gradually, reaching full strength by about 30 days. Typical unconfined compressive strengths for various materials are given in Table 7.1. It can be seen that many fresh, unweathered rocks are considerably stronger than the highest strength concrete, although they might be less durable due to chemical instability, and this needs to be considered, for example, when selecting armourstone for a breakwater. Geological materials are generally classified primarily according to predefined ranges of unconfined compressive strength, as set out in Appendix B. Compressive strength is not a relevant concept for purely frictional materials such as sand, which must be confined to develop shear resistance.

Rocks do not actually fail in compression, despite the apparent loading condition, but either in tension or in shear or in some hybrid mode including both. If the rock contains any weak and adversely dipping discontinuities or fabric, then these will cause a sample to fail at lower strength than would isotropic intact rock. Strengths can often be estimated quite adequately without laboratory testing by trying to break core by hand or hitting a sample with a geological hammer as set out by Hack & Huisman (2002) and discussed in Box 7.1. UCS can also be estimated using point load testing, the advantage being that they are quick and easy, although variable correlation factors are reported. The Schmidt hammer is sometimes used to estimate UCS and uses standard impact energy to measure rebound from a rock or concrete surface. It is very sensitive to surface finish and to any fractures behind the impact location, which cause low readings. It is also insensitive to strength over about 100 MPa. It is considered unsuitable for testing rock core, which will be broken or will bounce in its cradle, giving incorrect results. In the author's view, its main use in engineering geology is as an index test – used, for example, to help differentiate between different grades of weathered rock, as discussed in Chapter 6.

BOX 7.1 TO TEST OR NOT TO TEST?

Many ground investigations are wasteful in that they do not target or identify critical geological features and laboratory tests are commissioned without real consideration of whether or not they will be useful.

EXAMPLE 1

Figure B7.1.1 shows the formation level (foundations) for the Queen's Valley Dam, Jersey which was completed in 1991. The dam was to be an earth dam which exerts relatively low stresses on its foundations compared to a concrete dam such as an arch or gravity dam. With maximum height of 24 m and an assumed unit weight of 20kN/m^3, the bearing pressure might be of the order of 500 kPa. The author, who was mapping the foundations, was asked to select samples of core to be sent to the laboratory for uniaxial compressive strength testing.

Rock over much of the foundation was rhyolite that was extremely difficult to break with a geological hammer and had an estimated compressive strength of more than 300

Figure B7.1.1 The formation level (foundations) for the Queen's Valley Dam, Jersey, which was completed in 1991. The dam was to be an earth dam, which exerts relatively low stresses on its foundations compared to a concrete dam such as an arch or gravity dam. With maximum height of 24 m and an assumed unit weight of 20 kN/m^3, the bearing pressure might be of the order of 500 kPa. The author, who was mapping the foundations, was asked to select samples of core to be sent to the laboratory for uniaxial compressive strength testing.

Figure B7.1.2 Extremely strong rhyolite. Hammer and clip board for scale.

MPa. The rhyolite however contained numerous incipient fractures (Figure B7.1.2), which would mean that the mass strength was somewhat lower and, more significantly, would cause samples to fail prematurely in the laboratory. The author argued that if the samples were sent to the laboratory the reported result would simply be scattered with a range from 0 to 300 MPa and what would that tell us that we didn't already know? The allowable bearing pressure for rock of this quality (Chapter 3) would be at least 5 times the bearing pressure exerted by the dam. In the event the samples were still sent off to the laboratory for testing (because they had already been scheduled by the design engineers) and the money was duly wasted.

EXAMPLE 2

The Simsima Limestone is the main founding stratum in Doha, Qatar and is found extensively across the Middle East. It is a highly heterogeneous stratum including calcarenite, dolomite and breccia. The rock is often vuggy and re-cemented with calcite. RQD can be very high with sticks of core a metre or more in length without a fracture; elsewhere the RQD is zero. An example is shown in Figure B7.1.3 together with unconfined compressive strengths measured.

The properties of the stratum are clearly important for design of foundations and for other projects such as dredging as discussed in Chapter 3. UCS test data tend to be very scattered in part because the integral flaws in many samples lead to early failure. If a strongly indurated sample with few flaws is tested then it can give UCS strength of 60 or

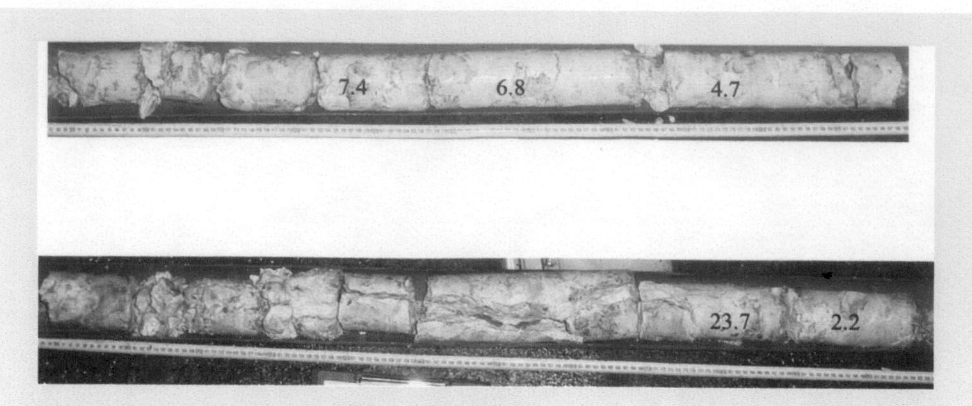

Figure B7.1.3 Example of core through Simsima Limestone.

70 MPa (higher than structural concrete). Samples of inherently weaker material (as could be estimated from scratch testing) or containing vugs or other flaws will fail at much lower strengths. If smaller intact pieces of dolomitised limestone are point load tested selectively they will, of course, err towards the higher strength of the rock mass. As a consequence, conversion factors from point load test to UCS for this rock are usually taken empirically as 8 to 9 (Khalaf, personal communication). For more uniform rocks with less flaws, elsewhere in the world conversion factors of about 22 are more commonly applied (Brook, 1993). If such a factor was to be used for the Simsima Limestone then it would imply strength for the intact limestone, without flaws, up to about 200 MPa.

Given this very wide range of possible strengths it would seem unwise simply to rely on a statistical testing campaign for characterising the rock mass as is commonly done in geological baseline reports (Essex, 2023). Far better to try first to characterise the rock geologically into units based on the strength of rock materials and then mass characteristics including flaws, degree of cementation and degree of fracturing. In this case, index tests (hammer, knife) combined with visual logging and selective testing of typical facies are likely to give a far better indication of mass properties than UCS testing alone. To obtain parameters for the large scale (say foundations) then *in situ* tests such as plate loading and perhaps seismic tests would help as would full scale instrumented pile testing. Where rock mass strength is very important as for the selection of dredging equipment then it would be very unwise to take UCS data at face value (as a statistical distribution). As for many tests, there are numerous reasons why values measured in the laboratory might be unrepresentative of conditions *in situ*, often too low, and considerable judgement is required if the parameters are critically important.

Lesson: Compressive strength of most rocks can often be estimated adequately by hitting with a hammer and the use of other index tests; if the material cannot be broken by a hard blow of a hammer then its strength probably exceeds that of any concrete structure to be built upon it. Where strength is critical, as in the selection of a tunnelling machine or choice of dredging equipment then any test data must be examined critically. If laboratory test samples contain flaws such as discontinuities then measured intact strength may be too low. Of course at the mass scale, the flaws and joints will be extremely important but their contribution cannot be properly assessed by their random occurrence and influence on laboratory test results.

Figure 7.13 Tensile tests on rock. Left, Brazilian test squeezing between loading platens; right, pulling apart sample of sandstone.

Source: Shang et al., 2016.

7.4.2 Tensile strength

Rocks usually fail in tension rather than compression although tensile strength is rarely measured directly or used in analysis or design, compressive strength being the preferred parameter for rock mass classifications and empirical strength criteria (see later). Exceptions are in research using either indirect methods such as the Brazilian test in which a disc of rock is squeezed axially until it breaks or by direct pulling apart in tension as shown in Figure 7.13 (Shang et al., 2016).

Tensile strength of rock and concrete is relatively low, typically about 1/10th of UCS. It is because of the weakness of concrete in tension that reinforcing steel needs to be used wherever tensile stresses are anticipated within an engineering structure.

7.4.3 Shear strength

Shear strength is a very important consideration for many geotechnical problems, most obviously in landslides where a volume of soil or rock shears on a sliding surface out of a hillside. It is also important for the design of foundations and in tunnelling (Chapter 3). There are two main types of tests used to measure shear strength in the laboratory – direct shear and triaxial testing. There are many other *in situ* tests used to measure shear strength parameters, either directly (e.g. vane test) or indirectly (e.g. SPT and static cone penetrometer tests), and these have been introduced in Chapter 6.

For persistent (continuous) rock discontinuities, direct shear testing is the most appropriate way of measuring shear strength. Details are given in Hencher & Richards (1989, 2015) and the interpretation is discussed by Hencher (1987). Because of the inherently variable roughness of different natural samples, dilation needs to be measured and results normalised, as discussed later. If this is not done then, in the author's opinion, the tests are usually a total waste of time. The details of a shear box capable of testing rock discontinuities and weak rocks with controlled pore pressures is described by Barla et al. (2007).

Direct shear tests are also carried out on soil and are much easier to prepare and conduct than tests on rock discontinuities, although the stress conditions are not fully defined in the test, which can cause some difficulties in interpretation (Atkinson, 2007). This is one reason why triaxial testing is preferred for most testing of soils. Other advantages are that factors like drainage and pore pressure measurement can be carefully controlled. A disadvantage is that the soil may well become disturbed during trimming and preparation for the test as well as during back saturation and loading/unloading, but that is a problem for all testing.

In a triaxial test, the cylindrical sample is placed inside a rubber membrane, in a cell, and then an all-around fluid pressure applied (σ_3). This stress is pre-selected according to the desired series of tests, for example, a series of tests might be run with σ_3 of 100, 200 and 400 kPa. This remains the constant minimum principal stress throughout the test. Drainage might be allowed or prevented. Once the sample is in equilibrium, the sample is gradually compressed axially (the axial stress is σ_1) whilst the confining stress remains the constant σ_3. The process is illustrated graphically using Mohr's stress circles in Figure 7.14.

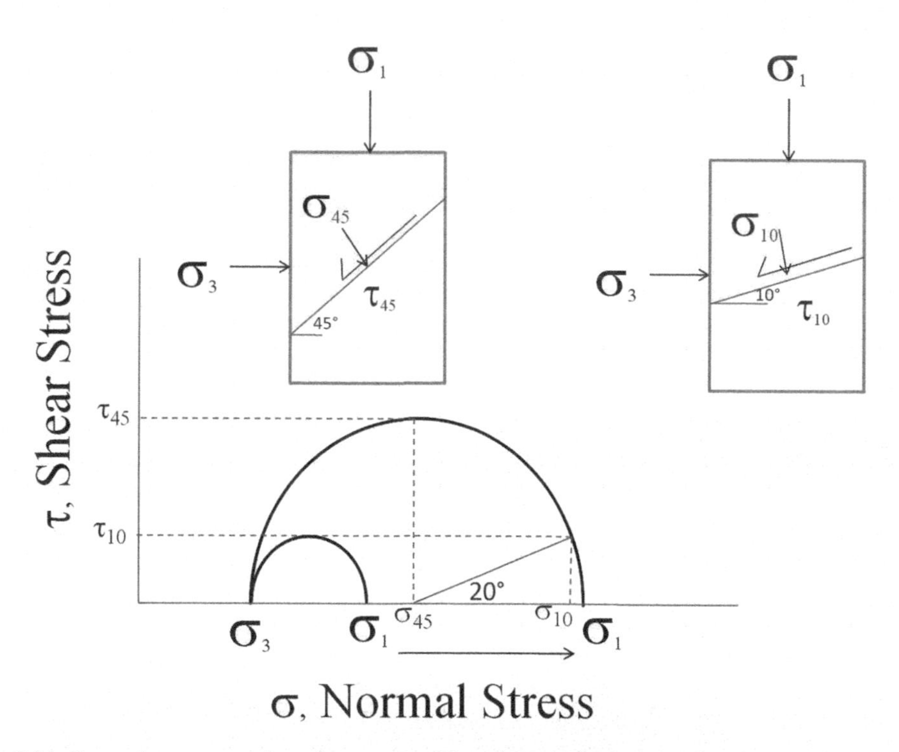

Figure 7.14 General representation of stress conditions in an individual sample.

The circles (only hemispheres shown) express the stress state on any plane drawn through the sample. The test proceeds from the state where the all-round state of stress is σ_3. Then, σ_1 is increased (hemispheres advance towards the right) until the sample eventually fails. The horizontal axis is normal stress on any plane; the vertical axis gives the shear stress on the same plane. So, at the start of the test, at the σ_3 position, the normal stress (on a vertical plane through the sample) is σ_3 and the shear stress is zero. Note that within the sample, the angle between σ_1 and σ_3 is 90 degrees, but in Mohr's circle presentation, this stress field is expressed graphically as a hemisphere (180 degrees). On ay other plane through the sample (other than vertical or horizontal), the stress state can be defined by a normal stress σ_n and a shear stress). For a plane inclined at 10 degrees (shown as 20 degrees graphically within Mohr's circle) the normal stress acting across that plane is σ_{10} and at 45 degrees (shown as 90) it is σ_{45}, with the corresponding shear stress (τ), as indicated, *at the same time*. Mohr's circle represents the state of stress, measurable on *any plane* simultaneously. As σ_1 is gradually increased, eventually the sample will fail, with a failure plane physically developed at some angle ($\theta/2$ degrees) to the horizontal and expressed as θ in Mohr's circle graph.

Mohr's stress circle representing the complete stress state at that stage is shown in Figure 7.15 for a single test. Further tests would be carried out on other similar samples at different starting confining stresses and used to define a strength envelope (a line joining the stress states at which all samples failed). Commonly, the envelope for a set of samples can be defined in terms of friction (gradient of line) and apparent cohesion, *c*, which is the intercept on the shear stress axis at zero normal stress.

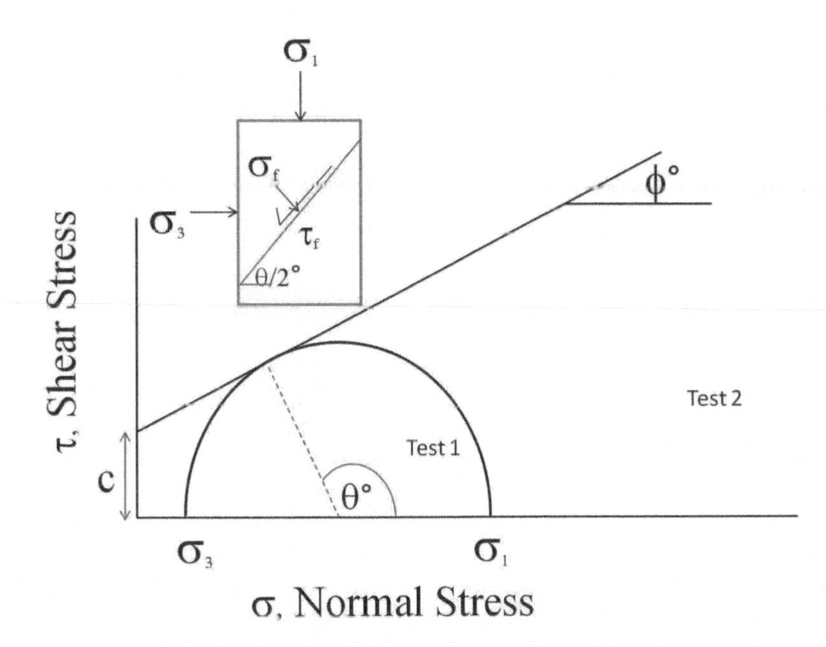

Figure 7.15 Mohr's circle at shear failure condition.

7.4.3.1 True cohesion

The nature, origin and even existence of cohesion – strength at zero normal load – causes considerable debate and confusion. This is partly because it can be either "apparent" (the result of dilation and dependant on confining stress) or "true" and due to cementation, grain bonding or impersistence of discontinuities in the case of rock. Quite often both factors contribute to the measured strength in the same test, for example, if shearing intact rock. In artificial, re-moulded soil, apparent cohesion is a function of the density of packing of the soil grains relative to the confining stress, and a theory of critical state soil mechanics has been developed, primarily for clay, that unifies shear strength with deformation characteristics (Roscoe et al., 1958; Schofield, 2006). Burland (2008) comments: "The 'critical state' framework has brought together a most valuable coherence to the understanding of the mechanical behaviour of soil for remoulded, overconsolidated fine grained soil". Burland goes on, however, to discuss the importance of geological history to natural soils, with the development of bonding and fabric leading to true cohesional, non-dilatational and stress-independent strength. While Burland was really discussing relatively young soils, it has been demonstrated earlier (Figure 7.1) how, with time, true cohesion can become very high and far outweigh the contribution of friction to shear strength. Conversely, as rock is gradually weathered to a soil it is primarily the cohesional strength that is lost – friction stays essentially constant.

7.4.3.2 Residual strength

After high shear displacement, all cohesion is lost, and shearing continues at a residual friction level. This is non-dilational friction and, in nature, residual friction angles can be lower than the "critical state" – also non-dilational – because of change in structure with, for example, flattening and alignment of particles in a clay or the development of highly polished shear surfaces. Such residual angles of friction can be very low and very significant, especially for landslides (see discussion of Carsington Dam and the Queensland failures in Chapter 8). To test residual strength, ring shear boxes are used, in which an annulus-shaped sample is prepared and then rotated until a constant low strength is obtained.

7.4.4 Deformability

When stress is applied to a rock or soil, it changes shape. For rock, the key parameter is Young's modulus (E) and is expressed as stress/strain (with units of stress). It is a key parameter for predicting settlement of a structure and therefore needs to be defined on a mass scale rather than at a laboratory scale. Because of the difficulties in determining E from first principles or testing, it is common to rely on empirical published data. Soils are similarly deformed by applied stresses and their response is often gradual as water is squeezed out slowly if the permeability is low, as in clay. It may take several years for a building to fully settle where built on clay. Samples of soil are tested in consolidation devices (oedometers) and measurements taken of the deformation against time. The main derived parameters are m_v, which is actually an inverted expression of E, i.e. strain over stress, and C_c, which is a measure of rate of consolidation. For normally consolidated clay that has been simply buried by overlying sediment, there will be quite a steep curve, as per Figure 7.9. If, however, the soil has become "overconsolidated" because of its geological history, it will be stiffer (have a higher modulus) on initial loading than normally consolidated soil. This high measured stiffness will only apply up to the loading level corresponding to its earlier "pre-consolidation" stress state. Once that pressure is exceeded, the stiffness will revert to the natural consolidation curve. At very small strains, overconsolidated clay can be much stiffer than at higher strain levels, and this can have consequence in realistic modelling of excavations (Jardine et al., 1984; Clayton, 2011). Geophysical testing can be used to interpret stiffness parameters from velocities of wave propagation through soil, and values are again on the high side compared to static tests at relatively high strains (Mathews et al., 2000). The same is true of rock masses – interpretation of compressional or shear velocities tend to give higher stiffness values than do static loading tests, and this probably reflects the low strain nature of loading from transient dynamic waves (Ambraseys & Hendron, 1969).

7.4.5 Permeability

Permeability is an intrinsic parameter of soil and rock, relating to rates of fluid flow through the material and strictly varies according to the fluid concerned – e.g. oil, water or gas. It has dimensions of area (L^2). In hydrogeology and geotechnical engineering, the term permeability is generally used to be the same as hydraulic conductivity and is the volume of water (m^3) passing through a unit area (m^2) under unit hydraulic gradient (1 m head over 1 m length) in a unit of time (per second), and this reduces to m/s. For low permeability rock suitable for a nuclear waste repository, the permeability, k, might be 10^{-11} m/s. For an aquifer of sandstone suitable for water extraction, it might be 10^{-6} m/s and for clean gravel 10^{-1} m/s. Typical values for other soils are given in BS 8004 (British Standards Institution, 1986).

In some soil, such as alluvial sand, the material permeability could be similar to the mass, so laboratory testing might be relevant, but for many ground profiles, water flow might involve natural pipes, fissures and open joints or faults. Field tests are then generally necessary to measure mass-scale permeability, as outlined in Chapter 5. Large-scale pumping tests from wells with observational boreholes at various distances can give reliable parameters for aquifer behaviour but localised testing in boreholes, as defined in BS 5930 (British Standards Institution, 1999), can be unreliable (Black, 2010).

7.5 SOIL PROPERTIES

7.5.1 Clay soils

As Skempton (1970) showed (Figure 7.9), for clay soil deposited offshore at rates of perhaps 2 m per thousand years, consolidation behaviour due to self-weight is fairly well-defined. As the clay is compressed by the overlying soil and water squeezed out, so strength increases and deformability reduces, even in the absence of other diagenetic processes. Hawkins et al. (1989), for example, show a consistent linear increase in shear strength with depth over 20 m at a test site in Bothkennar, Scotland, based on vane tests. Cone test data from the same site are very similar to other sites in the UK, confirming the trend. Similar results have been achieved from other sites worldwide, with a typical relationship:

$$S_u = 10 + 2.0d$$

where S_u = undrained shear strength, kPa.; d = depth below ground, m.

Elsewhere, values can be somewhat lower; for the Busan Clay in Korea, the gradient is closer to 1.0 (times depth) (Chung et al., 2007). Nevertheless, the trend is similar so for design in soft to firm clay, it is usual practice to carry out a series of vane tests down boreholes (SPT tests are not really appropriate for clay and especially for soft clay), or cone penetrometer soundings, and then try to define a relationship of increasing strength with depth that can easily be input to numerical simulations. Relationships are published both for shear strength and modulus of clay interpreted from SPT tests, and these are reviewed in Clayton (1995). Most of the values obtained from field tests are necessarily undrained and expressed as a value of apparent cohesion with no frictional component. Undrained shear strength of clay can also be obtained from undrained tests in the laboratory and is estimated during field description using index tests like resistance to finger pressure or in a rather more controlled way using a hand penetrometer. Undrained strength is useful for assessing the fundamental behaviour of clay empirically, for example, in designing foundations (Table 7.1). It is also used for numerical analysis in soils of low permeability immediately after or during construction. Drained conditions apply where excess pore pressures have dissipated following construction or where they dissipate relatively rapidly during construction. For design of structures in clay under drained conditions, effective stress parameters are required – friction and possibly some cohesion where there has been some geological bonding. These parameters are generally obtained from triaxial testing, in which pore pressures are monitored and corrected for throughout the test (Craig, 1992).

Laboratory tests are relied upon for characterising natural clay soils far more than for any other geological materials, because reasonably undisturbed samples can be taken and the small grain size relative to testing apparatus means that scale effects are not evident. An exception is in settlement analysis, where it is found that standard oedometer tests give lower stiffness than larger-scale plate load tests or as evident from back analysis of the construction of a structure. Specialised testing is necessary to simulate low strain deformation (e.g. Atkinson, 2000). Another exception is for older clay deposits such as the London Clay and within the Lambeth Group, where discontinuities (fissures) play a major role in strength and permeability assessments (Hight et al., 2004; Entwisle et al., 2013).

As noted earlier, for some active and ancient landslides, the strength along the slip plane through clay/mudstone is reduced below the critical state friction angle to a residual friction angle well below 20 degrees, even for clay of relatively low plasticity such as kaolinite or illite (Skempton et al., 1989). Such low values can be measured in the laboratory using ring shear boxes and back-analysed from landslide case histories.

Clays include some groups of very problematical soils. Quick clays are mostly formed of detrital materials (rock flour of clay size, produced by glacial scour), weakly cemented by salt, which can become disturbed and then flow, sometimes to disastrous effect. The Rissa, Norway, landslide in 1978 was filmed, flowing rapidly across flat ground, indicating the sensitivity of such materials. Videos are readily available on the internet. Other clays such as black cotton soils swell and shrink dramatically with changes in moisture, which causes damage to roads and other structures. The clay mineral group smectite (montmorillonite/bentonite) is most commonly associated with volume change causing damage to foundations and is typically identified by X-ray testing. Its presence is also indicated from high liquid limits and high plasticity indices in Atterberg limit tests as discussed in Chapter 6. These clay minerals can have very low shear strengths. Starr et al. (2010) describe a creeping major rock slope failure where the underlying rock is smectite-rich and for which, the operating residual friction angle, at depths of tens of metres, was only about 7 degrees, as established by numerical back-analysis and confirmed from laboratory tests.

7.5.2 Granular soils

The behaviour of granular soils such as silts, sand and gravel can be examined in the laboratory but for design, geotechnical parameters are generally determined by *in situ* testing, because of the difficulties of (a) obtaining and transporting undisturbed samples and (b) the problems of scale effects in testing samples of large grain size.

The most common test for assessing coarse soils such as sand and gravel is the SPT, as discussed in Chapter 6. Measured resistance needs to be corrected for various influences, including overburden pressure and the silt content of sand. Resistance may be affected by water softening in the base of a borehole. Details are given in Clayton (1995). SPT "N" values are used to infer a range of properties, including density (unit weight), friction angle and deformability and from there directly in the design of many types of structure, including foundations, retaining walls and slopes. CPT tests can also be used in this way and have proved particularly useful for the design of offshore structures (Robertson, 2009).

7.5.3 Soil mass properties

Usually, properties of intact soils of sedimentary origin are taken to be representative of the larger soil mass layer or unit. This can, however, be an oversimplification in that even quite recent soils can contain fractures and systematic joints and many are layered with different layers having different properties. In the latter case, permeability parallel to bedding might be orders of magnitude higher than at right angles to bedding, and there are many geotechnical situations where such a condition would be highly important. Collins & McGowan (1974) discuss origins of fractures in soil and how they might be dealt with when assessing geotechnical properties. The older the soil and more complex its geological history, the more likely that it will contain a significant number of fractures. The origin of discontinuities is discussed in some detail in Chapter 6. London Clay, for example, contains many fissures that are not random but can be interpreted using structural geological techniques (Fookes & Parrish, 1969). Chandler (2000) describes the significance of bedding parallel flexural-slip surfaces extending at least 300 m in London Clay. Similar features are discussed by Hutchinson (2001).

7.6 ROCK PROPERTIES

7.6.1 Intact rock

7.6.1.1 Fresh to moderately weathered rock

A piece of core of 52 mm diameter, of fresh to moderately weathered rock, cannot be broken by hand by definition (Chapter 4 and Appendix B). This equates to an unconfined compressive strength of at least 12.5 MPa and will not fail in a man-made slope, in the absence of discontinuities, almost irrespective of height and steepness.

The strength of fresh rock is a function of its mineralogy, internal structure of those minerals (cleavage), grain size, shape and degree of interlocking, strength of mineral bonds, degree of cementation and porosity. Some rocks have intact strength approaching 400 MPa – these might include quartzite, welded tuffs and fine – and medium-grained igneous rocks such as basalt and dolerite. Corresponding intact moduli can be as high as 1×10^6 MPa (1×10^3 GPa) (Deere, 1968).

Compressive strength is measured most accurately using very stiff servo-controlled loading frames, whereby, as the rock begins to fail, so the loading is paused to limit the chance of explosive brittle failure. Such test set-ups allow the full failure path to be explored, which can be important in underground mine pillars where, despite initial failure in one pillar, there may be sufficient remnant strength after load is transferred to adjacent pillars, so that overall failure of the mine level does not occur. For most civil engineering works, UCS values measured by less sophisticated set-ups are adequate. Nevertheless, the specification for UCS testing is onerous, particularly regarding test dimensions and flatness of the ends of samples. If these requirements are not followed, local stress concentrations can cause early failure. If samples are too short, then shear failure might be inhibited. As noted in Box 7.1, there are alternative, cheap ways of estimating UCS that might be adequate for the task at hand.

UCS is the starting point for many different empirical assessments of rock masses, including excavatability by machinery such as tunnel boring machines. Other parameters that might need to be quantified include abrasivity and durability. Appropriate tests are specified in the ISRM series of recommended methods (Ulusay & Hudson, 2007).

Intact rock modulus is rarely measured for projects and is not usually an important parameter for design. An exception is in numerical modelling of fractured rock mass, e.g. using UDEC (Itasca), where this parameter is required, but for this purpose, values are typically estimated from published charts or even selected to allow the model to come to a solution within a reasonable time.

7.6.1.2 Weathered rock

Weathered rock (highly to completely, Grades IV and V), by definition, can be broken down by hand, albeit for that at the top of Grade IV this is very difficult and the strength is similar to that of weak concrete. It is not until Grade V that samples of the rock will disaggregate when placed in water, which, for some projects, might be taken as the boundary between soil and rock.

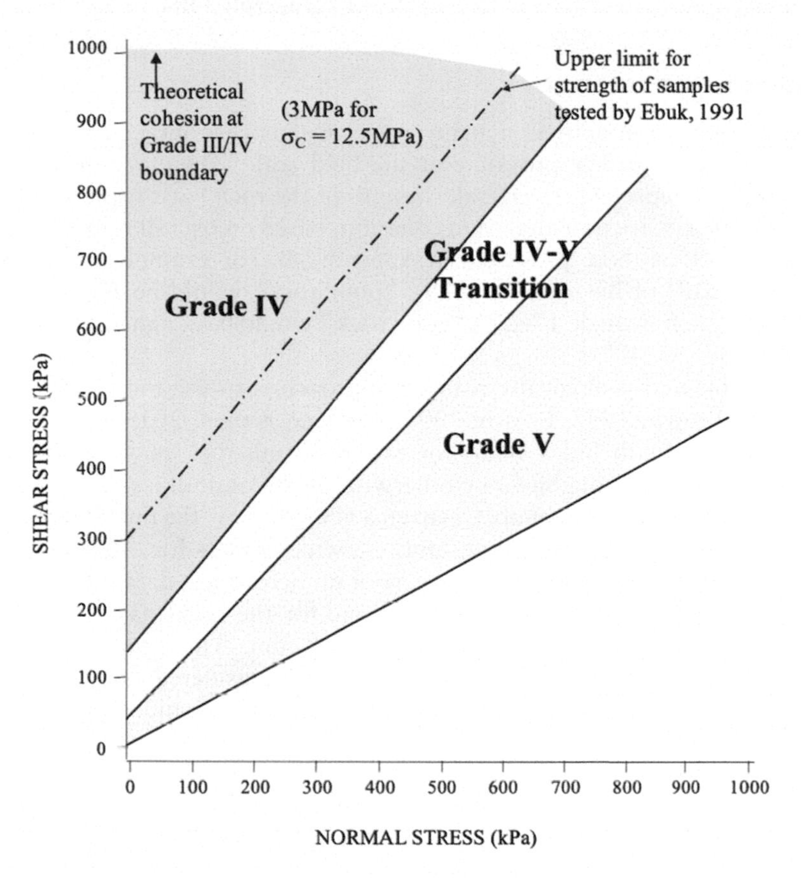

Figure 7.16 Measured strength envelopes for Grades IV, V and VI granite.
Source: Based on Ebuk, 1991.

Intact weathered rock has true cohesion from relict mineral bonding. In some cases, there may be secondary cementation, especially from iron oxides and the re-distribution of weathering products within the rock framework. At the strong end of Grade IV (in any rock), with unconfined compressive strength of about 12.5 MPa, cohesion of about 3 MPa might be anticipated (Hencher, 2006). In practice, such high values have never been reported. Ebuk, who tested a range of weathered rocks in direct shear, measured a maximum cohesion of 300 kPa for weaker end, Grade IV samples (Figure 7.16).

For design, parameter values for weathered rock are often estimated from SPT "*N*" data. Tests are often continued to 100 or even 200 blows, which is questionable practice for many reasons, not least damage to equipment. In terms of rock mass modulus, a typical relationship adopted for design is (Hencher & McNicholl, 1985):

$$E = 1.0 \text{ to } 1.2 \text{ "N" (MPa)}$$

For foundation design, parameters such as side friction and end bearing are also often estimated from empirical relationships linked to SPT data. Full discussion of practice in Hong Kong is given in Geotechnical Engineering Office (2006).

7.6.2 Rock mass strength

The presence of discontinuities in many rocks means that intact rock parameters from the laboratory are inappropriate at the field scale. Therefore, many attempts have been made to represent the overall strength of the rock mass using simple Mohr-Coulomb parameters, friction angle and cohesion, based on overall rock quality, using classifications such as those presented in Appendix B. For example, using the Rock Mass Rating (RMR) of Bieniawski (1989), "poor rock" would be assigned cohesion 1–200 kPa with friction angle 17–27°; "good rock" would be assigned cohesion 3–400 kPa, friction angle 37–47°.

A more flexible and geologically realistic approach is to use the Hoek-Brown criteria (Hoek & Brown, 1997; Brown, 2008; Hoek & Brown, 2019), which is linked to a Geological Strength Index (GSI) for rating overall rock mass conditions such as "blockiness" and the roughness or otherwise of discontinuities. The GSI chart is presented and discussed in Appendix C. Given a GSI estimate, the uniaxial compressive strength for the rock blocks and a constant, m_i, which differs for different rock types and has been derived empirically from review of numerous test data (Hoek & Brown, 1980), one can calculate a full-strength envelope for the rock mass. A programme, RockLab, is downloadable from www.rocscience.com. The programme provides values for cohesion and friction but these need to be considered carefully, especially with respect to the appropriate stress level for the problem at hand. Figure 7.17 shows a steep cut slope in weathered tuff. The question is whether it needs to be cut back or otherwise reinforced or supported. The rock mass is severely weathered. There are corestones of very strong tuff but these are separated and surrounded by highly and completely weathered materials which are much weaker. There are many joints and some of these have kaolin infill. In this case, there are no structural mechanisms for translational failure along daylighting joints, and it is a clear candidate for where a Hoek-Brown/GSI approach might help the assessment.

From the GSI chart, one might best characterise the mass as "very blocky" with "poor" joint surfaces. The rock type is tuff, so the m_i value is 17 (for granite it would be 33). The difficult parameter is intact strength. In this case, the corestones have UCS values in excess of 100 MPa, but for this assessment I have taken the strength of the weakest material making up this slope and, on balance, an average of 7 MPa is considered conservative. Using a spreadsheet modified to plot the H-B strength envelope, appropriate to the stress level of the problem, the curve shown in Figure 7.18 is obtained.

On that basis, for a potential slip surface at a depth of about 10 m (vertical stress, say 0.27 MPa), appropriate strength parameters might be c = 80 kPa and ϕ = 46 degrees, as shown. It must be emphasised that this assessment is for the rock mass, where there is no structural control by adverse geological fabric. Carvalho et al. (2007) discuss the assessment of rock mass strength where the intact rock has relatively low uniaxial compressive strength.

I was involved with a rock mass assessment in Turkey for the design check of a major suspension bridge. The original ground investigation contractor for the northern

Figure 7.17 Cut slope through weathered volcanic tuff where one might use the Hoek-Brown criteria to assess the mass strength.

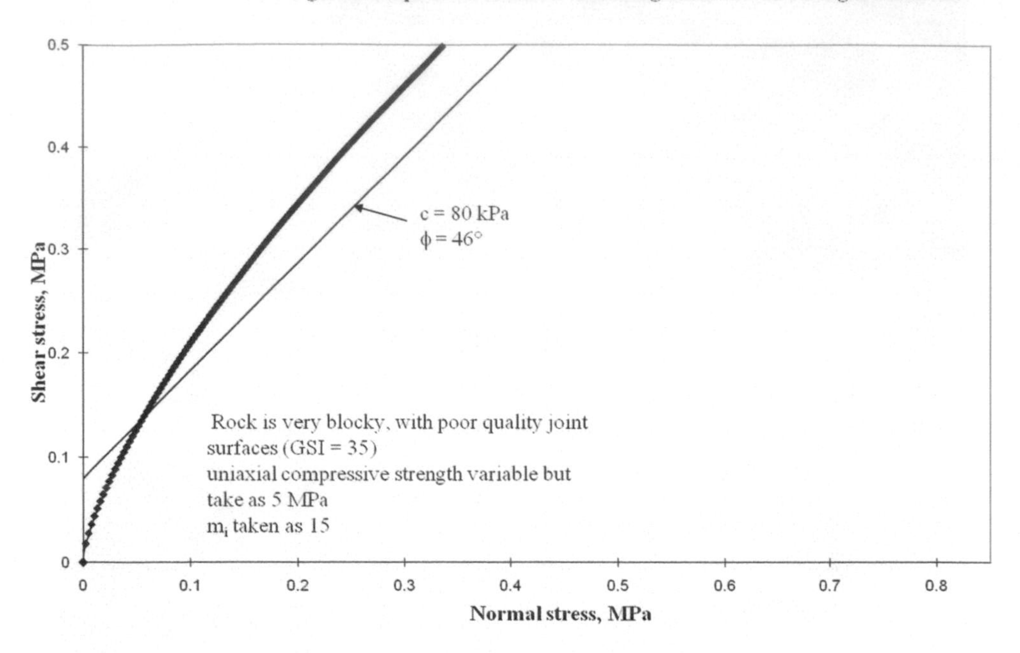

Figure 7.18 Strength envelope for slope in Figure 7.17, based on Hoek-Brown criteria (see text).

anchorages had simply mapped each exposure in the region using GSI directly, and then drawn conclusions regarding strength on that basis. In fact, there *were* adverse discontinuities present at the site, and we had to instigate a second investigation using boreholes and site investigation trial trenches together with a series of direct shear tests to prove that the rock anchorage was stable. This case is described in detail in Hencher (2015).

7.6.3 Rock mass deformability

Rock mass modulus is very difficult to predict with any accuracy, and measurements in boreholes or even by large *in situ* tests need to be considered critically and certainly should not be used directly in design without due consideration of the rock qualities of the zone tested (including relaxation) vs. the larger mass volume. Back calculations have been made from large projects, including dams and tunnels, and these data provide the main database for prediction (e.g. Gioda & Sakurai, 2005). Generally, poor-quality, highly fractured rock (up to RMR = 70) will have a rock mass modulus increasing from soil-type values of perhaps 700 MPa to about 20 GPa with decreasing fracture spacing and increasing intact compressive strength. As the rock mass quality improves, so the modulus increases markedly, up to values of 60 GPa or so for good-quality rock with RMR = 80. Many authors have attempted correlations between a variety of rock mass classifications (RMR and Q especially) and rock mass modulus, but with considerable scatter. This is perhaps not surprising given the inherent difficulties of (1) trying to represent an often complex, heterogeneous geological situation as a single "quality" number and (2) the non-uniform loading conditions of any project vs. the measurement system (deficiencies of data).

Hoek & Diederichs (2006) carried out a detailed review and proposed optimised equations linked to the GSI classification. The best-fit equation obtained was

$$E_{mass}(\text{MPa}) = 10^5(1 - D/2)/\left(1 + e^{((75+25D-GSI)/11)}\right)$$

where GSI is as taken from the chart in Appendix B (Table 11). The factor $D = 0$ for undisturbed masses, 0.7 for partially disturbed and 1.0 if "fully disturbed". Hoek & Diederichs present a more refined version of this equation using site-specific data for intact strength and modulus, but in many situations the rock mass will not be uniform, so considerable judgement is necessary anyway. Richards & Read (2007) tried applying the Hoek-Diederichs equations to the Waitacki Dam in New Zealand, which was founded on greywacke, and found that the mass modulus was considerably underestimated for a judged GSI of 20, but examination of their data shows how sensitive any prediction is on the GSI adopted. As discussed elsewhere, features like joint spacing and continuity are extremely difficult to measure and characterise and very risky to extrapolate from field exposures because of variations with weathering and the structural regime. This all reinforces the need for considerable judgement and engineering geological expertise in establishing ground models, and caution when applying any empirical relationships.

Large-scale pile loading tests yield data on rock mass deformation (see Hill & Wallace, 2001). They found that published correlations based on RMR and *Q* classifications overestimated the *in situ* modulus for deep foundation design by up to one order of magnitude, but this was only a significant consideration where the Rock Mass Rating was below 40 (poor and very poor rock masses), and in such cases site-specific testing might be required. As discussed in Chapter 3, the increasing use of Osterberg-type jacks embedded in large-diameter bored piles will no doubt provide very useful data in the future for assessing deformability of rock masses and this, combined with sophisticated numerical modelling, is allowing refinements to the empirical approaches currently in use.

7.7 ROCK DISCONTINUITY PROPERTIES

7.7.1 General

The majority of rocks, and some soil masses near the Earth's surface, contain many discontinuities and these dominate mass properties, including strength, deformability and permeability. Discontinuities include fissures, bedding planes, cleavage, lithological boundaries, faults and joints. The origins, nature and development of discontinuities are discussed in detail in Chapter 6. Many discontinuities are initiated geologically as incipient weakness directions and only with time do they develop as full mechanical discontinuities, as illustrated in Figure 7.19 (Hencher & Knipe, 2007; Hencher, 2013; Shang et al., 2018).

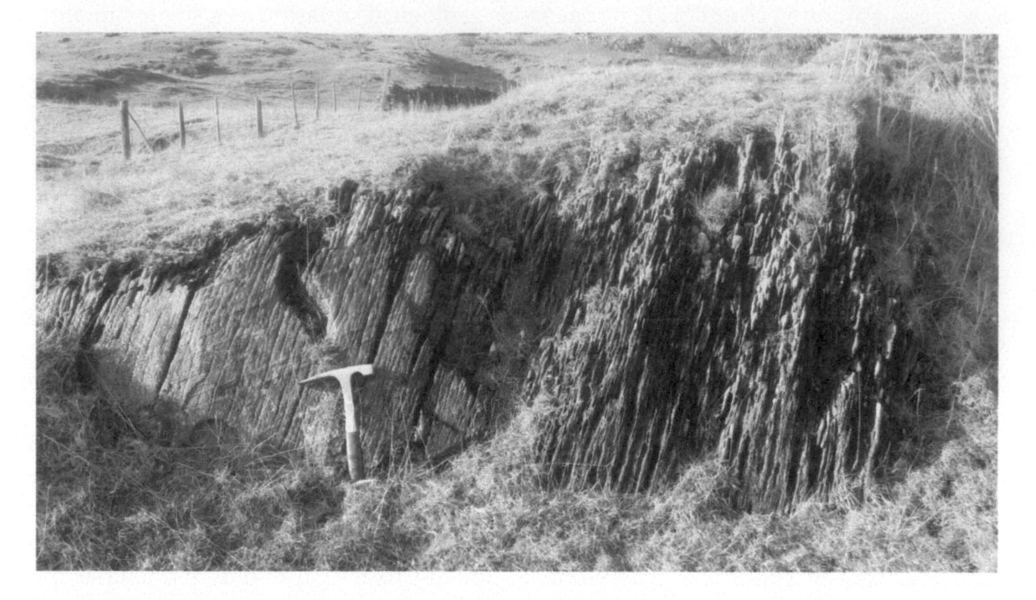

Figure 7.19 Incipient jointing, dipping to the left, and cleavage, opening up on the right, due to weathering. Horton in Ribblesdale, Yorkshire, UK.

At intermediate stages, before rock joints become full mechanical fractures, sections of incipient fractures are cohesive and will contribute strongly to shear strength and shear stiffness along the discontinuity plane. This is illustrated in Figure 7.20.

The persistence of rock joints is very challenging to measure or even estimate (Elmo et al., 2022). Rawnsley (1990) tried to relate joint properties such as style and persistence to geological origin. He concluded, after studying numerous rock outcrops of wide geological age, that whilst persistence can be typified at the scale of joint sets, it is far less predictable at smaller scales (Rawnsley et al., 1990). Zhang & Einstein (2010) review the situation and make some suggestions based on measurement, modelling and theory. Hencher (2013) proposed that incipient joints (with some tensile strength) should be logged in the field or as core as either open, weak, moderate or strong, relative to the strength of the parent rock as illustrated in Figure 7.21. This scheme has been adopted in trying to quantify rock impersistence for the assessment or rock mass strength, in field mapping of rock masses in Austria by Gerstner et al. (2023).

7.7.2 Geotechnical parameters

The main properties of rock joints that need to be measured or estimated are shear strength, normal and shear stiffness and permeability/hydraulic conductivity. These properties depend on the geometry of the joints, including roughness, the nature, strength and frictional properties of the wall rock and any infill between the walls, and their tightness or openness.

Shear and normal stiffness of rock joints are not normally required for civil engineering design but are needed as inputs when carrying out numerical simulations

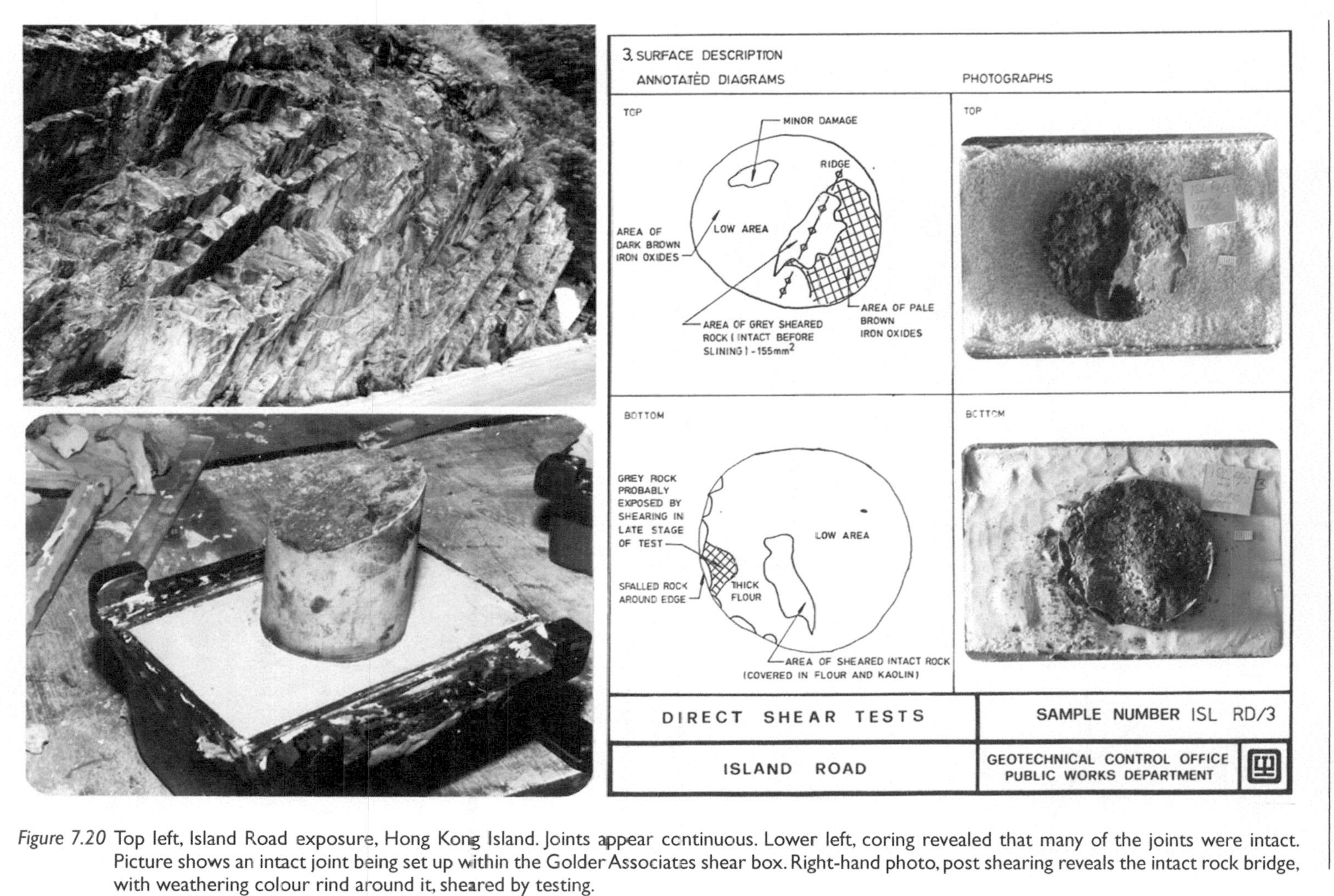

Figure 7.20 Top left, Island Road exposure, Hong Kong Island. Joints appear continuous. Lower left, coring revealed that many of the joints were intact. Picture shows an intact joint being set up within the Golder Associates shear box. Right-hand photo, post shearing reveals the intact rock bridge, with weathering colour rind around it, sheared by testing.

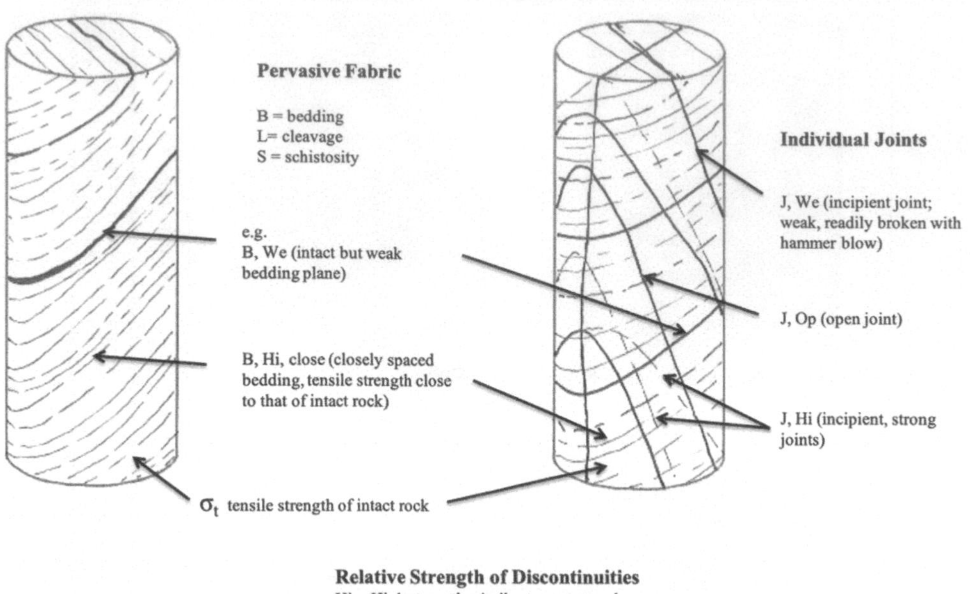

Figure 7.21 Logging scheme in which open joints (Op) are differentiated from incipient joints according to their strength (weak, moderately, high strength).

Source: After Hencher, 2013.

of jointed rock masses where each joint is modelled discretely using software such as UDEC. A review and guidance are given in the UDEC manuals (Itasca, 2004). Permeability of joints depends on their openness, tortuosity and connectivity. It is a very difficult but important subject area, especially for nuclear waste disposal considerations and tunnel inflow assessments (see Chapter 8 for discussion with reference to case examples). A recent review is given by Black et al. (2007). The rest of this section addresses shear strength of rock joints.

7.7.3 Shear strength of rock joints

7.7.3.1 Laws of friction

Amontons (1699) defined two main laws of friction:

1. Frictional resistance increases proportionally with normal load, and
2. Frictional resistance is independent of the apparent area of contact – a brick will provide the same frictional resistance whether slid on its end or its face.

Bowden & Tabor (1950) established that the answer to the second law was that "true" area of contact increases proportionally with normal load through measurements of electrical resistance through metal contacts during shear.

7.7.3.2 Friction of planar rock joints

Terzhagi (1925), developed an adhesion theory to explain the second law of Amontons as it applies to rock joints, arguing that friction was provided by chemical attractions between bodies in contact. He went on to reason that superimposed upon spherical Hertzian contacts may be minor asperities that do not deform according to Hertz's formula. The adhesion theory was proved by establishing that contacts changed proportionally in area with applied normal load using measuring electrical resistance between metal surfaces by Bowden & Tabor (1950, 1964) and for rock-like materials using graphite powder with a phenolic resin binder, as described by Power & Hencher (1996) and Power (1998).

Proof of the straight-line relationship between shear strength and normal load for rocks is shown by the upper line in Figure 7.22, which shows results from a series of direct shear tests on saw-cut surfaces of Darleydale Sandstone by Ross-Brown & Walton (1975). The samples were prepared consistently, being ground between runs with grade 60 carborundum powder. The linear expression of shear strength

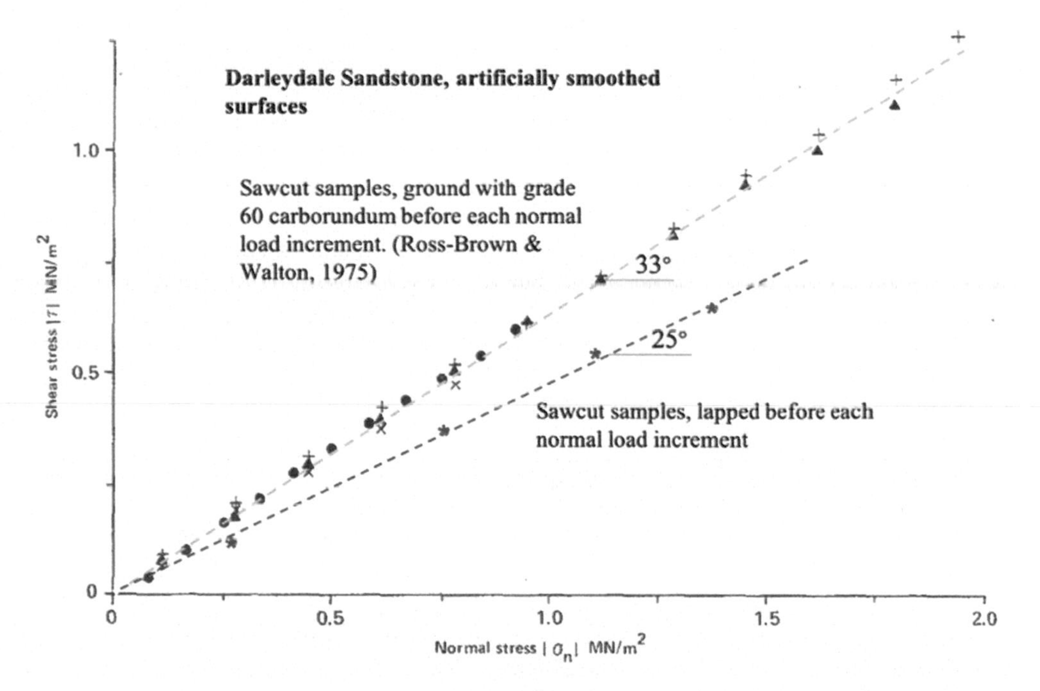

Figure 7.22 Shear strength of saw cut surfaces of Darleydale Sandstone, with different degree of polishing.

with respect to normal load confirms that Amontons first law applies to planar rock surfaces.

The lower line in Figure 7.22 similarly applies to planar (non-dilatant) surfaces of Darleydale Sandstone, but this time lapped using a finer polish. The line is inclined at a friction angle of 25°.

Clearly both lines demonstrate frictional behaviour, *but* the value of the frictional angle depends not only on *adhesion* at points of contact but *also* on the surface textural roughness.

> **"The most exciting phrase to hear in science, the one that heralds new discoveries, is not 'Eureka!' but 'That's funny'...."**
>
> **Isaac Asimov**

Whilst sitting in his cubicle, at Imperial College, London, preparing to carry out a series of tests on dynamic tilt testing using a specially designed test rig on bearings and pulled backwards and forwards, horizontally using strong springs, the author realised that he would first need to carry out static tilt tests to establish how the friction varied with displacement. He first tested rectangular blocks of wood, easily prepared, and sitting on a larger, planar piece of the same wood inclined at 20 degrees in his test device. The results were extremely interesting and showed that geometry of the block affected the angle of sliding. Relevant parts of my thesis can be accessed at hencherassociates.com

Back to the rocks, he had anticipated that there should be a "basic friction", which might decrease slightly with displacement.

What he later found, led him to discuss friction with the then Professor of Heavy Electrical Engineering at Imperial College, Eric Laithwaite who had published similar results in his Royal Institution Christmas Lectures (1974). Over a cup of coffee, he explained that physics equations were generally simplifications making the subject suitable for examination. He was not in the slightest bit surprised by my data!

Firstly, it must be recognised that the stresses beneath the block are not straightforward. When the block is placed horizontally, the weight is equally distributed (I KNOW that is even more complicated when you consider contact mechanics, but bear with me...). As the block is slowly tilted, then the stress state changes and is distributed towards the leading edge of the block as in Case 1 of Figure 7.23. At a particular point, the stress towards the rear of the block diminishes to zero (Case 2). By the time that the block is inclined as shown in Case 3, a piece of paper can be readily slid in and out under the rear of the block!

The results from one of the very many tests he conducted is shown in Figure 7.24.

This shows the results of sliding steel weighted, saw-cut samples of Darleydale Sandstone[1], where the rock flour was blown away between runs using a photographer's bulb and brush. This test showed that the angle of sliding could be reduced from about 32.5 to almost 12 degrees after 3 metres of sliding, which is approaching the low mineral values of Horn and Deere (main text). This reduction in strength was linked to reduced textural roughness as observed using a Talysurf surface profiling machine.

After reaching the lower strength (12.5 degrees), he took the sliding block and placed it elsewhere on the surface (not polished). The sliding angle was 24.5 degrees – which established the importance of polishing history to the sliding friction angle.

Case 1 –stress distributed beneath block ($\tan\beta < \frac{1}{3h}$)

$$\sigma_A = \frac{W}{l.b.\cos\beta}\left(1 - \frac{3h.\tan\beta}{l}\right)$$

$$\sigma_B = \frac{W}{l.b.\cos\beta}\left(1 + \frac{3h.\tan\beta}{l}\right)$$

$\beta =$ angle of inclination of plane

$l, b, h =$ length, breadth, height of block

$w =$ weight of block

Case 2 –Zero stress at rear of block ($\tan\beta = \frac{1}{3h}$)

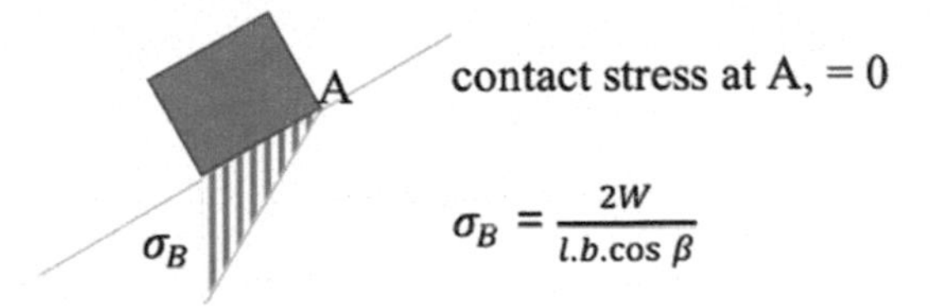

Case 3 –Zero stress towards rear of block
($\tan\beta > \frac{1}{3h}$)

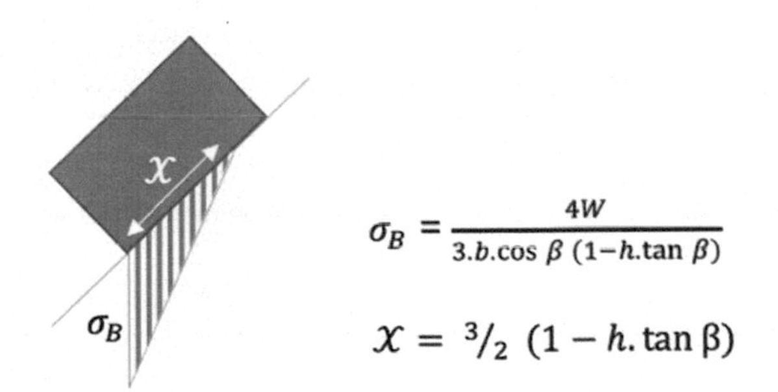

$$X = {}^{3}/_{2}\ (1 - h.\tan\beta)$$

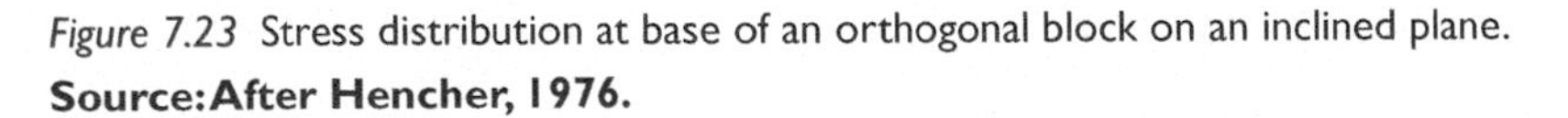

Figure 7.23 Stress distribution at base of an orthogonal block on an inclined plane.
Source: After Hencher, 1976.

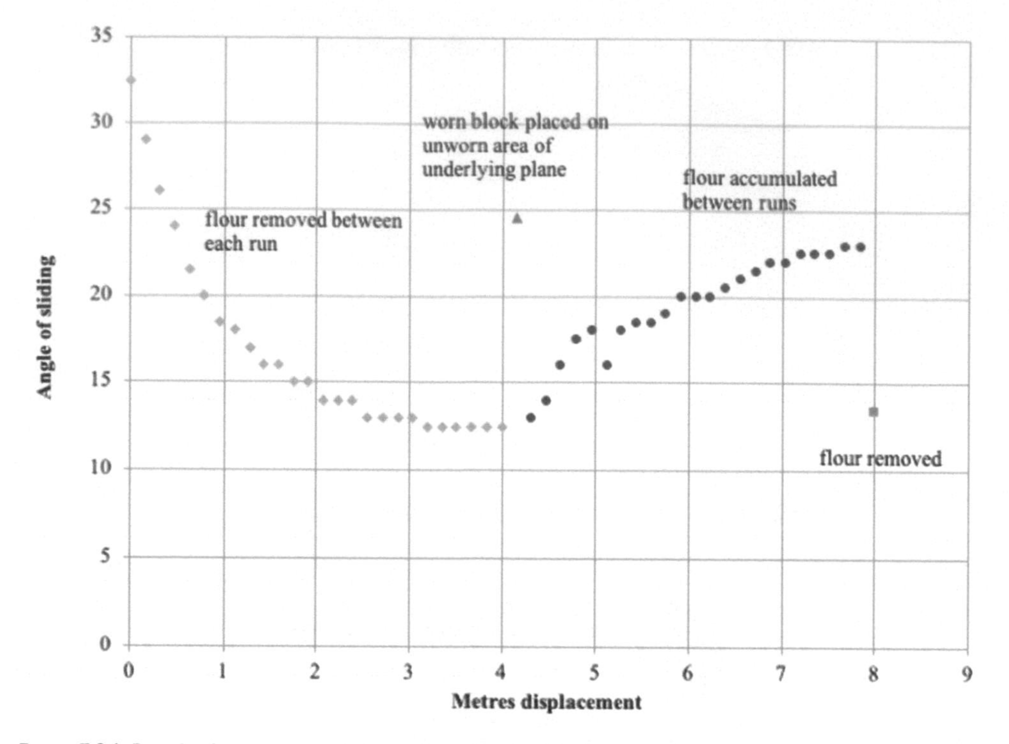

Figure 7.24 Results from repeated tilt tests on Darleydale Sandstone.

Then, placing the block in its original position, and allowing rock flour to accumulate between runs, the sliding angle was seen to increase.

The following conclusions can be made from the test:

(1) at 12.5 degrees, the surfaces were *still* rough enough to wear and produce rock flour, and,
(2) that the sliding angle representing the shear strength of the rock flour coated surfaces was higher than for the underlying polished rock surface (up to 23 degrees).

After 8 metres of sliding, the rock flour was blown away and the sliding angle reduced from 23 degrees to 13 degrees

Gonzalez et al. (2014) confirmed Hencher's findings of 1976. Figure 7.25 shows a test on a saw cut surface of gneiss. Initially the tilt test gave the friction angle as 24 degrees. Then, on repeated slides the angle reduced to 10.5 degrees. He then allowed flour to accumulate and the friction angle rose to roughly the same point as where he started the test, 20 metres earlier. He then blew away the flour and the sliding angle dropped sharply down to 9 degrees. Similar results were reported for tests on migmatite and serpentinised dunite, with flour blown from the surfaces or accumulated during the test.

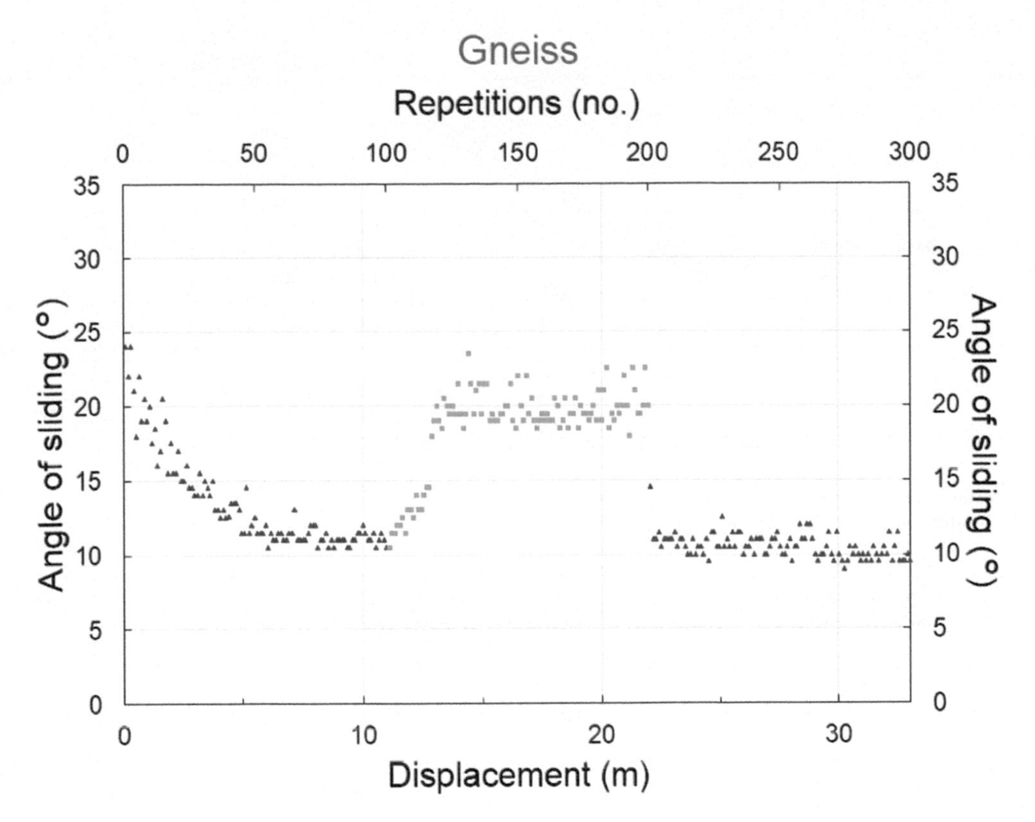

Figure 7.25 Tilt test data from Gonzalez et al. (2014).

The following can be concluded from these tilt tests:

1. There is a lower bound basic friction of about 7 to 10 degrees, due to adhesion between contact points.
2. This is added to by contact roughness, which leads to ploughing and deformation at contact points with the production of rock flour. This is a non-dilational process and is described by Scholz (1990). It is quite remarkable that the textural finish results in locking, which is proportional to normal load as shown in Figure 7.22 (following Amontons laws).

The origins of this "basic fiction" are illustrated in Figure 7.26.

Natural joints have variable roughness due their origins and degrees of impersistence related to their weathering state that result in dilation during shear and added apparent cohesion which cause additional strength.

7.7.3.3 Shear strength of rock joints in rock engineering

When faced with a problem such as checking the stability of a large cut slope, there are a couple of options for assessing the strength of the potentially sliding mass:

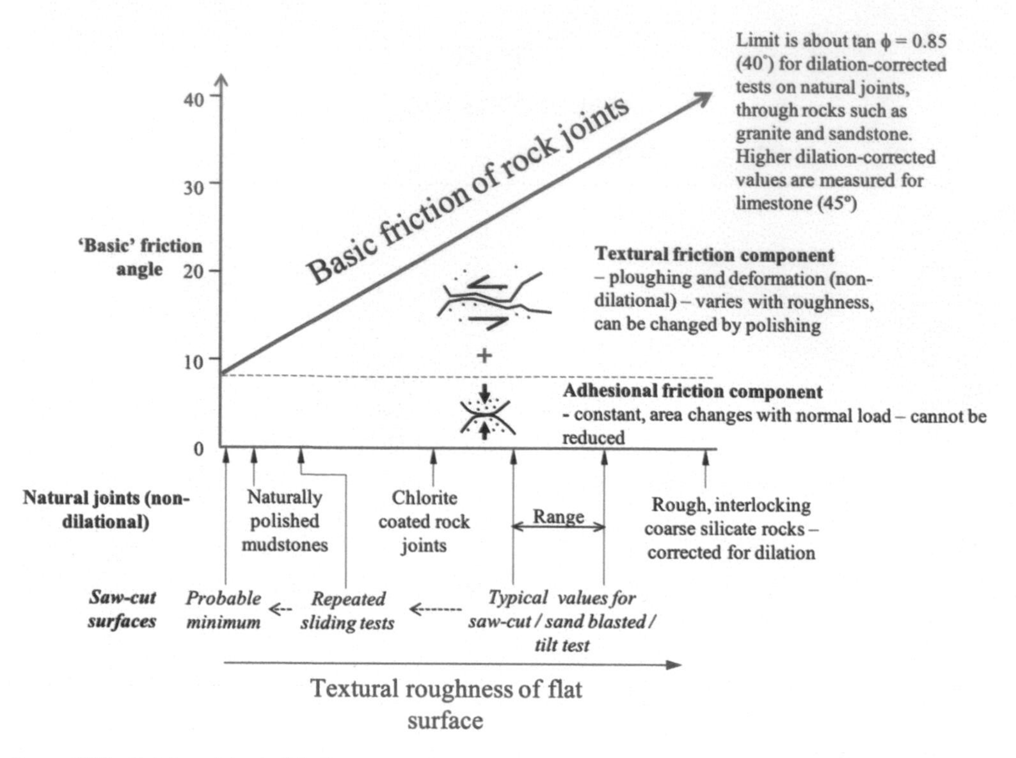

Figure 7.26 Origins of basic friction.

Firstly, for a large slope with no obvious adverse structural geological defect, but containing multiple fractures and weak zones that might coalesce to form a slip surface, then one would approach the problem using the Hoek-Brown relationship as discussed in Section 7.6.2. This involves factoring down the intact strength of rock, according to the degree of joints, weathering and alteration.

However, if the problem includes adversely oriented discontinuities allowing elements of planar or wedge failure, then the strength of individual joint planes will play a part, which one must investigate. There are two main options:

1. To use empirical methods (Barton & Choubey, 1977), which can also be used in numerical modelling as in UDEC-BB[2] ("BB" stands for Barton-Bandis), or
2. To use a realistic approach in which one measures the basic strength of actual, natural discontinuities, by dilation-corrected direct shear testing. This value which might be anywhere between 7 and 50 degrees, is then combined with roughness of the joints in the field, adjusted to the stress levels (as observed and measured in the shear tests), and the influence of any impersistence of the joints (rock bridges).

The first of these is easy to use, and provides reasonable estimates for many rough joints through igneous rocks. It is less reliable however, where dealing with polished faults or with rocks that have lower "basic" friction values, for example in shale horizons or mudstones generally, which have lower friction angles.

The second method requires much more skill in testing, measurement and theoretical understanding, but gives results which can be argued and justified, scientifically.

Empirical methods for estimating shear strength

Patton (1966) came up with the concept of "basic friction" (whereby he assumed that there is a value below which discontinuities would not slide). This value is aided by the effects of roughness and interlocking, which Patton differentiated into first order (small scale roughness) and second order (larger scale waviness):

$$\tau = \sigma \tan\phi + i$$

where τ is shear strength, σ is normal stress (vertical) and ϕ is the angle of basic friction and "i" represents the combined effect of the 1st and 2nd order roughness, which reduces to zero at some particular stress (with some apparent cohesion).

Barton (1973) carried out many direct shear tests on rock joints and concluded that shear strength, τ, as defined by Patton, could be represented by an equation in which the dilation angle due to roughness, i, is calculated by reference to a Joint Roughness Coefficient ("JRC"), which is dependent on the state of stress (σ_n) relative to the compressive strength of the asperities, represented by Joint Compressive Stress ("JCS"):

$$\tau = \sigma_n \left[\phi_r + \text{JRC} \times \log_{10} \frac{\text{JCS}}{\sigma_n} \right]$$

The "basic" friction angle was estimated to be between 28.5 and 31.5 degrees by Barton & Bandis (1990), and 30 degrees by Barton et al. (2023), but is identified here as just a reference number that can be used in his equation.

Estimating shear strength using a scientific approach involving measurement, testing and analysis

A more robust approach, which can be applied to all joints, regardless of their initial roughness and polishing involves field measurement methods including discontinuity orientation measurement and discontinuity profiling at different scales.

Field stress levels first need to be predicted. For example, for a large slope in granite, where unit weight is 27 kN/m^3, and depth of sliding is predicted to be 20 m, then the stress for testing would be around 540 kN/m^2 (kPa) so a test range of 200 to 700 kPa would be appropriate, to see if the asperities survive the testing.

Then it is necessary to take samples of the joints in question, and then to measure their shear strength, corrected for dilation. This is easier said than done!

Firstly, samples are selected from borehole core, preferably matching joints, and then cut to fit into the shear box as described in Hencher & Richards (1989). Record the roughness of the joint using physical methods ranging from a simple decorator's steel pin profiler to 3D photographic imaging, describe the samples and select the direction of shear to test whether particular asperities will survive or be crushed or modified, at the same stress levels as in the field. Note that the roughness determination

Figure 7.27 Golder Associates direct shear box set up at Leeds university (fully instrumented).

here is a reference, not for use in analysis, but for judging the roughness that should be adopted in the field situation, depending on the behaviour during testing.

In the box illustrated in Figure 7.27, the normal load is applied as a constant, dead load, through a lever arm, with the load passing through a ball bearing on the upper box. The shear load is applied through a jack and measured by a load cell, as well as by strain gauges on the shear yoke in this case. Shear displacement and normal displacement are recorded by LVDTs into a computer. In a more modern version of the box, the shear load is applied by a motorised drive.

7.7.4 Golder Associates' direct shear box set up at Leeds University (fully instrumented)

When a strong, rough sample is sheared, the upper sample will dilate, or open up, to overcome the roughness as shown in Figure 7.28. In the right image, the stress (normal and shear) at the asperity is clearly *many* times the amount calculated from the apparent area of contact, nevertheless it is the ratio of shear stress to normal stress that is important!

Work has been done in lifting the normal load over the roughness asperity and this is corrected for by calculating the stresses caused by dilation as illustrated in Figure 7.29.

The calculation is carried out using the following equations (reverse signs if compressive, going downhill):

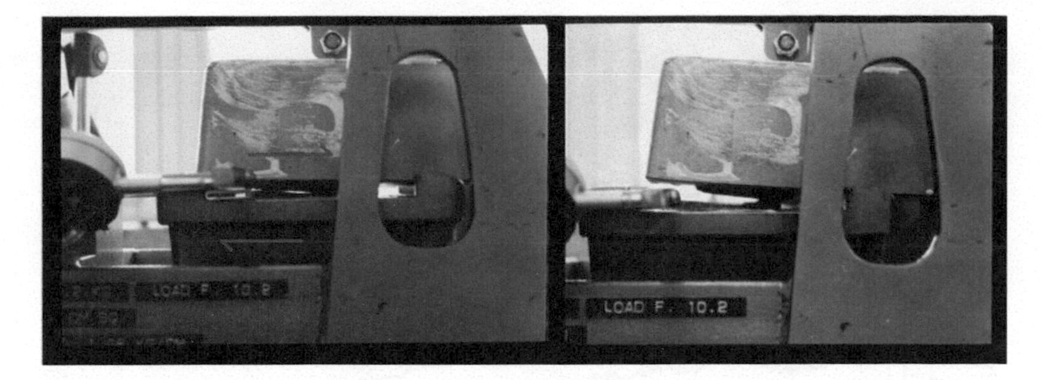

Figure 7.28 Dilation of joint during shear test.

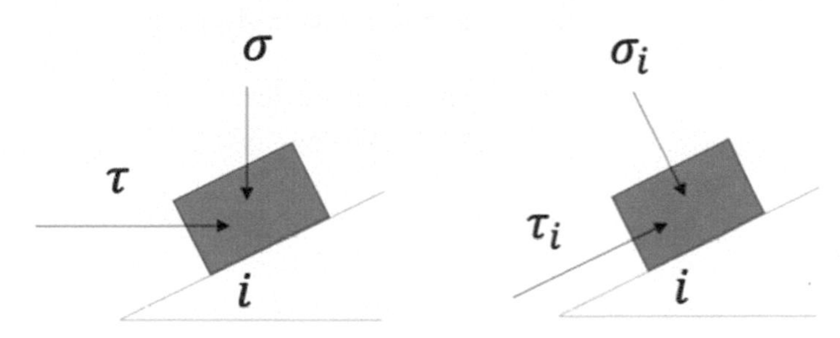

Figure 7.29 Illustration of dilation of rock joint during shear.

$$\tau_i = (\tau \cos i - \sigma \sin i)\cos i$$

$$\sigma_i = (\sigma \cos i + \tau \sin i)\cos i$$

where τ is shear stress measured horizontally, σ is normal stress measured vertically and i is the instantaneous dilation angle, measured over a short horizontal distance (typically 0.2 mm seems to be adequate to minimise reading errors from the LVDTs). An example of data – uncorrected and corrected for dilation is presented in Figure 7.30.

A second example is a four-stage test shown in Figure 7.31. The first stage was carried out on the fresh, matching joint, at a nominal normal stress of 150 kPa, increasing throughout as the apparent area of contact reduced. It can be seen that the measured shear stress rose to about 280 kPa, with an uncorrected friction angle of about 62 degrees[3]. Correcting for dilation however reduced the corrected strength to about 38 degrees. The asperities survived pretty-much undamaged at this stress level as shown by behaviour at the next stage of the test. The second stage was run at nominal normal stress of 300 kPa, after clearing off debris and re-setting the joint in its original position. The uncorrected shear stress rose to approximately 520 kPa, with an uncorrected friction angle of about 58 degrees; the corrected data again define a "basic"

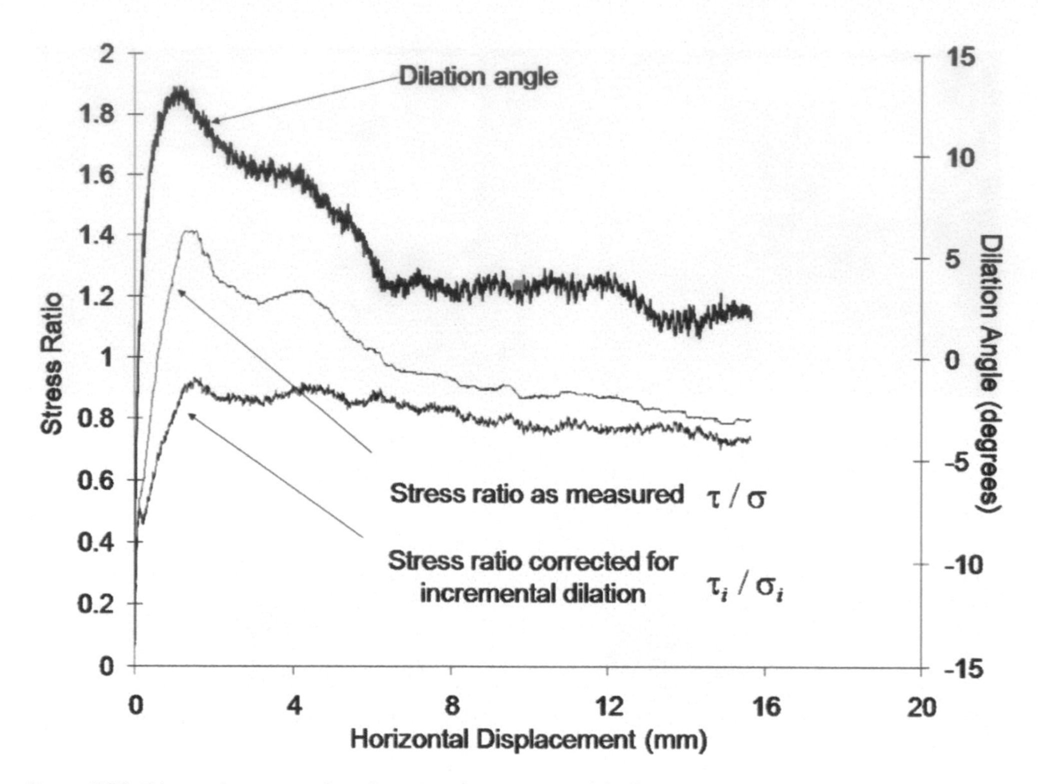

Figure 7.30 Direct shear test data from single run on tensile fracture through Darleydale Sandstone. Peak strength was measured as about 54 degrees (stress ratio 1.4, in central line) but this was accompanied by a dilation angle of approximately 12 degrees (upper line). Corrected strength (lower line) was approximately 42 degrees.

(corrected) friction angle of 38 degrees. After clearing off debris and re-setting the joint in its original position, a third-stage was run at a normal stress of 400 kPa. For the partially damaged joint, the shear strength rose to about 43 degrees, the dilation corrected value being 38 degrees. Finally, a fourth stage was run at 100 kPa, confirming the damage caused by the previous runs but also confirming the "basic" friction angle of 38 degrees for a non-dilatant joint.

Barton (1990) suggested that the dilation-corrected basic friction angle might be partly scale-dependent, as assumed for the asperity damage component in the Barton-Bandis model (Bandis et al., 1981), but further research using the same testing equipment as Bandis (1980), but with better instrumentation, indicates that this is not so (Hencher et al., 1993; Papaliangas et al., 1994). Dilation-corrected basic friction is independent of the length of the sample. Scale effects do need to be accounted for in design but as a geometrical consideration when deciding on an appropriate field scale *i* value.

Many silicate rocks are found to have a dilation-corrected, basic friction, ϕ_b, of approximately 40° (Papaliangas et al., 1994; Hencher & Richards, 2015), and Byerlee (1978) found a similar strength envelope (41°) for a large number of direct shear tests

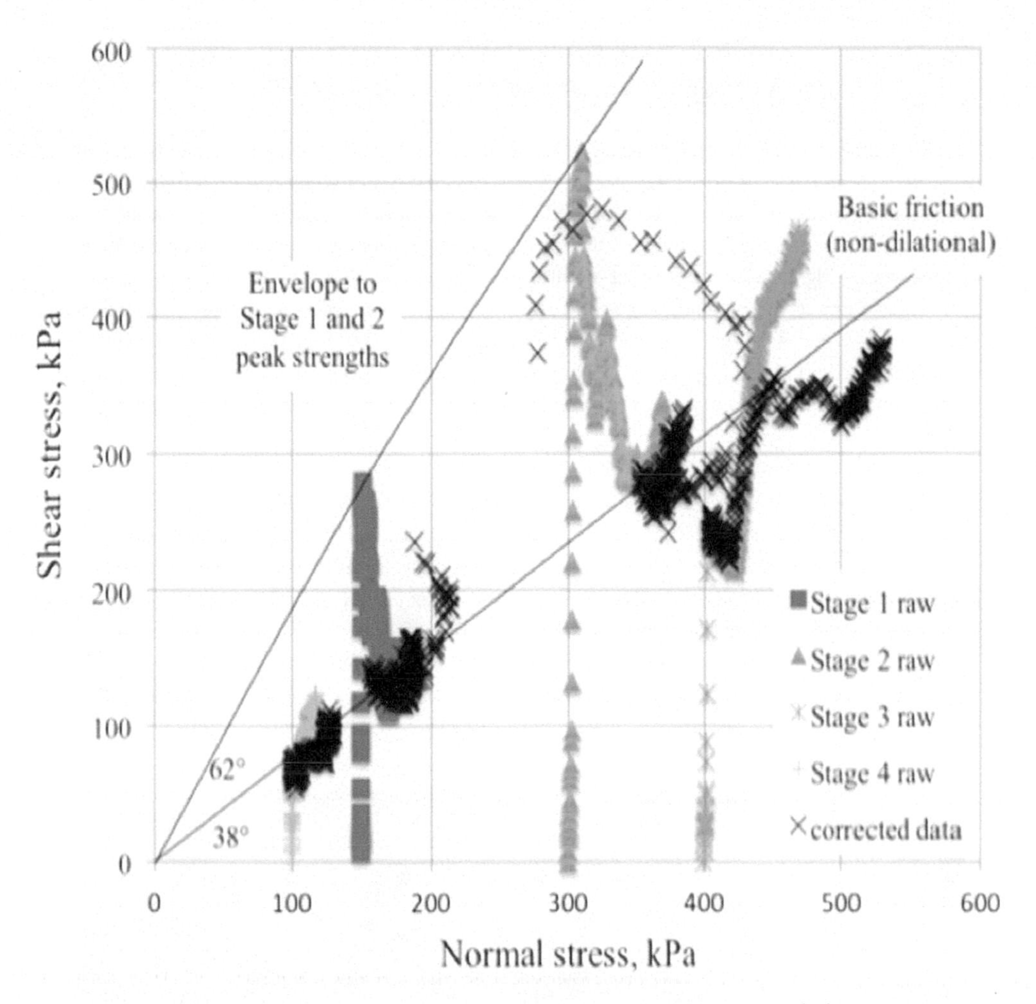

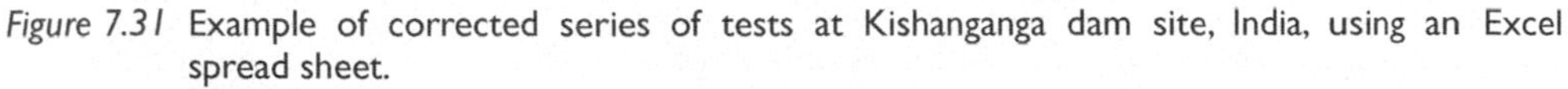

Figure 7.31 Example of corrected series of tests at Kishanganga dam site, India, using an Excel spread sheet.

Source: Available at www.hencherassociates.com.

on various rock types, where dilation was constrained by using high confining stresses. Empirically, it seems to be about the highest value for basic friction achievable for natural joints in many silicate rocks and applicable specifically to joints that are forced to dilate during shear or where dilation is suppressed because of the high normal load. Conversely, much lower basic fiction angles can be measured for natural joints where they are planar and where the surface texture is very fine, polished or coated with low-friction minerals, as illustrated by a case example in Box 7.2. The author has measured values of 10° to 17° for naturally polished joint surfaces through Coal Measures mudstones and coal of South Wales, UK.

BOX 7.2 YIP KAN STREET LANDSLIDE

AN EXAMPLE OF USE OF DIRECT SHEAR TESTING

The Yip Kan Street landslide occurred in July 1981 on a dry Sunday night. It mainly involved large blocks of rock of up to 10 m^3, which slid on persistent joint planes dipping at only 22 degrees out of the slope (Hencher, 1982). The total failure volume was estimated to be 1,235 m^3. The 8 metres high, near vertical slope was cut in very strong, slightly decomposed, coarse grained igneous rock (quartz-syenite). The upper part of the slope was in saprolite. The failure occurred next to a construction site where blasting had been carried out recently before the failure but not over the weekend. There had been intense rainfall a week before the failure. The slope had been deteriorating in the days preceding the failure with cracks in chunam cover in the weathered part having been repaired 5 days before failure.

Because of the low angle of sliding as illustrated in Figure B7.2.1, it was decided to investigate in some detail. Blocks were collected – both matching discontinuities and mismatched. It was noted that some blocks were coated with red iron oxides and others with green chlorite (a hard, thin coating). Each sample was carefully described and then tested multi-stage in a Golder Associates direct shear box. At each stage the test proceeded until peak strength was reached and then for another mm or two following which the normal stress was increased without re-setting the sample in some cases. For some tests complete runs of about 15 mm shear displacement were conducted and in

Figure B7.2.1 One of the sliding planes with debris removed.

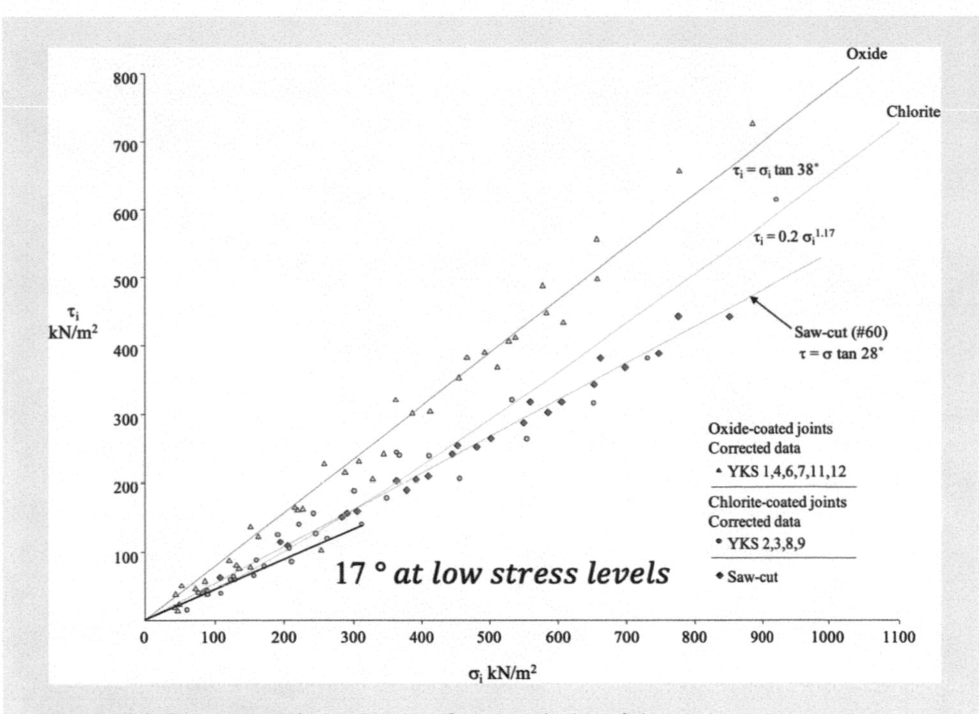

Figure B7.2.2 Shear strength data, Yip Kan Street rock slope failure.

one test the sample was tested at the highest stress level first, which was then reduced in stages incrementally. Samples were photographed, roughness measured and damage described carefully. For reference, a series of tests were conducted on saw-cut samples, ground with grade 60 carborundum powder

Results from the tests are presented in Figure B7.2.2. All tests on natural joint surfaces were corrected for dilation incrementally. It can be seen that the saw-cut surfaces gave a friction angle of about 28 degrees, which is about what might be expected.

The tests from natural joints fall into distinct groups. The data from joints coated with iron oxides define a friction angle of 38 degrees, which is the same as one finds for many weathered rocks (Hencher et al., 2010). The data for the chlorite coated joints were much lower however and unexpectedly so. At low stress levels especially, values were very low, below that of the saw-cut joints as can be seen from the inset figure and about the same as the angle of dip of the planes along which the failure took place ($\phi \equiv 17°$ at the lowest stress levels). Field scale roughness was measured at 5° using a 420 mm diameter plate and 9° using a base plate of 80 mm. It was concluded that the failure was progressive, probably having been exacerbated by blasting and previous rainfall and that the initial movements overcame the field scale roughness. The eventual failure was explained by the presence of persistent chlorite-coated joints with inherently low frictional resistance (Brand et al., 1983).

Figure 7.32 The problem of roughness, at Skipton Quarry, UK.

Roughness at the field scale

Roughness is expressed as an anticipated dilation angle, $i°$, which accounts for the likely geometrical path for the sliding slab during failure (deviation from mean dip). There are two main tasks for the geotechnical engineer in analysing the roughness component: firstly, to determine the actual geometry of the surface along the direction of likely sliding at all scales (Figure 7.32) and secondly to judge which of those roughness features along the failure path will survive during shear and force the joint or joints to deviate from the mean dip angle. This is the most difficult part of the shear strength assessment, not least because it is impossible to establish the detailed roughness of surfaces that are hidden in the rock mass. Considerable judgement is required and has to be balanced against the risk involved. Hack (1998) gives a good review of the options, and the difficulties in exercising engineering judgement are discussed in an insightful way by Baecher & Christian (2003).

In Figure 7.32, measurement is being carried out of roughness (variability of dip and dip direction) on a grid pattern is being carried out, using plates of different size as originally described by Fecker & Rengers (1971), adopted in the ISRM Suggested Methods (Brown, 1981) and described in Richards & Cowland (1982). The problem can be seen, that at the small scale the blocks are rough, but at a larger scale, the

Figure 7.33 Attempts to characterise roughness at different scales.

roughness is zero. The problem is to define which scale of roughness will be operative at the stress level (this is where the observations from direct shear testing come into play). Figure 7.33 illustrates methods used to determine roughness parameters, from the top left clockwise: using plates of different size, on a grid pattern using a Clar compass, by surveying along different profiles, and using a tooth impression device.

Spatial variability may be an important issue; the important first-order roughness represented by major wave features may vary considerably from one area to another, as of course also might the mean dip of the plane (Hencher & Richards, 2015). At one location, a block might be prevented from sliding by a wave in the joint surface causing a reduction in the effective down-dip angle along the sliding direction; elsewhere, a slab of perhaps several metres length may have a dip angle steeper than the mean angle for the joint as a whole because it sits on the down-slope section of one of the major waves as shown in Figure 7.34.

Defining the scale at which roughness will force dilation during sliding, rather than being sheared through, requires considerable judgement. Some assistance is provided by Schneider (1976) and by Goodman (1989) who indicate that for typical rough joint surfaces, where slabs are free to rotate during shear, as the length of the slab increases (at field scale), the dilation angle controlling lifting of the centre of gravity of the upper block will reduce. The problem cannot be finessed by improved analytical methodology. There is no substitution to careful engineering geological inspection,

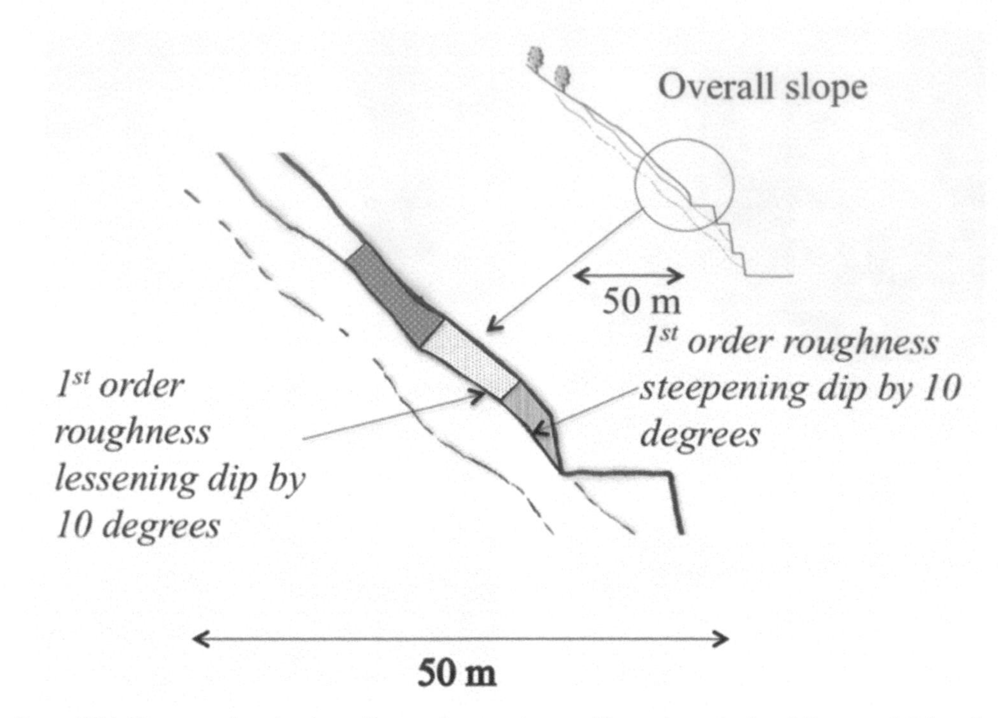

Figure 7.34 Diagram showing how first-order roughness affects the calculated Factor of Safety for a large block – compare the lowest segment of the potential sliding block with the block above.

investigation, characterisation of the ground model and judgement based on experience of similar joints and geological settings, and an appreciation of the fundamental mechanics controlling the potential failure.

7.7.5 Infilled joints

The two walls of a joint might be separated by a layer or pockets of weaker material which may reduce shear strength. A similar situation arises from preferential weathering along a persistent joint. The effect of the "infill" is a function of the relative height of roughness asperities in the wall rock vs. the thickness of weaker material (Papaliangas et al., 1990). If persistent and the infill is of low strength, the consequences can be serious. Cut slopes on the A77 at Rhuallt, North Wales (Figure 7.35), failed by sliding on bedding-parallel thin clay "infilled" discontinuities with faults acting as release surfaces (Gordon et al., 1996). The mechanism had not been anticipated from ground investigation prior to the failure, which involved more than 187,000 m^3 of rock.

In some slopes, incremental movement may take place over many years before final detachment of a landslide and, following each movement, sediment may be washed in to accumulate in dilated hollows on the joint (Figure 7.36). The presence

Figure 7.35 Cut slope at Rhuallt, North Wales, with continuous soft clay infill which led to failure (on wedges).

of such infill might cause alarm during ground investigation but in many cases is confined to local down-warps and probably plays little part in decreasing shear strength, other than in restricting drainage (Halcrow Asia Partnership, 1998b). It may, however, be taken as a warning that the slope is deteriorating and approaching failure (Hencher, 2000).

7.7.6 Dynamic shear strength of rock joints

There is some evidence that frictional resistance for rock joints is dependent on loading rate, and this may be significant for aseismic design and for understanding response to blasting. For a block of rock sitting on an inclined plane, given a value for static friction, one can calculate the horizontal acceleration necessary to initiate movement and when the block should stop, given a particular acceleration time history, as illustrated in Figure 7.37. This type of calculation is the basis of the Newmark (1965) method of dynamic slope stability analysis, which is used to calculate the distance travelled.

Hencher (1977) carried out a series of experiments and found that initiation of movement was generally later than anticipated (or did not occur), meaning that the peak friction angle was higher than predicted from static tests (Figure 7.38). The effective friction angle for initiation increased with the rate of loading. The implication is that if the loading is very rapid and reversed quickly (as in blast vibrations),

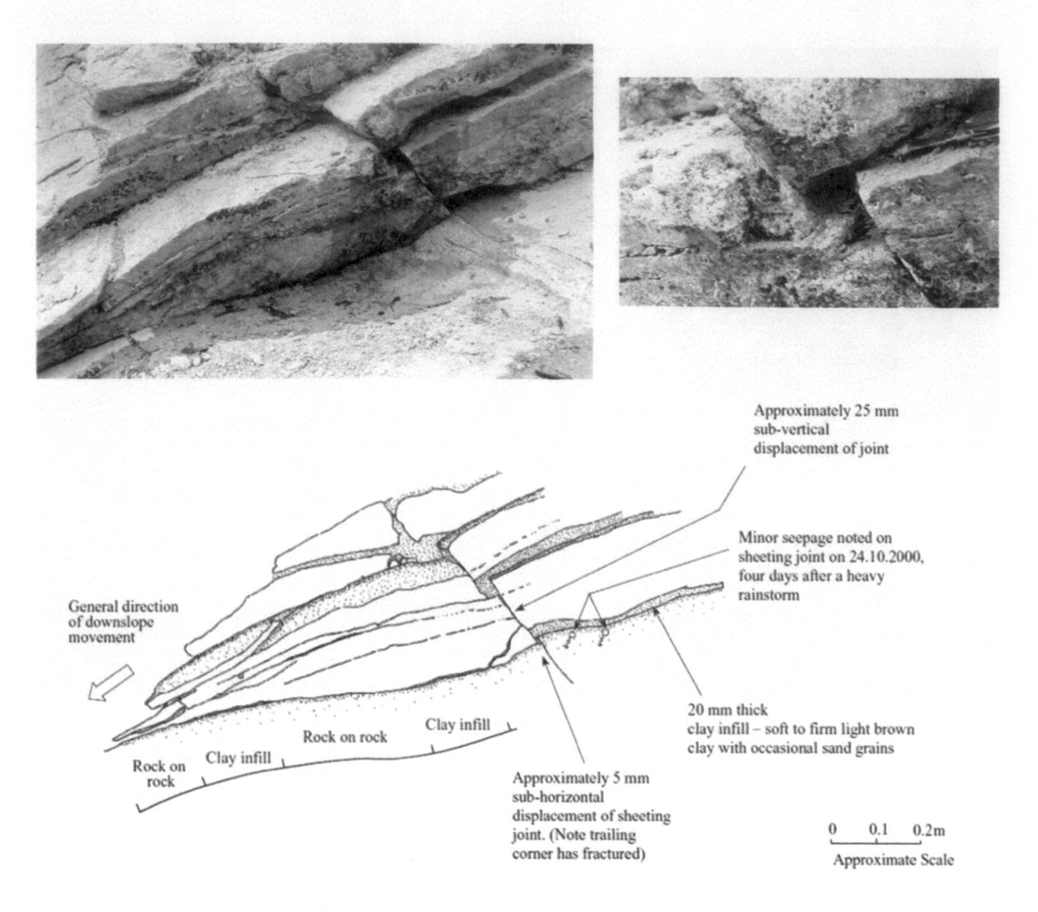

Figure 7.36 Detailed view of failure above Leung King, showing evidence of partial disintegration, prior to final collapse.

shear displacement might not occur, despite the supposed "critical" acceleration being exceeded. However, once movement *was* initiated, Hencher found that the distance travelled was higher than anticipated from static strength measurements and interpreted this as reflecting rolling friction and the inability of strong frictional contacts to form during rapid sliding. Hencher (1981a) suggested that for Newmark-type analysis, residual strength should be used for calculating displacements. Recent work confirms low sliding friction angles post-failure (Lee et al., 2010).

7.8 ROCK-SOIL MIXES

It has long been recognised that mixes of soil and rock, such as illustrated in Figure 7.39, can often stand safely at steeper angles than if the slope were comprised only of the soil fraction. From testing on soils together with theoretical studies, the point at which the hard inclusions start to have a strengthening effect is about 30% by volume.

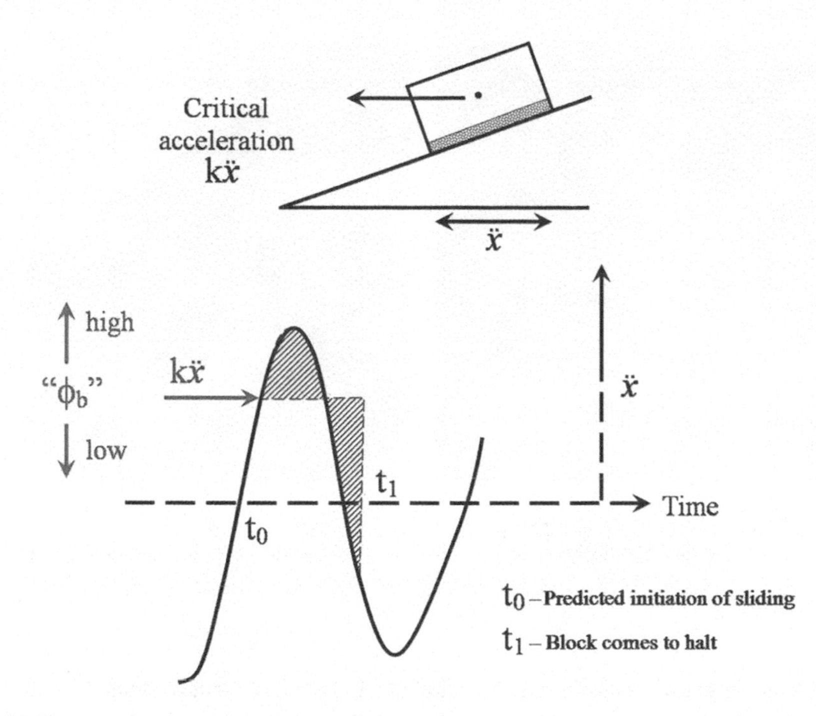

Figure 7.37 Transient loading of block on a plane. As the acceleration reaches a critical value, $k\ddot{x}$, which depends upon the dynamic friction angle, at time t_0, the block begins to slide. It continues down the slope, accelerating and then decelerating, and comes to a halt at time t_1.

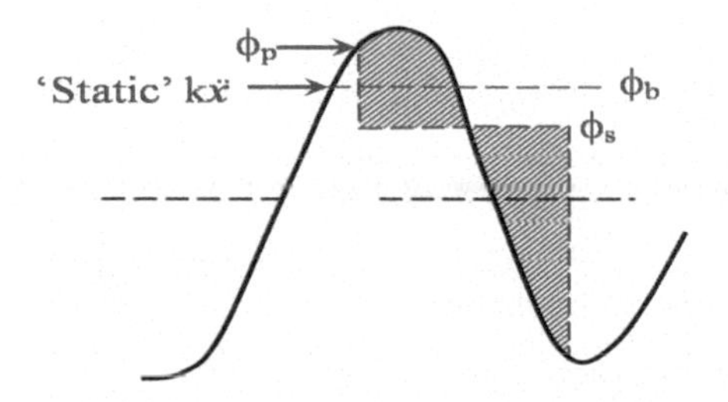

1. **Sliding commences later than predicted (ϕ_p rather than ϕ_b)**
2. **Velocity developed greater than predicted (ϕ_s)**

Figure 7.38 Initiation of sliding occurs later than predicted (or not at all at higher frequencies of acceleration, as during blasting) (after Hencher, 1977; 1981a). Once sliding begins, the frictional coefficient is lower than for static sliding, presumably because of the interactions required to adhere at contact points on the surface.

Figure 7.39 Slopes with considerable boulders affecting strength. Left-hand photo is of Chai Wan road failure in Hong Kong comprising colluvium overlying completely decomposed volcanic rock. Right-hand photo shows unprotected slope standing at about 70 degrees in South Africa.

7.8.1 Theoretical effect on shear strength of included boulders

Hencher (1983c) and Hencher et al. (1985) report on the back-analysis of a landslide involving colluvium containing a high percentage of boulders, in which an attempt was made to estimate dilation angles on the basis of the coarse fraction percentage estimated in the field and measurements taken from idealised drawings. These estimated field dilation angles were added to the strength for the matrix, determined from laboratory testing. West et al. (1992) took this further and identified several different ways that included boulders might influence shear strength, based on physical modelling and back analysis of slopes. Factors included the following: boulders preventing failure along an otherwise preferred failure path, failure surface forced to deviate around a boulder, and a failure zone incorporating the boulder. Triaxial tests reported by Lindquist & Goodman (1994) similarly concluded that boulders increase the mass strength. Additional review is provided by Irfan & Tang (1993) and by Khorasani et al. (2019, 2022).

Practical methods for addressing the strength of mixed soils and rocks remain difficult. One of the main problems is that such masses can be highly heterogeneous and difficult to characterise realistically. The other is that whilst trends of increasing mass strength with percentage of rock clasts and boulders are clear, general rules have not yet been formulated. Further advances will probably be by numerical modelling and could be done using PFC3D (Itasca). Whilst the largely intractable geological characterisation nature of the problem would remain, the problem could probably be resolved parametrically in a similar way and with a similar level of success for prediction as the Hoek-Brown model for fractured rock masses.

7.8.2 Bearing capacity of mixed soil and rock

Mixed soil and rock deposits include sedimentary deposits like colluvium and glacial boulder clay, but also some weathered rocks. As for assessing shear strength, there are considerable difficulties for sampling and testing and there can also be significant problems for construction (e.g. Weltman & Healy, 1978). The conservative position for design is to take the strength and deformability of the matrix as representative of the mass, but allowance might be made for the included stiffer and stronger clasts by rational analysis, perhaps backed up by numerical modelling.

7.9 ROCK USED IN CONSTRUCTION

Crushed rock and quarried or dredged sand and gravel are important materials used in making concrete and construction generally, perhaps as fill. Rock is also used as armourstone, for example, in protecting earth dams from wave action or for forming harbours. It is also cut and polished as "dimension stone" to be used as kitchen work surfaces or as cladding on the outside of prestigious buildings. Engineering geologists are often required to identify sources of aggregate, either from existing quarries but sometimes from new "borrow area" in the case of sand for reclamation or new quarries for a remote project such as a road. Some of the properties that are important for their use are the same as in much of geotechnical design: strength, unit weight and porosity, but there are other properties that need to be tested specifically.

7.9.1 Concrete aggregate

For concrete, the aggregate must be "sound", durable and chemically stable. Materials to be avoided include sulphates and sulfides (e.g. gypsum and pyrites), clay and some silicate minerals such as opal and volcanic glass, which can cause a severe reaction and deterioration of the concrete if present in the wrong proportions (see case example of Pracana dam in Chapter 8). Tests are available and should be used to ensure that the aggregate being sourced is suitable. These include mortar bar tests whereby a test mix of concrete is formed and observed to see if it expands with time. Other factors might include the need for light – or heavy-weight concrete, fire resistance and overall strength. Concrete mix design for a large project may require a research programme to optimise the aggregate specification and type of cement to use. For smaller projects or where the demands are less onerous, then cost may be the controlling factor; aggregates and quarries have "place value", which is a matter of the quality of aggregate at a particular quarry together with the costs of transport to the project site. A useful review of the factors to be considered in specifying concrete aggregate is given by Smith & Collis (2001).

7.9.2 Armourstone

Armourstone or "riprap" is used to protect structures primarily from wave action and is often made up of blocks of rock of several tonnes. Generally, the rock must be durable and massive. If it softens or discontinuities open up with time, then the function is lost. Massive crystalline limestone often works well, as do many igneous and metamorphic rocks. Usually, durability (and availability and cost) is all-important but see the case history of Carsington Dam in Chapter 8 where the choice of limestone as riprap contributed to adverse chemical reactions and environmental damage. Weak or fractured rocks are obviously not appropriate. For many coastal defence works in the east of England, large rock blocks are brought by barge from Scandinavia because of a lack of suitable local rock. Where suitable rock is not available then concrete tetrapod structures known as "dolosse" are used in the same way, piled on top of one another and interlocked, to protect coasts and structures by dissipating wave energy.

7.9.3 Road stone

Aggregate is used in road construction in many different ways – as general fill or in the sub-base, as drainage material and in the wearing course. There are many different standard tests to be applied in road construction, and these are described in Smith & Collis (2001). The most demanding specification is for wearing course material, within the "hard top". Rock must be strong and durable but also must resist polishing as it is worn by traffic. This requires the rock used to comprise a range of different minerals that are strongly bonded but wear irregularly. Rocks like limestone are generally unsuitable (the "polished stone value" is too low). Rocks like Ingleton "granite", which is really an arkose, have excellent properties and therefore very high place values – worth quarrying and transporting large distances – even from a National Park (Figure 7.40). It might be come as a surprise that the rock shown in Figure 7.19 is a

Figure 7.40 Drygigg Quarry in 2023, just south of Horton-in-Ribblesdale in the Yorkshire Dales National Park, UK.

weathering product (fractured and cleaved) of the same rock type being quarried and transported great distances...

7.9.4 Dimension stone

The term "dimension stone" is used for rock quarried to be used directly in building, construction or even sculptures. Typical rocks quarried in this way include marble, granite and slate for roofs. Rocks are generally chosen for their colour and appearance – the quarry at St Bees headland, Cumbria, UK (a fairly ordinary sandstone), was re-opened temporarily in the 1990s to provide rock for shipping to New York to repair buildings faced with sandstone carried by ships as ballast in the 19th Century – because of its appearance. Dimension stone must also be resistant to wear, frost and chemical attack. This can be difficult to determine from direct testing, so experience of the long-term performance of a particular rock from a particular quarry may be the best clue. Generally, a lack of visible joints is a very important attribute.

Notes

1 See Hencher (1976) to see results for three rock types; see Hencher (1977) for full details of testing including areas of wear and theoretical analysis of stress distribution.
2 "BB" stands for Barton-Bandis. Bandis (1980) carried out some revolutionary testing involving scaled models at Leeds University under the supervision of Alistair Lumsden, published as Bandis et al. (1981).
3 Note that the true area of contact is much smaller, and stresses applied to the asperities would be much greater, as illustrated in Figure 7.28, but so would it be in the field situation, so any observations made in the laboratory will also apply in the field.

Chapter 8

Unexpected ground conditions and how to avoid them: case examples

> As failure is exactly what engineers do not want, it is essential that we learn lessons, when it does happen.
>
> *Blockley, 2011*

8.1 INTRODUCTION

Generally, multiple causative factors contribute to failure (Reason, 1990). When unexpected ground conditions are a major factor, the consequences can be delays, extra costs or even abandonment of contracts. Sometimes the adverse ground conditions were truly onerous and unpredictable but it is often clear that the problem is exacerbated by the way the project was set up, managed and contracted (Muir Wood, 2007; Baynes, 2007). Quite often, good practice, which is set out in standards and the literature is simply not followed because of lack of knowledge, experience or application in the engineering teams or for commercial reasons. Where unexpectedly difficult conditions are encountered during a project, for whatever reason, the consequences can be minimised provided the attitudes of the various parties are to work together to solve the issues. This is often a simple matter of good professional practice on both sides but can be actively encouraged in contracts as discussed in Chapter 2.

8.2 GROUND RISKS

First it is worth considering where ground risks arise. Clayton (2001) divides them essentially into three:

1. technical;
2. contractual;
3. managerial.

Of the "technical", essentially geotechnical risks, these were split down by McMahon (1985) and Trenter (2003) into

1. unknown geological conditions and
2. using incorrect design criteria.

DOI: 10.1201/9781003348894-8

The first of these reflects inadequate investigation or incorrect geological modelling at some scale or other – including the perhaps 5% of sites described as "unforgiving" in Chapter 7. The second is due to ignorance of how to design works and the parameters to use, for a given scenario of loading.

If one examines failures forensically, however, often the causes are far more complex and it is the interaction of the *various* pre-disposing conditions at a site and other construction factors that caused the problems. Very often mismanagement is fundamental to why critical factors are missed, overlooked or not dealt with properly as discussed for tunnels by Muir Wood (2000). For the following discussion, case examples are grouped according to the predominant cause following the "geology" plus "environment" plus "construction" approach introduced in Chapter 7 (Table 7.1). This is not always easy to do because it is often a combination of factors that led to failure or to the extent of the failure. A fourth cause introduced here is "systemic" – where fundamental decisions or concepts seem to have played a major role in what happened.

8.3 GEOLOGY: MATERIAL SCALE FACTORS

Geotechnical hazards occur at a full range of scales from micro (porosity, mineralogy, friction) to macro (earthquakes, volcanic hazards, typhoons). Hazards at the material scale are associated with the physical and chemical nature and properties of the various geological materials making up the site and used in construction, including their durability. It is the scale of most laboratory and *in situ* tests.

For example, liquefaction caused by an earthquake is definitely, fundamentally, a material scale geological problem in that the soil-type is susceptible to liquefaction due to its looseness and grading (silty-sand) and degree of saturation (environmental hazard), combined with a larger-scale environmental hazard (dynamic earthquake loading – environmental hazard), which lasts some considerable time. All of these matters should be considered during site investigation by the engineering geologist (Boulanger & Idriss, 2014). The material scale factors (strength and grading) determine whether or not the site would be susceptible to liquefaction.

8.3.1 Chemical reactions: Carsington Dam, UK

Carsington Dam failed in July, 1984 during construction when it had almost reached full design height and this incident is discussed later. One of the lesser known problems with the Carsington Dam project however concerned chemistry of the materials making up the dam, which should have been anticipated. The dam was constructed of locally-derived rock fill with a clay core. The rock included black shale with pyrite (FeS) and the riprap was limestone ($CaCO_3$) (Figure 8.1). Sulphuric acid was generated as a result of the geological materials present and this acid polluted local stream courses. Carbon dioxide, because it is heavier than air, collected in underground inspection chambers and four workers died as a consequence. The various chemical reactions and processes are discussed in detail by Pye & Miller (1990). The lesson is that potential chemical reactions should be considered for any project as described by Cripps et al. (2019). The project is also an example of failure due to geological mass factors as considered in Section 8.4.1 below.

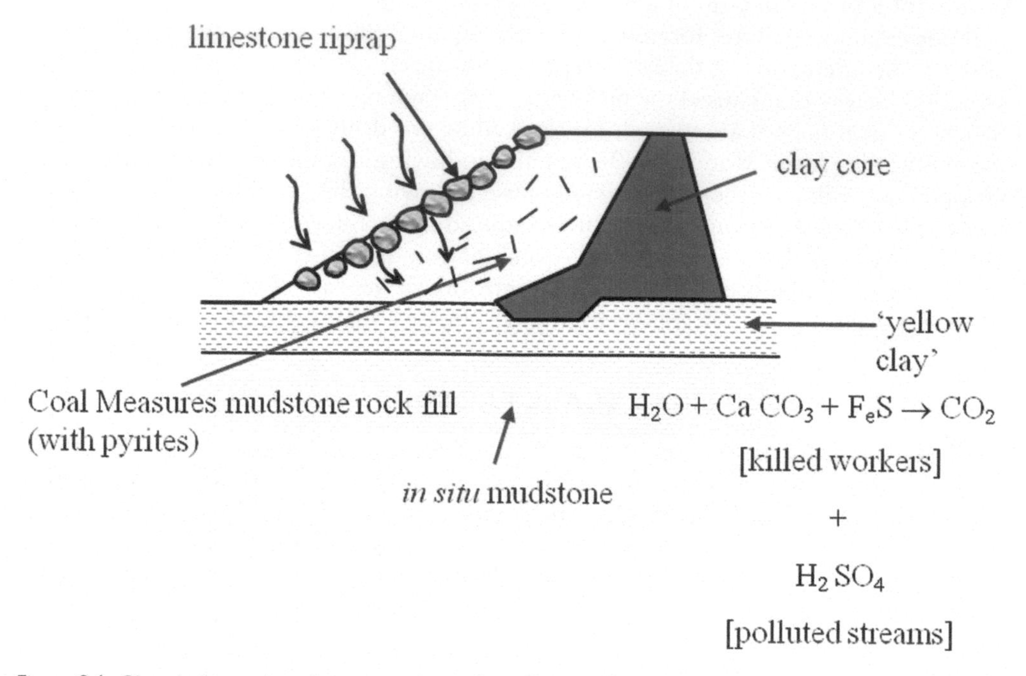

Figure 8.1 Chemical reactions between materials making up Carsington Dam.

8.3.2 Strength and abrasivity of flint and chert: Gas storage caverns Killingholme, Humberside, UK

Another example of a material scale factor that should have been anticipated is shown in Figure 8.2, which shows a cavern in chalk under construction for the storage of liquefied gas beneath South Killingholme, Yorkshire. The road header pictured had been specified for the excavation of the caverns and had been dismantled and lowered down a narrow shaft for this purpose. This proved impractical economically because of wear caused by the presence of bands of extremely strong and abrasive flints and chert in the chalk. Blasting had to be used instead to excavate the caverns (Anon, 1985). Perhaps the presence of bands of flint might have been anticipated in the chalk, even if it had not been logged during site-specific drilling (probably due to inadequate sampling).

8.3.3 Abrasivity: TBM Singapore

Shirlaw et al. (2000) provide examples of the high abrasion that can result from high quartz content in soils, weathered and fresh rock during tunnelling. As part of the construction of the North East Line of the Mass Rapid Transit in Singapore, two tunnels were driven from the east bank of the Singapore River to Clarke Quay Station on the

Figure 8.2 Killingholme LNG caverns under construction.The road-header is being used to trim blasted caverns rather than excavate them as had been the original intention.

west bank. The total length of each drive was only 80 m. A single 6.53 m diameter machine (Figure 8.3) was used to construct the tunnels with the TBM turned around after the first drive. The tunnelling passed from weathered sandstone to marine clay on the southbound tunnel and the reverse on the northbound tunnel. To maintain the face pressure in the weathered sandstone the EPB shield needed to be operated at almost maximum torque, which resulted in extensive wear to the cutting tools and dangerously high temperatures in the slurry and spoil. Almost all of the discs had to be replaced after 80 m before re-launching the shield for the northbound drive. Furthermore the sandstone was broken down to abrasive slurry which, in EPB mode, fills the chamber and the area between the cutting head and the excavated face causing further damage to the machine, including the screw conveyor. In a separate case, for an EPBM tunnelling mainly through Old Alluvium in Singapore, the pressure bulkhead (40 mm thick steel), became so abraded that it failed, resulting in a major loss of ground (Marshall & Flanagan, 2007). The incident caused about 280 mm of settlement and closure of a lane of the central expressway for 12 hours (Shirlaw, personal commnication).

Similar problems have been reported and should be anticipated for other soils and rock with high silica content including granite and soils derived from granite and can be predicted using Cerchar Abrasivity Index tests. The abrasion can be lessened by appropriate use of additives. In the example above, for the southbound and most of the northbound drives either water or polyacrylamide were used as a conditioning

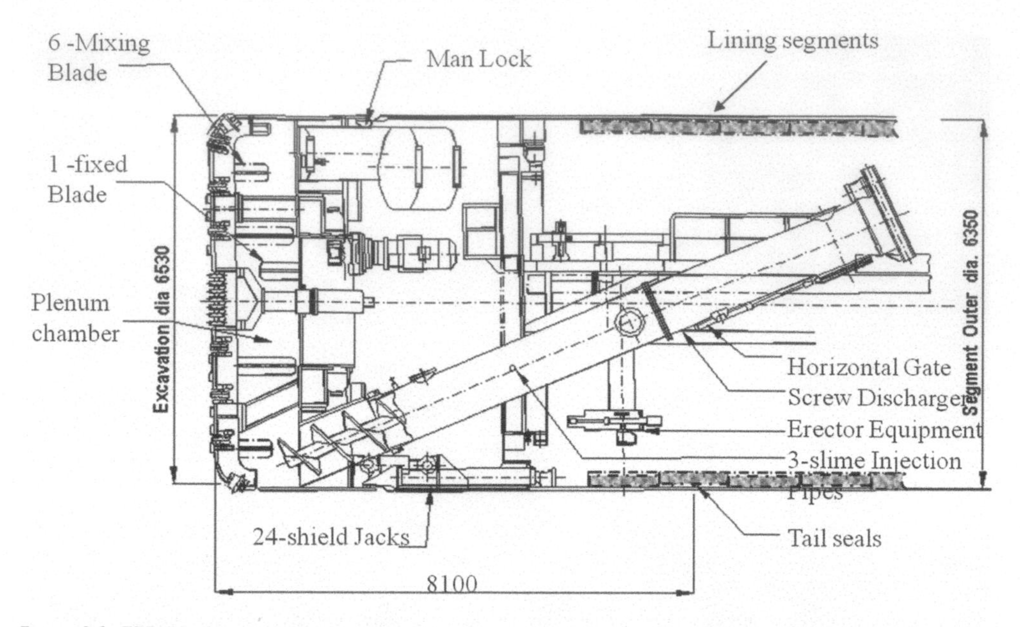

Figure 8.3 EPB Machine in cross section. The pressure at the face is maintained by the rate at which broken down material is removed in the Archimedes screw.

agent. For the last 10 m of the northbound tunnel the contractor switched to foam. This had a dramatic effect in reducing the torque required to rotate the cutting head, and the temperature of the spoil.

8.3.4 Concrete aggregate reaction: Pracana Dam, Portugal

Pracana Dam is a 60 m high concrete gravity buttress dam built between 1948 and 1951 and located in the centre of Portugal, on the Ocreza River, a tributary of the Tagus River. In 1962 cracks were detected in some buttresses and, in 1964 seepage through the dam increased suddenly (Gomes et al., 2009). Cracking further developed together with progressive upward movement of the crest and downstream movement of the dam. Cores were drilled into the dam and samples taken to analyse deposits in cracks and the aggregate and to conduct expansion tests. It was established that alkali–silica reactivity (ASR) of the locally-derived aggregate including quartzite, milky quartz, feldspar, granite and shale/greywacke, was probably the cause for the swelling phenomenon. A photograph of the crazing resulting from the ASR is given in Figure 8.4. Remedial works included sealing of individual cracks, resin grouting and placing a 2.5 mm PVC geomembrane with geotextile on the upstream side of the dam. As illustrated by this case, alkali–silica reactivity in concrete can cause severe deterioration and necessitate expensive remediation and yet is readily avoided if the mineralogy of the aggregate is considered properly and appropriate reactivity and expansion tests conducted when the aggregate is selected for the project (Smith & Collis, 2001).

Figure 8.4 Cracks in face of the Pracana Dam, Portugal, due to alkali–silica reactivity from use of reactive aggregates in concrete (reproduced with kind permission from Mr. Ilidio Ferreira, Head of Dam Safety Department at EDP, Portugal).

8.4 GEOLOGY: MASS SCALE FACTORS

8.4.1 Pre-existing shear surfaces: Carsington Dam Failure

Carsington Dam is infamous because it collapsed over a long length during construction in June 1984 (Figure 8.5). The project was delayed by 8 years as a result. A major factor was that the foundation materials were mistakenly considered to comprise undisturbed residual clay derived from *in situ* weathering of the underlying mudstone whereas they were later identified as periglacial head deposits (Figure 8.6). The head deposits contained pre-existing slip planes along which the shear strength was far lower than had been assumed for the original design (Skempton & Vaughan, 1993). In hindsight probably, the clay above bedrock should have been excavated prior to construction of the dam as was done for the reconstruction of the dam (Banyard et al., 1992). As noted in Chapter 4, pre-existing shear surfaces are recognised commonly in quite young soils and particularly in the London Clay where they have been associated recently with landslides.

Figure 8.5 Failure of Carsington Dam. Black area behind water intake structure is tension crack due to failure over 400 m length of the dam crest. White area in front is limestone riprap.

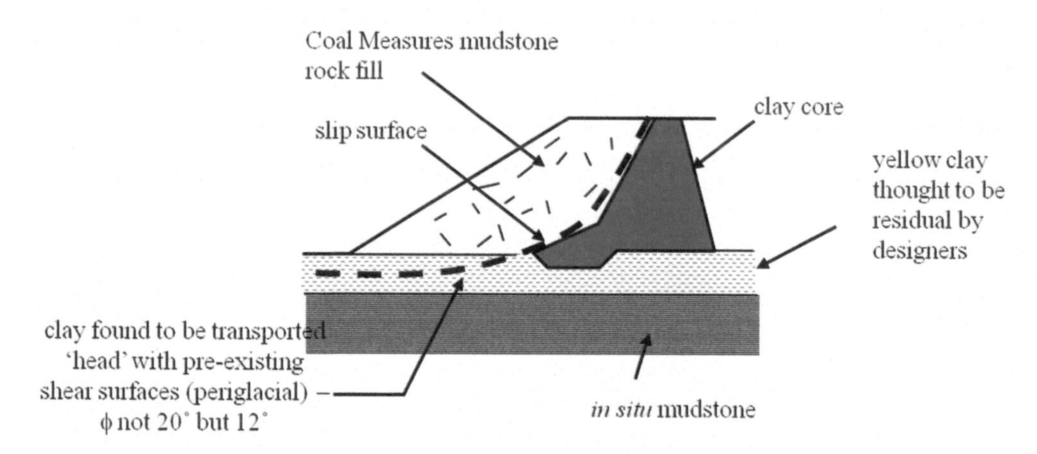

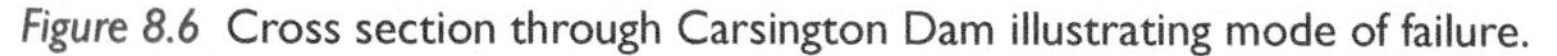

Figure 8.6 Cross section through Carsington Dam illustrating mode of failure.

8.4.2 Faults in foundations: Kornhill Development, Hong Kong

Figure 8.7 shows foundations for high-rise structures under construction at Kornhill, Hong Kong. The presence of a major fault (weathered zone in line with the valley) meant that foundations had to be taken locally 10s of metres deeper than adjacent foundations. Clearly the valley was indicative of the potential for poor ground conditions. That said, all valleys are not associated with faults and all major faults are not associated with valleys. At Kornhill some faults that had been anticipated caused no difficulties whilst other, unpredicted faults were discovered during construction (Muir et al., 1986).

8.4.3 Faults: TBM collapse, Halifax, UK

A tunnel was to be constructed in Northern England through Carboniferous mudstones and a tunnel boring machine (TBM) was selected for the construction. The disc cutters of the machine are seen in Figure 8.8. Normally you should not be able to photograph this view of the machine until the tunnel has been completed and the TBM has entered a reception excavation. Unfortunately, in this case the tunnel had collapsed during construction when the TBM encountered a fault (Figure 8.9). The TBM came to a halt as material collapsed around it. The only way of advancing the tunnel was to sink a shaft in advance of the TBM and excavate a larger tunnel by hand, back

Figure 8.7 Fault in valley at Kornhill, Hong Kong. Poor-quality rock in the fault zone resulted in foundations being taken 10s of metres deeper than in better rock away from the fault zone.

Figure 8.8 TBM exposed in fault zone, Northern England.

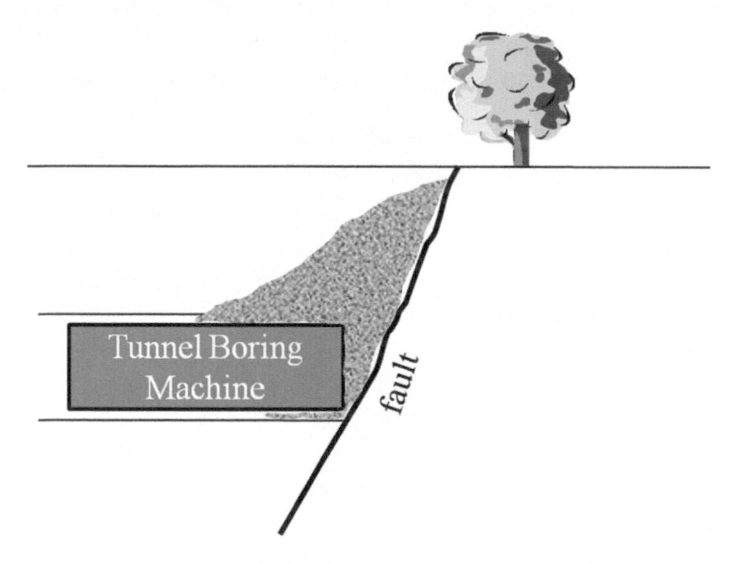

Figure 8.9 Collapse of fault zone around TBM – exacerbated by attempt to move TBM backwards, which necessitated emergency backfill and grouting above the tunnel.

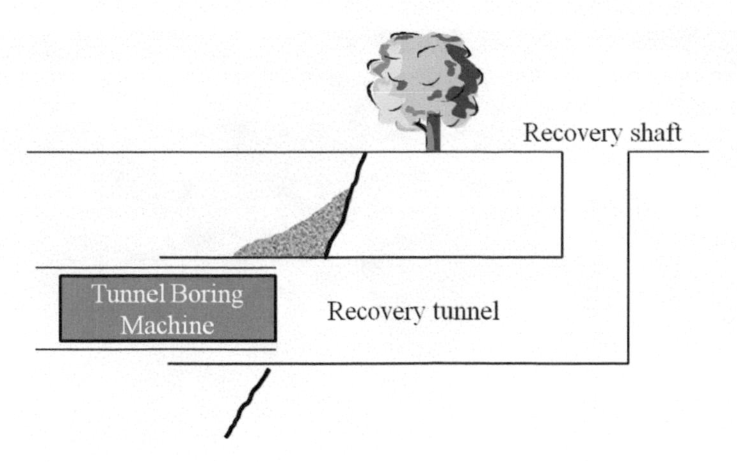

Figure 8.10 Solution – sink shaft in advance of TBM and excavate tunnel by hand back to TBM allowing it to be protected and freed up.

to the collapse and freeing the machine (Figure 8.10). This of course was expensive and led to some dispute. The Contractor (constructing the tunnel and whose TBM had got buried) claimed that he did not expect such poor ground. An expert for the Client said that he should have done so because a fault was clearly indicated by the ground investigation. The Contractor agreed that there was a fault shown from the boreholes but argued that the GI did not indicate the degree of disturbance that had led to the collapse. Another expert was called to give evidence. He was asked "was such poor ground an unusual occurrence with such faults?" The expert answered "yes, very unusual" – [case going for the Contractor]. He then went on to say "it is so interesting that I bring my students each year to examine it in the quarry near the site where the collapse occurred". Case dismissed. [The New Engineering Contract (Institution Civil Engineers, 2005) states that in judging the physical conditions the Contractor is assumed to have considered:

- The site information
- Publicly available information referred to in the site information
- <u>Information available from a visual inspection of the site</u>, and
- Other information that an experienced contractor could reasonably be expected to have or to obtain.]

8.4.4 Geological structure: Ping Lin Tunnel, Taiwan

The Ping Lin Tunnel in Taiwan (now called the Hsuehshan Tunnel) was eventually completed after numerous delays, collapses and deaths. The plan was to excavate a pilot tunnel using a 4.8 m diameter rock TBM followed by two 11.84 m diameter TBMs for the main tunnels but the anticipated tunnelling rates of up to 360 m/month/machine proved hopelessly optimistic because of adverse ground conditions. The geology included a syncline of sandstone with several faults. The pilot tunnel TBM

Figure 8.11 River of water out of pilot tunnel TBM, Ping Lin, Taiwan.

soon ran into trouble as slow seepage from a fault draining the saturated aquifer above rapidly increased and flooded the tunnel (Figure 8.11). Attempts were made to construct by-pass tunnels and to advance each of the TBMs using pre-grouting to improve the rock mass but much of the tunnelling had to be done using drill & blast methods rather than using the purpose built TBMs. Details of this project are described by Barla & Pelizza (2000).

8.4.5 Deep weathering and cavern infill, Tung Chung, Hong Kong

The new town at Tung Chung is situated close to Hong Kong International Airport and for a large part was built on off-shore reclamation. Prior to the planning of Tung Chung and the formation of the reclamation, only limited site investigations and geophysical surveys were undertaken. As a consequence, the geology of the substrate below the seabed was essentially unknown, and extrapolation of the geology from the on-shore rock outcrops proved to be misleading.

Initial site investigations on the newly formed reclamation indicated that the subsurface bedrock geology was composed of rhyolite dykes, marble, meta-sedimentary rock and skarn. The thickness of weathered bedrock varied greatly across the reclaimed area from less than 50 m to over 150 m. At the site of a proposed 50 storey residential tower block a very steep gradient in the rockhead surface was identified.

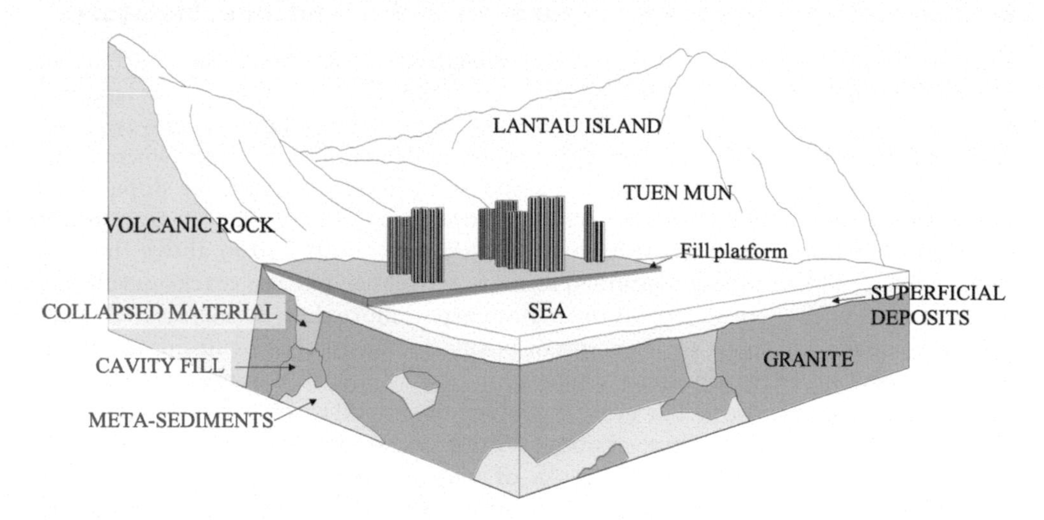

Figure 8.12 3D geological model, Tuen Mun, Hong Kong. [Courtesy of Dr Chris Fletcher (Fletcher, 2004).]

In addition, cavities up to 12 metres in thickness and of unknown lateral extent were recorded in the drill logs, deduced from zones of no sample recovery and sudden drops of the drill string. Using this information, a foundation design consisting of 2.5 m diameter bored piles socketed into rock at depths in excess of 120 m was proposed. Ground conditions are illustrated in a summary way in Figure 8.12. However, the costs and risks involved with this design were considered to be too high and further ground investigations were undertaken to investigate the geology more fully and to determine whether the cavities were actually present or were filled with soil. The new boreholes used polymer drilling fluids and improved sampling techniques, followed by down-hole electrical cylinder resistivity, gamma density and sonar surveys. No open cavities were identified in the second phase of drilling and a new geological model was proposed. The foundation design was then re-assessed, but concern over the mention of open cavities in the original borehole logs still remained and in the event the tower block was never built.

The key points to be learnt from this project are the following:

- The town planning of Tung Chung should have considered the sub-surface geology, so that areas with problematic foundation conditions could have been avoided, if at all possible.
- The use of geophysical techniques, such as microgravity surveys, prior to the formation of the reclamation would have greatly assisted the formulation of a realistic geological model, thereby maximising the potential of the site and reducing costs.
- Site investigations in areas of complex ground, deep tropical weathering and the presence of calcareous meta-sedimentary rocks require advanced drilling techniques and high levels of soil recovery.

8.4.6 Pre-disposed rock structure: Po Selim landslide, Malaysia

The Pos Selim landslide is a currently active landslide in Malaysia. Some details are provided by Malone et al. (2008). The landslide occurred in one of the many large and steep cut slopes along the new 35 km section of the Simpang Pulai–Lojing Highway project and it is pertinent to ask the question why it occurred there rather than somewhere else?

Failure occurred early on in the cut slope and affected the natural slope above the cutting (Figure 8.13). Progressively the slope was then cut back in response to further failures until the works reached the ridgeline about 250 m above the road (Figure 8.14). The slope has continued to move with huge tension cracks developing near the crest with vertical drop at the main scarp of more than 20 metres in 3 years.

Clearly at the site there are some predisposing factors that are causing instability whereas many other equally steep slopes along the 35 km of new highway show no similar deep-seated failures. The general geology of the site is schist but the main foliation actually dips into the slope at about 10 degrees so the common mode of failure associated with such metamorphic rocks of planar sliding on daylighting, adverse schistosity or on shear zones parallel to the schistosity (Deere, 1971) is not an option to explain this landslide. Following detailed face mapping by geologists, review of displacement data and examination of the various stages of failure, a model was derived that can be used to explain the nature of the failure, the vectors of movement and the fact that it has not yet failed catastrophically but is bulging at one section of the toe (Figures 8.15 & 8.16). Key aspects of the geology are frequent joints that are oriented roughly orthogonal to the schistosity, three persistent faults cutting across the failure

Figure 8.13 View of Po Selim landslide, Malaysia.

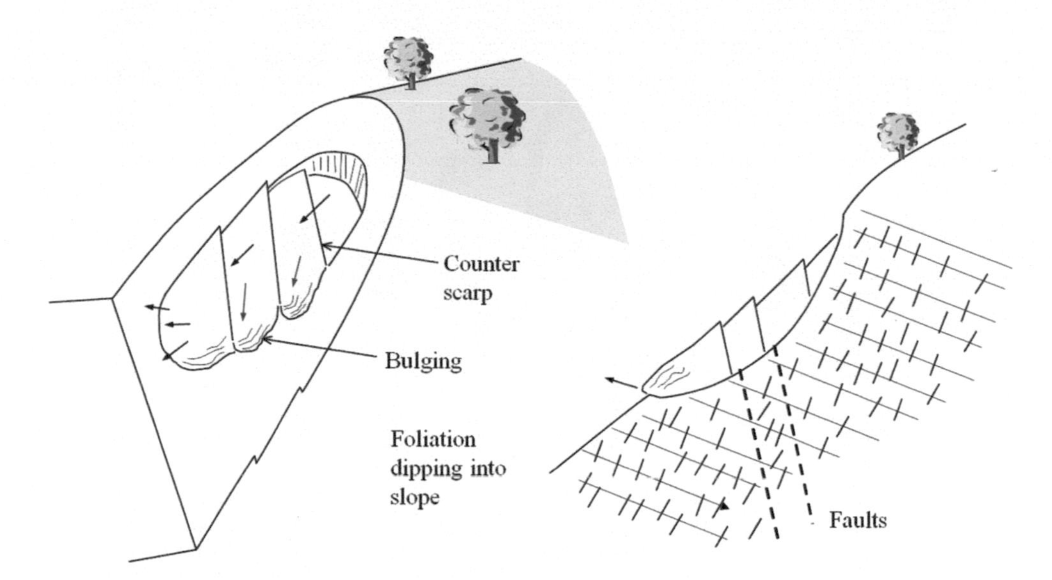

Figure 8.14 3D model of postulated failure mechanism, Po Selim.

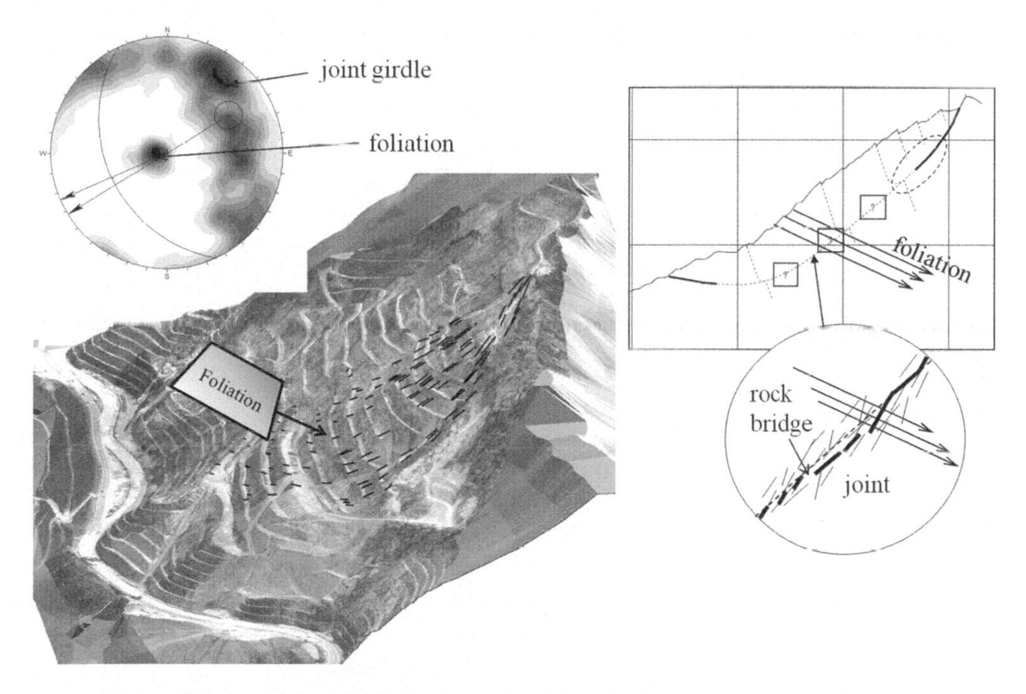

Figure 8.15 (a) Digital model of Po Selim landslide with movement vectors compiled by comparing coordinates of prominent points on a 2005 orthophoto and in the 2003 as-built CAD drawing. (b) Stereo plot with concentration of poles near the centre representing the foliation dipping back into the slope at about 10 degrees. Most joints form a girdle at about 90 degrees to the mean pole of the foliation. (c) Explanation for main body of landslide – exploiting jointing and kicking out at the toe and upwards, along foliation.

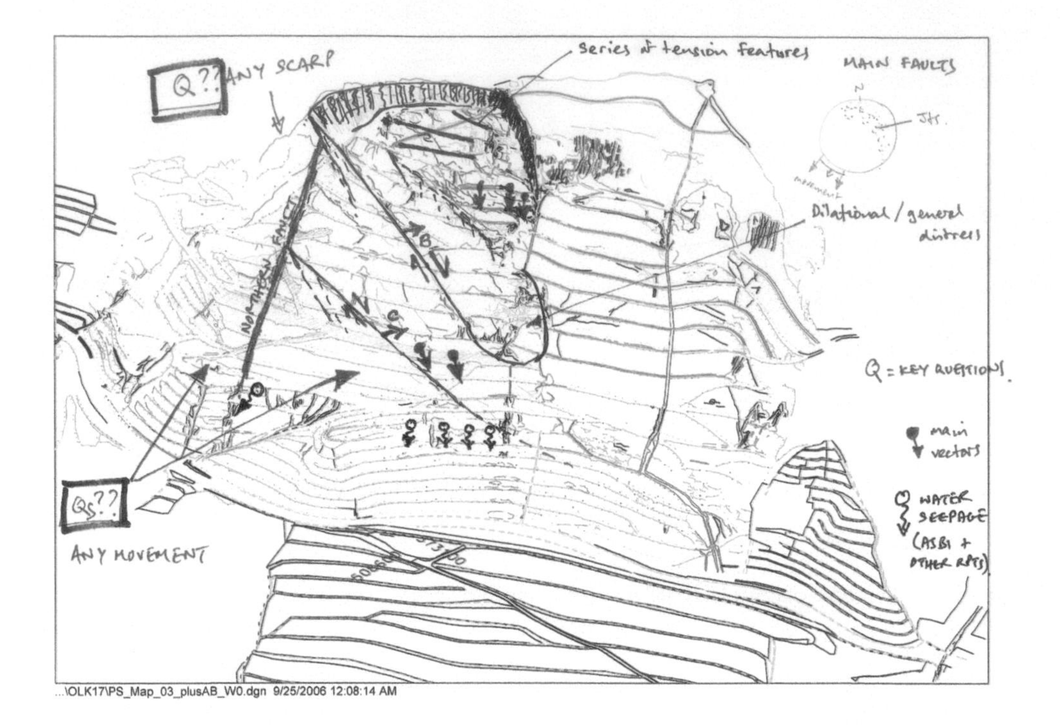

Figure 8.16 Model of Po Selim failure.

and another major fault to the north of the landslide area. The derived model is of a mechanism of sliding on the short, impersistent joints that combine with offset sections of schistose fabric to form a basal rupture surface. The shearing forces are largely balanced by sliding friction on joints, along the transverse faults and schistosity in one part of the toe where the failure is kicking out. Resistance is also provided by the dilating mass towards the right-side toe of the slope (facing). One possible option for remediation that can be derived from this model therefore involves strengthening that toe area by anchoring or otherwise buttressing.

It is to be noted that this model is not numerical but could certainly be used as the basis for a numerical model that would, indeed work for some realistic set of parameters. Without this understanding of geological mechanism, it would be impossible even to begin to design successful remedial measures.

Currently there is some minor evidence that groundwater is playing a part in the failure (seasonal movements and seepage) and therefore it has been recommended that long, trial raking drains be installed at points of seepage in such a way that they also allow water pressures to be monitored within the slope (a cost-effective combination of ground investigation and remedial measure).

8.5 GENERAL GEOLOGICAL CONSIDERATIONS

One of Terzaghi's principles as reported by Goodman (2002) was to "assume the worst configuration of properties and boundary conditions consistent with the data from site investigations", i.e. within the confines of an appropriate ground model and two examples illustrate the consequences of not so doing.

8.5.1 Tunnel liner failure at Kingston on Hull, UK

The failure of a tunnel at Kingston on Hull is reported by Grose & Benton (2005). The tunnel was constructed with a TBM through a sequence of saturated quaternary sediments. During construction, water and then soil migrated through one of the already constructed segmental liner joints and the tunnel had to be abandoned temporarily as the situation deteriorated. A subsequent investigation failed to come up with a definitive answer but it was clearly a matter of soil-structure interaction and possibly construction defects. The discussions by Hartwell (2006) and Shirlaw (2006) are instructive and they come up with various ideas to do with problems with grouting of the liner to explain the failure that seem feasible. The bottom line seems to be that the tunnel design and construction methodology was not robust for the ground conditions. In this case, whatever the geological variability that ultimately brought about the failure, variability was to be anticipated and the design and construction methodology should have coped with this.

8.5.2 Major temporary works failure: Nicoll Highway collapse, Singapore

A major failure of temporary works occurred during the construction of part of the Circle Line of the MRT, Singapore on 20 April 2004. Four persons were killed (Magnus et al., 2005). The post-failure investigations were presented to a meeting of the British Geotechnical Association in (Hight, 2009).

At the time of collapse, excavation was taking place of a 34 m deep excavation for a cut and cover tunnel between two diaphragm walls. As the excavation was lowered, the vertical walls were supported by a system of steel struts, waler beams and kingposts. By the time of the failure there were nine levels of struts and some of these were instrumented with load cells and strain gauges. About 6 hours before the final failure some of the struts began to lose load rapidly whilst others took on more load. This has been interpreted as brittle failure (rapid and unrecoverable loss of strength) of some of the strutting. A detail of the strut beams/waling connection was considered a major factor in the collapse (Ng, 2021).

The excavation was in an area of reclaimed land but most of the excavation was through very soft and soft to firm predominantly clays of the Kallang Formation, which is an extensive and well-investigated stratum in Singapore (Bird et al., 2003). At the meeting, it was argued by Hight (op cit) that the failure also related to the fact that the temporary works were designed to a FoS of 1.2 yet contained brittle elements including the steel strutting/waling connection and a concrete strut formed at depth using jet piling. It was suggested that the trigger for the failure was due to ground conditions deviating from the design assumptions in that the undrained strength profile was lower than assumed and a "complex geology", which involved significant variations in stratigraphy between the diaphragm walls. The comment regarding undrained strengths actually does not relate to incorrectly measured or anticipated strengths but an error in the way that strength was dealt with in numerical modelling

(Magnus et al., op cit). The comment on complex geology apparently relates to a local deepening of the Kallang Clay but such occurrences are commonplace in Singapore (Bird et al., op cit).

In this case the more significant of the two design errors was in the design of the structural connection between struts and waling beams that was under-designed. As reported from the BGA Meeting, "the waling detail that yielded and underwent brittle failure had a direct load capacity that was only marginally lower than the load that was predicted would be applied to it". Perhaps if a higher overall FoS (say 1.4) had been adopted for the design of the walls it might have compensated for the error in structural design detail but if it had done so it would have been essentially by chance. Blockley (2011) in his paper on engineering safety discusses a Swiss cheese model (Reason, 1990) where the components controlling safety are expressed as a series of layers such as management, design, ground model, each with defects (holes) in them including unsafe acts. These are dynamic – some of the holes move around during construction or generally with time. Failure occurs where the holes line up. The Nicoll Highway collapse is an example where there were several such layers with defects and it just happens that perhaps three out of five or so lined up at this particular location and these were enough to initiate failure. The fact that others lined up elsewhere on site (including differing ground conditions and incorrect analysis) contributed to the scale of the failure. The FoS or partial factor approaches of Eurocode 8 cannot be expected routinely to cope with structural design faults nor errors in ground models and analytical mistakes. At the BGA meeting, the site was described as "unforgiving" but this seems outside the sense of the term as used in Chapter 6 where it is taken to describe particularly adverse geological conditions that would be very difficult to anticipate or to investigate using routine approaches.

8.5.3 General failings in ground models

As a general point, when things do go wrong and a detailed examination is made of the ground model assumptions vs. what caused the failure it is often found that the ground model was inadequate or incorrect. Sometimes this can be because the GI was poorly designed or conducted but also because practice was poor. Some of the common errors and poor practices are set out in Table 8.1.

8.6 ENVIRONMENTAL FACTORS

Environmental factors include hydrogeological conditions, *in situ* stresses and earthquake shaking. Such factors should be considered in preparing the ground model for a site. The environmental factors to be accounted for depend largely on the nature, sensitivity and design life of structures and the consequence of failure.

8.6.1 Incorrect hydrogeological ground model and inattention to detail: Landfill site in the UK

A quarry in Gloucestershire, UK was used for disposing domestic refuse. The quarry had been used to extract moderately strong to strong Great Oolite limestone with

Table 8.1 Errors in foreseeing ground conditions

	Common failings	*Mitigating factors/comments*
General	Failure to carry out and interpret desk study	Commonplace. Can prove disastrous to a project – for example, old mine workings for High Speed Railway, Korea.
	Failure to carry out site reconnaissance, examine exposures, make measurements and draw appropriate conclusions. Failure to interpret the landscape – e.g. colour of soil, topography, vegetation, seepage. Failure to anticipate geological associations, e.g. - Link between topography and geological history such as glacial whereby one would anticipate soils and distributions - Presence and angularity of boulders–alluvial, colluvial, weathering, erratic etc. - Metamorphism and hydrothermal associated with igneous plutons - Structural variation in discontinuity networks - Lateral and vertical variation of soil and rock quality, especially in weathered terrain	These are really tasks for a trained and experienced engineering geologist/geomorphologist, after examining desk study data. If delegated to a geotechnical engineer who has not had the training and experience then he may miss crucial clues.
Site-specific investigation data	Inadequate or incompetent investigation	Investigations can almost always be criticised in hindsight when something goes wrong although sometimes the difficulties are so enormous that even the best investigation would have been inadequate.
	Failure to examine samples (designer)	Commonplace; might be constraint from Client trying to limit costs. Sometimes the Project Manager, within the design team, controlling budget and looking to maximise profit, fails to allow the geotechnical team adequate time. It is up to the engineering geologist EG/GE to advise the PM where this is the case.
	Misinterpretation	Commonplace. Over-reliance of designer on interpretation of ground model by ground investigation contractor (incorrectly relying on logs and test results as reliable and representative).
	Ignoring significance of lost samples	There is a poor practice by GI contractors automatically to make assumptions in logs as to what lost core might have been. The designer might not readily appreciate that this has been done.

(*continued*)

Table 8.1 (*Cont.*)

	Common failings	*Mitigating factors/comments*
	Designer fails to see significance of the ground information	Most soil and rock textbooks say very little about geology and training in university civil engineering courses is often shallow. Consequently many engineers have only very simple ideas regarding possible ground models. Rock mass classifications generally do not cope with "minor" geological features that turn out to be of "major" significance.
	Contractor fails to examine samples or site	Tender restrictions might constrain accessibility but if prevented from so doing the Contractor should document this.
	Interpretation of ground conditions incorrect by designer (Reference Ground Conditions inaccurate)	Contractor needs to be aware of this possibility – there will be contractual implications and risks for the tendering process and for carrying out the works.

well-defined bedding planes dipping shallowly at about 4 degrees to the northeast. The rock also had many joints orthogonal to bedding, many of which were open, partly as a result of dissolution and partly due to blasting damage. The limestone overlies Fullers Earth, which is a clay-rich formation. The landfill operatives had been advised that the risk of leachate (liquid-rich in waste products) migration from the quarry was low but it turned out that this was based on an incorrect interpretation of local geology. The landfill was then operated using internal clay bunds and a series of drains and lagoons but, crucially, these were constructed overlying a thin layer of limestone that had been left in place in the floor of the quarry as an operating surface for vehicles both during quarrying and landfill operations. When the quarry was about ½ full of refuse, leachate emerged at a spring in an adjacent valley about a kilometre away. This polluted a stream, which then impacted a fish hatchery downstream as well as water supply. Re-appraisal of the geological model using available, published maps indicated that the leachate was probably passing laterally through the lower stratum of jointed limestone that had been left in the base of the quarry and then channelled down a fault to the spring. Tracer tests were commissioned and these confirmed the link between the quarry source and the polluted spring (Smart, 1985). Various options were considered to improve the situation, including grouting but the preferred remedial measures involved the excavation of landfill along the downstream margin, excavating a trench through the lower limestone and into the underlying Fullers Earth clay and then using geotextile membrane on the inside and an impermeable sheet on the outside wrapped around a drain falling to a sump where leachate could be collected regularly for separate treatment. Upstream of the quarry, a similar membrane and drain system was keyed into the Fullers Earth, collected groundwater throughflow and channelled it around and away from the quarry. As the landfill was completed it was to be capped to prevent direct rainfall ingress. The works led to significant reductions in migration of leachate and improvement in the quality of water emerging from the spring.

It is to be noted that at the time the quarry was operated a "dilute and disperse" approach was generally adopted to land filling in the UK and world-wide whereby it was assumed that the overall migration of leachate would be insignificant as it mixed with a large volume of groundwater. This case illustrates that where the leachate is transported by local channel flow rather than volumetric dispersion through a porous mass, dilution cannot be relied upon. Nowadays in most countries the base and sides of landfills will be sealed using a 0.6 to 1.0 m layer of low permeability soil followed by flexible geomembranes, which again will be covered by a layer of soil to prevent puncturing by traffic or other means. Levels of leachate will also be controlled by internal drainage. While these systems can seldom be completely leak-proof, pollution levels should generally be low.

8.6.2 Corrosive groundwater conditions and failure of ground anchors: Hong Kong and the UK

In the 1970s and early 1980s many ground anchors were installed to stabilise cut slopes in Hong Kong but this practice largely came to a halt with the GCO embargo of 1979 (Rodin, 1979) and failures of anchors in slope such as that shown in Figure 8.17. The problem was recognised following the explosive failure of concrete anchor head covers as the bar anchors broke. Investigation showed that whilst the

Figure 8.17 Anchored slope below Clearwater Bay Road, Hong Kong.

Rusted dynamometer and face plate. This
detail was supposed to be protected by a
polythene cap
Whole face plate
assemblage missing
(fallen off)

Figure 8.18 Deteriorated rock bolt with proving ring. Jeffrey's Mount, above M6 near Tebay, UK.

steel anchor bars were protected from chemical attack along most of their lengths either by grease or by cement grout, close to a steel coupler there was an air space and inadequate protection, which allowed rusting and failure. Similar deterioration has been observed in the UK on the A685, above the M6 near Tebay where rock bolts were installed, some with built in dynamometers to allow loads to be checked periodically (Edwards, 1971). The bolts, with lengths up to 9.15 m, had fixed anchor lengths varying from 0.31 to 0.61 m (epoxy resin). The rest of the length was grouted with either cement or bitumastic following a nominal loading of 50 kN. Several of the bolts have now lost their anchor plates and therefore are unable to carry their design loads due to deterioration of the rock mass and general rusting of the assemblages (Figure 8.18). Designers and owners need to recognise long-term maintenance requirements including regular inspection and testing and the likelihood that anchors will need to be replaced periodically. Edwards (1971) identified the need for corrosion protection and long-term maintenance but this appears not to have been done. Even in more recent projects and despite stringent standards in force for design and construction including corrosion protection, anchors fail. The design of rock cut slopes at Glyn Bends on the A5 in North Wales required 4,500 m of ground anchors stressed to between 400 and 600 kN (Green & Hawkins, 2005). Within 10 years of construction the new section of road had to be closed after two anchor heads fell off and other anchors failed on testing (North Wales Geology Association, 2006). Anchorages also fail due to corrosion in other situations including holding down anchors for concrete dams (Cederstrom et al., 2005). Similar problems can arise with corrosion of steel piles or of steel reinforcement within concrete piles especially in aggressive environments such as at the coast and these issues need to be addressed by the designer.

8.6.3 Explosive gases: Abbeystead, UK

In 1984 a methane gas explosion destroyed a valve house at Abbeystead waterworks in Lancashire. Eight people were killed instantly by the explosion and others died subsequently. The inquiry into the disaster concluded that the methane had seeped from coal deposits 1,000 m below ground and had built up in an empty pipeline. Following a trial and appeal, the designers were held negligent for failing to exercise "reasonable care" in assessing the risk of methane in the finished structure. A review of the various opinions as to which party (Client, Contractor and/or Engineer) should have foreseen the accumulation of methane is given by Abrahamson (1992). Other details are given by Orr et al. (1991).

8.6.4 Resonant damage from earthquakes at great distance, Mexico and Turkey

One of the interesting aspects of earthquakes is the damage that is caused at great distances through resonant affects. One of the classic cases is the earthquake that killed an estimated 10,000 people in Mexico City in September 1985. The earthquake was large (8.1 magnitude) but the epicentre was more than 350 km away from the city so the peak bedrock acceleration was not very high at about 0.04g (as would be anticipated from attenuation equations discussed in Chapter 6). Consequently, in areas of the city built on rock, the earthquake was hardly felt. However, the dominant period of the incoming waves matched the natural period of a basin of lakebed sediments on which part of Mexico City is constructed (about 2.5 seconds) and the ground motion was amplified by the long duration ground motion. That ground motion in turn matched the natural resonant frequencies of some buildings and almost all damage was caused to structures between 6 and 15 stories in height.

In a similar manner, whilst most damage from the M6.8 Erzincan earthquake 1992 occurred close to the epicentre and near to the North Anatolian Fault, anomalous damage occurred at great distances including complete failure of a six-storey reinforced concrete structure at about 40 km, the very top level of a minaret at 80 km (Figure 8.19) and large landslides at more than 100 km distance. This damage almost certainly was caused by long-period wave resonance affects (Hencher & Acar, 1995).

8.6.5 Geological history of a river in Borneo

The Mahakam river in Indonesia is slowly moving its location through meandering and landsliding and this lesson is provided for engineering geologists who believe (wrongly) that the world is not forever changing (see Chapter 5). The river at the site in question is almost a mile wide. One of many mines on its eastern bank, feeds coal to a couple of moorings on piles in the river. Barges pull up several times a day to

Figure 8.19 Damage to top of Minaret at about 80 km from epicentre, Sularbasi.

be loaded with huge amounts of poor-quality coal that then drift downstream to be loaded up to Chinese freighters to take away to their coal-fired power stations. One day, the slope failed massively into the river, destroying one of the moorings and this ended up in arbitration. We investigated and found that inland of the moorings was a clay-filled lake, with pieces of wood at depth that could be dated over 5 thousand years of gradually filling in. The lake had no current source of water supply implying that the paleogeography had changed, which certainly made me stop and think. The river profile changed overnight locally, from hummocky with an underwater hill in the middle, 36 metres high, to a smooth bed, feeding sediment down to the Mahakam delta at the mouth. Further details are given in Hencher (2015) and more generally on the river development by Vermeulen et al. (2014).

8.7 CONSTRUCTION FACTORS

The third verbal equation of Knill and Price (Knill, 1976) addresses the interaction between the geological and environmental conditions at a site and the construction and operational constraints and interactions. The systemic review and investigation of site geology and environmental factors, discussed in Chapter 4, needs to be conducted with specific reference to the project at hand – how construction is to be achieved and the long-term performance of the structure.

8.7.1 Soil grading and its consequence: Piling at Drax Power Station, UK

It is often during construction that things go seriously wrong. An example is provided from the 20,000 piles driven to support the 2nd Phase of Drax Power Station in Yorkshire, UK. The ground conditions appeared to be straightforward with about 18 m of firm, varved silt and clay overlying 2 to 3 m of sand that was typically dense and in turn overlying Triassic sandstone. This profile was proved to be consistent across the site by numerous shell and auger boreholes with many SPT tests in the sand. Piles were therefore expected to come to a halt during driving at depths of about 18 to 19 m. In the event the piles were driven to depths, which varied unexpectedly and unpredictably by up to about 4 m. Figure 8.20 shows some of the piles; the holes in the front are where piles disappeared below carpet level without reaching a satisfactory resistance (set) during driving. Where piles were driven to carpet level without achieving the specified resistance they had to be re-driven contractually. Invariably it was found that on re-driving the piles could not be advanced at all. This was not through design or choice and a costly problem because piles had to be manufactured to cater for longer lengths of penetration to avoid re-drives that were expensive and caused delays to the programme. An investigation was carried out using driven sample tubes into the sand as well as additional cone penetrometer soundings in groups of piles where they came to an early set and where the depth of penetration was much greater

Figure 8.20 Driven piles at Drax Power Station Completion. Dark holes in the foreground are where pile heads have been driven below carpet level without achieving an adequate set and need re-driving.

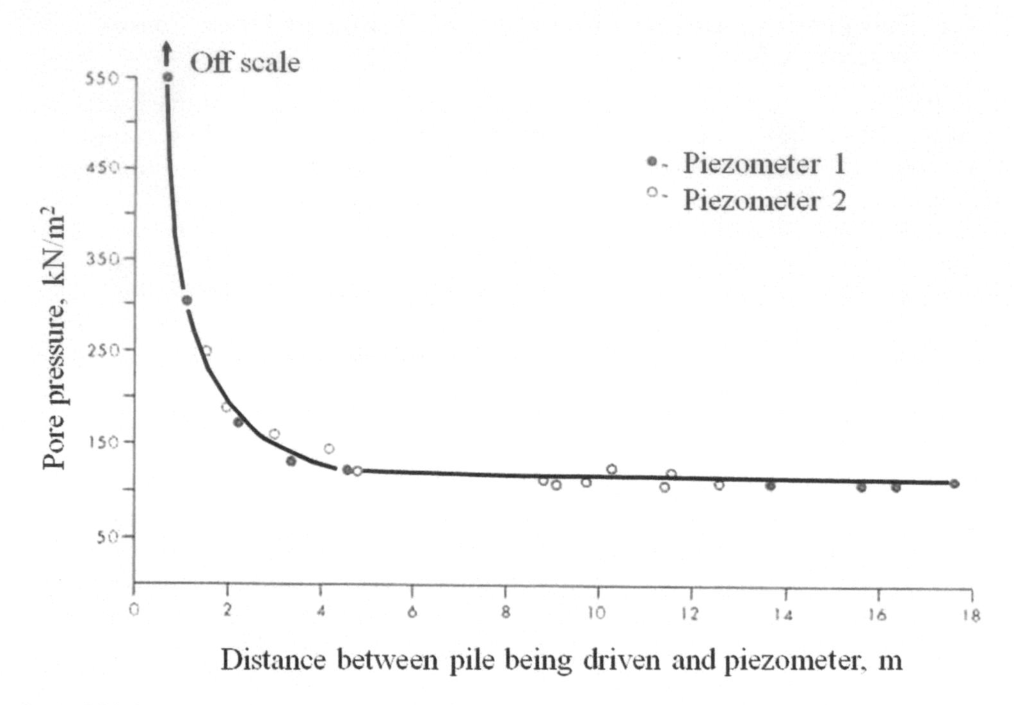

Figure 8.21 Piezometric pressure vs. horizontal distance between pile and instrument.

respectively. In addition, some pneumatic piezometers were installed within the sand horizon and readings of groundwater pressure taken during pile driving as the piling front approached the instrument positions. Data are presented in Figure 8.21. It can be seen that for piles more than 4 m away from the piezometers, there was no influence from the pile driving. However, as the piles approached the piezometer positions, water pressures increased markedly and for the closest piles were off scale at more than twice overburden pressure. Clearly, we were dealing with an effective stress problem. Examination of samples established that an explanation for the different behaviour was in the soil grading. In areas where the piles were driven to refusal at unexpectedly high levels the sand was uniform and clean. Where they penetrated to deeper levels the sand had higher contents of silt and clay and clearly had lower permeability. Evidently the lower permeability allowed higher water pressures to develop, reduced effective stress and greater penetration. If the driving was stopped, the water pressure dissipated and the piles could not be advanced however hard they were hit. This was proved by driving some piles to a target depth next to boreholes, 1.5 m into the "sand" rather than to set or to carpet level. Piles that had not reached set on initial drive were then re-driven (they did not advance) and some were tested without re-driving to 1.5 times working load, which was carried perfectly satisfactorily. In hindsight the original GI had been too simple; the standard clack valve type sampling of the sand horizon typical of UK practice was not adequate for establishing the true nature of the soil. The variability across the site was due to the outwash fan origin of the "sand" with local pools of clay and silt-rich soils in the glacial landscape. These difficulties added perhaps

10% to the cost of pile production (Hencher & Mallard, 1989). The problem had not been anticipated and was related entirely to the method of construction. If bored piles had been used instead of driven piles for the Drax foundations, which would have been feasible, then the problem would not have arisen. What was particularly galling was when a director, who had been involved in the construction of the first Phase of Drax power station, visited the site and said "Oh you have found that again have you?" when informed of what was happening.

8.7.2 Construction of piles in karstic limestone, Wales, UK

Fookes (1997) discusses a number of examples of "unexpected" ground conditions where he believes that the situation could have been anticipated prior to tender and includes several dealing with karstic limestone. One that is described is for a river crossing in Wales and Fookes argues that karstic conditions could readily have been identified from local evidence and boreholes specific to the project. The case is rather more complicated than that. The karst was indeed recognised prior to tender by all parties and as is good practice (Cole, 1988), a ground improvement contract was let to wash out and grout voids in the limestone around pile locations prior to the piling contract. That being so, the conditions anticipated by the piling contractor were as shown in Figure 8.22. In the event, for various reasons the grouting was not comprehensive, in particular because some arbitrary rockhead was adopted at some depth below true rockhead (see Chapter 6), which rather negated the point of the grouting.

During the piling contract, where the rock mass was of good quality (either naturally or due to successful grouting) piles were installed without difficulty. Elsewhere the contractor encountered soil overlying a very irregular rock profile containing voids some of which were open, others infilled with soil and some grouted. Cased holes collapsed and there was loss of concrete from the holes once formed, with a risk of pollution to the adjacent river. The karstic conditions were not unforeseen as suggested by Fookes but the lack of ground improvements was. Of course, other foundation schemes might have been conceived that would have placed less reliance on the pre-piling grouting contract but as things stood there was a valid claim.

8.7.3 Inappropriate excavation method for tunnels: Singapore

A major collapse occurred during the construction of the launch chamber for a TBM tunnel in Singapore. It started as the tunnel was blasted from rock into saprolite, below the water table. The collapse was initially gentle, with the workers trying to shotcrete the upper corner of tunnel, but then accelerated, with the tunnel eventually filled with debris. The problem was that the engineer had had no idea of the ground model and assumed that the weathered rock, below the water table, would be stable after the blast. The true situation was investigated by drilling and taking Mazier samples back to Hong Kong where they were subjected to a novel test whereby the sample was pressurized with water, which was then released suddenly from the base. As in the field, the sample rapidly collapsed under gravity and the imbalanced, internal water pressure as illustrated in Figures 8.23a & b. The collapse led to arbitration.

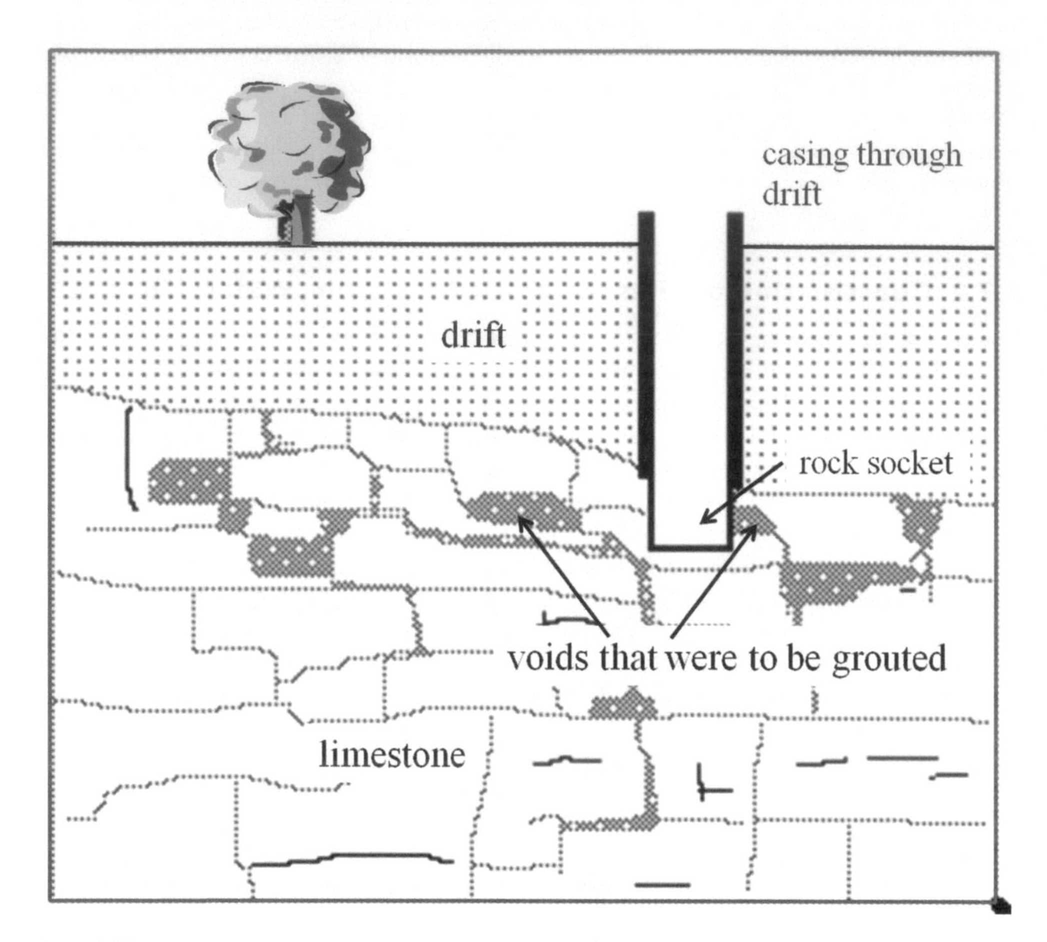

Figure 8.22 Anticipated conditions for piling in karstic limestone, following ground improvement contract.

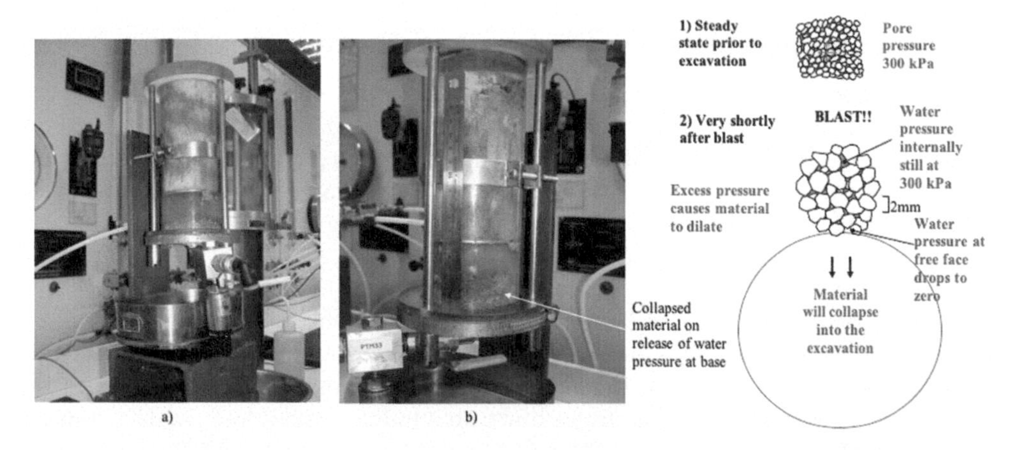

Figure 8.23 Novel experiment to investigate the consequence of sudden release of pressure at the base of saturated Grade V granite by blasting. (a) before; (b) after release of water pressure at base.

8.8 SYSTEMIC FAILING

Sometimes things go wrong because of some fundamental misunderstanding of the proper approach to follow or due to some miscommunication or lack of management of the process. Muir Wood (2000) compares the design of a tunnel to the performance of a symphony. Whereas in the concert hall, all the players are using the same music sheets and are watching a conductor, all too often in engineering projects the control is inadequate and all the players have their own agenda. The result, as Muir Wood puts it, can end up cacophonic. Often when things go wrong it is argued that ground conditions were unexpected but on analysis it is shown that ground characterisation of the site was fatally incorrect or there were failings in management of the project.

Regarding the Heathrow Express tunnel collapse discussed below this was described by the Health and Safety Executive, UK (2000) as an "organisational accident" with a "multiplicity of causes". Other cases below could be described in similar fashion.

8.8.1 Heathrow Express tunnel collapse

One case from which many lessons can be learned is the well-documented collapse of a series of tunnels and chambers in London Clay under construction for the Heathrow Airport Express (HEX) in October 1994. This case is described in detail, both technically and regarding management and contractual issues by Muir Wood (2000) who was appointed as part of the team investigating the collapse by the Health and Safety Executive (HSE, 2000). The monetary cost of the failure was at least £400 million whereas the original contract value was £60 million. The reputation of civil engineering in the UK was also badly damaged overseas.

The project was to be conducted by the New Austrian Tunnelling Method (NATM), which was a novel approach for tunnelling in London Clay although it had been used successfully for trial tunnels at Terminal 4. As described in Chapter 3, the traditional concept of NATM relies on the rock mass "locking up" as joints and interlocking blocks of rock interact and dilate during convergence towards the excavation. The mass often forms a natural arch within the surrounding rock mass as it relaxes and no or little support is needed. Optimising support requirements involves prediction of likely convergence rates, making measurements as excavation proceeds and then applying any support necessary to prevent collapse. Generally, shotcrete is used to prevent loss of loose blocks and wedges, which would destabilise the arch and maybe lead to ravelling failure. Further active support options include the use of bolts, lattice girders and steel arches as appropriate. The principle is that the rock mass carries most of the stress and by waiting until the appropriate time the engineered works are kept to a minimum. After the rock mass has stabilised an inner liner can be constructed, with or without waterproofing.

There is some debate over whether the method adopted for HEX really was or should be termed NATM. Muir Wood prefers the term "informal support" to describe the use of a primary shotcrete liner for the Heathrow Express project; ICE (1996) uses the term Sprayed Concrete Liner (SCL). The common feature between NATM in rock and SCL in weaker ground is the use of shotcrete as a preliminary support but in weaker ground the shotcrete is used to minimise the deformation and settlement of the ground which "effectively reverses the original NATM principle of encouraging controlled ground deformation" (ICE, 1996). It is a subtle but important difference. Even before the

Heathrow tunnel project, considerable question marks had been raised over whether the London Clay (an, extremely weak, fissured rock mass) would behave as a jointed rock mass, appropriate to the use of an NATM approach. In fact the design concept was to use sprayed concrete (with steel mesh and lattice girders) to form a load-bearing closed ring (without any bolts or dowels as might be used for rock) to limit settlement – a quite different concept to NATM in rock. One of the features of the failure at HEX was cracking and perhaps repair works of that damage prior to a complete folding in of the shotcrete liner and this has been blamed as a contributory factor although it was argued, post-collapse by experts that even with part of the invert completely removed, "the tunnel, as designed, should still have stood up for up to 80 days" (Wallis, 1999).

Muir Wood summarises the factors he considers as contributing to the collapse of the Heathrow tunnels. He notes that if any one of these had been addressed competently, in all probability the collapse would not have occurred.

Management: An unfamiliar system of project management based on the New Engineering Contract (NEC) (ICE, 2005) with self-certification by the Contractor of a design by others. Muir Wood advocates a more partnering approach.

Control of Works: It was assumed that the specialist consulting engineers advising the Contractor would bring particular knowledge and expertise to the project but commercial factors limited his presence on site and in reality he had limited power to exercise control as should have been done for a design-led system of construction.

Compensation Grouting: Due to higher than anticipated ground settement, compensation grouting was carried out beneath one of the buildings adjacent to the works. Muir Wood describes this as "in reality grout-jacking, requiring pressures in excess of pre-existing vertical ground stress". The grouting probably loaded the tunnel and contributed to the collapse.

Lack of Reaction to Instrument Data: There is no evidence that there was any reaction by the Contractor and his advisors to the data indicating circumferential movement of the tunnel lining together with depression of the crown. Muir Wood argues that there was no acceptable explanation for this phenomenon other than a weakness of the invert. This should have been clear from distress that had been seen in the invert. The advisors had not established any criteria for acceptable deflections or movement of the lining so there was no quantitative "trigger" to warn that action needed to be taken. Data were presented in figures and diagrams without any commentary or discussion. Nevertheless, according to Muir Wood, the data caused the Client (British Airport Authorities) to question the integrity of the tunnelling, a suggestion that was dismissed by the Contractor.

In addition, an important factor was probably the presence of several parallel openings at the time of the failure, which would have allowed failure mechanisms to develop more readily than in the trial tunnels at Terminal 4 (Karakuş & Fowell, 2004).

Little is said in the various papers dealing with the Heathrow collapse about the geology of the site. According to an expert for the prosecution by the HSE "the London Clay is a well-documented, largely homogeneous, uniform and extensive body of over consolidated sedimentary clay with very few discontinuities and none identified in the area of the collapse that could have caused a landslip..." (Wallis, op cit). This description is somewhat at odds with the observations of Skempton et al. (1969) and others of joints and fissures in London Clay including numerous polished and slickensided ones. The Contractor's expert argued that there must have been "some unforeseeable and completely unpredictable behaviour or geotechnical mechanism in the clay body is the only explanation for the collapse..." (Wallis, op cit) linked to the opinions of other experts that the excavated tunnels should have stood up for up to 80 days. It is noted that in subsequent years several authors have highlighted the importance of low shear strength flexural shears in the London Clay (Chandler, 2000; Hutchinson, 2001), which have led to landsliding including near to Heathrow but no such surfaces were apparently observed in the post-collapse investigations at HEX (Norbury, personal communication; Health and Safety Executive, 2000).

8.8.2 The extensive collapse of a tunnel at Glendoe in Scotland

The collapse of the Glendoe tunnel in Scotland is an interesting one, not least regarding whether to classify it as a rock mass failure or a systemic failure. In the end, I believe that it was systemic, in that the rock mass was all clear and exposed, so cannot be explained as unexpected, through the Conagleann major fault zone, and mapped in quite a lot of detail (Bricker et al., 2021). The complex geology illustrated in Figure 8.24 was essentially ignored as a hazard, and a length estimated to be up to 71 metres eventually collapsed, and the original tunnel was abandoned (Hencher, 2019).

Figure 8.24 Complex geology in extensive fault zone, tunnelled using a TBM, Glen Doe, Scotland.

The tunnel was designed by Jacobs to be constructed by drill and blast, and to be lined 100% for 6.2 km, as would be appropriate for the surging and intermittent forces caused by water flow along a head race tunnel (Brox, 2019). It was well-understood that the tunnel was to be driven through a major fault zone for more than 2 km. One of the tendering contractors however came up with an alternative design, using a TBM to construct the tunnel and this was accepted. They also proposed a "rock mass classification" for designing the concrete liner, blah blah blah.... To cut a long story short, what had been proposed as a drill and blasted, 100% concrete-lined tunnel, ended up as a TBM tunnel, with smooth "rifle bore" sides that was fully-lined with shotcrete for less than 1%. One of the excuses is that drill and blast exposes and emphasises weak areas and discontinuities as overbreak zones and that such weak zones may be less visible in a TBM tunnel wall surface according to the expert advisors. This begs the question of whether cutting rock smoothly reduces the risk of failure compared to a rough drill and blasted one – evidently it does not, in the long-term.

8.8.3 Planning for a major tunnelling system under the sea: SSDS Hong Kong

The Strategic Sewerage Disposal Scheme (SSDS) Stage 1 was an ambitious project to construct a series of shafts and deep tunnels at typical depths of 120 to 145 m leading from a number of catchments around Victoria Harbour in Hong Kong to a central treatment works constructed at Stonecutter's Island. The layout of the scheme is set out in Figure 8.25 and details are given by McLearie et al. (2001).

These tunnels were to be the first bored tunnels under the sea in Hong Kong and the general thinking amongst the design engineers was that if the tunnels were deep enough, in rock that was Grade III or better, then water inflow would be low and ground conditions suitable for excavating with open face rock TBMs. In the event, most boreholes put down for the project and available at the time of tender did not prove ground conditions to the depths of the tunnel alignment that was eventually adopted. Buckingham (2003) summarises the project as follows:

> When the SSDS project was conceived, site investigation was undertaken along the planned alignment to levels determined by this alignment. This revealed ground conditions at the planned depth to be worse than expected, the tunnels were lowered by several tens of metres below the depth of these boreholes with the idea that they would be below the poor ground. The tender was based upon this assumption that the tunnels would be in better rock with minimal water ingress.

There was a tender competition and the successful contractors elected to use mostly refurbished open face rock tunnel boring machines (TBMs) with diameters of between 3.2 and 4.3 m. These TBMs had very limited ability to grout ahead of the tunnels and a low level of shielding for electrical and mechanical devices, which reflected the general assumption by all parties – the HK Government, the design engineers and the bidding contractors that water inflow would be low. The Contractor was required to accept

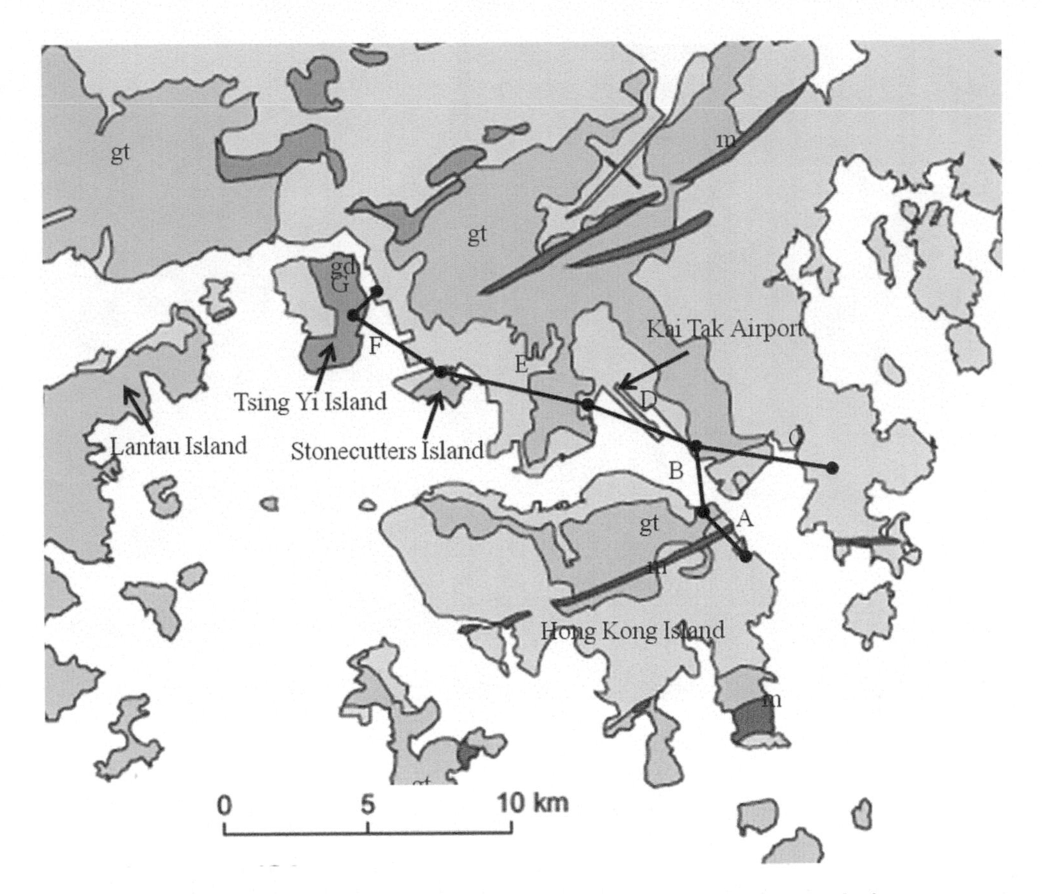

Figure 8.25 Layout of SSDS Stage I tunnels A to G, Hong Kong harbour. Darker shades are granitic rock – g_t = granite, g_d = granodiorite, m = monzonite. Most of the other rocks are mainly of volcanic origin.

all geotechnical risks such that there was no mechanism for additional payment in the event of "unexpected conditions".

In the event, once tunnelling commenced severe water inflow conditions were encountered in tunnels F and C especially. In tunnels A/B there was almost zero water inflow, due to the massive nature of the rock. Fairly onerous conditions had been set for the original contract including a limit for groundwater inflow of 200 l/min/1,000 m length of tunnel. This compares to a final inflow rate of 5,300 litres/minute for the completed tunnel F, which equates to 1,400 l/min/1,000 m (McLearie et al., op cit) despite extensive ground treatment and re-letting of the contract as discussed below.

The original contractor had great difficulty in grouting ahead of the tunnels because the machines and ancillary drilling rigs were not designed for such poor conditions. The original specification was for 120 degree drilling capability rather than 360 degrees, which is more appropriate for stability improvement than preventing water ingress. In tunnel F, measured water pressures – to be overcome by grouting – was

Figure 8.26 View along tunnel F, 1995. Note flow weir in invert.

up to 14 bars. Post-excavation grouting behind the TBM proved ineffective in meeting the requirements of the contract. When water was stopped at one location it re-appeared somewhere else. Figure 8.26 shows water conditions in tunnel F at one stage.

The original contractor halted work in June 1996 and took advice on the likely time to complete the tunnels from specialist consultants brought in to assess the situation. Following negotiations between the contractor and the Government the Contractor was removed from the SSDS project in December 1996 and this was followed by arbitration.

In 1997 the project was re-let as three separate contracts, which were completed by November 2000. The contractual arrangements are unknown in terms of acceptance of ground risks but what is known is that considerable difficulties were encountered including tunnelling through major fault zones (as had been predicted by the specialists advising the original Contractor at the time that he stopped work). In tunnel F the Tolo Channel Fault Zone comprised 268 m of mostly poor-quality rock, which required up to 20 grout holes of typically 54 m length around the full perimeter with grout quantities averaging 5,000 kg/m of tunnel through most of the fault zone (McLearie et al., op cit). Some details of the additional works and additional costs resulting are given in Legislative Council Panel on Environmental Affairs, Hong Kong (2000).

Other problems associated with the re-let tunnelling contracts included settlement in an area of recent reclamation about 1 km away from tunnel C. Inflows into tunnel C from a discrete highly fractured fault zone peaked at 10,400 l/min (compared to a permitted limit of about 1,000 l/min for the full length of the 5.3 km tunnel C) and this led to considerable drawdown, settlement and damage to several housing developments (Kwong, 2005; Maunsell, 2000).

The final breakthrough of the tunnels was 4 years late and US$ 200 million over budget (Wallis, 2000).

Buckingham (op cit) states "Inadequate thought and planning during the site investigation stage lead to poor equipment and method selection. Also, the fact that the contractor was open to all ground condition risks, which eventually led him to pull off the job resulting in lengthy and expensive arbitration, created additional problems".

It was argued at the time of the failed Stage 1 works, in defence of the Govt.'s case that the investigation for the SSDS tunnels was the most extensive seen in Hong Kong (albeit that most boreholes did not reach eventual tunnel depth, failed to sample or intersect the major faults along the route and provided almost no data on permeability conditions). That argument resonates in some ways with a quote regarding nuclear works (Nirex application see below) highlighted by Green & Western (1994):

> If a problem is too difficult to solve, one cannot claim that it is solved by pointing to all the efforts made to solve it.
>
> *Bulletin of Atomic Scientists (1976)*

8.8.4 Inadequate investigations and mismanagement: The application for a rock research laboratory, Sellafield, UK

An application by Nirex to construct an underground rock research laboratory at Sellafield to investigate the site's suitability for disposal of nuclear waste was rejected following a public inquiry in 1996. This essentially brought a halt to investigations in a dramatic way. Considering the considerable cost of investigations up to that time and the consequences for the nuclear industry and Britain's energy policies, this can be considered a major failed project. Moreover, the failure was basically a matter of ground modelling and interpretation.

Nirex in the 1990s were given the task of developing a site for the disposal of waste in the UK. The Government set strict safety guidelines that would need to be met for a final application for a repository. Following a high-quality ground investigation that was "the most expensive 'single' geological investigation carried out in Britain other than the North Sea oil projects" (Oldroyd, 2002), Nirex decided that part of their studies should include an underground rock research laboratory. Other parties were not convinced that Nirex were ready for this stage because of doubts regarding

1. the general suitability of the site at Sellafield;
2. the capability of the Nirex team to control the necessary science;
3. an opinion that the base line hydrogeological conditions had not yet been established and that these would be disturbed irrevocably by the construction of a laboratory. The base line conditions would be a crucial part of establishing a safety case in the future.

It was argued that there was a major risk that construction of the underground laboratory would itself damage the site irretrievably. A research laboratory should be constructed elsewhere to investigate and refine the science even if the final disposal site were to be at Sellafield. Furthermore, various parties were suspicious that the Rock Research Laboratory would be a "Trojan Horse". Once considerable money had been invested in a rock research laboratory, it would be very difficult for the UK Government to argue that Sellafield was fundamentally unsuitable as a repository site. Many of the arguments put forward at the Public Inquiry as expert evidence are published by Haszeldine & Smythe (1996).

Some authors have interpreted the failure at the Public Inquiry to represent a fundamental ruling on the unsuitability of the Sellafield site. Others are less convinced that that point was established, simply that Nirex were not ready at that time to make the case to proceed with an underground laboratory at the site where waste might be disposed of. Even at the time of the Public Inquiry various people were of the opinion that the fundamental problem was the Government's insistence on "disposal" (such that the site would not need to be monitored or any provision made for retrieval), rather than stored underground whilst a safety case was established and tested, possibly over many years. Warehousing waste underground would reduce many of the risks of storage at the ground surface although it would not be the final solution desired by the politicians. There are arguments for disposal, if it can be achieved, not least to remove a burden from future generations, but perhaps that is simply unrealistic at the moment. It seems highly likely that radioactive waste might be regarded as less dangerous in the future due to advances in medical science and there may be ways of modifying or even using the waste in the future as an energy source. There certainly will be improved methods for investigating the geology and hydrogeological conditions at a site.

A year before the Public Inquiry – which brought the whole process to a halt – Green & Western (1994) wrote on behalf of Friends of the Earth that Government should ensure (amongst other things) the following:

1. Radioactive wastes are held in interim, retrievable and monitorable storage until scientific knowledge has advanced to enable permanent solutions to be adopted.
2. An on-going and comprehensive research programme is initiated for all waste streams into waste conditioning, retrievability and long-term management and the long-term behaviour of radioactive wastes in the environment.

These recommendations still seem eminently sensible and, if that advice had been accepted, then the investigations at Sellafield would have proceeded but with a different focus, programme and at a different location. Oldroyd (op cit) comments "in retrospect, the whole Nirex enterprise appears to have been in too much of a hurry". As a result, rather than an ongoing, well-planned and managed approach to deal with the radioactive waste disposal problem, the UK efforts seem to have come to a halt due to fundamental mismanagement.

The latest state of play is reported by Turner et al. (2023). It is still the case that the post-closure period over which the England and Wales independent regulators will require a safety case to demonstrate the long-term containment and isolation capabilities of a geological disposal facility (GDF) is up to 1 million years (OECD, 2009). I would remind the reader that this is the same period which has involved approximately 10 ice ages in Britain, with the sea level rising and falling through more than 120 metres, again and again (as illustrated in Figure 5.17). In that time the ice thickness over the UK is predicted to be in excess of 2 km as it was last time. There will be zero population in Britain during each ice age unless we figure out some drastic new measures for feeding and keeping warm, the population, (beneath the ice). Just imagine the problems that this will cause with millions of people trying to migrate south (opposite of now), let alone the hazards associated with leaking copper drums, filled with toxic waste.

The authors (2023) identify three uncertainties (which are much the same as around the time of the Public Inquiry in 1996):

- *Rock availability – host rock depth, thickness, areal extent and compartmentalization;*
- *Properties and behaviour of the deep environment – required to produce a robustoperational and post-closure safety case; and*
- *Constructability and operability – suitable rock properties that enable constructionand operation of a GDF, including its surface to sub-surface access ways.*

"Quantification of uncertainty generally requires expert judgment. There are well-developed techniques for accessing expert judgment and incorporating it into decision-making procedures, for design and safety case development and strategic project decisions."

These "expert judgements", no doubt will be the same "well-developed" means that were used previously to judge the permeability of the Brockram beds (incorrectly). I would remind the authors of the quote from Baecher and Christian (2003) regarding the value of expert judgement in Box 3.4.

Regarding the problem of groundwater flow and the presence of highly permeably fracture zones, not intercepted by borehole, they state:

> The current UK site selection programme will benefit from the considerable progress that has been made in refining methods for upscaling borehole data to produce discrete fracture network and continuous porous medium models of fractured oil and gas reservoirs (e. g. James, 2007).

Again, I would remind all readers, of the experience from BP modellers of oil and gas reservoirs in North Africa where they admitted in 2012 that their geological models are "geological guesswork" and that, when they find that the model does not work, they simply add another fault to see if that improves matters (Hencher, 2013).

The authors reveal that of the four sites under consideration, three are just north of Sellafield.

Without wishing to inflict the waste problem onto future generations (Tondel & Lindahl, 2019), can I suggest that the heart-felt Government plea to "**dispose**" of the nuclear waste is changed into the design of a "**storage facility**" for nuclear waste to reduce the risk by, say 90%, so that we can rely on long-term, detailed monitoring over possibly hundreds of years, rather than "expert judgement" now.

Can I say further, that the idea of "disposing" of waste underground so that it cannot be dug up in 100,000+ years seems ludicrous. We **cannot** contemplate that length of time in any realistic way. If the waste was a button that when pressed, would destroy the world completely, would we contemplate putting it underground where someone might well access it, say after the next ice age? That seems an argument to stop nuclear power as a source of energy completely, unless we can render it unharmful to the planet (including people). I know that there are other uses (medical, military), but we should be reducing the quantity of waste towards zero and phasing out nuclear power.

8.8.5 Landslide near Busan, Korea

The failure of slopes and the subsequent costs of remedial works are often the result of insufficient geological investigation and inadequate interpretation of ground conditions prior to design. This is compounded by poor investigations into the causes of failures and systemic problems associated with poorly defined responsibilities for the stability of cut-slopes. This was illustrated by the repeated failure of a large slope in Korea (Figure 8.27; Lee & Hencher, 2009). The original ground investigation and design were deficient particularly considering the predictable complexity of the geological conditions. Subsequent investigations were similarly deficient. As a consequence, the slope failed 6 times despite nine re-assessments by various professional engineers and the implementation of several different remedial schemes over a period of 8 years up to a disastrous failure in 2002. During the history of design, failure and re-assessment the height of the cut-slope increased from 45 m to 155 m and the cost increased from 3.3 million to 26 million US dollars.

The investigation, design and management of excavation of this cut-slope can all be strongly criticised. There were many warnings that the slope was not safe and yet opportunities over a 4 year period, to prevent the final failure were not taken.

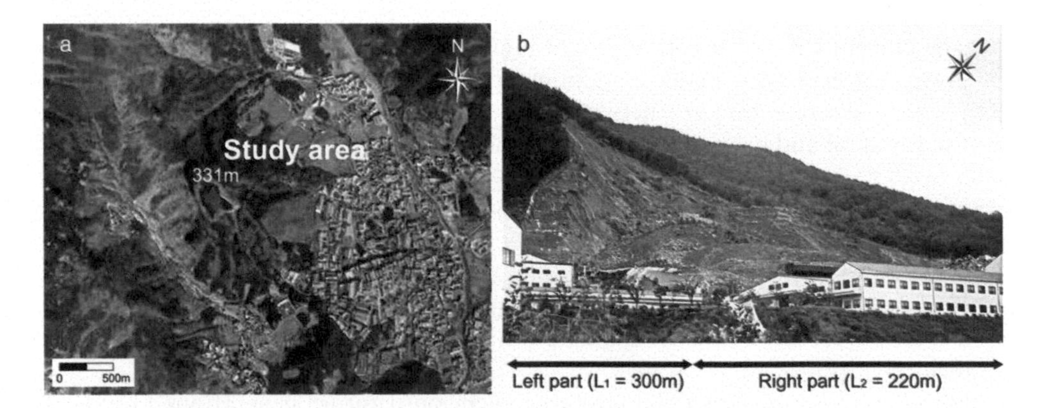

Figure 8.27 Aerial photograph and general landslide from 2009.

Investigations, instrumentation, monitoring and design fell far short of international good practice and failed to meet the then current, Korean (rather poorly specified) guidelines. It appears that all (or nearly all) engineers involved in investigations and reviews, right through to and even after the final collapse, thought that failure could only occur in soil overlying rock and that proving a metre of rock was enough for modelling purposes. There was neither appreciation nor understanding by the engineers that they were dealing with a weathered rock slope with structural control of mass shear strength and hydrogeology. The slope was modelled and analysed as if it comprised layers of cheese, albeit "soil mechanics" cheese overlying stainless steel rock, which could never fail. This certainly indicates a lack of engineering geological thinking and may reflect some fundamental problems with training and the sometimes unhealthy compartmentalisation of geotechnics into rock mechanics and soil mechanics. This is not just a Korean problem. As noted in Chapter 6 in many situations sites are "forgiving" in that an incorrect model does not inevitably lead to disaster. The geotechnical engineer proceeds in normal fashion, ignorant of the real geological conditions, but gets away with it. Unfortunately, this was one of those "unforgiving" sites with adverse geological structure where a superficial and approach is simply not enough.

8.8.6 A series of landlides on Ching Cheung Road, Hong Kong, which occurred several days after heavy rain

One section of Ching Cheung Road in Hong Kong has been the focus for a series of large and unusual landslides since it was first built in 1963–68 (see Figure 6.7).

In the early 1970s two major landslides occurred on Ching Cheung Road. The HK Government engaged Consultants to study the two landslides, partly because they were unusual in that both occurred on dry days. The smaller of the two landslides occurred 4 days after a major rainstorm, the larger 10 days after the rainstorm. The investigation concentrated on a theoretical consideration of the infiltration and storage characteristics of the ground but no explanation could be given for the delayed nature of the failures. The consultant designed the slopes to be re-profiled but during these works it was noted that one of the slopes was issuing water and that movements were occurring. A pragmatic, innovative solution was recommended to install a series of raking drains and this seems to have been effective in stopping the movements. The two slopes subsequently survived effectively for several years without any major event. In 1982 another major, delayed failure occurred on the section of the road away from the area that had been cut back with raking drains installed in the 1980's. As part of the investigation for that failure the early photographs were found and interpreted showing the pre-disposing poor condition of the hillside. The delayed nature of the failure was explained conceptually through a delayed rise in the main groundwater table (Hencher, 1983b; Hudson & Hencher, 1984).

Following the 1982 failures, the section of Ching Cheung Road was selected for investigation and upgrading as necessary under the Landslide Preventive Measures Programme (LPM) programme. Several boreholes were put down, analysis conducted

Figure 8.28 The setting for the 1997 failure at Ching Cheung Road with conditions for short-term shallow failure and later delayed failure of the whole slope, with development of pipes.

and a programme of cutting back and vegetating of the slopes instigated. In so doing, all the previous channels leading from the series of raking drains, installed after the 1982 failures, were stripped away as part of the change from hard covering to a vegetated finish. In essence, the drains and their function had been forgotten about. In 1993, another failure occurred, shallow, and by chance, I happened to be flying into Kai Tak airport in 1995 and photographed the failure, shaped rather like a Playboy bunny, with two ears (Figure 8.28).

In 1997 another large failure occurred essentially at the same location as one shown in the 1940's photographs and the 1993 failures.

The geology of the landslide is illustrated in Figure 8.29. It can be seen that the surface material was Grade V granite, completely decomposed but with original fabric remaining, and samples slaked (fell apart on submersion in water)

Photographs of the failure show raking drains, hanging out of the slope as shown in Figure 8.30. These drains were the same ones that were installed in 1972 to stabilise the slope. (Halcrow Asia Partnership, 1998). Post-failure investigations showed evidence of extensive natural pipes that were almost certainly associated with previous movements in the slope. Trial pits similarly showed evidence of previous movement with staining along the truncated pieces of granite (see example trial pit log in Appendix C at C6). Details are given in Hencher (2006) and the case is illustrated in Figure 8.31.

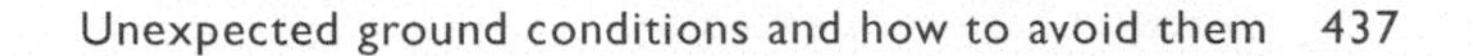

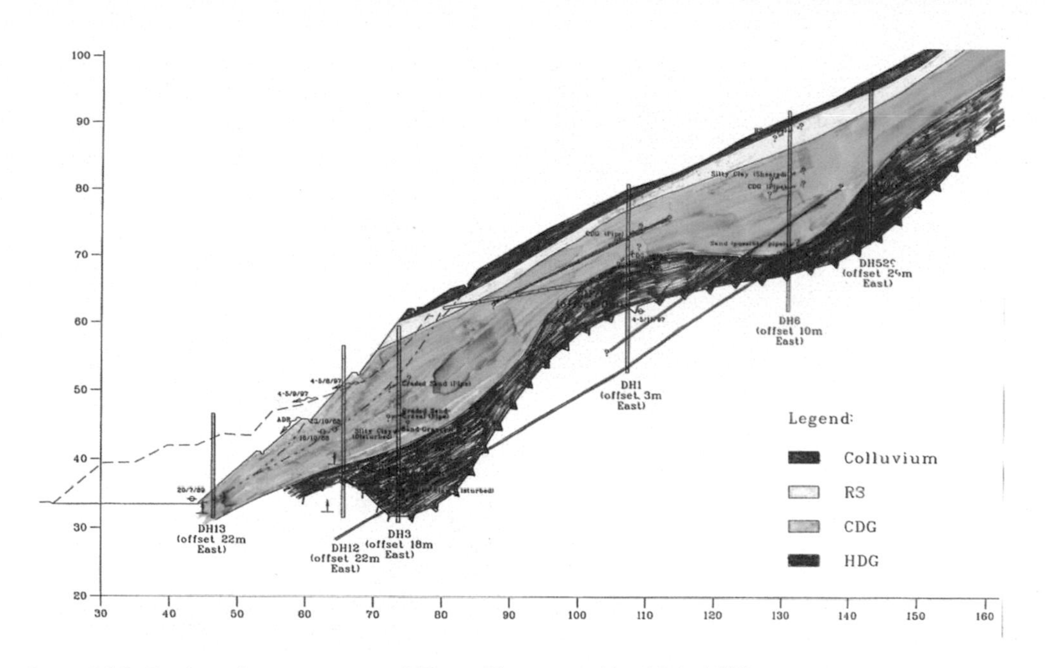

Figure 8.29 Geological cross-section of Ching Cheung road landslide 1997.

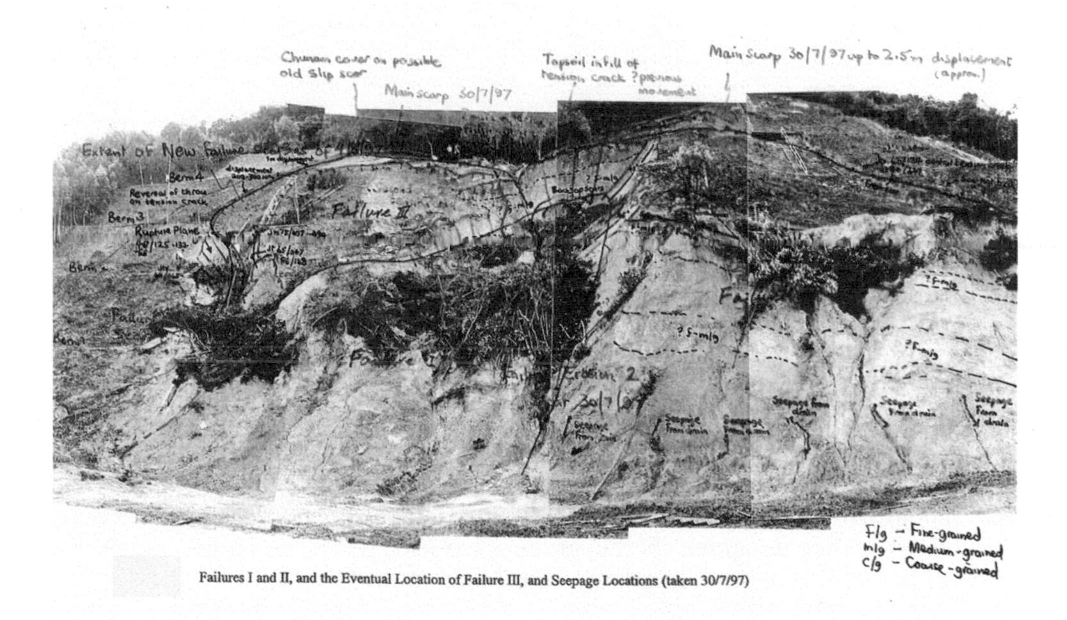

Figure 8.30 View of slope on 30th July 1997 showing raking drains hanging from the face.

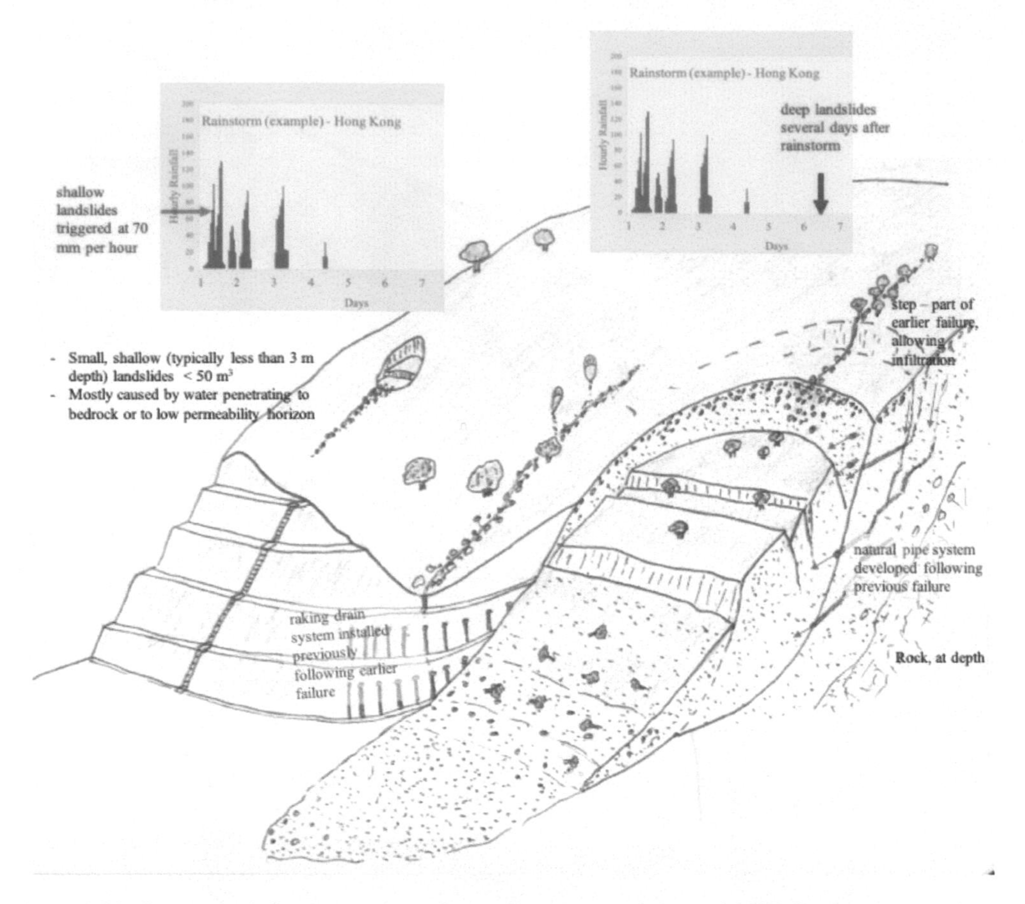

Figure 8.31 Generalized sketch based on Ching Cheung road failure of 1997. Shallow landslides are seen at the surface during heavy rainfall (typically 100 mm/hr). The Ching Cheung road failure was deep-seated with pipes contributing water from depths of about 40 metres and occurred on a dry day almost a week following the major storm.

There are two systemic failings illustrated by this example. Firstly, all of the subsequent failures might have been avoided by proper site investigation prior to construction of the road as advocated in Chapter 6. The photographs illustrating the pre-disposing factors were obviously available in Government but were not consulted perhaps because API was not routinely used for engineering projects in the 1960s. The second failing was a management error in disregarding the previously successful engineering works when designing the "upgrading works", which probably made things worse rather than better and may have contributed to the 1997 landslide.

8.9 FRAUD

Fraud causes failures either due to the incompetence of a contractor unable to complete the task, or just due to money. There are several cases that I know about, from Hong Kong, where I acted as Expert Witness in prosecutions. In one of these, everybody seemed to have been up to their ears in it, taking money – from the structural engineer, who falsified his test pile records to try to make the piles "fit" (that had already been driven) – to the site agent who signed off as delivered, segments of piles, that were not actually in the ground (Hencher et al., 2005).

The short piles scandal shows that people can be pretty devious – I remember one case in Central district, Hong Kong where the designers, Arup became suspicious about pile length and ordered the contractor to core out the piles. This he did over the weekend when there were no supervisors around, drilling again and again into the concrete then arranging the core length-wise into a core box!

The second case concerns piles that were simply not constructed properly. The case came to light when it was found that lifts could not be installed because the buildings were not vertical (Figure 8.32).

Daido piles as used at the site, are manufactured in lengths of up to 2 metres in a factory then transported to the site. At the site they are driven into the ground and then a new length welded on top (Figure 8.30).

Investigations were carried out and it was found that pile segments had either not been installed at all (Figure 8.31a) or had not been connected properly (Figure 8.31b).

Figure 8.32 Tower blocks at Tin Shui Wai, Hong Kong that were found to be non-vertical when the lifts were installed.

Figure 8.33 Daido pile segments being welded together before being driven into the ground (Shenzhen, China).

Figure 8.34a Foundation conditions found beneath tower block. (a) Daido pile segment installed beneath foundation with nothing beneath it. Clear fraud. (b) Shows segment of Daido pile, above cut section of pile (unwelded). Clear fraud.

Figure 8.34b (continued)

Appendix A

Training, institutions and societies

A.1 TRAINING

Thanks are due to professional colleagues world-wide who kindly found the time to advise on situations re training and professional matters. I have left this section essentially unchanged from the 1st edition, because I think that junior geologists, starting out on their careers, probably need some sort of guidance, whatever the gaps and errors in coverage. Again, apologies for non-inclusivity – things vary continually from country to country, as do "standards" and "codes" as in "Eurocodes" in Europe (including the UK perhaps)[1] and "Geoguides" issued by GEO and "Foundation Guidelines" issued by Buildings Department, in Hong Kong – there is no way that I can keep up with it all..., however, I can and do comment on some of it, especially the incorrect bits, as in Appendix B and elsewhere in the book.

A.1.1 United Kingdom

Most engineering geologists from the UK begin their careers with an undergraduate degree in geology or some closely related subject and there is no real short cut for gaining an adequate understanding of geological processes. Portsmouth University has for many years offered a practical undergraduate course in engineering geology, which provides graduates with a suitable entry level of training for the profession but inevitably some of the basic geology teaching has to be omitted in such a course. Furthermore there is considerably more fundamental geology to learn even for graduates whose undergraduate courses offer a much high modular content in geology.

Generally the best way to gain the next level of knowledge is through formal training via an MSc course but even after that it will take some years of experience before the engineering geologist becomes a person that can contribute fully to a geotechnical team. Geotechnical engineers would normally follow a similar career path, starting with a degree in civil engineering and then taking a specialist MSc course. Many MSc engineering geology courses will fulfil the need for specialist training for both geologists and civil engineers – but the career route for the MSc graduate will really follow from his initial training as a geologist or engineer. In the case of a geologist graduating with an MSc in Engineering Geology, he will still know that he is out of his depth if asked in his first year in employment to check the structural design of a strutted excavation. Civil engineers who have proceeded to take an MSc in

Engineering Geology are similarly unlikely to have gained enough knowledge of geological processes and relationships to identify realistic geological models other than for simple situations. It is important that even an MSc-qualified individual follows a period of training and this is often arranged in a formal and structured manner by large consulting or contracting companies or by Government Departments. In the UK and several other countries such as Hong Kong, the aim of the individual is to become either a Chartered Geologist (CGeol) through accreditation with the Geological Society of London or to become a member of one of the Institutions of Engineering and a Chartered Engineer (CEng). A chartered status indicates that the individual has gained adequate experience in various facets of the profession. As stated earlier, many engineering geologists have careers in civil engineering and become skilled in geotechnical input to the design of structures and should then aspire to becoming members of engineering institutions and to become chartered engineers. That can only be achieved through further study, possibly formal exams in engineering subjects and/or extensive proven experience. Further and continuing study (self taught, reading technical and scientific journals and attending lectures and seminars) is a formal requirement of membership of most institutions. Some of the career routes are set out in Chapter 1, Box 1.2 and details of Institutions and Learned Societies, what they do and offer, and routes for membership are presented later.

A UK Register of Ground Engineers (the term has been specifically chosen in preference to "geotechnical engineers") has been established and includes engineering geologists. The scheme was drawn up by the ICE, Geological Society (Engineering Geology Group) and IoM3 and administered by the ICE. The scheme is open to Chartered members from the three professional bodies. Applicants are required to demonstrate their competence on six specific topics (innovation, technical solutions, integration, risk management, sustainability and management) and there are three levels of Registrant – Professional, Specialist and Adviser. As at 2011, the scheme does not have any particular legal status. The emphasis is for applicants to demonstrate "competence" in the various areas.

A.1.2 Mainland Europe

In continental Europe higher education is moving towards the UK model. Through the Bologna Declaration most European countries (including the UK) agreed that university systems and degrees should be the same with the same standards within the so-called "European Higher Education Area". The background of the declaration is that it facilitates the employment and study of European citizens in any European country, which is one of the main goals of the European Union.

Not every country has implemented the Declaration to the same level however. In The Netherlands, the first degree for an engineering geologist will now generally be a BSc in Civil and Geo-Engineering, Geology, Geography or indeed any subject that has some relevance to geotechnical engineering or engineering geology. Depending on any deficiencies in the first degree, additional subjects may be required to be studied before or during an MSc degree.

The German university system has changed to a system comparable to The Netherlands and the UK with introduction of a split system of first and second degrees so the situation for engineering geology education is now broadly the same as in the UK and The Netherlands. Other EU countries are at various stages in implementing the Bologna Declaration.

The European Federation of Geologists has adopted a system of multi-lateral recognition between affiliated national associations, which is incorporated in the professional title European Geologist (EurGeol). As with CGeol status in the UK, the title EurGeol is open to all geologists whether they work in government, academia or industry and therefore gives no indication of competence in engineering geology.

A.1.3 United States of America

As in the UK, generally, engineering geologists from the United States start off with an undergraduate degree in geology or geological engineering. Typically, most engineering geologists are initially trained and educated in geology, primarily obtaining undergraduate degrees in geology. A number of universities also offer Bachelor of Science (BS) degrees in geological engineering. These programmes provide the student with a general background in fundamental geology and geophysics, and geological engineering design including such subjects as soil and rock engineering, geological and geophysical exploration, geological hazard evaluation, groundwater hydrology, geographic information systems (GIS), hazardous waste management and environmental science. The undergraduate student choosing this field of study will learn to apply geologic principles to engineering solutions related to design of geotechnical/civil infrastructure such as tunnels, dams, bridges, excavations and waste disposal sites but as with similar undergraduate degree courses in the UK, this is inevitably at some expense regarding the depth of geology learned. The same applies to others aspiring to become engineering geologists whose initial degree contains a relatively small amount of geological training (e.g. physical geographers and even many "earth scientists" whose geological modules may make up perhaps only 30% of the course). Such individuals obviously have other skills and learning that will help them in their careers but they will find that there is still a lot of geological topping up to do as well as all the engineering during their early years in employment (Box 1.1).

In the USA, a minimum of a Master of Science (MS) is now normally required for persons seeking employment in geological or geotechnical engineering fields. For most engineering geologists the MS degree programme includes studies in rock and soil mechanics, geotechnics, groundwater, hydrology, strengths and permeability of soil and rock, and civil engineering design. Once in employment, geological engineers are encouraged to obtain their professional engineering license or registration, particularly if working for a smaller consulting firm or are working for a state agency. Currently, 31 states have Geologist Licensing Boards (California, Florida, Oregon and Washington to name a few). Typically, education requirements consist of graduation from an accredited college or university with a degree in geology, hydrogeology or engineering geology. To be licensed as a geologist, typically one must have at least 5 years (Oregon requires 7 years under direct supervision of a Registered Geologist) of documented and verifiable professional geological practice or, if applying for a specialty such as Engineering Geologist, 3 to 5 years of specialty practice that is acceptable to the review board. In some states, an undergraduate degree and/or each year of graduate study may count as one year or more of experience. Geological research or teaching at the college or university level may be credited year for year if, in the opinion of the board, it is comparable to experience from practice of geology or a specialty. In most states, applicants must also pass a geologist and/or specialty geologist examination.

In the states of Oregon and California, the licensure title is Certified Engineering Geologist, and in Washington State the title is Licensed Engineering Geologist. There are some states that have reciprocity (i.e. California Board for Geologists and Geophysicists and the Washington State Geologist Licensing Board agree to reciprocity). Applicants requesting licence through reciprocity must however have certified proof from the state where they are licensed.

A.1.4 Canada

In Canada, engineering programmes must be accredited by the Canadian Council of Professional Engineers. Accreditation is normally evaluated every 6 years by a visiting team, some of whom evaluate the general programme and university environment. Students graduate with the subject matter that is required for professional registration. Graduates may register as Engineers in Training, but require 4 years of experience, supervised by a professional engineer, before they can become professional engineers in their own right. They have to pass a Professional Practice Exam, which covers engineering law and engineering ethics, before the P.Eng is conferred. The accreditation review is completed by a national organisation (the Canadian Council of Professional Engineers), but the Law and Ethics exams, as well as the evaluation of the applicant's file are administered by the provincial organization (in Ontario, for example, the organisation is called Professional Engineers Ontario).

Some Universities in Canada, which offer earth sciences, have put together a package of courses that is expected to meet the course requirements for registration as Professional Geologists. While Geoscientists Canada publishes national guidelines as to the courses required, these are not binding, and programmes are not directly accredited. Rather, the experience and course review by each provincial organisation is completed for each individual applicant. Three available electives are considered within the Professional Geologist designation: Geology, Geophysics and Environmental Geoscience. Each applicant must pass a Professional Practice and Ethics exam.

In both cases (P.Eng and P.Geol), because registration is completed on a provincial level, the geoscientist or geological engineer must become registered in all of the provinces where they plan to work. Furthermore, in both cases, where an applicant is missing core courses, they must complete subject area exams to demonstrate their technical proficiency in the subject.

In summary, engineering geologists in Canada have a legal requirement to register as professional geoscientists in each province where they wish to work but there is no specific professional qualification for engineering geologists. P.Eng applies specifically to engineers but engineering geologists with long experience in civil engineering can register – i.e. hold both P.Geol and P.Eng qualifications.

A.1.5 China

China's universities provide courses in both geology and engineering geology as 1st degrees. Some level of professional status as a Geologist or Engineering Geologist is provided by membership of the Geological Society of China (GSC). To become a member requires 5 years' experience after graduation or 2 years' experience after completing a postgraduate MSc or MEng. Engineering geologists can become members of the Engineering Geology Committee (EGC) (the China National Group of IAEG), which is a professional committee within the GSC. Engineering geologists' status in the workplace is identified by positional title, a ranking system – junior to senior. This professional ranking is reviewed and entitled by a committee called the "positional title audit panel", organised by local government, or within some big institutes/companies authorised by the government. Salaries and allowances are largely determined by this positional system.

Currently in China there are professional qualifications of Registered Civil Engineer, Registered Geotechnical Engineer, Registered Structural Engineer, Registered Mining Engineer, and so on and it is likely that the role of Registered Geologist will be established and recognised in China soon. Currently some engineering geologists do achieve the status of Registered Geotechnical Engineer by taking professional examinations.

A.1.6 Hong Kong

Hong Kong, which is a Special Administered Region (SAR) of China, has its own Institution of Engineers with an equivalent status to the Institution of Civil Engineers in the UK. Engineering geologists, qualified as Chartered Engineers through one of the UK institutions can become members of HKIE through mutual recognition of societies or may become members through normal routes requiring graduation from a recognised course with adequate engineering input or taking additional professional examinations. This initial step must be followed by a period of additional training and professional experience over a period of typically 4 or 5 years. After at least a year's relevant professional practice in Hong Kong, members of HKIE can apply to become Registered Professional Geotechnical Engineers RPE(G). Many other geologists and engineering geologists in Hong Kong achieve Chartered Geologist (CGeol) status through the Geological Society of London and this requires the individual to follow a prescribed course of training and experience as detailed below. As noted earlier however the qualification CGeol can be achieved by all geologists who are members of the Geological Society of London and does not in itself indicate knowledge or experience of engineering practice.

A.2 INSTITUTIONS

A.2.1 Introduction

There are a number of Professional Institutions in the UK that govern professional practice to some degree by setting out training routes and requirements for their members, thereby setting standards. Membership is conferred and recognised by letters that can be appended to the member's name as in MICE (Member of the Institution of Civil Engineers). Most institutions also act as learned societies, publishing journals, books and offering their support to conferences and meetings. Similar bodies exist in different countries around the works and offer reciprocal recognition of qualifications, allowing a member from one country to practice professionally in another. All the UK institutions are open to foreign members and there are well-established regional groups in different countries (and throughout the UK).

For geologists, in the UK, direct membership is possible to the Institution of Geologists, which has its home within the Geological Society of London. Through the Institution of Geologists one can become a Chartered Geologist (CGeol). Other institutions that the engineering geologist might wish to join because of their importance to the industry are the Institution of Civil Engineers and the Institution of Mining, Metallurgy and Minerals (which is obviously more attuned to mining and the extractive industries). Through both of these institutions a suitably trained and experienced engineering geologist can become a Chartered Engineer (CEng) – a title conferred by the Engineering Council and which applies to many types of engineers – e.g. mechanical, aeronautical or structural.

A.2.2 The Geological Society of London

The title Chartered Geologist (CGeol) is awarded to suitable geological graduates with a period of training and experience on a par with that required of engineers to become Chartered Engineers (CEng). It is open to all geologists however, not only those working in engineering geology.

Candidates must, via a Professional Report, supporting documentation and through a Professional Interview, prove their competence against each of the following criteria:

1. Understanding of the complexities of geology and of geological processes in space and time in relation to their speciality.
2. Critical evaluation of geoscience information to generate predictive models.
3. Effective communication in writing and orally.
4. Competence in the management of Health and Safety and Environmental issues, and in the observance of all other statutory obligations applicable to their discipline or area of work.
5. Clear understanding of the meaning and needs of professionalism, including a clear understanding of the Code of Conduct and commitment to its implementation.
6. Commitment to Continuing Professional Development throughout their professional career.
7. Competence in their area of expertise.

Usually the training period prior to successful application is several years. Large companies and Government organisations run in-house training schemes under a nominated supervisor that encourages the junior geologist gets the range of experience he needs. Further details can be found from the Geological Society web page www.geolsoc.org.uk or by writing to the Geological Society at

The Geological Society
Burlington House
Piccadilly
London

W1J 0BG
UK

A.2.3 The Institution of Civil Engineers (ICE)

The Institution of Civil Engineers is primarily the governing body for practicing, graduate civil engineers but other disciplines including engineering geologists can join if they can demonstrate sufficient engineering in their education, take additional professional exams and/or have a proven track record of experience. The usual route is first degree followed by a period of training to meet a range of achievements, usually to be "signed off" by an Engineering Supervisor within the employing company. After the period of training the candidate will apply and need to demonstrate his competence in a professional interview. There are two main grades that an engineering geologist might aim for:

(a) **Member**
Membership of the Institution of Civil Engineers (MICE) can be awarded to a wide range of engineers practising in the broad area of civil engineering.

(b) **Fellow**
Fellow is the highest grade of membership of the Institution and may be awarded to those engaged in a position of responsibility in the promotion, planning, design, construction, maintenance or management of important engineering work.

Chartered Engineer (CEng) status can be awarded by ICE or other engineering institutions. Applicants must be able to demonstrate their professional engineering experience and managerial skills. They must have practical knowledge and understanding of the engineering principles relevant to the disciplines of the Institute. They must also demonstrate the use of such knowledge to contribute to the design, manufacture, maintenance, testing and safety of components, devices and structures or the control of process plant. Details of the Engineering Council that administers the scheme can be found at www.engc.org.uk/

Full details of the ICE, benefits and membership requirements are given at www.ice.org.uk/Membership

Alternatively write to

The Institution of Civil Engineers
One Great George Street
Westminster
London SW1P 3A
UK

A.2.4 Institution of Materials, Minerals and Mining (IOM^3)

As for ICE there are two main grades that would interest engineering geologists: Professional Member and Fellow. As for ICE, Members and Fellows can become Chartered Engineers (CEng) if suitably qualified. Other grades are for those with appropriate technical qualifications and experience.

The IOM3 can be contacted at www.iom3.org or the address below.

The Institute of Materials, Minerals and Mining
1 Carlton House Terrace,
London, SW1Y 5DB
UK

A.2.5 Other countries

Many other countries have their own Institutions that govern practice, maintain standards and act as learned societies in the same way as UK Institutions; many such as the American Society of Civil Engineers (ASCE) is open to foreign membership. Obviously the way to find these is by searching on the internet.

A.3 LEARNED SOCIETIES

A.3.1 Introduction

Membership of Professional Institutions provides some certification that the member is competent to practice in a particular field and that he follows some Code of Conduct in his professional activities. Engineering geologists might also find benefit from joining various societies, most of which produce journals, newsletters and organise meeting for their members. The societies sometimes take it upon themselves to provide advice on practice although generally this has no legal status and historically different societies have offered different advice on the same subject, which can be somewhat confusing.

The following is a list of those societies that might be of particular interest to an engineering geologist. Several are really UK-focussed but similar groups can be found world-wide and often are very active and good fun organising social events as well as dealing with local practice. In recent years, the IAEG, ISRM and ISSMFE, which are international in nature, have forged better links between themselves with Joint Commissions looking at different aspects of good practice and research, which may go some way to avoiding the overlaps (and blinkered approaches) between the concepts of engineering soil and rock as addressed in Chapter 1.

A.3.2 Geological Society of London

The Geological Society of London (also known as The Geological Society or Geol. Soc.) is a learned society, based in the UK with the declared aim of "investigating the mineral structure of the Earth". It is the oldest national geological society in the world and the largest in Europe with over 9,000 Fellows entitled to the title FGS (Fellow of the Geological Society) – over 2,000 of whom are also Chartered Geologists (CGeol). Membership is open to any geology graduates (greater than 25% earth science subjects) but also to other graduates who are particularly interested in Geology or where they work in a profession where geology is a core subject. FGS is therefore open to most geotechnical engineers who are particularly interested in geology. The

Geol Soc produces the Quarterly Journal of Engineering Geology and Hydrogeology along with many other journals and books dealing with specialist interests.

A.3.3 International Association for Engineering Geologists and the environment

The International Association for Engineering Geology and the environment (IAEG) was founded in 1964.

According to www.iaeg, the aims of the International Association for Engineering Geology and the environment are

- to promote and encourage the advancement of Engineering Geology through technological activities and research;
- to improve teaching and training in Engineering Geology; and
- to collect, evaluate and disseminate the results of engineering geological activities on a worldwide basis.

A journal the Bulletin of Engineering Geology and the Environment is produced and the IAEG runs regular conferences and organises Commissions with the aim of improving practice.

A.3.4 British Geotechnical Association (BGA)

According to its web site (http://bga.city.ac.uk), the British Geotechnical Association (BGA) is the principal association for geotechnical engineers in the United Kingdom. It performs the role of the ICE Ground Board, as well as being the UK member of the International Society for Soil Mechanics & Geotechnical Engineering (ISSMGE) and the International Society for Rock Mechanics (ISRM).

The BGA organises renowned events like the annual Rankine Lecture, as well supporting young engineers and professional and technical initiatives throughout the field of geotechnics. Membership includes a copy of the monthly magazine *Ground Engineering*, which generally has a practical bias with articles on topical subjects and case studies.

A.3.5 Association of Geotechnical and Geoenvironmental Specialists

The Association of Geotechnical and Geoenvironmental Specialists (AGS) is a non-profit making trade association established to improve the profile and quality of geotechnical and geoenvironmental engineering. Information can be found at www.ags.org.uk. The membership comprises UK organisations and individuals having a common interest in the business of site investigation, geotechnics, geo-environmental engineering, engineering geology, geochemistry, hydrogeology, and other related disciplines. The AGS is also active in Hong Kong.

The AGS produces guidelines on what it considers good practice and organises meetings. It has played a particularly important role in defining ways for transfer of geotechnical data electronically.

A.3.6 International Society for Rock Mechanics

As noted above the ISRM and sister society ISSMGE in the UK are taken under the wing of the British Geotechnical Association which is, in turn, provided support by the ICE.

The following is taken from the ISRM webpage www.isrm

The International Society for Rock Mechanics (ISRM) is a non-profit scientific association supported by the fees of the members and grants that do not impair its free action. The Society has some 5,000 + members. The field of Rock Mechanics is taken to include all studies relative to the physical and mechanical behaviour of rocks and rock masses and the applications of this knowledge for the better understanding of geological processes and in the fields of Engineering (ISRM Statutes).

The main objectives and purposes of the Society are

- to encourage international collaboration and exchange of ideas and information between Rock Mechanics practitioners;
- to encourage teaching, research, and advancement of knowledge in Rock Mechanics;
- to promote high standards of professional practice among rock engineers so that civil, mining and petroleum engineering works might be safer, more economic and less disruptive to the environment.

The ISRM holds rock mechanics congresses every 4 years and also sponsors and supports other conferences internationally.

A.3.7 International Society for Soil Mechanics and Geotechnical Engineering

The aim of the ISSMGE is the promotion of international co-operation amongst engineers and scientists for the advancement and dissemination of knowledge in the field of geotechnics, and its engineering and environmental applications (which of course overlaps with other societies).

Like the ISRM and IAEG it produces a regular newsletter and organises Congresses and supports other conferences. Its webpage is at www.issmge.org

Note

1 See Box A.1.

Appendix B

Soil and rock terminology for description and classification for engineering purposes

B.1 INTRODUCTION

When teaching the MSc in Engineering Geology, at Leeds University in the UK, I always started my course on soil and rock description by asking the class of mostly geologists to classify a piece of rock. They thought about this, and came back with a wide variety of names for the fine-grained, grey rock that I had given them. "No" I would say, imperiously, in an all-knowing way, "no-one knows the answer because you cannot see the mineral grains". But then went on to demonstrate how through careful description, you could go some way towards defining the rock – which might be a basalt, a tuff or a sedimentary greywacke, for example.

In this Appendix, I set out standards for soil and rock description, or at least, commonly-used terms and classifications, throughout the world.

My preferred guide to soil and rock description is Geoguide 3 (Geotechnical Control Office, 1988), from Hong Kong, largely because this was based on BS 5930: 1981, which had been well thought through, but with many of the failings of that document improved upon. I am impressed that it has lasted a long time with little need to change. I would estimate that it is about 95% workeable (a good A+ mark). In particular, I like the material scale classification for logging hand-sized samples of core, which is kept separate from that for the rock mass. This scheme has made it into BS5930 (1999), which was also good, then out again in (2015), then in again (2020)[1] Note that Geoguide 3 does not tally with various ISRM and ISMFE standards (particularly in terms of rock strength classification), but provided you know this, then there are few difficulties.

Engineering geologists and geotechnical engineers, working in any country need to comply with local regulations and this will often apply to standards for soil and rock description. This can be difficult and many will feel uncomfortable at using strange and perhaps poorly defined terms. One option sometimes is to avoid use of classifications and to give actual data. We might disagree internationally over what "strong" or "highly weathered" rock is but UCS = 45 MPa is recognisable and, a metre is a metre, everywhere. That said, there are many different testing methods and different capabilities re testing so even "absolute" numbers must be treated with some caution.

B.2 INTRODUCTION AND HISTORY

Standardisation of soil and rock description has been an aspiration of many individuals and working groups over the last 50 years. A brief review is given below.

Much of the soil description and classification terminology has been fairly standardised since Casagrande (1965). Deere (1968) made various proposals re rock strength classification and discontinuity spacing and introduced the Rock Quality Designation (RQD) and much of the terminology he recommended continues to be used in some parts of the world, not least the USA (Hunt, 2005). One of the earliest attempts to provide fairly comprehensive guidance for logging rotary core through rock was the Geological Society Engineering Group Working Party Report, Anon (1970). That report and another on Maps and Plans (Anon, 1972) set many standards that were then followed, particularly in the UK. Unfortunately, conflicting use of terms was already a feature, especially regarding rock strength as illustrated later. Other bodies (ISRM and IAEG in particular), were meantime setting up their own working groups and coming up with sets of terms to describe rock features that were in conflict with those suggested by others. The ISRM publication on Suggested Methods for the Quantitative Description of Discontinuities (International Society for Rock Mechanics, 1978) is a particularly well-illustrated and useful guide, but some of the terminology is unique to itself.

This was followed in the UK by the preparation of fuller guidance on the description and classification of both soil and rock in the BS 5930: 1981 Code of Practice for Site Investigations. Following its publication, a conference was held to review the BS and papers and discussion were published as Geological Society Engineering Geology Special Publication No. 2 (Hawkins, 1986). The Geotechnical Control Office (1988) published Geoguide 3 on Soil and Rock Descriptions, which largely followed British practice but with some distinct differences especially regarding the description and classification of weathered rock, which is of particular importance in Hong Kong.

BS 5930 was revised and republished in 1999 and is a better document. Most recently amendments have been made as part of introduction of the Eurocode and some of the changes are not necessarily improvements as discussed by Hencher (2008). Meanwhile other countries have adopted their own schemes (e.g. USA, Australia, New Zealand and China) and whilst there are common aspects, often the same terms (and certainly the same properties) are re-defined in different ways, which is confusing to say the least. In China there are currently separate standards for site investigation for different industries: transportation, railway, houses and buildings, oil and gas, mining, hydro-electricity, geological survey and for exploration. US practice is illustrated by Hunt (2005) and CALTRANS (2010).

B.3 SYSTEMATIC DESCRIPTION

Systematic description is essential in ground investigation – logging boreholes and exposures. Descriptions should be thorough and unambiguous so that the end user, perhaps in a design office, will know what has been observed on site. The scope and style of routine description of soil and rock for engineering purposes in logging are well established. Examples are presented in Appendix D and serve to illustrate the differences in practice internationally.

B.3.1 Order of description

No preferred order is given in EN ISO 14688 (British Standards Institution, 2002) or 14689 (British Standards Institution, 2003) so, for the UK, the BS 5939 recommendations should be adopted in logging. In essence, the order of description is the same for soil and rock and this is largely to encourage the logger to consider all aspects.

B.3.1.1 Soil

BS 5930:1999 states that soils should be described in the following order:

(A) Mass characteristics

1. density/compactness/field strength;
2. discontinuities;
3. bedding.

(B) Material characteristics

1. colour;
2. composite soil types: particle grading and composition; shape and size;
3. principal soil type (name in capitals, e.g. SAND), based on grading and plasticity;
4. shape.

(C) Stratum name: geological formation, age and type of deposit; classification (optional)

Examples presented in BS 5930 are as follows:

> Firm, closely-fissured, yellowish-brown CLAY (LONDON CLAY FORMATION).
> Loose, brown, sub-angular, fine and medium flint GRAVEL (TERRACE GRAVELS).

It is advised that materials in interstratified beds may be described as follows:

> Thinly interbedded dense yellow fine SAND and soft grey CLAY (ALLUVIUM).

and that any additional information or minor details should be placed at the end of the main description after a full stop, in order to keep the standard main description concise.

B.3.1.2 Rock

For rock the BS recommended order is as follows:

(A) Material characteristics

1. strength;
2. structure;
3. colour;
4. texture;
5. grain size;
6. rock name (in capitals, e.g. GRANITE).

{B) General information

1. additional information and minor constituents;
2. geological formation.

(C) Mass characteristics:

1. state of weathering;
2. discontinuities.

BS 5930: (British Standards Institution, 1999; 2020) should be consulted for all terms and definitions as used in British practice and a detailed commentary is given by Norbury (2010). Other countries have their own terms and practice guidance so the engineering geologist needs to be aware of local usage wherever he is working.

Most of the terms and definitions used in "standards", such as colour and bedding thickness, are not discussed further here; the text instead concentrates on difficulties, discrepancies and practical advice.

B.4 SOIL DESCRIPTION

For "soil", strength (compactness or consistency), colour, particle shape and composition, structure, state of weathering, discontinuities and any additional geological information. Geological origin should be stated if known but simply interpreting this from local published geological maps is not recommended without confirmation that the map is correct. Often geological maps are not sufficiently detailed to be interpreted at a site-specific scale. Soil type or group is named according to commonly accepted conventions regarding distribution and proportions of different materials (clay grade, silt, sand and so on). Clearly for heterogeneous soil masses such as colluvium or many glacial moraine deposits, mass characteristics need to be addressed in some detail – this cannot be done using standard proportional descriptive terminology. Tables B1 and B2 set out common terminology to be adopted for soil description and Table B3 presents common terminology for describing the strength and compactness of detrital sediments. Such terms and index tests may not be appropriate or adequate for describing and classifying "soil"-like materials derived by *in situ* weathering.

Table B1 Basic soil types

Soil type	*Particle size (mm)*		*Identification*
BOULDERS	—	> 200	Only seen complete in pits or exposures.
COBBLES	—	60–200	Often difficult to recover by drilling.
GRAVEL	Coarse Medium Fine	20–60 6–20 2–6	Visible to naked eye; little or no cohesion but where cemented should state so; particle shape and grading can be described.
SAND	Coarse Medium Fine	0.6–2 0.2–0.6 0.06–0.2	Well-graded means wide range of grain sizes. Poorly-graded is the opposite. Can use the terms "uniform" or "gap-graded" as appropriate.
SILT	Coarse Medium Fine	0.02–0.06 0.006–0.2 0.002–0.006	Coarse silt is barely visible to naked eye – easily seen with hand lens; exhibits little plasticity (can't roll into a cohesive sausage) and marked dilatancy (when wet and squeezed, it increases in volume so that water will disappear); slightly granular or silky to the touch. Disintegrates in water; lumps dry quickly; may possess cohesion but can be powdered between fingers. Silt is often detrital quartz (not clay minerals).
CLAY	—	<0.002	Dry lumps can be broken by hand but not powdered between the fingers. Disintegrates in water more slowly than silt; smooth to the touch; exhibits plasticity but no dilatancy; sticks to the fingers and dries slowly; shrinks appreciably on drying, usually showing cracks. These properties more noticeable with increasing plasticity. Quartz of clay size (often glacial rock flour) has very different properties than true clay minerals.
ORGANIC CLAY, SILT OR SAND	—	varies	Contains much organic vegetable matter; often has a noticeable smell and changes different colours because of oxidation (red /yellow) or reducing environment (green/blue/black).
PEAT	—	varies	Predominantly plant remains; usually dark brown or black in colour; often with distinctive smell; low bulk density.

Source: Based on Geotechincal Control Office (1988) but generally compatible with much international practice.

B.5 ROCK DESCRIPTION AND CLASSIFICATION

For rock, key issues are intact rock strength, nature of discontinuities, weathering and rock mass classification.

B.5.1 Strength

Intact strength of rock material is very different from mass strength as addressed in Chapter 6 and usually to a greater degree than for soils albeit that soil also contains joints, shear planes and general fissures and fractures that influence mass behaviour.

Strength can be estimated quite readily in absolute terms (MPa) by simply hitting a piece of rock with a hammer or trying to break it by hand and quite often no further

Table B2 Dealing with composite soil types

Principal soil type	*Terminology sequence*	*Term for secondary constituent*	*% of secondary constituent*
Very coarse (BOULDERS & COBBLES) (>50% of soil > 60 mm)	Secondary constituents (finer material) after principal[2]	With a little	< 5
		With some	5–20
		With much	20–50
Coarse (GRAVEL & SAND) (>65% gravel & sand sizes)	Secondary constituents before principal (excluding cobbles & boulders)	Slightly (silty, clayey or silty/clayey)[3]	< 5
		Silty, clayey or silty/clayey	5–15
		Very (silty, clayey or silty/clayey)	15–35
		AND/OR	
		Slightly (gravelly or sandy)	< 5
		Gravelly or sandy	5–20
		Very (gravelly or sandy)	20–50
Fine (SILTS & CLAYS) (> 35% silt & clay sizes)	Secondary constituents before principal (excluding cobbles & boulders)[4]	Slightly (gravelly or sandy or both)	< 35
		(gravelly or sandy)	35–65

Source: Based on Geotechnical Control Office (1988) but generally compatible with much international practice.

Notes:

Examples: Slightly silty/clayey, sandy GRAVEL. Slightly gravelly, sandy SILT. Very gravelly SAND. Sandy GRAVEL with occasional boulders. BOULDERS with much finer material (silty/clayey, very sandy gravel).

For fine soils, plasticity terms should also be described where possible, viz: "non-plastic" (generally silts), "intermediate plasticity" (lean clays), "high plasticity" (fat clays).

testing will be required (see Box 7.2). Despite this, the geotechnical community has been inconsistent in its terminology for 50 years (Bieniawski, 1984). Some of the issues and definitions used world-wide are addressed in Tables B4 and B5. The engineering geologist must be cautious especially where using an empirical guideline say for allowable bearing pressure or rippability of rock. Refer to the actual UCS in MPa and not the descriptive term (such as "weak") that could mean different things depending on who has logged the sample and in which country.

B.5.2 Joints and Discontinuities

ISRM (1978)
Describes them as follows:

B.5.3 Joint

A break of geological origin in the continuity of a body of rock along which there has been no visible displacement. Joints can be open, filled or healed.

Table B3 Guide to soil strength terminology

Soil type	*Term*	*Identification*
Very Coarse (COBBLES & BOULDERS)	Loose	By inspection of voids and particles packing in the field.
	Dense	
Coarse (SAND & GRAVEL)	Very loose	SPT "N" value 0–4
	Loose	"N" value 4–10; can be excavated with spade, 50 mm peg easily driven.
	Medium dense	SPT 10–30.
	Dense	"N" value 30–50, requires pick for excavation; 50 mm peg hard to drive.
	Very dense	"N" value > 50.
Fine (CLAY& SILT)	Very soft	Undrained shear strength (S_u) < 20 kPa; exudes between fingers when squeezed in hand.
	Soft	S_u 20–40 kPa; moulded by light finger pressure.
	Firm	S_u 40–75 kPa; can be moulded by strong finger pressure
	Stiff	S_u 75–150 kPa; cannot be moulded by fingers; can be indented by thumb.
	Very stiff or hard	S_u > 150 kPa; can be indented by thumbnail.
Organic (ORGANIC CLAY, SILT SAND & PEAT)	Compact	Fibres already compressed together.
	Spongy	Very compressible and open structure.
	Plastic	Can be moulded in hand and smears fingers.

Source: Based on Geotechnical Control Office (1988) but generally compatible with much international practice.

Comments:

The terms "open" and "filled" are clear although many joints are only open or infilled locally but elsewhere incipient. The meaning of infill is sometimes stretched to include zones of weathering. The term "healed" implies a secondary or contemporaneous mineralisation process as described by Ramsay & Huber (1987) and Miller et al (1994) but is a misnomer for incipient joint planes which have bridges of intact rock Figures B.1.1 and B.1.2 (see also incipient joints in Figures 4.30 to 4.32 and Figure 7.20).

B.5.4 Discontinuity

The general term for any mechanical discontinuity in a rock mass *having zero or low tensile strength* (emphasis added). It is the collective term for most types of joints, weak bedding planes, weak schistosity planes, weakness zones and faults.

Comment:

The definition does not allow for somewhat stronger visible traces of rock fabric, which may still play an important role in engineering performance and fluid flow. It is common practice to record all visible traces of discontinuity in outcrop mapping, regardless of strength, a practice that which does not tally with ISRM definition (Figure B.1.3).

BS EN ISO 14689-1:2003 (European standard)

Table B4 Intact rock strength. The left-hand column classification is used in several textbooks and standards for foundation design (e.g. Tomlinson, 2001; BS 8004, 1986). The changes in the amended BS5930 are inconsistent; note in particular the very different definitions of the term weak. Care must be taken when interpreting boreholes to check the governing standard at the time of the works. Other terms are used in different standards in different countries

BS5930:1999 and Geotechnical Control Office (1988) – still valid for description (2022)		*BS EN ISO 14689-1:2003; ISRM (1981); amended BS 5930:2010, 2015*		*BS 5930:2015+A1*	
Term & UCS (MPa)	*Identification*	*Term & UCS (MPa)*	*Identification (amended wording in BS5930:2015)*	*Term & UCS (MPa)*	*Identification*
Extremely weak <0.5	Easily crumbled by hand; indented deeply by thumbnail.	Extremely weak <1.0	Indented by thumbnail	Extremely weak 0.6–1.0	Scratched by thumbnail, breaks in brittle manner, gravel sized lumps can be crushed between finger and thumb.
Very weak 0.5–1.25	Crumbled with difficulty; scratched easily by thumbnail; peeled easily by pocket knife.				Scratched by thumbnail, lumps can be broken by heavy hand pressure, can be peeled easily by pocket knife, hand-held specimen crumbles under firm blows with the point of a geological hammer.
Weak 1.25–5	Broken into pieces by hand; scratched by thumbnail; peeled by pocket knife; deep indentations (to 5 mm) by point of geological pick; hand-held specimen easily broken by single light hammer blow.	Very weak 1–5	Crumbles under firm blows with point of geological hammer, can be peeled with pocketknife	Very weak 1–5	

Moderately weak 5–12.5	Broken with difficulty in two hands; scratched with difficulty by thumbnail; difficult to peel but easily scratched by pocket knife; shallow indentations easily made by point of pick; hand-held specimen usually broken by single light hammer blow.	Weak 5–25	Can be peeled by a pocket knife with difficulty, shallow indentations made by a firm blow of geological hammer	Weak 5–12.5	Thin slabs, corners or edges can be broken off with hand pressure, can be peeled by a pocket knife with difficulty, easily scratched by pocket knife, shallow indentations made in hand-held specimen by firm blow with the point of a geological hammer.
	Scratched by pocket knife; shallow indentations made by firm blow with point of pick; hand-held specimen usually broken by single firm hammer blow.			Moderately weak 12.5–25	Thin slabs, corners or edges can be broken off with heavy hand pressure, can be scratched with difficulty by pocket knife, hand-held specimen can be broken with a single firm blow of a geological hammer.
Moderately strong 12.5–50		Medium Strong 25–50	Cannot be scraped or peeled with a pocket knife, specimen can be fractured with single firm blow of geological hammer.	Medium Strong 25–50	Cannot be scraped with a pocket knife, specimen on a solid surface can be fractured with a single firm blow of a geological hammer.

(*continued*)

Table B4 (Cont.)

BS5930:1999 and Geotechnical Control Office (1988) – still valid for description (2022)		*BS EN ISO 14689-1:2003; ISRM (1981); amended BS 5930:2010, 2015*		*BS 5930:2015+A1*	
Strong 50–100	Firm blows with point of pick cause only superficial surface damage; hand-held specimen requires more than one firm hammer blow to break.	Strong 50–100	Specimen requires more than one blow of geological hammer to fracture it	Strong 50–100	Specimen requires more than one blow of a geological hammer to fracture it.
Very strong 100–200	Many hammer blows required to break specimen.	Very Strong 100–250	Specimen requires many blows of geological hammer to fracture it.	Very Strong 100–250	Specimen requires many blows of a geological hammer to fracture it.
Extremely strong >200	Specimen only chipped by hammer blows	Extremely Strong >250	Specimen can only be chipped with a geological hammer.	Extremely Strong >250	Specimen can only be chipped with a geological hammer.

Table B5 Examples of other "standard" terminologies in use for intact rock strength

New Zealand Geotechnical Society (2005)			*Australian Standard AS1726*	
Strength	*UCS MPa*	*Is (50) (MPa)*	*Strength*	*Is (50) (MPa)*
Extremely weak	<1.0		Extremely Low	Generally N/A
Very weak	1–5	<1	Very Low	<0.1
Weak	5–20		Low	0.1–0.3
Medium strong	20–50	1–2	Medium	0.3–1
Strong	50–100	2–5	High	1–3
Very strong.	100–250	5–10	Very High	3–10
Extremely strong	> 250	>10	Extremely High	>10

Figure B.1.1 Discntinuous joints with flowing water, Chai Wan, Hong Kong.

B.5.5 Discontinuity

Surface, which breaks the rock material continuity within the rock mass and that is open or may become open under the stress applied by the engineering work.

Comment:
This definition requires interpretation by the observer and assumes an awareness of stress levels and engineering application, which may cause difficulties for logging.

ASTM (2008)

Figure B.1.2 Discontinuous joints through granite, northern Seoul, South Korea.

B.5.6 Structural discontinuity

An interruption or abrupt change in a rock's structural properties. Such changes might include strength, stiffness or density, usually occurring across internal surfaces or zones, such as bedding, parting, cracks, joints, faults or cleavage.

Comments:
This particular ASTM (2008) addresses the use of rock mass classifications, including RMR, Q and RQD. The definition of a structural discontinuity encompasses pervasive rock fabric and incipient features that might have high tensile strength and conflicts with the ISRM definitions. Most rock mass classifications assume that all discontinuities are mechanical fractures with zero tensile strength (as in the definition of RQD) so the ASTM definition might be regarded as inconsistent.

Joints, faults and fractures are of major importance to rock engineering and yet are rather poorly defined.

Discontinuities are synonymous with fractures – mechanical breaks which intersect the soil or rock.

Figure B.1.3 Bedding with discontinuous joints, Durdle Door, Dorset, UK.

Joints are breaks in the continuity of a body of rock along which there has been no visible displacement.

Fissures are exactly the same as joints but the term is reserved for soil.

Incipient fractures are natural fractures which retain some tensile strength and so may not be readily apparent on visual inspection.

This largely tallies with ISRM (1978).

BS EN ISO 14689-1:2003 does not say what a joint is, but defines the term discontinuity.

Discontinuity: surface which breaks the rock material continuity within the rock mass and that is open or may become open under the stress applied by the engineering work; the tensile or shear strength across or along the surface is lower than that of the intact rock material.

In other words, whereas a discontinuity as defined by Norbury and ISRM is a mechanical fracture, the European standard broadens this to include incipient fabric (such as cleavage or bedding planes) that might open up under stress. The implication is that a feature might be described as a discontinuity in some logs but not in others.

B.5.7 Consequence

The distinctions might appear trivial but where it matters is when characterising the rock mass. Rock mass classifications discussed later mostly incorporate discontinuity or joint spacing as a fundamental parameter. Rock Quality Designation (RQD) is a measure in boreholes of the proportion of pieces of "sound" core more than 100 mm in length between discontinuities in the sense of mechanical fractures (natural and not drill induced). If one were to include incipient fabric such as intact bedding and schistosity then the whole measure of RQD would be changed and this has knock on effects for rock mass classifications systems such as Q and RMR.

It is best to follow the Norbury/ISRM definition that a discontinuity is a mechanical break (and this should be used in RQD). Other fabrics that retain tensile strength (including bedding and cleavage) should be described as incipient and not counted in RQD. Incipient discontinuities should be logged and characterised as high tensile strength (close to that of intact rock), intermediate or low (readily split). Geologically however incipient fabric such as cleavage, where it has little effect on intact strength would be called "faint" as in faintly defined cleavage.

Geometrical description and measurement of discontinuities is covered in the ISRM guidelines. Extrapolation should not be made from one exposure or sampling point to another without careful consideration of possible geological change, including structural regime and degree of weathering.

B.5.8 Weathering

As discussed in Chapter 4, many rocks are weathered to great depths, especially in tropical and sub-tropical areas of the world. Weathering affects should be described and recorded and may be interpreted directly from changes in colour, discontinuity spacing, infill on joints and intact strength. In some circumstances, weathering classifications are useful to characterise rock at the scale of an intact, uniform sample or at the mass scale. The classification provides a shorthand description, which is often treated synonymously with mass strength.

Unfortunately, there are many different classifications some of which use the same terminology to describe different phenomena and profiles; others describe the same phenomena and profiles using different terms all of which is very confusing (Martin & Hencher, 1986; Hencher, 2008).

1. Material weathering classifications

Having used weathering classifications a great deal in practice, the author is convinced that material classifications such as that used in Hong Kong and presented in Table C6 are the most useful for dealing with rock that weathers from a strong condition progressively to a soil so that thick profiles of saprolite are sometimes found. It has been used for logging many tropically weathered igneous and sedimentary rocks and

Table B6 Material classification: Geoguide 3 (Geotechnical Control Office, 1988). This classification from Geotechnical Control Office (1988) is applicable to uniform samples of weathered igneous and volcanic rocks and other rocks of equivalent strength in fresh state. It is broadly compatible with recommendations of Anon (1995), BS5930:1999 and used in other standards such as CP4 (2003) in Singapore. It has stood the test of time (from Moye, 1955) as a useful tool in logging core and describing thick weathered profiles. The classification should be supplemented by other descriptive terms and it is often helpful to qualify with index tests such as Schmidt hammer readings on exposures (not applicable for core logging)

Decomposition term	*Grade symbol*	*Typical characteristics*
RESIDUAL SOIL	VI	Original rock texture completely destroyed; can be crumbled by hand and finger pressure into constituent grains.
COMPLETELY DECOMPOSED	V	Original rock texture preserved; can be crumbled by hand and finger pressure into constituent grains; easily indented by point of geological pick; slakes in water; completely discoloured compared with fresh rock.
HIGHLY DECOMPOSED	IV	Can be broken by hand into smaller pieces; makes a dull sound when struck by a hammer; not easily indented by point of pick; does not slake in water; completely discoloured compared with fresh rock.
MODERATELY DECOMPOSED	III	Cannot usually be broken by hand; easily broken by a hammer; makes a dull or slight ringing sound when struck by a hammer; completely stained throughout.
SLIGHTLY DECOMPOSED	II	Not broken easily by a hammer; makes a ringing sound when struck by a hammer; fresh rock colours generally retained but stained near joint surfaces.
FRESH ROCK	I	Not broken easily by a hammer; makes a ringing sound when struck by a hammer; no visible signs of decomposition (i.e. no discolouration).

even in temperate climates and is essentially a strength rating. The Australian standard weathering scheme is presented in Table B7 and is essentially the same as that in Table B6 except that different terminology is used and "distinctly weathered" includes a very wide range of strengths (including Grades III and IV from Table B6). In practice, the boundary between Grades III and IV is often taken as distinguishing between rock-like and soil-like behaviour for slope analysis, is readily identifiable and is often used in specifying target depths for piling so the advantage of grouping them together is not obvious, The Australian and Hong Kong example borehole logs presented in Appendix C both illustrate the use of material weathering grades in practice. Larger scale mass exposures such as those with core stone development can be described according to the distribution of the various material classification grades to characterise weathering profile.

2. Mass weathering classifications

Formal mass schemes are commonly prescribed in different standards because they are readily drawn on paper (percentages of soil and rock) but in practice are often difficult to apply, inflexible and not particularly useful. The scheme used in Hong Kong is

Table B7 from Australian Standard AS 1726 (1993). In the Australian Standard a material weathering classification is prescribed (but no mass classification). The various grades are essentially the same definition as in Table C4 (Geoguide 3) but using some different terms. DW includes material ranging from the strength of "fresh" rock to material that falls apart in water

Term	*Symbol*	*Description*	*Comments*
RESIDUAL SOIL	RS	Soil developed on extremely weathered rock; the mass structure and substance fabric are no longer evident; there is a large change in volume but the soil has not been significantly transported.	Same as Geoguide 3
EXTREMELY WEATHERED ROCK	XW	Rock is weathered to such an extent that it has "soil" properties, i.e. it either disintegrates or it can be remoulded in water. However, it retains rock structure.	Same definition as completely weathered rock (Moye, 1955) and Geoguide 3
DISTINCTLY WEATHERED ROCK	DW*	Rock strength usually changed by weathering. The rock may be highly discoloured, usually by iron staining. Porosity may be increased by leaching, or may be decreased due to deposition of weathering products in pores.	
SLIGHTLY WEATHERED ROCK	SW	Rock is slightly discoloured but shows little or no change of strength from fresh rock.	
FRESH ROCK	FR	Rock shows no sign of decomposition or staining.	

* Covers highly weathered and moderately weathered classes commonly used internationally. HW and MW classes may be used if noted in Explanatory Notes.

presented in Table B8. This was prepared following the work of Martin & Hencher (1986) and was largely a response to the perceived failings of the scheme used in the then BS 5930:1981 and similar schemes adopted by the ISRM. The zones have some sense to them and it seems to work reasonably well in practice although conditions are found where the application is simply impossible and/or unhelpful. Describing profiles according to material grade distributions is the preferred method of the author. Elsewhere the ISRM recommendation is still to use a mass scheme by preference – the old scheme of BS 5930:1981 – despite it being impossible to use this in logging (though some people try). The more recent move in this sad story is that Eurocode 7 has reverted to a mass scheme for classification, which is essentially the same as the ISRM but with different zone numbers and even poorer definitions (Table B9). For most countries this is relatively unimportant but where the engineering geologist finds himself dealing with severely weathered rock he will probably find it useful to adopt the approaches used in Hong Kong and in Singapore [refer to Geotechnical Control Office (1988), Anon (1995), BS 5930:1999 and Singapore Standard CP4 (2003)].

Other schemes are used in different countries; many of these have been prepared from agricultural or soil science perspectives and some are discussed by Selby (1993). In Japan and Korea, distinctions are made between hard rock, soft rock and weathered rock. In China, a pragmatic classification is used based on the ability to cut material with a shovel (Table C10).

As discussed at length in Anon (1995) and adopted by BS5930:1999 and Eurocode 7, weathering in limestone needs special consideration as do rocks that weather in a

Table B8 Mass weathering zones (Geotechnical Control Office, 1988). Comment: Largely similar to scheme recommended in Anon (1995) and BS5930:1999. This classification was developed for classifying thick, heterogeneous weathered profiles. The zone boundaries make good sense in terms of engineering significance. Principles are often useful starting points for differentiating engineering units in a ground model

Term	*Zone symbol from BS 5930:2020*	*Zone symbol*	*Typical characteristics*
RESIDUAL SOIL		RS	Residual soil derived from *in situ* weathering; mass structure and material texture/fabric completely destroyed: 100% soil.
PARTIALLY WEATHERED ROCK MASS (PW)		PW 0/30	Less than 30% rock. Soil retains original mass structure and material texture/fabric (i.e. saprolite). Rock content does not affect shear behaviour of mass, but relict discontinuities in soil may do so. Rock content may be significant for investigation and construction.
		PW 30/50	30% to 50% rock. Both rock content and relict discontinuities may affect shear behaviour of mass.
		PW 50/90	50% to 90% rock. Interlocked structure.
		PW 90/100	Greater than 90% rock. Small amount of the material converted to soil along discontinuities.
UNWEATHERED ROCK		UW	100% rock. May show slight discolouration along discontinuities.

relatively uniform way such as chalk and some mudstones. Norbury (2010) discusses these schemes comprehensively.

B.6 ROCK MASS CLASSIFICATIONS FOR ENGINEERING PURPOSES

This review is not comprehensive but considers the most commonly used Rock Mass Classification systems. A particularly useful review, mostly with respect to tunnelling and underground structures is given by Professor Hoek at www.rocscience.com/hoek/corner/3_Rock_mass_classification.pdf

B.6.1 RQD

Rock Quality Designation (RQD) was proposed by Deere (1968) as an index of rock quality by which a modified core recovery percentage is measured by counting only pieces of "sound and hard" core 4 inches (100 mm) or greater in length as a percentage of core run. This measure is usually recorded on core logs and has stood the

Table B9 International Society for Rock Mechanics (1978) vs. Eurocode 7 schemes. See Hencher (2008) for further discussion.
The Eurocode 7 scheme is the 1978 ISRM scheme but with different grade numbers. This type of scheme can be applied as descriptive shorthand in areas where weathering is a relatively minor consideration as illustrated for the greywacke and schists of Southern Portugal (Pinho et al, 2006) and can also be used for more deeply weathered heterogeneous zones. It is difficult however to log core using this scheme and this causes difficulties for many workers. Deere & Deere (1988) redefine "sound and hard" rock for RQD as Fresh or Slightly Weathered (ISRM); if the rock core is, Moderately Weathered, they say that RQD should be annotated RQD*. This overlooks the dilemma for a logger that Grade III (ISRM, 1978) can include up to 50% "soil" by definition

Term	*ISRM (1978)*		*Eurocode 7*	
	Description		*Description*	
FRESH (100% ROCK)	No visible sign of rock material weathering; perhaps slight discolouration on major discontinuity surfaces.	I	No visible sign of rock material weathering; perhaps slight discoloration on major discontinuity surfaces.	0
SLIGHTLY WEATHERED (100% ROCK)	Discolouration indicates weathering of rock material and discontinuity surfaces. All the rock material may be discoloured by weathering and may be somewhat weaker than its fresh condition.	II	Discoloration indicates weathering of rock material and discontinuity surfaces.	1
MODERATELY WEATHERED (>50% ROCK)	Less than half of the rock material is decomposed and/or disintegrated to a soil. Fresh or discoloured rock is present either as a discontinuous framework or as corestones.	III	Less than half of the rock material is decomposed or disintegrated. Fresh or discoloured rock is present either as a continuous framework or as core stones.	2
HIGHLY WEATHERED (<50% ROCK)	More than a half of the rock material is decomposed and/or disintegrated to a soil. Fresh or discoloured rock is present either as a discontinuous framework or as corestones.	IV	More than half of the rock material is decomposed or disintegrated. Fresh or discoloured rock is present either as a discontinuous framework or as core stones.	3
COMPLETELY WEATHERED (100% SOIL)	All rock material is decomposed and/or disintegrated to soil. The original mass structure is still largely intact.	V	All rock material is decomposed and/or disintegrated to soil. The original mass structure is still largely intact.	4
RESIDUAL SOIL (100% SOIL)	All rock material is converted to soil. The mass structure and material fabric are destroyed. There is a large change in volume, but the soil has not been significantly transported.	VI	All rock material is converted to soil. The mass structure and material fabric are destroyed. There is a large change in volume, but the soil has not been significantly transported.	5

test of time, partly because it is simple (like the SPT test for soils and weak rock). Many authors have pointed out inconsistencies (e.g. Pells et al, 2017) and suggested modifications, for example, over core lengths to be considered and have questioned the definition of "sound and hard" but most of these have not been widely adopted. Basically, one adds the lengths of intact core greater than 100 mm in length, along the centre line of core and expresses this as a percentage of core run. If there is 100% recovery and all sticks between natural discontinuities are >100 mm in length the rock has an RQD of 100%. If all sticks of core were 95 mm the RQD would be 0%. It is a crude measure but used in practice and is a key parameter in other rock mass classifications RMR and Q as discussed below. Details of measurement are given in BS 5930:1999 and most other standards including Geoguide 3 (Geotechnical Control Office, 1988). Various authors have used RQD directly to correlate with rock mass parameters such as Young's modulus or to estimate bearing capacity for foundations (Peck et al., 1974). It should be borne in mind that rock with 100 mm spacing of open joints (100% RQD) is in fact severely fractured rock that would be described as closely spaced (60 mm to 200 mm). According to BS 8004 (British Standards Institution, 1986) any rock with joint spacing less than 100 mm would require "tests on rock" as discussed in Chapter 7 to assess allowable bearing pressure.

Palmström (1982) suggests that RQD can be estimated from the number of discontinuities per unit volume based on visible discontinuity trace surface exposures using the following relationship:

$$\text{RQD} = 115\text{–}3.3\, J_v$$

where J_v is the sum of the number of joints per unit length for all discontinuity sets and known as the "volumetric joint count". There is a practical difficulty here in that traces of discontinuities do not necessarily equate with natural breaks in core (the incipient joint problem as discussed in Chapter 4). The engineering geologist must beware of counting all visible traces in defining RQD otherwise the ground will be assigned a much lower quality rating than is really justified. This can have major consequences in assessing the potential of tunnel boring machines or "road headers" to make progress when cutting rock.

B.6.2 More sophisticated rock mass classification schemes

Various rock mass classifications have been developed largely as methods of estimating support requirements for underground excavations linked to case histories although GSI (covered below) is aimed instead at predicting engineering parameters rather than engineering performance and support requirements. As discussed in Chapter 4 these classifications really come into their own whilst tunnelling and where a decision has to be made quickly, on mucking out, as to the level of support that is required to stabilise the rock mass. In a drill and blast operation there is no time for the engineering geologist to take sophisticated measurements and carefully weigh up the potential modes of failure and rock loads whilst the drilling crew (no doubt on a bonus linked to advance rate) wait patiently...

However, not all tunnel engineers are great fans of rock mass classification schemes. Sir Alan Muir Wood (2000) comments "Currently, much time and effort tend to be wasted in assembling prescribed data, often painstakingly acquired at the tunnel face, to enable calculation of a RMC algorithm which is then filed in a geological log book but not applied to serve any further purpose" and "for weak rocks, the contribution from RMC is more limited… based on material… as well as discontinuities. Attempts to base support needs for weak rocks on RMC have been notably unsuccessful". He continues "RMC are inadequate to provide reliable information on failure modes" and "the way ahead must be to identify important parameters for a particular situation and to present these in a multidimensional way".

B.6.3 RMR

The Rock Mass Rating (RMR) of Bieniawski (1976, 1989) has also stood the test of time as a useful classing system despite many question marks over definitions. The five parameters for the RMR system and their ranges of assigned points are as follows:

1. Uniaxial compressive strength of rock material (0–15).
2. Rock Quality Designation (3–20).
3. Spacing of discontinuities (5–20).
4. Condition of discontinuities (0–30).
5. Groundwater conditions (0–15).

Points are assigned for each parameter and then summed to rate the rock as very good to very poor. It is to be noted that degree of fracturing and condition of those fractures covers 70% of the rock mass rating and that there is a high level of double counting between factors 2 & 3. There is also somewhat of a conundrum in that rock with RQD of 90–100% is allocated the full 20 points but rock with joint spacing 60 to 200 mm is only allocated 8 out of 20 points.

Orientation of discontinuities is used to adjust the summed rating according to whether discontinuities are adverse relative to the engineering project.

RMR is used (as is RQD by itself) to correlate with rock mass parameters, including rock mass strength and deformability.

B.6.4 Q SYSTEM

The *Q* system of Barton et al (2002) is commonly used to classify the quality of rock mass – as a predictive tool in estimating tunnel support requirements for a planned tunnel or cavern, for judging whether ground conditions during tunnelling were as expected or not (contractual issues) and for making decisions on temporary and permanent support requirements during tunnelling (Chapter 3). Barton (2000, 2005) also discusses ways of using the *Q* system in predicting TBM performance. *Q* has a value range from 0.001 to 1000.

$$Q \text{ (quality)} = \text{RQD}/J_n \times J_r/J_a \times J_w/\text{SRF}$$

Table B10 Chinese Standard GB50021-2001 (2009 Edition). Classification is essentially a practical mass scheme, which does not deal with weathered rock masses that develop as corestone-rich profiles. Not applicable for core logging in boreholes

Term	*Description (simplified from original)*	*Comment*
RESIDUAL SOIL	Structure fully destroyed. Easily dug with a shovel.	Actually, many residual soils are quite strong, cemented with iron oxides especially, so can be quite difficult to dig.
FULLY WEATHERED	Structure basically destroyed but still recognisable. Residual structural strength existing. May be dug with a shovel.	
HIGHLY WEATHERED	Majority of structural planes destroyed. May be dug with a shovel.	From experience in Hong Kong, highly decomposed (weathered) rock often shows the highest degree of fracturing – they become healed by the completely weathered and residual soil stages (Hencher & Martin, 1982)
MODERATELY WEATHERED	Structure partially destroyed; secondary minerals along joints; weathered fissures developed. Hard to cut with a shovel.	
NON-WEATHERED	Structure basically unchanged.	

where
RQD is Rock Quality Designation
J_n is joint set number
J_r is joint roughness number
J_a is joint alteration number
J_w is joint water reduction factor

and SRF is a Stress Reduction Factor (relating to loosening of rock, rock stress in competent rock and squeezing conditions in incompetent rock).

Ranges and descriptions for each parameter are given in the original publications by Barton (op cit) but also in Hoek & Brown (1980) and at Hoek's Corner referred to earlier.

B.6.5 GSI

The Geological Strength Index (GSI) of Hoek (1999) provides a means of estimating rock mass strength and deformability through broad classification based on rock type and quality of rock mass as illustrated in Table B11. Ways in which the GSI can be used to estimate rock mass parameters are dealt with in Chapter 7.

Table B11

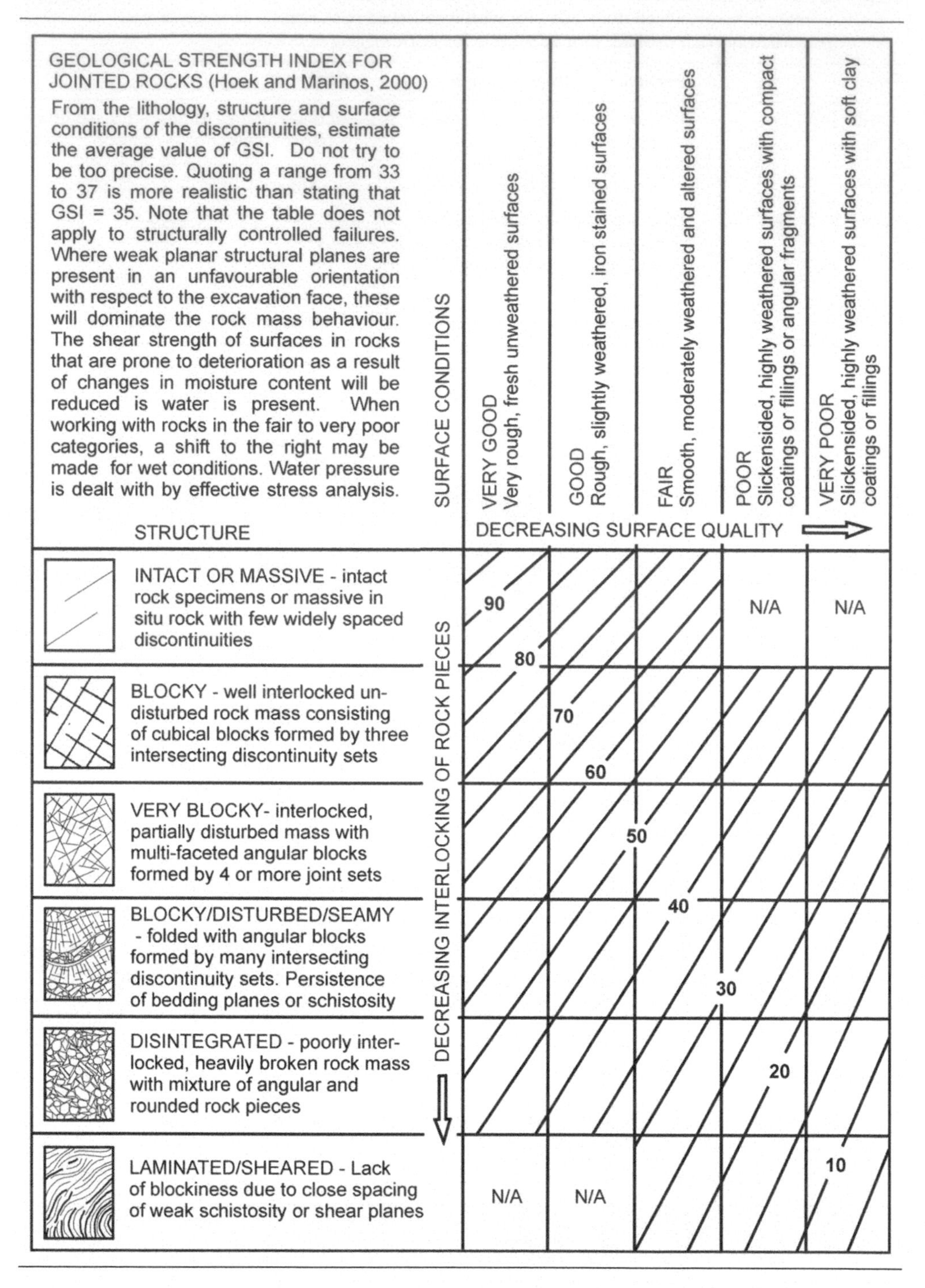
GEOLOGICAL STRENGTH INDEX FOR JOINTED ROCKS (Hoek and Marinos, 2000)
From the lithology, structure and surface conditions of the discontinuities, estimate the average value of GSI. Do not try to be too precise. Quoting a range from 33 to 37 is more realistic than stating that GSI = 35. Note that the table does not apply to structurally controlled failures. Where weak planar structural planes are present in an unfavourable orientation with respect to the excavation face, these will dominate the rock mass behaviour. The shear strength of surfaces in rocks that are prone to deterioration as a result of changes in moisture content will be reduced is water is present. When working with rocks in the fair to very poor categories, a shift to the right may be made for wet conditions. Water pressure is dealt with by effective stress analysis.
SURFACE CONDITIONS
VERY GOOD
Very rough, fresh unweathered surfaces
GOOD
Rough, slightly weathered, iron stained surfaces
FAIR
Smooth, moderately weathered and altered surfaces
POOR
Slickensided, highly weathered surfaces with compact coatings or fillings or angular fragments
VERY POOR
Slickensided, highly weathered surfaces with soft clay coatings or fillings
STRUCTURE
DECREASING SURFACE QUALITY
INTACT OR MASSIVE - intact rock specimens or massive in situ rock with few widely spaced discontinuities
BLOCKY - well interlocked un-disturbed rock mass consisting of cubical blocks formed by three intersecting discontinuity sets
VERY BLOCKY- interlocked, partially disturbed mass with multi-faceted angular blocks formed by 4 or more joint sets
BLOCKY/DISTURBED/SEAMY - folded with angular blocks formed by many intersecting discontinuity sets. Persistence of bedding planes or schistosity
DISINTEGRATED - poorly inter-locked, heavily broken rock mass with mixture of angular and rounded rock pieces
LAMINATED/SHEARED - Lack of blockiness due to close spacing of weak schistosity or shear planes
DECREASING INTERLOCKING OF ROCK PIECES
90
80
70
60
50
40
30
20
10
N/A
N/A
N/A
N/A

B.6.6 Slope classifications

Various authors have devised soil and rock mass classifications as a means of judging slope stability (Romana, 1991; Selby, 1993; Hack, 1998). In the experience of the author these are less commonly applied than are rock mass classifications for tunnels and caverns, perhaps because there is not usually the same degree of urgency to make decisions and need to link observations to empirical experience. When assessing a slope one can usually take some time to examine and investigate the geological conditions in some detail and carry out rational analysis and design – as of course should also be done wherever time allows for underground openings as discussed in Chapter 4. Slope stability assessment classifications can be useful where, for example, carrying out a rapid comparative hazard and risk survey along a highway.

Notes

1. In the 2020 version, there are two typographic errors. For Zone 3 for mixed soils, it should say 30–50% Grades III or better; for Zone 6 it should be 100% soil of Grades IV to VI not just Grade IV...
2. Full name of finer materials should be given (see examples).
3. Secondary soil type as appropriate; use "silty/clayey" when a distinction cannot be made between the two.
4. If cobbles or boulders are also present in coarse fine soil, this can be indicated by using one of the following terms relating to the very coarse fraction after the principal: "with occasional" (< 5), "with some" (5–20), "with many" (20–50), where figures in brackets are % very coarse material expressed as a fraction of the whole soil (see examples below).

Appendix C

Examples of borehole and trial pit logs

In this Appendix, examples are presented of borehole (or "drillhole" used synonymously) and trial pit logs from real examples in the UK, Hong Kong and Australia albeit that locations have been omitted in some examples. They contain a great deal of information not only as records of the ground conditions encountered but also on the way in which the investigation was conducted, the machinery used, the tests conducted and their results and the groundwater conditions. These serve to illustrate the variety of use of descriptive terms and classifications in description as well as typical techniques used in ground investigation.

C.1 CONTRACTOR'S BOREHOLE LOGS

Contractor's logs are carried out following standards and codes in the country where the work is carried out. As discussed in Chapter 4 they tend to provide somewhat simplified descriptions of ground conditions. The Engineer who, unlike the Contractor, has designed the GI, should have carried out a thorough desk study and be aware of the factors that will be crucial to the success of a project. He needs to examine samples himself to ensure that key features at a site have been correctly identified, described and highlighted.

C.1.1 UK example

The borehole log Figure C.1.1 is courtesy of Geotechnics Ltd and describes the materials encountered and methods used, for a hole taken to 20 m depth. Descriptions of the soil and rock are to BS 5930:1999 (and effectively Geoguide 3 from Hong Kong).

All of the following detail (and much more) can be read directly from the log without any report or further explanation.

The first 1.2 m was excavated as an inspection pit. This is usual practice just in case there might be services such as electric cables or pipes not identified on plans or using detecting equipment. Water was encountered at 0.4 m.

A cable percussion rig was then used to advance a hole of diameter 0.15 m. The hole was advanced essentially following the strategy outlined in Figure 6.24. Sample types are *B* (bulk or bag sample of the arisings), *E* (environmental for contamination testing), *W* (water sample) and *D* (small disturbed tub or jar sample, generally taken by a split tube sampler in an SPT test). SPT results (*N* values) are given in a separate column (S5 and so on) and more details of the SPT tests are given in a separate sheet.

BOREHOLE RECORD - Cable Percussion and Rotary

Project SAMPLE PROJECT

Engineer GEOTECHNICS LIMITED

Borehole CP5/RC5

Project No PC060000

Client GEOTECHNICS LIMITED

Local Grid Coordinates 9826.53 E 1424.53 N

Ground Level 30.48 m OD

Scale 1:50

Sampling / Properties

Depth	Sample Type	Depth Cased & (to Water)	Strength kPa	w %	SPT (FI)
0.00- 1.00	B			38	
0.50	E				
1.00	D				
1.00	E				
1.20- 1.70	B				
1.20- 1.65	D	1.00 (0.40)			S5
2.00	D			38	
2.00	E				
2.20- 2.70	B				
2.20- 2.65	D	2.00 (DRY)			S8
3.00	D			45	
3.20- 3.70	B				
3.20- 3.65	D	3.00 (DRY)			S6
4.00	D			63	
4.20- 4.70	B				
4.20- 4.65		4.00 (3.70)			S16
5.00	D				
5.20- 5.70	B				
5.20- 5.65		5.20 (4.65)			C53
6.00	D				
6.20- 6.70	B				
6.20- 6.39		6.20 (4.70)			C50/70
7.00	D				
7.00	W				
7.20- 7.70	B				
7.20- 7.65	D	7.20 (5.70)			S35
8.00- 8.30	D	7.20 (5.70)			S50/ 146

Core Run Core Dia	Depth Cased	TCR/SCR %	Length Max/Min	RQD %	SPT (FI)
8.30- 9.00	8.30	90 28	0.05 0.01	0	
9.00-10.50	8.30	50 22	0.05 0.01	0	(>20)

Strata

Description	Depth	Level m OD
	G.L.	30.48
Very soft dark grey brown gravelly sandy clay with occasional rootlets. Gravel is angular to subrounded fine to coarse brick, mudstone and slag. [MADE GROUND]	1.40	29.08
Very soft brown and dark brown slightly sandy gravelly clay. Gravel is subangular brick and slag. [MADE GROUND]	2.10	28.38
Very soft grey slightly sandy organic CLAY. [ALLUVIUM] Below 3.20m, with occasional plant remains and occasional black speckling.	3.90	26.58
Very soft grey slightly sandy slightly gravelly CLAY. Gravel is subrounded fine to coarse sandstone and mudstone. [ALLUVIUM]	4.20	26.28
Medium dense brown slightly clayey sandy GRAVEL with a low cobble content of sandstone. Gravel is subrounded fine to coarse sandstone. [RIVER TERRACE DEPOSITS] At 5.00m, recovered as very soft slightly sandy very gravelly clay. Below 5.20m, becoming very dense.	7.00	23.48
Very stiff grey mottled brown slightly sandy CLAY with many fine to medium gravel sized mudstone lithorelicts. [COAL MEASURES] Between 7.00-7.20m, recovered as soft.	8.30	22.18

Continued by Rotary techniques

General	Detail
Extremely weak light grey thinly laminated MUDSTONE. Extremely closely to very closely spaced subplanar slightly polished discontinuities with slight clay smear dip 60-70 degrees to core axis. [COAL MEASURES]	Between 8.30-8.45m and 10.06-10.22m, stepped subvertical discontinuity with some orange staining penetrating up to 1mm. Between 9.00-9.75m, assumed zone of core loss.

Boring

Depth	Dia	Technique	Crew
1.20		Inspection Pit	JS/CS
8.30	0.15	Cable Percussion	JS/CS
20.00	0.12	Rotary Core	GC

Progress

Depth of Hole	Depth Cased	Depth to Water	Date	Time
G.L.			03/12/09	08:00
8.30	7.20	5.70	03/12/09	18:00
8.00	7.20	5.50	04/12/09	08:00
16.50	8.00	7.40	04/12/09	18:00
16.50	8.00	1.40	07/12/09	08:00
20.00	8.00	11.90	07/12/09	18:00

Groundwater

Depth Struck	Depth Cased	Rose to	in Mins	Depth Sealed	Remarks on Groundwater
0.40		0.40	20	1.50	
4.20	4.00	3.70	20		Slow.
11.10	8.00	6.40	20		Fast.

Remarks

Inspection pit hand excavated to 1.20m depth.
A 50mm standpipe was installed to 11.00m with a slotted section from 1.00m to 11.00m with flush lockable protective cover. Detail as follows from base of hole: bentonite seal up to 12.00m, gravel filter up to 2.00m, bentonite seal up to 0.20m, concrete up to ground level.
Chiselling: 6.30-6.60m for 60 minutes and 7.50-8.00m for 60 minutes.
Flush: 8.00-20.00m, Water, 80% returns.

Symbols and abbreviations are explained on the accompanying key sheet.

All dimensions are in metres.

Logged in accordance with BS5930:1999

Logged by NG/LJ
Checked by TNH
Figure 1 of 2
17/12/2010

geotechnics

Figure C.1.1

In this example no "undisturbed" samples were taken. If they had been they would have been reported as *U* (open 102 mm tube) or *UT* (thin walled open drive sample) together with a record of the number of blows taken to drive the sample (for general information only as the driving force from the hammer is not usually standardised). The cable percussion driving was continued to a depth of 8.30 metres by which time the investigation had encountered weathered Coal Measures mudstone (described as very stiff slightly sandy CLAY with lithorelicts of the parent rock). An SPT was attempted in the weathered rock but only penetrated 146 mm (rather than 300 mm) for 50 blows of the hammer, which is standard practice in the UK. In the cable percussion section of hole, water was encountered close to the ground surface but then, as casing was installed, the hole proceeded "in the dry" until the base of alluvium at 4.20 m depth. Water rose in the borehole to 3.90 m.

From 8.30 m the hole was advanced using rotary drilling using water flush with a diameter of 0.12 m. Advance rates are given at the bottom of page 1. The first drilling run was only 0.5 m but after that 1.5 m runs were adopted up to 18 m when a 2 m run was carried out. Total core recovery (TCR) was not bad, generally over 80% but from 10.50 to 12.00 m 47% of core was lost (TCR = 53%).[1] Solid Core Recovery (SCR) is defined as percentage of core with full circumference. RQD is percentage of core (on a drilling run basis – NOT rock type) in full sticks of "sound" rock > 100 mm in length. Fracture index (no. per metre) is also recorded. Note that core run lengths are not consistent with geological changes – no reason why they should be unless the driller noted some sudden change in advance rate or loss of flushing fluid perhaps, which might cause him to stop and extract the drill string to investigate the cause. In the Contractor's own style of log, he has chosen to split the description into "general" and "detail" columns which is helpful. Water was encountered in the rock at a depth of 11.10 m and rose to 6.40 m. And when the hole was complete a standpipe piezometer was installed – details of the installation are on the log. Also noted is chiselling time – this is useful information for the designer (perhaps if he is considering using driven piles) but is also a record for payment purposes. As a comment note that there is no attempt to describe weathering state (the strength consequence is clearly recorded so there is no need) – this contrasts with the HK and Australian examples presented later. Initials of the crew who carried out the boring, who logged the materials and prepared the log and who checked the whole report are given.

C.1.2 Hong Kong example

The drillhole record Figure C.1.2 is provided courtesy of Gammon Construction Ltd, HK for a hole taken to 14.78 m depth. Descriptions of soil and rock are to Geotechnical Control Office (1988).

As for the UK example, the hole commenced with an inspection pit to 1.50 m to check for services. After that the hole was advanced by rotary drilling using water flush. PX casing with an outside diameter (OD) of 140 mm was installed to 3.50 m and then HX (OD 114 mm) to 6.50 m at the top of rock head from which depth the hole was uncased. To a depth of 8.50 m, an intermittent sampling strategy was adopted with Mazier samples and SPT tests. To 5.70 m the rock is all described as Grade V, completely decomposed granite. At 5.70 there is a 0.8 m thick basalt stratum,

Gammon Construction Limited
Ground Engineering Department
DRILLHOLE RECORD

HOLE No. **ID**
SHEET 1 of 2
CONTRACT NO. 1234

Project Title

METHOD Rotary	CO-ORDINATES	PROJECT No. 1234
MACHINE & No. 123	E N	DATE from DD/MM/YYYY to DD/MM/YYYY
FLUSHING MEDIUM WATER	ORIENTATION Vertical	GROUND LEVEL mPD

Depth (m)	Grade	Description
0.00 – 1.50		Orangish brown, very silty fine to coarse SAND with some angular to subangular fine to coarse gravel sized rock fragments and rootlets. (COLLUVIUM)
1.50 – 2.60	V	Extremely to very weak, orangish brown, mottled white, completely decomposed GRANITE. (Very sandy angular to subangular medium to coarse GRAVEL sized rock fragments)
2.60 – 3.50	V	Extremely weak, reddish brown, completely decomposed GRANITE. (Very silty fine to medium SAND)
3.50 – 5.70	V	Extremely weak, light brown, completely decomposed GRANITE. (Very silty fine to medium SAND)
5.70 – 6.50	V	Extremely weak, orangish brown, completely decomposed BASALT. (Very stiff, slightly sandy SILT)
6.50 – 9.20	III	Moderately strong, light pink, mottled brown, straked black, moderately decomposed fine grained GRANITE. Joints are medium to closely, locally very closely spaced, rough planar and undulating, tight to very narrow, iron and manganese stained, dipping 0°-10°, 10°-20°, 30°-40°, 50°-60° and 60°-70° 6.50 - 6.70m: Moderately weak to moderately strong, orangish brown, moderately decomposed and non intacted. 7.05 - 7.37m: Orangish brown. 8.55 - 8.72m: Orangish brown. 8.72 - 9.00m: Moderately weak, orangish brown, moderately decomposed and highly fractured.
9.20 – 9.83	II	Strong, pinkish grey, spotted black, slightly decomposed fine grained GRANITE. Joints are medium to closely, locally very closely spaced, rough and smooth, planar and undulating, tight
9.83 – 10.00	III	

Samples: A 0.45–0.50; B 0.95–1.00; C 1.45–1.50 (INSPECTION PIT); 1; 2 2.50; 3 2.60; 4 3.00–3.05; 5 3.50; 6 4.50; 7 4.60; 8 5.60; 9 5.70; 10 6.10–6.15; T2101 6.50–6.70, 8.00, 9.50.

Tests: SPT 2,2 3,4,4,5 N=16; SPT 5,8 16,20,24,31 N=91.

Casing: PX; PX 3.50 HX; HX 6.50.

Water depth: 08:00; Dry at 18:00; Dry at 08:00; 4.20m at 18:00; 8.74m at 08:00. Water recovery 70.

Core runs: Total core recovery 90, 50, 70, 100, 100, 91, 100; Solid core recovery 0, 100, 87, 80; R.Q.D. 0, 100, 73, 72; Fracture index N.I., 3.5, >20, 8.2, 12.7.

Depths: 1.50, 2.60, 3.50, 5.70, 6.50, 6.70, 7.05, 7.37, 8.55, 8.72, 9.00, 9.20, 9.83, 10.00

Key: ● Small disturbed sample; SPT liner sample; U76 undisturbed sample; U100 undisturbed sample; Mazier sample; Piston sample; ↓ Standard penetration test; Vibrocore sample; ▲ Water sample; Piezometer / Standpipe tip; Permeability test; Water absorption (Packer) test; Impression packer test; Acoustic Televiewer Survey Test; V In-situ vane shear test; Pressuremeter Test

LOGGED XXX DATE 13/11/2010
CHECKED YYY DATE 15/11/2010

REMARKS
1. Inspection pit was dug to 1.50m depth.
2. Piezometer was installed at 6.30m depth.

Figure C.1.2

completely decomposed to very stiff, slightly sandy SILT (also Grade V). The basalt was sampled in an SPT test which gave an *N* value of 91 (in the UK the test would normally have been terminated at 50 blows).

From 6.50 m the drilling was continued using a Craelius T2-101 double-tube core barrel with an outside diameter of 101 mm and core diameter of about 84 mm. The rock recovered is all described as moderately strong or strong, moderately (Grade III) or slightly decomposed (Grade II) granite. Recovery in the rock is generally good and RQD high. There is some information about discontinuities, their nature, closeness and orientations but no real detail. If this drillhole was for a slope stability assessment then more information would be required, especially regarding the dip direction of the joints. If it was for a foundation design then such information would probably not be necessary. Water levels have been recorded each morning and evening but these would have been affected by the drilling process and water flush so little can be read from these data. A piezometer was installed at a depth of 6.30 m (at the basalt horizon). No details are given on this log but probably the piezometer would have been installed in a sand/gravel pocket extending above the basalt into the sandy decomposed granite.

C.2 CONSULTANT'S BOREHOLE LOG

The drillhole record Figures C.1.3a is provided courtesy of Golder Associates (GA) / SinclairKnightMerz, Brisbane and with acknowledgements to Department of Transport and Main Roads, Queensland for permission. Only the first three of seven sheets are presented. Soil and rocks are described according to Australian Standards (1993). Shorthand descriptors and other terms are given in Figure C.1.4 (terms).

GA are consulting engineers responsible for investigating the site where this borehole has been put down. It is the site of a complex landslide so cores have been logged not only by the GI contractor (not presented here) but also by GA. In the original report, the logs are accompanied by high-quality photographs. The borehole diameter is 96 mm and core diameter 76 mm. The log only provides information on the rock recovered; for information on how the borehole was conducted, how long it took, casing used, information on flush and ground water encountered etc. one would need to examine the GI contractor's log for the same hole. The factor that distinguishes this log from the HK and UK examples is the attention to detail in describing each discontinuity with information on roughness, infill and mineral coatings as applicable. That said, this log is really only a preliminary step. After considering other details regarding the site, not least results from instruments monitoring ground movements and water pressures then sections of core where movement is suspected were examined and logged in even greater detail, with samples taken for strength testing, microscopic examination and chemical analysis.

C.3 CONTRACTOR'S TRIAL PIT LOGS

Examples from the UK and from Hong Kong are presented in Figures C.1.5 and C.1.6, respectively. Descriptions and classification used reflect local standards and codes. The UK example is essentially the same as a borehole log (one dimensional). The HK example is more graphic and shows all four sides of the pit. As is common practice in

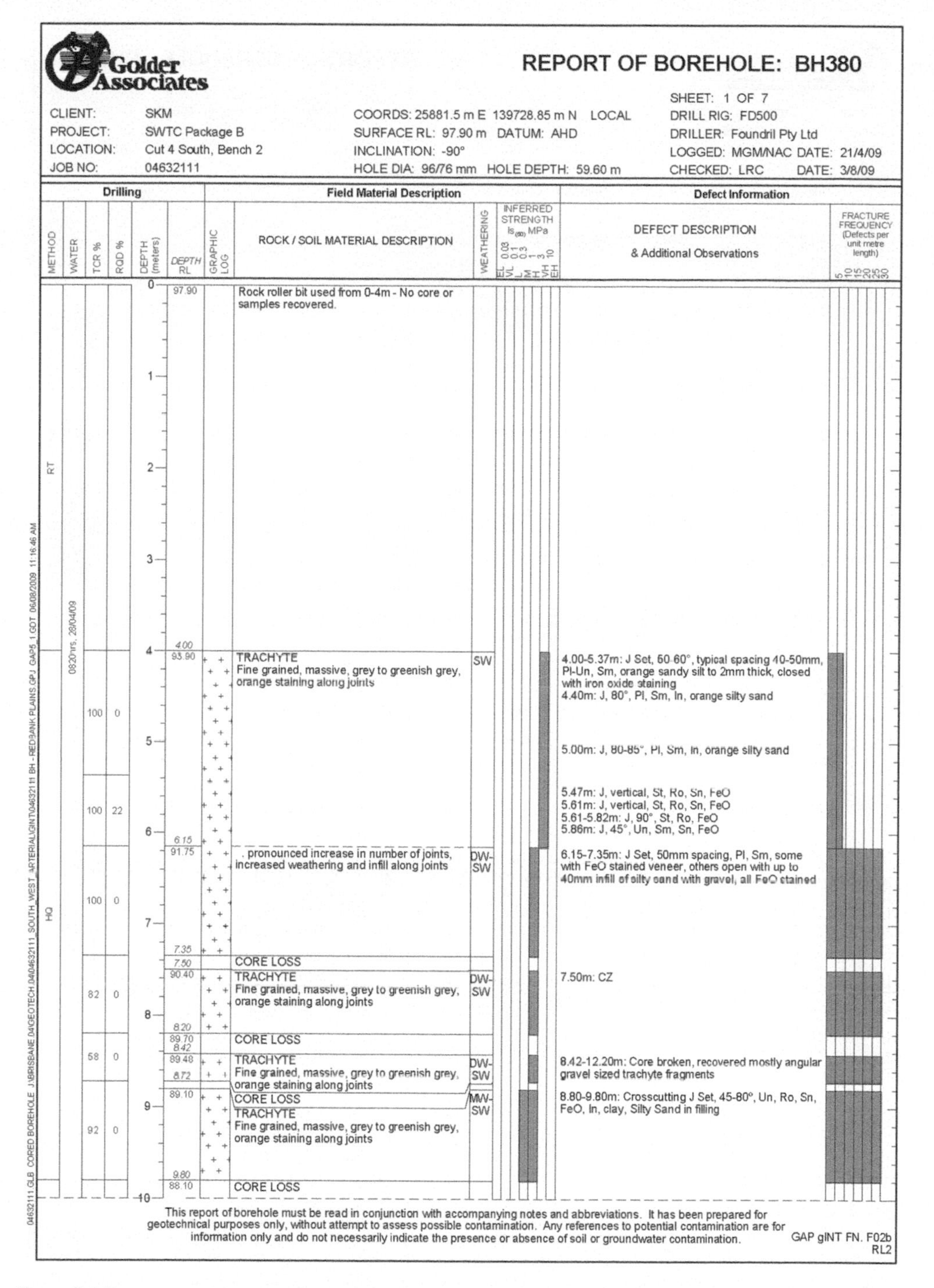

Golder Associates

REPORT OF BOREHOLE: BH380

CLIENT: SKM	COORDS: 25881.5 m E 139728.85 m N LOCAL		SHEET: 1 OF 7
PROJECT: SWTC Package B	SURFACE RL: 97.90 m DATUM: AHD		DRILL RIG: FD500
LOCATION: Cut 4 South, Bench 2	INCLINATION: -90°		DRILLER: Foundril Pty Ltd
JOB NO: 04632111	HOLE DIA: 96/76 mm HOLE DEPTH: 59.60 m		LOGGED: MGM/NAC DATE: 21/4/09
			CHECKED: LRC DATE: 3/8/09

METHOD	WATER	TCR %	RQD %	DEPTH (meters)	DEPTH RL	ROCK / SOIL MATERIAL DESCRIPTION	WEATHERING	DEFECT DESCRIPTION & Additional Observations
RT				0	97.90	Rock roller bit used from 0-4m - No core or samples recovered.		
HQ	0820hrs, 28/04/09	100	0	4	4.00 / 93.90	TRACHYTE Fine grained, massive, grey to greenish grey, orange staining along joints	SW	4.00-5.37m: J Set, 50-60°, typical spacing 40-50mm, Pl-Un, Sm, orange sandy silt to 2mm thick, closed with iron oxide staining 4.40m: J, 80°, Pl, Sm, In, orange silty sand 5.00m: J, 80-85°, Pl, Sm, In, orange silty sand
		100	22					5.47m: J, vertical, St, Ro, Sn, FeO 5.61m: J, vertical, St, Ro, Sn, FeO 5.61-5.82m: J, 90°, St, Ro, FeO 5.86m: J, 45°, Un, Sm, Sn, FeO
		100	0	6	6.15 / 91.75	. pronounced increase in number of joints, increased weathering and infill along joints	DW-SW	6.15-7.35m: J Set, 50mm spacing, Pl, Sm, some with FeO stained veneer, others open with up to 40mm infill of silty sand with gravel, all FeO stained
				7	7.35 / 7.50	CORE LOSS		
		82	0		90.40	TRACHYTE Fine grained, massive, grey to greenish grey, orange staining along joints	DW-SW	7.50m: CZ
				8	8.20 / 89.70 / 8.42	CORE LOSS		
		58	0		89.48 / 8.72	TRACHYTE Fine grained, massive, grey to greenish grey, orange staining along joints	DW-SW	8.42-12.20m: Core broken, recovered mostly angular gravel sized trachyte fragments
		92	0	9	89.10	CORE LOSS TRACHYTE Fine grained, massive, grey to greenish grey, orange staining along joints	MW-SW	8.80-9.80m: Crosscutting J Set, 45-80°, Un, Ro, Sn, FeO, In, clay, Silty Sand in filling
				10	9.80 / 88.10	CORE LOSS		

INFERRED STRENGTH $Is_{(50)}$ MPa: EL VL L M H VH EH (0.03, 0.1, 0.3, 1, 3, 10)

FRACTURE FREQUENCY (Defects per unit metre length): 5 10 15 20 25 30

This report of borehole must be read in conjunction with accompanying notes and abbreviations. It has been prepared for geotechnical purposes only, without attempt to assess possible contamination. Any references to potential contamination are for information only and do not necessarily indicate the presence or absence of soil or groundwater contamination.

GAP gINT FN. F02b RL2

04632111 GLB CORED BOREHOLE J:\BRISBANE 04\GEOTECH 04\04632111 SOUTH WEST ARTERIAL\GINT\04632111 BH - REDBANK PLAINS.GPJ GAP5_1.GDT 06/08/2009 11:16:46 AM

Figure C.1.3

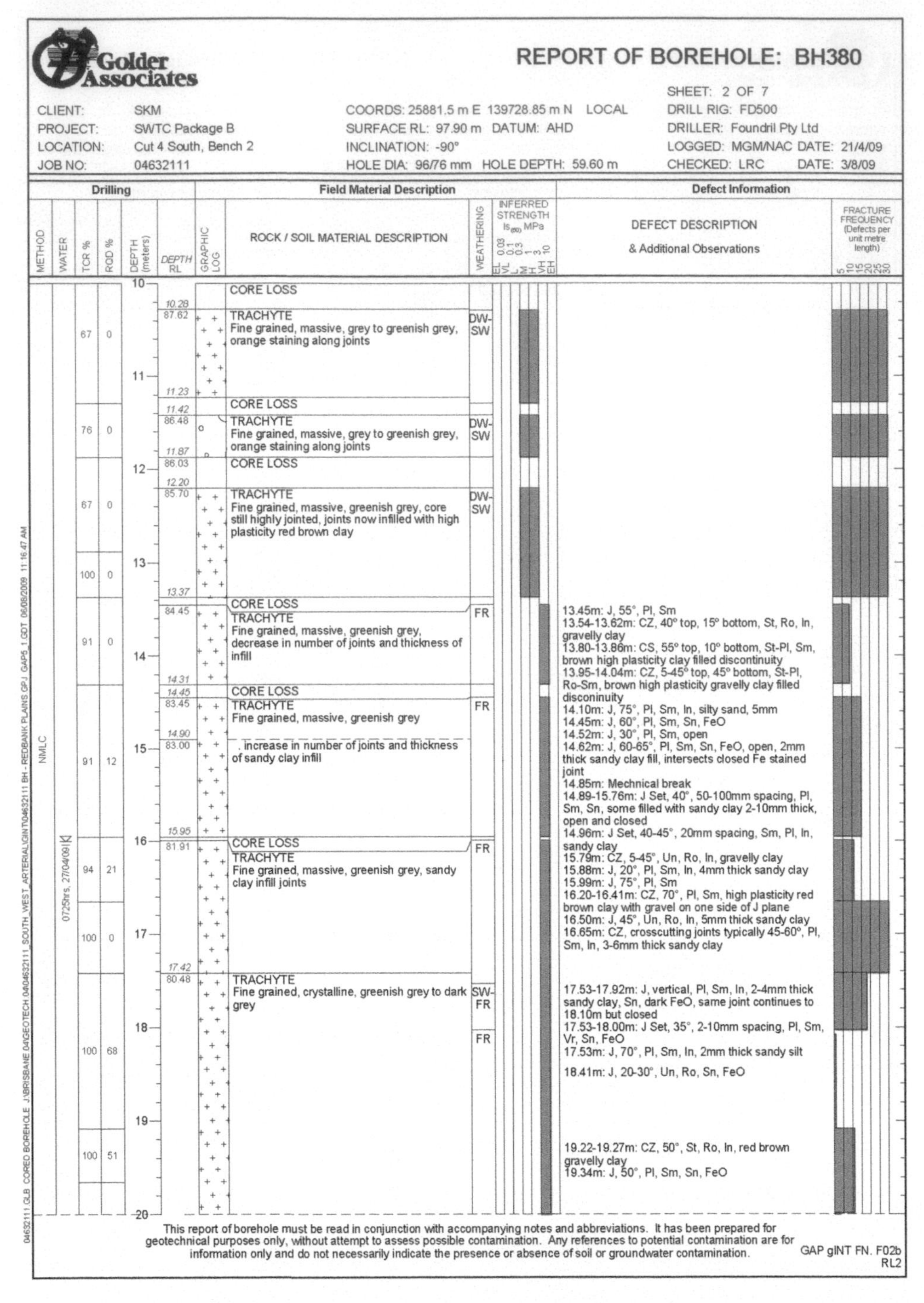

Golder Associates

REPORT OF BOREHOLE: BH380

CLIENT: SKM	COORDS: 25881.5 m E 139728.85 m N LOCAL	SHEET: 2 OF 7	
PROJECT: SWTC Package B	SURFACE RL: 97.90 m DATUM: AHD	DRILL RIG: FD500	
LOCATION: Cut 4 South, Bench 2	INCLINATION: -90°	DRILLER: Foundril Pty Ltd	
JOB NO: 04632111	HOLE DIA: 96/76 mm HOLE DEPTH: 59.60 m	LOGGED: MGM/NAC DATE: 21/4/09	CHECKED: LRC DATE: 3/8/09

Drilling: METHOD | WATER | TCR % | RQD % | DEPTH (meters) | DEPTH RL
Field Material Description: GRAPHIC LOG | ROCK / SOIL MATERIAL DESCRIPTION | WEATHERING | INFERRED STRENGTH $Is_{(50)}$ MPa (EL 0.03, VL 0.1, L 0.3, M 1, H 3, VH 10, EH)
Defect Information: DEFECT DESCRIPTION & Additional Observations | FRACTURE FREQUENCY (Defects per unit metre length) 5 10 15 20 25 30

Method: NMLC. Water: 0725hrs, 27/04/09

TCR %	RQD %	Depth / RL	Rock / Soil Material Description	Weathering
		10.00 – 10.28	CORE LOSS	
67	0	10.28 / 87.62 – 11.23	TRACHYTE Fine grained, massive, grey to greenish grey, orange staining along joints	DW-SW
		11.23 – 11.42	CORE LOSS	
76	0	11.42 / 86.48 – 11.87	TRACHYTE Fine grained, massive, grey to greenish grey, orange staining along joints	DW-SW
		11.87 / 86.03 – 12.20	CORE LOSS	
67	0	12.20 / 85.70 – 13.37	TRACHYTE Fine grained, massive, greenish grey, core still highly jointed, joints now infilled with high plasticity red brown clay	DW-SW
100	0			
		13.37 / 84.45	CORE LOSS	
91	0	– 14.31	TRACHYTE Fine grained, massive, greenish grey, decrease in number of joints and thickness of infill	FR
		14.31 – 14.45	CORE LOSS	
91	12	14.45 / 83.45 – 14.90	TRACHYTE Fine grained, massive, greenish grey	FR
		14.90 / 83.00 – 15.95	. increase in number of joints and thickness of sandy clay infill	
		15.95 / 81.91	CORE LOSS	
94	21	– 17.42	TRACHYTE Fine grained, massive, greenish grey, sandy clay infill joints	FR
100	0			
100	68	17.42 / 80.48 – 20.00	TRACHYTE Fine grained, crystalline, greenish grey to dark grey	SW-FR, FR
100	51			

Defect Description & Additional Observations:

13.45m: J, 55°, Pl, Sm
13.54-13.62m: CZ, 40° top, 15° bottom, St, Ro, In, gravelly clay
13.80-13.86m: CS, 55° top, 10° bottom, St-Pl, Sm, brown high plasticity clay filled discontinuity
13.95-14.04m: CZ, 5-45° top, 45° bottom, St-Pl, Ro-Sm, brown high plasticity gravelly clay filled disconinuity
14.10m: J, 75°, Pl, Sm, In, silty sand, 5mm
14.45m: J, 60°, Pl, Sm, Sn, FeO
14.52m: J, 30°, Pl, Sm, open
14.62m: J, 60-65°, Pl, Sm, Sn, FeO, open, 2mm thick sandy clay fill, intersects closed Fe stained joint
14.85m: Mechnical break
14.89-15.76m: J Set, 40°, 50-100mm spacing, Pl, Sm, Sn, some filled with sandy clay 2-10mm thick, open and closed
14.96m: J Set, 40-45°, 20mm spacing, Sm, Pl, In, sandy clay
15.79m: CZ, 5-45°, Un, Ro, In, gravelly clay
15.88m: J, 20°, Pl, Sm, In, 4mm thick sandy clay
15.99m: J, 75°, Pl, Sm
16.20-16.41m: CZ, 70°, Pl, Sm, high plasticity red brown clay with gravel on one side of J plane
16.50m: J, 45°, Un, Ro, In, 5mm thick sandy clay
16.65m: CZ, crosscutting joints typically 45-60°, Pl, Sm, In, 3-6mm thick sandy clay

17.53-17.92m: J, vertical, Pl, Sm, In, 2-4mm thick sandy clay, Sn, dark FeO, same joint continues to 18.10m but closed
17.53-18.00m: J Set, 35°, 2-10mm spacing, Pl, Sm, Vr, Sn, FeO
17.53m: J, 70°, Pl, Sm, In, 2mm thick sandy silt
18.41m: J, 20-30°, Un, Ro, Sn, FeO

19.22-19.27m: CZ, 50°, St, Ro, In, red brown gravelly clay
19.34m: J, 50°, Pl, Sm, Sn, FeO

This report of borehole must be read in conjunction with accompanying notes and abbreviations. It has been prepared for geotechnical purposes only, without attempt to assess possible contamination. Any references to potential contamination are for information only and do not necessarily indicate the presence or absence of soil or groundwater contamination.

GAP gINT FN. F02b RL2

04632111 GLB CORED BOREHOLE J:\BRISBANE 04\GEOTECH 04\04632111_SOUTH WEST ARTERIAL\GINT\04632111 BH - REDBANK PLAINS.GPJ GAP5_1.GDT 06/08/2009 11:16:47 AM

Figure C.1.3 (Continued)

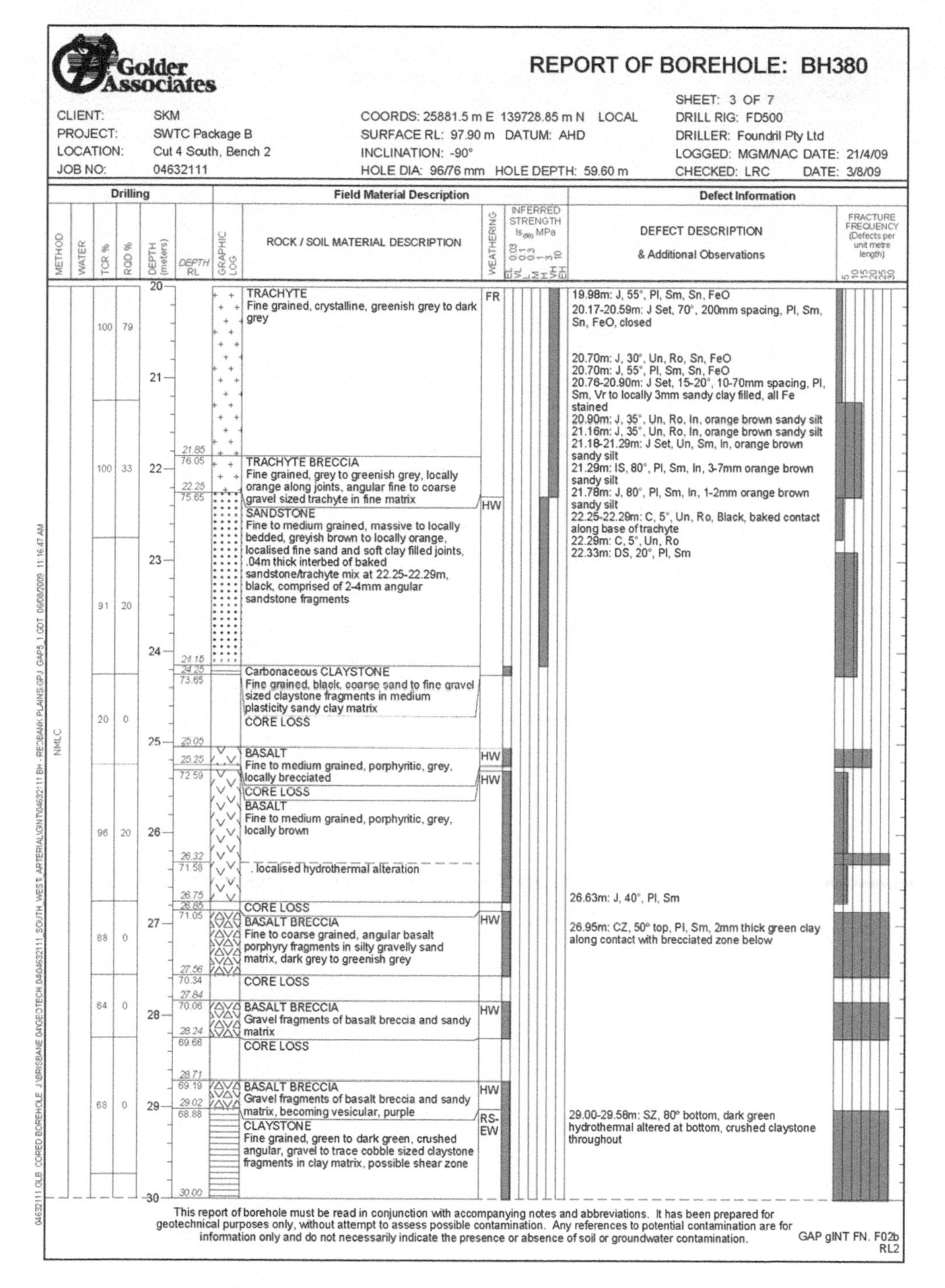

Golder Associates

REPORT OF BOREHOLE: BH380

CLIENT: SKM			SHEET: 3 OF 7
PROJECT: SWTC Package B	COORDS: 25881.5 m E 139728.85 m N LOCAL		DRILL RIG: FD500
LOCATION: Cut 4 South, Bench 2	SURFACE RL: 97.90 m DATUM: AHD		DRILLER: Foundril Pty Ltd
JOB NO: 04632111	INCLINATION: -90°		LOGGED: MGM/NAC DATE: 21/4/09
	HOLE DIA: 96/76 mm HOLE DEPTH: 59.60 m		CHECKED: LRC DATE: 3/8/09

Drilling: METHOD (NMLC), WATER, TCR %, RQD %, DEPTH (meters), DEPTH RL, GRAPHIC LOG. Field Material Description: ROCK / SOIL MATERIAL DESCRIPTION, WEATHERING, INFERRED STRENGTH $Is_{(50)}$ MPa (EL 0.03, VL 0.1, L 0.3, M 1, H 3, VH 10, EH). Defect Information: DEFECT DESCRIPTION & Additional Observations, FRACTURE FREQUENCY (Defects per unit metre length) 5 10 15 20 25 30.

TCR %	RQD %	Depth / RL	Rock / Soil Material Description	Weathering
100	79	20–21.85	TRACHYTE Fine grained, crystalline, greenish grey to dark grey	FR
100	33	21.85 / 76.05	TRACHYTE BRECCIA Fine grained, grey to greenish grey, locally orange along joints, angular fine to coarse gravel sized trachyte in fine matrix	
91	20	22.25 / 75.65	SANDSTONE Fine to medium grained, massive to locally bedded, greyish brown to locally orange, localised fine sand and soft clay filled joints, .04m thick interbed of baked sandstone/trachyte mix at 22.25-22.29m, black, comprised of 2-4mm angular sandstone fragments	HW
20	0	24.15 / 24.25 / 73.65	Carbonaceous CLAYSTONE Fine grained, black, coarse sand to fine gravel sized claystone fragments in medium plasticity sandy clay matrix; CORE LOSS	
		25.05	BASALT Fine to medium grained, porphyritic, grey, locally brecciated	HW
96	20	25.25 / 72.59	CORE LOSS; BASALT Fine to medium grained, porphyritic, grey, locally brown	HW
		26.32 / 71.58	. localised hydrothermal alteration	
		26.75 / 26.85	CORE LOSS	
88	0	71.05	BASALT BRECCIA Fine to coarse grained, angular basalt porphyry fragments in silty gravelly sand matrix, dark grey to greenish grey	HW
		27.56 / 70.34	CORE LOSS	
64	0	27.84 / 70.06	BASALT BRECCIA Gravel fragments of basalt breccia and sandy matrix	HW
		28.24 / 69.66	CORE LOSS	
68	0	28.71 / 69.19	BASALT BRECCIA Gravel fragments of basalt breccia and sandy matrix, becoming vesicular, purple	HW
		29.02 / 68.88	CLAYSTONE Fine grained, green to dark green, crushed angular, gravel to trace cobble sized claystone fragments in clay matrix, possible shear zone	RS-EW
		30.00		

Defect Description & Additional Observations:

19.98m: J, 55°, Pl, Sm, Sn, FeO
20.17-20.59m: J Set, 70°, 200mm spacing, Pl, Sm, Sn, FeO, closed
20.70m: J, 30°, Un, Ro, Sn, FeO
20.70m: J, 55°, Pl, Sm, Sn, FeO
20.76-20.90m: J Set, 15-20°, 10-70mm spacing, Pl, Sm, Vr to locally 3mm sandy clay filled, all Fe stained
20.90m: J, 35°, Un, Ro, In, orange brown sandy silt
21.16m: J, 35°, Un, Ro, In, orange brown sandy silt
21.18-21.29m: J Set, Un, Sm, In, orange brown sandy silt
21.29m: IS, 80°, Pl, Sm, In, 3-7mm orange brown sandy silt
21.78m: J, 80°, Pl, Sm, In, 1-2mm orange brown sandy silt
22.25-22.29m: C, 5°, Un, Ro, Black, baked contact along base of trachyte
22.29m: C, 5°, Un, Ro
22.33m: DS, 20°, Pl, Sm
26.63m: J, 40°, Pl, Sm
26.95m: CZ, 50° top, Pl, Sm, 2mm thick green clay along contact with brecciated zone below
29.00-29.58m: SZ, 80° bottom, dark green hydrothermal altered at bottom, crushed claystone throughout

This report of borehole must be read in conjunction with accompanying notes and abbreviations. It has been prepared for geotechnical purposes only, without attempt to assess possible contamination. Any references to potential contamination are for information only and do not necessarily indicate the presence or absence of soil or groundwater contamination.

GAP gINT FN. F02b RL2

04632111 GLB COREO BOREHOLE J:\BRISBANE 04\GEOTECH 04\04632111_SOUTH WEST ARTERIAL\GINT\04632111 BH - REDBANK PLAINS.GPJ GAP5_1.GDT 05/08/2009 11:16:47 AM

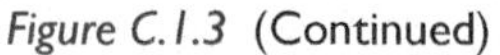

Figure C.1.3 (Continued)

Golder Associates

TERMS FOR ROCK MATERIAL STRENGTH & WEATHERING AND ABBREVIATIONS FOR DEFECT DESCRIPTIONS

STRENGTH

Symbol	Term	Point Load Index, $Is_{(50)}$ (MPa)	Field Guide
EL	Extremely Low	< 0.03	Easily remoulded by hand to a material with soil properties.
VL	Very Low	0.03 to 0.1	Material crumbles under firm blows with sharp end of pick; can be peeled with knife; too hard to cut a triaxial sample by hand. Pieces up to 30 mm can be broken by finger pressure.
L	Low	0.1 to 0.3	Easily scored with a knife; indentations 1 mm to 3 mm show in the specimen with firm blows of pick point; has dull sound under hammer. A piece of core 150 mm long by 50 mm diameter may be broken by hand. Sharp edges of core may be friable and break during handling.
M	Medium	0.3 to 1	Readily scored with a knife; a piece of core 150 mm long by 50 mm diameter can be broken by hand with difficulty.
H	High	1 to 3	A piece of core 150 mm long by 50 mm diameter cannot be broken by hand but can be broken with pick with a single firm blow; rock rings under hammer.
VH	Very High	3 to 10	Hand specimen breaks with pick after more than one blow; rock rings under hammer.
EH	Extremely High	>10	Specimen requires many blows with geological pick to break through intact material; rock rings under hammer.

ROCK STRENGTH TEST RESULTS

▼ Point Load Strength Index, $I_s(50)$, Axial test (MPa)

◀ Point Load Strength Index, $I_s(50)$, Diametral test (MPa)

Relationship between $I_s(50)$ and UCS (unconfined compressive strength) will vary with rock type and strength, and should be determined on a site-specific basis. UCS is typically 10 to 30 x $I_s(50)$, but can be as low as 5.

ROCK MATERIAL WEATHERING

Symbol		Term	Field Guide
RS		Residual Soil	Soil developed on extremely weathered rock; the mass structure and substance fabric are no longer evident; there is a large change in volume but the soil has not been significantly transported.
EW		Extremely Weathered	Rock is weathered to such an extent that it has soil properties - i.e. it either disintegrates or can be remoulded, in water.
DW	HW	Distinctly Weathered	Rock strength usually changed by weathering. The rock may be highly discoloured, usually by iron staining. Porosity may be increased by leaching, or may be decreased due to deposition of weathering products in pores. In some environments it is convenient to subdivide into Highly Weathered and Moderately Weathered, with the degree of alteration typically less for MW.
	MW		
SW		Slightly Weathered	Rock is slightly discoloured but shows little or no change of strength relative to fresh rock.
FR		Fresh	Rock shows no sign of decomposition or staining.

ABBREVIATIONS FOR DEFECT TYPES AND DESCRIPTIONS

Defect Type

B	Bedding parting
X	Foliation
C	Contact
L	Cleavage
J	Joint
SS/SZ	Sheared seam/zone (Fault)
CS/CZ	Crushed seam/zone (Fault)
DS/DZ	Decomposed seam/zone
IS/IZ	Infilled seam/zone
S	Schistocity
V	Vein

Coating or Infilling

Cn	Clean
Sn	Stain
Vr	Veneer
Ct	Coating or Infill

Planarity

Pl	Planar
Un	Undulating
St	Stepped

Roughness

Sl	Slickensided
Sm	Smooth
Ro	Rough

Vertical Boreholes – The dip (inclination from horizontal) of the defect is given.

Inclined Boreholes – The inclination is measured as the acute angle to the core axis.

GAP Form No. 7
RL6

Figure C.1.4

TRIAL PIT RECORD   *Trial Pit*

Project SAMPLE PROJECT   Engineer GEOTECHNICS LIMITED   Trial Pit TP8
Project No PC060000

Client GEOTECHNICS LIMITED   Ground Level 31.10 m OD

Scale 1:50

Samples and Tests: Depth	Type	Stratum No	Results	Strata: Description	Depth	Legend	Level m OD
					G.L.		31.10
				Soft dark brown slightly sandy silty clay with frequent roots and rootlets. [TOPSOIL]	0.20	1	30.90
0.50	D		mc=17%	Soft brown slightly gravelly sandy CLAY with frequent rootlets and a low cobble content of sandstone and quartzite. Gravel is subrounded to rounded fine to medium quartzite. [ALLUVIUM]		2	
1.00	B		mc=38%				
1.50	B			Brown slightly clayey sandy GRAVEL with a high cobble and boulder content of sandstone and quartzite. Gravel is angular to subrounded fine to coarse shale, mudstone, sandstone and quartzite. [GLACIAL DEPOSITS]	1.30	3	29.80
2.00	D						
				End of Excavation	2.30		28.80

Excavation

Plant JCB 3CX
Date 20/01/2010
Shoring None.
Stability Unstable during excavation.
Width (B) 0.80
Length (C) 2.20
Date Backfilled 20/01/2010

Groundwater

Depth Observed	Depth of Pit	Details
		None encountered during excavation.

Remarks AGS All sides of pit collapsing below 1.80m, unable to excavated below 2.30m depth.

Symbols and abbreviations are explained on the accompanying key sheet.

All dimensions are in metres.   Logged in accordance with BS5930:1999

Logged by DF
Checked by DRB
Figure 1 of 1
17/12/2010

geotechnics

Figure C.1.5

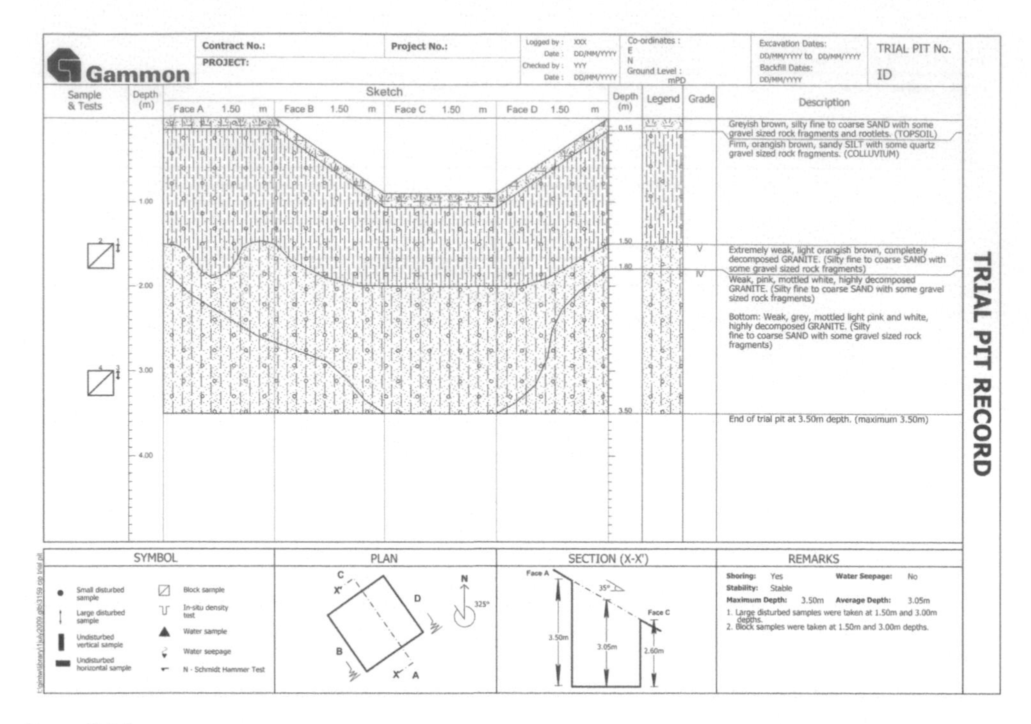

Figure C.1.6

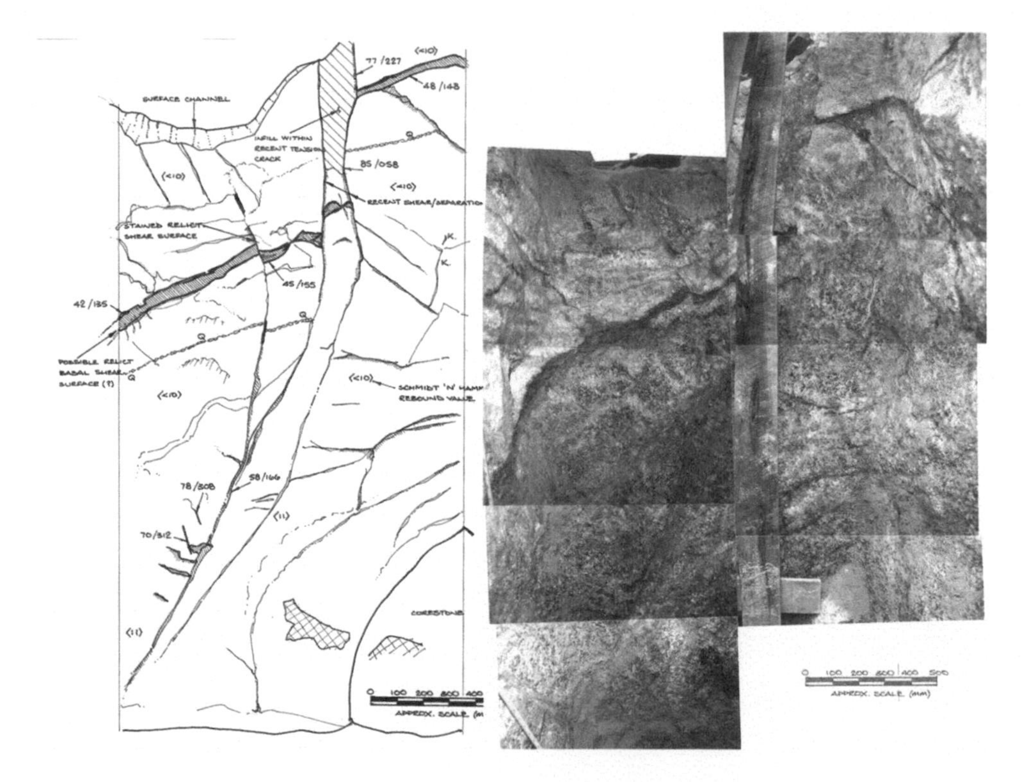

Figure C.1.7

HK, large block samples were taken at depths of 1.5 m and 3.0 m. These are cut by hand, covered in foil, waxed and placed in a wooden box for transportation to the laboratory where they are opened, described and prepared for testing as illustrated in Figure 4.12.

C.4 CONSULTANT'S TRIAL PIT LOG

Figure C.1.7 is an example log from a trial pit put down as part of a landslide investigation (Halcrow Asia Partnership, 1998b). In this case, the log is not a standard record but a geological and geotechnical sketch aimed at recording key factors about the landslide. Infill within recent tension crack relates to the recent failure. Discontinuity orientations are recorded as dip/dip direction. Rather than describe the strength of the weathered granite verbally, this is quantified by Schmidt hammer rebound values, from tests repeated at locations, mostly less than 10 but locally 11. This places the rock in at the weak end of highly decomposed (Grade IV) following Hencher & Martin (1982). A key observation and record is the presence of stained, displaced relict shear surfaces which indicates that the slope had a history of movement over many years prior to the eventual failure.

Note

1 Such a high core loss might well not be acceptable and might be dealt with by a clause of the contract, which specifies acceptable recovery as perhaps 95%.

Appendix D

Tunnelling risk

D.1 EXAMPLE OF TUNNELLING RISK ASSESSMENT AT PROJECT OPTION STAGE FOR YOUNG DONG MOUNTAIN LOOP TUNNEL, SOUTH KOREA

The Young Dong Railroad Relocation Project for the Korean National Railways (KNR) included a single-track railway tunnel in rock approximately 16.3 km long with a span of approximately 8 metres and is the longest tunnel in Korea. The tunnel had to be constructed as a large radius loop to limit the gradient of the track as illustrated in Figure AD.1.1. The maximum depth of the tunnel is approximately 400 m with most of the alignment being at depths in excess of 100 m.

The route for the tunnel was identified as having intrinsic hazards including the following:

- potentially high water pressures, up to 40 bars (4 MPa) pressure;
- fault zones, possibly associated with significant groundwater inflows;
- highly sheared and closely jointed rocks;
- some rocks with high strength and abrasively;
- possible cavernous limestone with groundwater;
- old mine workings (coal).

Halcrow was commissioned to advise a consortium of contractors tendering for the construction of the tunnel and as part of the brief carried out a thorough hazard and risk assessment of the route for both TBM and drill & blast options. Several state-of-the-art reports by Dr Graham Garrard of Halcrow were based on existing ground investigations along the route, mapping and further GI together with an in-depth review of tunnel case histories in similar terrain throughout the world. An example of a summary hazard and risk register for this project (TBM) is given at Appendix D.2.

A risk assessment method was developed using a risk matrix to make a quantitative and objective assessment of the construction methods of the tunnel. The risks associated with tunnel excavation are dependent on the hazards encountered and were defined for this project with respect to programme (rather than other issues such as safety or cost as might have been done). The likelihood of a hazard occurring was assigned one of three levels and consequence of each hazard assumed to be at one of five levels as set out in Table D.1.1 below.

Figure AD.1.1 The Young Dong (now Solan) Mountain Loop Tunnel Project, South Korea after Daewoo Corporation.

Source: Figure modified from Kim et al (2001).

The level of risk for each hazard can be determined by finding its likelihood of occurrence and considering its consequence. The level of risk associated with the hazard is then established conventionally as follows:

Level of Risk = Likelihood × Consequence

Once the level of risk has been ascertained, it can be compared with Table D.1.2 below to identify the action that should be taken to mitigate the risk.

Having made an assessment of the risk associated with each hazard, appropriate mitigation measures are considered. The residual risk remaining after mitigation is then assessed in the same way to determine acceptability or otherwise.

D.1.1 Risk assessment

The assessment of risks associated with the use of a shielded TBM was carried out separately from that for drill and blast excavation.

Table D.1.3 provides a brief summary of the hazards identified for a drill & blast option with level of probability of occurrence and likely consequence and hence risk. Possible ways of mitigating the risk are identified and the residual likely risk identified.

Table D.1.1 Tunnel risk assessment method

Likelihood			*Consequence*		
Title	*Description*	*Scale*	*Title*	*Description*	*Scale*
Probable	Likely to occur during the construction of the tunnel, possibly on more than one occasion	3	Catastrophic	Total loss of a section of tunnel	5
Occasional	Likely to occur at least once during construction of the tunnel	2	Critical	Major damage or delay to tunnel or major environmental impact affecting programme	4
Remote	Unlikely to occur during construction of the tunnel	1	Serious	Some damage or delay to tunnel or some environmental impact affecting programme	3
			Marginal	A routine maintenance repair to tunnel or minor hindrance	2
			Negligible	Of little consequence to programme	1

Table D.1.2 Comparative table used to assess consequence and need for mitigation of risk

Consequence / *Likelihood*	*Catastrophic*	*Critical*	*Serious*	*Marginal*	*Negligible*
Probable	15	12	9	6	3
Occasional	10	8	6	4	2
Remote	5	4	3	2	1

Score	
10–15	Very High Risk – not acceptable for tunnel construction – need to apply mitigation measures to eliminate or reduce risk
6–9	High Risk – apply mitigation measures to eliminate or reduce risk. Residual risk at this level indicates need for active management control and response plans to be well developed with well trained personnel, materials and plant readily available
1–5	Low Risk – may be accepted if mitigating measures are in place under active management control

The original reports go into far more detail of the nature of hazards, ways to mitigate them and the potential ways those mitigation measures might prove unsuccessful leaving residual risks.

D.1.2 Conclusions

It was found that the number of hazards and residual risks associated with a shielded TBM construction would be greater than for the drill and blast method. The principal reasons included the following:

Table D.1.3 Programme risk assessment for excavating tunnel by drill and blast method. (L = Likelihood, C = Consequence, R = L × C = Risk). Residual Risk is likely outcome <u>after</u> application of mitigation measures

No.	*Hazard*	*Risk*	*Risk level*			*Mitigation measures*	*Residual risk level*		
			L	*C*	*R*		*L*	*C*	*R*
1	Highly jointed rock mass (possibly in association with high pressure water) See Hazard 3 for water ingress specifically.	Ravelling ground, roof falls and sidewall and/or face instability with high amount of primary support.	3	4	12	1. Reduce length of excavation advance; face support and/or buttressing and/or partial face advance 2. Reduce powder factor to lessen blast damage. 3. Increase rock support and install rock support in the form of rock-bolts and steel fibre reinforced shotcrete without delay. 4. Probing and pre-injection.	2	2	4
2	Fault zones	Soft ground or mixed face conditions with potential roof falls and sidewalls instability requiring a high degree of primary support.	3	4	12	1. Reduce length of excavation advance; face support and/or buttressing and/or partial face advance. 2. Reduce powder factor. 3. Increase rock support and install rock-bolts, steel fibre reinforced shotcrete, lattice girders and spilling bars without delay. 4. Provision of probe drilling to identify these features ahead of the excavation face. 5. Provision of Tunnel Seismic Prediction (TSP) to identify fault zones ahead of the excavation face. 6. Provision of instrumentation to monitor movement to optimise support.	3	2	6
3	Water ingress, possibly under high pressure up to 40 bar (4 MPa)	Water in cavities, joints and fissures in the rock mass entering excavation and causing instability of ground. Difficulties with shotcrete application.	3	4	12	1. Tunnel drive to be up-grade to allow water to drain. (Not possible with all drives.) 2. Provision of pumps to cope with high flows and back-up systems to deal with pumps and power failures. 3. Provision of probe drilling to identify areas of high water flows and to carry out pre-injection grouting to stem the flow.	2	2	4

(continued)

Table D.1.3 (Cont.)

No.	Hazard	Risk	Risk level L	Risk level C	Risk level R	Mitigation measures	Residual risk level L	Residual risk level C	Residual risk level R
						4. Excavation equipment systems to be rated to IP68 or equivalent. 5. Use drainage channels to control inflows prior to shotcreting			
4	Cavities in the rock mass (including mine workings) possibly associated with water inflow.	Instability of tunnel face, roof fall and side wall instability. Flooding. Need for major structural work or infilling.	3	4	12	1. Provision of TSP to identify cavities in advance of excavation. 2. Provision of probe drilling to determine extent of cavities and provide means for grouting or other advance stabilisation measures. 3. Reduce length of excavation advance.	2	3	6
5	Tunnel atmosphere and ventilation including accumulation of explosive and noxious gases.	Explosion risk. Possible accumulation of explosive and or noxious gas. Methane, associated with coal or other sources is a flammable gas, lighter than air and can give rise to explosion. In large quantities it can also cause asphyxiation. Other gases such carbon dioxide, carbon monoxide sulphur dioxide and hydrogen sulphide are noxious.	3	5	15	1. Provision of adequate fresh air from the portal to the excavation face. 2. Provision of adequate and suitable atmospheric monitoring system. 3. Avoid the use of dry shotcrete mix. 4. Use explosive appropriate to tunnels prone to fire risk. 5. Standby generators to power fans.	1	4	4
6	Mechanical breakdown	Failure of key item of plant	3	3	9	1. Planned maintenance strategy. 2. Maintain spare plant items. 3. Maintain stocks of spares.	3	1	3
7	Use of Explosives	Premature detonation or uncontrolled explosion	2	5	10	1. Employ qualified staff. 2. Comply with safety regulations. 3. Use proper storage and transport facilities 4. Use non-electric detonators	1	5	5

- the relative inflexibility of mechanised excavation and lining systems to deal with conditions for which they may not have been specifically designed;
- the dependence of the tunnel progress entirely on the performance and reliability of a single item of mechanical plant which would require a high level of technological input for their successful operation and maintenance.

Further details are given by Kim et al (2001).

D.2 EXAMPLE OF HAZARD AND RISK PREDICTION TABLE

Example table of predicted hazards at option/design stage for TBM with potential intersection of old mine workings, karst, collapsed and weak zones; this is part of a more detailed report prepared for a specific project and was supported by similar tables for Drill and Blast options, detailed text, recommended methods of working, international case studies and discussion. The report was prepared by Dr Graham Garrard of Halcrow with inputs from many others including Dr Laurie Richards.

Hazard			*Consequence of event on tunnel*	*Possible Mitigation Measures prior to and during construction to reduce likelihood or consequence of hazard Notes * indicates technique more likely than others to be useful*	*Impact of mitigation measures and Risk*			*Summary of main residual risks*
Number	*Title*	*Example of circumstances*			*Likelihood (L)*	*Consequence (C))*	*Risk (L x C)*	
1	Inrush of water or unconsolidated ground	cavity, fault, area of collapsed ground, shaft, adit	H	1. GI[1]boreholes and packer tests 2. pre-construction geological desk study and risk assessment* 3. surface grouting ahead of tunnel 4. probe drilling ahead of and around the face through TBM[2] head. Monitoring of water volume and pressure* 5. inspection of probe drill cuttings for weak or unconsolidated material, monitoring of strata colour and odour for unanticipated changes* 6. TSP[3] to detect likely geological structures * 7. monitoring of water make from face and significant changes in volume or pressure* 8. ground probing radar to detect cavities, clay or high density of fractures 9. geological m apping to detect changes in structure, strata distress 10. grouting ahead of tunnel face* 11. Consider also likelihood of hazards: 2, 3, 5, 6, 7,8,9,10,11,16 and 17	L	H	H	1. failure to identify inrush conditions 2. blockage of probe holes with debris, or unconsolidated material resulting in underestimation of water pressure and volume 3. failure of grouting program 4. failure of EPBM[4] bulkhead 5. injury to workforce 6. partial or complete loss of tunnel drive
2	Inrush of flammable gas	as 1 above	M	1. GI boreholes 2. pre-construction geological desk study and risk assessment* 3. probe drilling ahead of and around the face through TBM head. Monitoring of character of flush return* 4. monitoring of tunnel atmosphere (gas monitoring at top of face) to detect significant changes in character or composition *	L	L	L	1. failure to identify presence of gas 2. explosion during probe drilling because of flush failure at the drill bit

				5. inspection of probe drill cuttings to identify weak or unconsolidated material, monitoring of strata colour and odour for unanticipated changes* 6. TSP to detect likely geological structures * 7. monitoring of tunnel atmosphere for significant changes in character or composition* 8. GPR to detect cavities, clay or high density of fractures 9. geological mapping to detect changes in structure, strata distress 10. provision of adequate ventilation (and if necessary methane drainage system)* 11. adoption of flammable gas working practices (no smoking policy, safety light policies, intrinsically safe plant and equipment 12. Consider also likelihood of hazards: 1, 3, 4, 5, 7,8,9,10,11,12 and 17				3. escape of gas and explosion in the driveage 4. injury to workforce 5. temporary evacuation of drive during remedial works or 6. temporary loss of drive
3	Noxious gases (eg CO_2, SO_2, H_2S	as 1 above	M	1. GI boreholes 2. pre-construction geological desk study and risk assessment* 3. geological mapping of face cuttings to detect changes in structure and strata, presence of organic material, sulphides etc* 4. monitoring of tunnel atmosphere (gas monitoring at top and base of face) to detect significant changes in character or composition * 5. probe drilling ahead of and around the face through TBM head. Monitoring of character of flush return 6. provision of adequate ventilation* 7. consider also presence of hazard 2	L	L	L	1. failure to identify gases 2. injury to workforce 3. temporary evacuation of drive during remedial works

(continued)

Hazard				*Possible Mitigation Measures prior to and during construction to reduce likelihood or consequence of hazard Notes * indicates technique more likely than others to be useful*	*Impact of mitigation measures and Risk*			*Summary of main residual risks*
Number	*Title*	*Example of circumstances*	*Consequence of event on tunnel*		*Likelihood (L)*	*Consequence (C))*	*Risk (L x C)*	
4	Radon	as 1 above	L	1. pre-construction geological desk study and risk assessment* 2. monitoring of tunnel atmosphere for significant changes in character or composition 3. provision of adequate ventilation* 4. consider also hazard 3	L	L	L	1. failure to identify radon at elevated levels 2. long term affects on workforce health 3. temporary evacuation of drive during remedial works
5	Excessive volume of groundwater	as 1 above plus permeable strata and fractured ground associated with 1 above	H	1. hydrological testing in GI boreholes* 2. pre-construction geological desk study and risk assessment, tunnel depth* 3. probe drilling ahead of face through control valve grouted into rock ahead of TBM cutters and monitoring of water volume and pressure* 4. packer testing to measure permeability ahead of face* 5. TSP to detect likely geological structures * 6. monitoring of water make from face and significant changes in volume or pressure* 7. ground probing radar to detect cavities, clay or high density of fractures 8. geological mapping to anticipate permeable or fractured structure, strata distress etc 9. grouting ahead of face* 10. advanced dewatering ahead of tunnel from the surface 11. ground freezing ahead of face 12. Consider also hazard 1	M	H	H	1. failure of grouting program / ground freezing 2. underestimation of inflow 3. failure of EPBM bulkhead 4. failure of pumps 5. temporary evacuation of drive or 6. temporary loss of drive

6	Excessive groundwater pressure	as 5 above	M	1. hydrological testing in GI boreholes 2. pre-construction geological desk study and risk assessment, tunnel depth* 3. probe drilling ahead of face through control valve grouted into rock ahead of TBM cutters and monitoring of water volume and pressure * 4. monitoring of water make from face and significant changes in volume or pressure* 5. TSP to detect likely geological structures 6. ground probing radar to detect cavities, clay or high density of fractures 7. geological mapping to anticipate permeable or fractured structure, strata distress 8. grouting ahead of face* 9. ground freezing ahead of face 0. advanced dewatering ahead of tunnel from the surface 11. use permeable liner to reduce pressure on liner 12. consider also hazard 1 and 2	M	H	H	1. failure of grouting program / ground freezing 2. underestimation of groundwater pressure 3. failure of EPBM bulkhead 4. failure of pumps 5. temporary evacuation of drive or 6. temporary or permanent loss of drive
7	Collapsed Ground	karst areas (boulder chokes), shafts, areas of collapsed mine workings	M	1. GI boreholes 2. pre-construction geological desk study and risk assessment 3. probe drilling ahead of and around face through TBM head. Monitoring of rates of drilling and character of drill cuttings (presence of clay, infill materials etc)* 4. monitoring of water make from face 5. TSP to detect change in reflection characteristic of strata 6. geological mapping to identify disturbed or fractured structure and strata distress 7. monitoring relative volume of spoil produced and driveage distance to identify caving ground	L	M	M	1. failure of grouting program or strata control foams 2. stability problems and damage to cutter head 3. high than anticipated loading on tunnel linings leading to premature failure

(*continued*)

Hazard				Possible Mitigation Measures prior to and during construction to reduce likelihood or consequence of hazard Notes * indicates technique more likely than others to be useful	Impact of mitigation measures and Risk			Summary of main residual risks
Number	Title	Example of circumstances	Consequence of event on tunnel		Likelihood (L)	Consequence (C))	Risk (L x C)	
				8. use of strata control (eg bentonite) foams ahead of face 9. grouting ahead of face 10. grouting behind liner to fill cavities 11. consider also hazard 1, 2, 8, 9, 10, 11, 12, 13, 14, 15, 16, 17				4. higher post construction maintenance 5. temporary or permanent loss of drive
8	Cavity << tunnel diameter	karst, old mine workings, possibly faults	M	1. GI boreholes 2. pre-construction geological desk study and risk assessment 3. probe drilling ahead of and around face through TBM head. Monitoring of rates of drilling and character of drill cuttings (presence of clay, infill materials etc)* 4. TSP to detect air or water filled cavity (unlikely to detect small cavities) 5. geological mapping to identify disturbed or fractured structure and strata distress, 6. consider also hazard 1, 2, 9, 10, 11, 12, 13, 14, 15, 16, 17	L	L	L	1. failure of grouting program 2. difficulty in maintaining tunnel alignment 3. higher risk if water or gas inrush (see above)
9	Cavity ~ tunnel diameter	karst, old partial extraction mine workings	M	1. GI boreholes 2. pre-construction geological desk study and risk assessment 3. probe drilling ahead of and around face through TBM head. Monitoring of rates of drilling and character of drill cuttings (presence of clay, infill materials etc)*	L	L	L	1. failure to detect cavity 2. failure of grouting program 3. difficulty in maintaining tunnel alignment

				4. TSP to detect cavity 5. geological mapping to identify disturbed or fractured structure and strata distress, 6. grouting ahead of face to fill cavities to create plug to improve driving conditions and help minimise TBM deviation* 7. grouting behind lining to fill cavities* 8. consider also hazard 1, 2, 8, 11, 12, 13, 14, 15, 16, 17				4. higher risk of water or gas inrush (see above)
10	Cavity >> tunnel diameter	karst, old partial extraction mine workings	H	1. GI boreholes 2. pre-construction geological desk study and risk assessment 3. probe drilling ahead of and around through TBM head. Monitoring of rates of drilling and character of drill cuttings (presence of clay, infill materials etc)* 4. TSP to detect cavity 5. geological mapping to identify disturbed, fractured or distressed strata 6. grout cavities ahead of tunnel face (may be very difficult)* 7. drain cavity, take TBM off line and construct dummy tunnel rings on tunnel line, grout remaining cavities around lining then advance through dummy tunnel. Note this option may not be possible due to water make* 8. consider also hazard 2, 3, 4, 8, 9, 10, 11, 12, 14, 15, 17 and particularly hazards 1 and 16 during construction	L	H	H	1. failure to detect cavity systems 2. failure of grouting program 3. unable to drain cavities 4. significant delay to program 5. significantly increased risk if water and gas inrush and settlement (hazards 1,2, 3 and 16)

(continued)

Hazard				*Possible Mitigation Measures prior to and during construction to reduce likelihood or consequence of hazard Notes * indicates technique more likely than others to be useful*	*Impact of mitigation measures and Risk*			*Summary of main residual risks*
Number	*Title*	*Example of circumstances*	*Consequence of event on tunnel*		*Likelihood (L)*	*Consequence (C))*	*Risk (L x C)*	
11	Variable face conditions	karst boulder beds and cave / mine working infill, fault zones and highly folded/ disturbed areas	M	1. GI boreholes 2. pre-construction geological desk study and risk assessment 3. probe drilling ahead of face through TBM head and monitoring rates of drilling and character of drill cuttings* 4. geological mapping to detect changes in structure* 5. TSP to detect likely geological structures which could be associated with variable face conditions 6. ground probing radar to detect clay or high densities of fractures 7. use TBM face with ability to deal with small amounts of soft material (tunnel alignment may be compromised* 8. Consider also likelihood of hazards: 2, 3, 5, 6, 7,8,9,10,11,16 and 17	M	L	M	1. failure to anticipate conditions 2. damage to cutters 3. stability problems and damage to cutter head 4. delays to program 5. increased risk of water or gas inrush
12	Metal obstructions	old mine workings, surface GI boreholes (lost drill strings) and lost bits from advance drilling	H	1. pre-construction geological desk study and risk assessment (presence of old mine worlkings) 2. maintenance of good GI drilling records and records of advanced drilling programme* 3. monitoring of face production rates* 4. stop machine draw back and use cutting or explosives to remove obstacles* 5. Consider also likelihood of 1, particularly 2, 7, 8, 9, 10, 11, 13, 14, 15, 16, 17	M	H	H	1. failure to identify possibility of metal in face 2. stability problems and damage to cutter head 3. unable to cut or remove metal from ahead of TBM due to face instability or water make 4. delays to program 5. difficulty in maintaining tunnel alignment

13	Subsidence along tunnel line	karst and old mine workings beneath tunnel line	H	1. GI boreholes, identification of karst or mineworkings 2. pre-construction geological desk study and risk assessment* 3. probe drilling ahead of face and beneath tunnel to detect areas of weak ground. Monitoring rates of drilling and character of drill cuttings* 4. identification of karstic or mining features* 5. geological mapping to detect disturbed or distressed strata* 6. grouting ahead and below tunnel to consolidate strata* 7. Post construction microgravity survey (gradiometry) along line of tunnel. to identify cavities and areas of untreated ground or to confirm consolidation undertaken satisfactorily* 8. Consider also likelihood of hazards: 1, 2, 5, 7, 8, 9, 10, 11 and 15	L	H	H	1. failure to identify unconsolidated ground 2. failure of grouting program 3. workings / karst too deep to consolidate from tunnel 4. settlement of TBM, delays to program and loss of line 5. failure of tunnel lining post construction
14	Ground Heave	proximity to faults, shear zones, old mine workings, folded ground and places where stress conditions atypical	M	1. GI boreholes / structural interpretation 2. pre-construction geological desk study and risk assessment 3. probe drilling ahead of face and beneath tunnel to detect areas of weak ground. Monitoring rates of drilling and character of drill cuttings* 4. identification of mining features 5. geological mapping to identify weak or disturbed strata* 6. use copy cutters in weak ground to increase tunnel diameter* 7. Consider also likelihood of hazards: 15	M	M	M	1. failure to identify ground conditions 2. difficulties in moving TBM forward 3. difficulties in placing lining 4. delays in program 5. incorrect lining selection 6. post construction lining failure 7. increased maintenance costs

(continued)

Hazard			*Consequence of event on tunnel*	*Possible Mitigation Measures prior to and during construction to reduce likelihood or consequence of hazard Notes * indicates technique more likely than others to be useful*	*Impact of mitigation measures and Risk*			*Summary of main residual risks*
Number	*Title*	*Example of circumstances*			*Likelihood (L)*	*Consequence (C))*	*Risk (L x C)*	
15	Squeezing ground	as 14 above	M	1. GI boreholes / structural interpretation 2. pre-construction geological desk study and risk assessment 3. probe drilling ahead of face and beneath tunnel to detect areas of weak ground. Monitoring rates of drilling and character of drill cuttings* 4. identification of mining features 5. geological mapping to identify weak or disturbed strata* 6. use of copy cutters in weak ground to increase tunnel diameter* 7. Consider also likelihood of hazards: 14	M	M	M	1. failure to identify ground conditions 2. difficulties in mowing TBM forward 3. difficulties in placing lining 4. delays in program 5. incorrect lining selection 6. post construction failure of tunnel lining 7. increased maintenance costs

16	Shock loading of tunnel lining	collapse of karst, old mine workings above, below or adjacent to tunnel lining. Earthquake, fault rupture	H	1. GI boreholes proving of extensive karst features or open mineworkings 2. pre-construction geological desk study and risk assessment* 3. seismic hazard assessment 4. detection of cavities >> tunnel diameter from probe drilling ahead of and around the face through TBM head. * 5. TSP to detect cavities in vicinity of tunnel bore * 6. use concrete rather than spheroidal steel lining 7. Post construction microgravity survey (gradiometry) along line of tunnel. to locate and determine size of cavities in vicinity (30m radius) of tunnel 8. Consider also likelihood of hazards: 1, 2, 3, 5, 6, 7, 8, 9, 10, 11, 16 and 17	L	H	H	1. failure to detect cavities 2. failure of grouting program 3. inability to grout workings due to depth / topography 4. tunnel collapse and temporary loss of drive 5. selection of incorrect tunnel lining 6. post construction damage to tunnel lining 7. higher risk of water inrush through damaged lining

(continued)

Hazard				*Possible Mitigation Measures prior to and during construction to reduce likelihood or consequence of hazard Notes * indicates technique more likely than others to be useful*	*Impact of mitigation measures and Risk*			*Summary of main residual risks*
Number	*Title*	*Example of circumstances*	*Consequence of event on tunnel*		*Likelihood (L)*	*Consequence (C))*	*Risk (L x C)*	
17	Contaminated ground	industrial and domestic effluents disposed of down mine shafts.	L	9. GI boreholes proving contamination 10. GI investigation identifying post mining uses of mine shafts and adits* 11. pre-construction geological desk study and risk assessment* 12. monitoring of spoil properties during tunnelling* 13. monitoring of tunnel atmosphere 14. Consider also likelihood of hazards: 1, 2, 3, 7, 8, 9, 10, 11, 12, and 16	L	L	L	1. failure to identify contaminants 2. higher risk of hazard 2 from petroleum wastes 3. injury (short and long term) to workforce 4. selection of incorrect tunnel lining 5. corrosion of tunnel lining

1 GI = Ground Investigation.
2 TBM = Tunnel Boring Machine.
3 TSP = Tunnel Seismic Profiling or Prediction.
4 EPBM = Earth Pressure Balance tunnelling Machine.

D.3 EXAMPLE RISK REGISTER

Table of typical tunnelling hazards and possible mitigation measures to be considered during design and construction (modified from Brown, 1999).

Hazard (alphabetical)	*Aspect*	*Mitigations/actions*
Access/egress	To site and work areas	Safe routes and methods
Biological health hazards	Burial sites, etc.	Site investigation Liaison with authorities Avoid disturbance/minimise Appropriate disposal
Confined spaces	Asphyxiation, explosion, flooding, heat, humidity: (a) Existing confined spaces (b) Confined spaces to be constructed/ maintained	Minimise need for working in confined spaces in design Reduce need to enter confined spaces Safe working practices
Contaminated land	Contaminated ground Ground gas	Site investigation to identify Avoid disturbance Ventilation and monitoring Appropriate disposal
Demolition and site clearance	To existing structures	Survey structures and condition Consider stability Plan and phase work to minimise disturbance Fencing and security
	New structures	Design to ensure practical and safe sequence Communication
Earthworks	Ground movements	Site investigation Minimise earthworks Consider effect on existing structures Adequate information to Contractor
Excavation	Collapse, or falls associated with ground movements	GI to identify design constraints Minimise deep excavation Consider existing structures Adequate information to Contractor
	Areas prone to flooding (cavernous limestone, old mine workings etc.)	Site investigation Liaise with authorities re flood potential Safety plan
	Contaminated ground, ground gas, old mine workings	See contaminated land above Probing ahead/geophysics
Fire/explosion	General construction Confined spaces and tunnels	Use non-flammable and benign materials Appropriate plant and machinery Emergency systems including liaison with Emergency Services
Mechanical lifting operations	Impacts and loads	Design works to minimise interaction with existing structures Design detailing and documentation
Maintenance	Access/egress to working environment	Incorporate into design High durability materials and details Allow for access Address residual risks
Manual handling		Minimise Design to ensure legislation compliance

Hazard (alphabetical)	*Aspect*	*Mitigations/actions*
Noise and vibration	General construction	Specification of methods and techniques Set limits and ensure compliance with Health & Safety requirements
Public safety		Minimise disruption and interaction by phasing works and design Security fencing and control Minimise influence of subsequent maintenance works
Services	Overhead and underground services	Agree with utility providers. Avoid diversions Avoid work near services Provide information on services and signage
Site plant/traffic		Ensure adequate site size Traffic control measures Phase works Maximise separation between plant and personnel
Substances hazardous to health		Identify hazards Eliminate or substitute Reduce exposure Proper handling
Temporary stability of structures	Existing structures affected by works	Consider existing structure stability Provide info to Contractor Condition surveys
Unexploded ordnance (bombs)		Site investigation including desk study Survey and use specialists Emergency plan

References

AASHTO. 2007. *LRFD Bridge Design Specification*, 4th edition. Washington DC: American Association of State Highway and Transportation Officials,.

Abrahamson, M.W. 1992. Abbeystead–a legal view. *Proceedings of the ICE, Civil Engineering*, 92, 98–99.

Allum, J.A.E. 1966. *Photogeology and Regional Mapping*. Pergamon Press, 107p.

Al-Sanad, H.A., Shaqour, F.M., Hencher, S.R. & Lumsden, A.C. 1990. The influence of changing groundwater levels on the geotechnical behaviour of desert sands. *Quarterly Journal of Engineering Geology*, 23, 357–364.

Ambraseys, N.N. & Hendron, A.J. Jr. 1969. Dynamic behaviour of rock masses. Chapter 7. In *Rock Mechanics in Engineering Practice*, Stagg, K.G. & Zienkiewicz, O.C. (eds.), New York: Wiley, 203–236.

Ambraseys, N.N. & Bilham, R. 2011. Corruption kills. *Nature*, 469, 153–155.

Ambraseys, N.N. & Srbulov, M. 1995. Earthquake-induced displacements of slopes. *Soil Dynamics & Earthquake Engineering*, 14, 59–72.

Ambraseys, N.N., Simpson, K.A. & Bommer J.J. 1996. Prediction of horizontal response spectra in Europe. *Earthquake Engineering & Structural Dynamics*, 25, 371–400.

Ameen, M.S. 1995. Fractography: Fracture topography as a tool in fracture mechanics and stress analysis. An introduction. In *Fractography: Fracture Topography as a Tool in Fracture Mechanics and Stress Analysis*, Ameen, M.S. (ed.), Geological Society London Special Publication No. 92, 1–10.

American Society for Testing and Materials. 1999. *Standard Practice for Rock Core Drilling and Sampling of Rock for Site Investigation*, ASTM, D 2113-99, 20p.

Amontons, G. 1699. *Mémoires de l' Académie Royale A*, 257–282.

Anderson, A.E., Weiler, M., Alila, Y. & Hudson, R.O. 2008. Dye staining and excavation of a lateral preferential flow network. *Hydrology and Earth System Sciences Discussions*, 5, 1043–1065.

Anderson, E.M. 1951. *The Dynamics of Faulting*. Edinburgh: Oliver and Boyd, 206p.

Anon, 1972. Working party report on the preparation of maps and plans in terms of engineering geology. Q. Jl Engng. Geol. 5, 293–381.

Anon. 1970. The logging of rock cores for engineering purposes. *Quarterly Journal of Engineering Geology*, 3, 1–14.

Anon. 1985. Liquefied petroleum gas caverns at South Killingholme. *Photographic Feature. Quarterly Journal of Engineering Geology*, 18, 1, ii–iv.

Anon. 1995. The description and classification of weathered rocks for engineering purposes. Geological society engineering group working party report. *Quarterly Journal of Engineering Geology*, 28, 207–242.

Association of Geotechnical & Geoenvironmental Specialists. 2022. 2022. *UK Specification for Ground Investigation*, Third edition, ICE Publishing, 328p

Atkinson, J.H. 2000. Non-linear soil stiffness in routine design. *Géotechnique*, 50, 487–508.

Atkinson, J.H. 2007. *The Mechanics of Soils and Foundations*, 2nd edition. Taylor & Francis, 442p.

Australian Standards. 1993. *Geotechnical Site Investigations*. AS-1726.

Baecher, G.B. & Christian, J.T. 2003. *Reliability and Statistics in Geotechnical Engineering*. John Wiley & Sons Ltd., 605p.

Bahat, D., Grossenbacher, K. & Karasaki, K. 1999. Mechanism of exfoliation joint formation in granitic rocks, Yosemite National Park. *Journal of Structural Geology*, 21, 85–96.

Bandis, S.C. 1980. *Experimental Studies of Scale Effects on Shear Strength, and Deformation of Rock Joints*. unpublished PhD thesis, the University of Leeds, 385p plus appendices.

Bandis, S.C., Lumsden, A.C. & Barton, N.R. 1981. Experimental studies of scale effects on the shear behaviour of rock joints. *International Journal of Rock Mechanics and Mining Sciences and Geomechanics Abstracts*, 18, 1–21.

Banyard, J.K, Coxon, R.E. & Johnston, T.A. 1992. Carsington reservoir–reconstruction of the dam. *Proceedings of the ICE–Civil Engineering*, 92, 106–115.

Barla, G. & Pelizza, S. 2000. TBM tunnelling in difficult conditions. *Proceedings of GeoEng 2000*, Melbourne, Australia, 329–354.

Barla, G. and Jarre, P. 1993. Tunnelling under an industrial waste landfill in Italy: Environmental controls and excavation procedures. *Safety and Environmental Issues in Rock Engineering, Eurock '93*, Lisbon: Balkema, 259–266.

Barla, G., Barla, M., Camusso, M. & Martinotti, M.E. 2007. Setting up a new direct shear testing apparatus. *Proceedings of the 11th Congress of the International Society for Rock Mechanics*, 1, 415–418.

Barla, G., Paronuzzi, P. The 1963 Vajont landslide: 50th Anniversary. *Rock Mechanics and Rock Engineering*, 46, 1267–1270.

Barton, N. 2002. Some new Q-value correlations to assist in site characterization and tunnel design. *International Journal of Rock Mechanics and Mining Science*, 39, 185–216.

Barton, N., Choubey, V. 1977. The shear strength of rock joints in theory and practice. *Rock Mechanics*, 10, 1–54.

Barton, N.R. 1973. Review of a new shear-strength criterion for rock joints. *Engineering Geology*, 7, 287–332.

Barton, N.R. 1990. Scale effects or sampling bias? *Proceedings of 1st International Workshop on Scale Effects in Rock Masses*, Löen, Norway, 31–55.

Barton, N.R. 2000. *TBM Tunnelling in Jointed and Faulted Rock*. Rotterdam: Balkema, 173p.

Barton, N.R. 2003. TBM or drill-and-blast. *Tunnels & Tunnelling International*, June, 20–22.

Barton, N.R. 2005. Comments on "A critique of QTBM". *Tunnels & Tunnelling International*, July, 16–19.

Barton, N.R. & Bandis, S.C. 1990. Review of predictive capabilities of JRC-JCS model in engineering practice. In *Rock Joints, Proceedings International Symposium on Rock Joints*, Loen, N., Barton, N. and Stephansson, O. (eds.), Rotterdam: Balkema, 603–610.

Batchelor, C.L., Christie, F.D.W., Ottesen, D. et al. 2023. Rapid, buoyancy-driven ice-sheet retreat of hundreds of metres per day. *Nature*, 617, 105–110.

Bates, A.D. 1981. Profit of loss pivot on pre-dredging surveys. *Dredging + Port Construction*, April 1981, UK: Intec Press. [Referenced in PIANC (2000) Site investigation requirements for dredging works, Report of Working Group No. 23, 32p.]

Baynes, F.J. 2007. Sources of geotechnical risk. *Engineering Geology in Geotechnical Risk Management*, Hong Kong Regional Group of the Geological Society of London, 1–12.

Baynes, F.J. & Parry, S. 2022. Guidelines for the development and application of engineering geological models on projects. *International Association for Engineering Geology and the Environment (IAEG) Commission 25 Publication No. 1*, 129p.

Baynes, F.J., Parry, S. & Novotný, J. 2020. Engineering geological models, projects and geotechnical risk. *Quarterly Journal of Engineering Geology and Hydrogeology*. https://doi.org/10.1144/qjegh2020-080

Beale, G. & Read, J. 2014.*Guidelines for Evaluating Water in Pit Slope Stability*. CRC Press, Taylor & Francis Group, A Balkema Book (CSIRO Publishing), 614p.

Beitnes, A. 2005. Lessons to be learned from Romeriksporten. *Tunnels & Tunnelling International*, June, 36–38.

Bell, D.K. & Davidson, B.J. 1996. Response identification of Pacoima Dam for the 1994 Northridge Earthquake. *Eleventh World Conference on Earthquake Engineering*, Oxford, Disc 2, Paper No. 774.

Bell, F.G. 1999. *Geological Hazards*. Taylor & Francis.

Benson, R. 1989. Design of unlined and lined pressure tunnels. *Tunnelling and Underground Space Technology*, 4, 2, 155–170

Bieniawski, Z.T. 1976. Rock mass classification in rock engineering. In *Exploration for Rock Engineering, Proceedings of Symposium*, 1, Z.T. Bieniawski (ed.), 97–106.

Bieniawski, Z.T. 1984. *Rock Mechanics Design in Mining and Tunnelling*. A.A. Balkema, Rotterdam.

Bieniawski, Z.T. 1989. *Engineering Rock Mass Classifications*. New York: Wiley, 251p.

Bird, M.I., Chang, C.H., Shirlaw, J.N., Tan, T.S., & Teh, T.S. 2003. The age and origin of the quaternary sediments of Singapore with emphasis on the Marine Clay. *Proceedings of Underground Singapore 2003*, Singapore, 428–440.

Bisci, C., Dramis, F. & Sorriso-Valvo, M. 1996. Rock Flow (Sackung). Chapter 7.2 in *Landslide Recognition*, Dikau, R, Brunsden, D, Schrott, L & Ibsen, M.L. (eds.), Wiley, 150–160.

Black, J.H. 2010. The practical reasons why slug tests (including falling and rising head tests) often yield the wrong value of hydraulic conductivity. *Quarterly Journal of Engineering Geology*, 43, 345–358.

Black, J.H., Barker, J.A. & Woodman, N.D. 2007. *An investigation of "sparse channel networks"*. Report No. R-07-35. *Svensk Kärnbränslehantering AB (SKB)*, 123p.

Blake, J.R., Renaud, J.-P. Anderson, M.G. and Hencher, S.R. 2003. Retaining wall impact on slope hydrology using a high-resolution finite element model. *Computers and Geotechnics*, 30, 6, 431–442.

Blockley, D.I. 2011. Engineering safety. *Proceedings of the Institution of Civil Engineers: Forensic Engineering*, 164, FE1, 7–13.

Blyth, F.G.H. & de Freitas, M.H. 1984. *A Geology for Engineers*, 7th edition. Edward Arnold, 325p.

Blyth, F.G.H. 1967. *A Geology for Engineers*. 5th Edition. Edward Arnold (Publishers) Ltd., 351p.

Bo, M.W., Na, Y.M. & Arulrajah, A. 2009. Densification of granular soil by dynamic compaction. *Proceedings of the Institution of Civil Engineers: Ground Improvement*, 162, 121–132.

Bock, H. 2006. Common ground in engineering geology, soil mechanics and rock mechanics: Past, present and future. *Bulletin Engineering Geology & Environment*, 65, 209–216.

Bolton, M.D. 1979. *A Guide to Soil Mechanics*, 1st edition, Palgrave, 456p.

Bolton, M.D. 1986. The strength and dilatancy of sands. *Géotechnique*, 36, I, 65–78.

Bommer, J.J. & Martinez-Pereira, A. 1999. The effective duration of earthquake strong motion. *Journal of Earthquake Engineering*, 3, 127–172.

Bond, A.J. & Harris, A. 2008. *Decoding Eurocode 7*. London: Taylor and Francis, 598p.

Bond, A.J. & Simpson, B. 2010. Pile design to Eurocode 7 and the UK National Annex. Part 2: UK National Annex. *Ground Engineering*, January, 28–31.

Boulanger, R.W. & Idriss, I.M. 2014. *CPT and SPT Liquefaction Triggering Processes*. University of California, Center for Geotechnical Modelling. Report No. UCD/CGM-14/01, 134p.

Bowden, F.P. & Tabor, D. 1950. *The Friction and Lubrication of Solids. Part I*. Oxford: Clarendon Press.

Bowden, F.P. & Tabor, D. 1964. *The Friction and Lubrication of Solids. Part II*. Oxford: Clarendon Press.

Bowles, J.E. 1996. *Foundation Analysis and Design*, 5th edition. McGraw Hill, 1175p.

Brady, B.H.G. & Brown, E.T. 2004. *Rock Mechanics for Underground Mining*, 3rd edition. Springer.

Brand, E.W., Hencher, S.R. & Youdan, J.D. 1983. Rock slope engineering in Hong Kong. *Proceedings* of the 5th International Rock Mechanics Congress, Melbourne, 17–24.

Bray, R.N., Bates, A.D. & Land, J.M. 1997. *Dredging. A Handbook for Engineers*, 2nd edition. Butterworth-Heinemann, 434p.

Bricker, S., Terrington, R., Dobbs, M., Kearsey, T., Burke, H., Anrhardt, R. & Thorpe, S. 2021. Urban Geoscience report–The value of geoscience data, information and knowledge for transport and linear infrastructure projects. *British Geological Survey Open Report*, OR/21/065. 67p.

Bridges, M.C. 1990. Identification and characterisation of sets of fractures and faults in rock. *Proceedings of the International Symposium on Rock Joints*, Loen, Norway, 19–25.

British Standards Institution. 1964. BS3618:1964, Sec. 5. *Glossary of Mining Terms, Geology*, 18p.

British Standards Institution. 1981. BS 6031:1981. Code of Practice for Earthworks.

British Standards Institution. 1982. BS 5930:1981. Code of Practice for Site Investigations (Formerly CP 2001), 147p.

British Standards Institution. 1986. BS 8004:1986. Code of Practice for Foundations.

British Standards Institution. 1989. BS 8081:1989. Code of Practice for Ground Anchorages.

British Standards Institution. 1990. BS 1377:1990. Methods of Test for Soils for Civil Engineering Purposes (9 parts).

British Standards Institution. 1999. BS 5930:1999. Code of Practice for Site Investigation. 206p.

British Standards Institution. 2000. BS EN 1537:2000. Execution of Special Geotechnical Work–Ground Anchors.

British Standards Institution. 2001. BS 10175:2001. Code of Practice for Investigation of Potentially Contaminated Sites.

British Standards Institution. 2002. EN ISO 14688-1. Geotechnical Investigation and Testing–Identification and Classification of Soil–Part 1: Identification and Description.

British Standards Institution. 2003. EN ISO 14689-1. Geotechnical Investigation and Testing–Identification and Classification of Rock–Part 1: Identification and Description.

British Standards Institution. 2004. EN 1997-1:2004. Eurocode 7: Geotechnical Design. Part 1: General Rules. Code of Practice.

British Standards Institution. 2019. BS 5975. The Management of Temporary Works in the Construction Industry.

British Standards Institution. 2020. BS 5930:2015+A1:2020 Code of Practice for Ground Investigation

British Tunnelling Society. 2003. The Joint Code of Practice for Risk Management of Tunnel Works in the UK, 18p.

British Tunnelling Society. 2005. *Closed-Face Tunnelling Machines and Ground Stability*. London: Thomas Telford.

Broch, E., Myrvang, A.M. &. Stjern, G. 1996. Support of large rock caverns in Norway. *Tunnelling and Underground Space Technology*, 11, 11–19.

Brodie, K., Fettes, D. & Harte B. 2007. Structural terms including fault rock terms. In *Metamorphic Rocks: A Classification and Glossary of Terms* D. Fettes & J. Desmonds (eds.), 24–31.

Broms, B.B. & Lai, P.H. 1995. The republic plaza in Singapore–Foundation design, *Bengt B. Broms Symposium on Geotechnical Engineering*, Singapore, 3–24.

Brook, N. 1993. The measurement and estimation of basic rock strength. *Comprehensive Rock Engineering*, 3, Hudson, J. (ed.), Pergamon Press, 41–66.

Brown, E. T. (ed.) 1981. *Suggested Methods for Determining Shear Strength.* In Rock Characterisation Testing and Monitoring, Pergamon Press, Oxford.

Brown, E.T. 2008. Estimating the mechanical properties of rock masses. *Proceedings, 1st Southern Hemisphere International Rock Mechanics Symposium*, 1, September, Y. Potvin, J. Carter, A. Dyskin & R. Jeffrey (eds.), Perth, 16–19, 3–22.

Brown, R.H.A. 1999. The management of risk in the design and construction of tunnels. *Korean Geotechnical Society, Tunnel Committee Seminar, 21st Seminar*, 21st September, 22p.

Brox, D. 2019. Hydropower tunnel failures: Risks and causes. *Tunnels and Undersground Cities. Engineering and Innovation meet Archaeology, Architecture and Art.* CRC Press

Brunsden, D. 2002. Geomorphological roulette for engineers and planners: Some insights into an old game. *Quarterly Journal of Engineering Geology & Hydrogeology*, 35, 101–142.

Buckingham, R.J. 2003. *Problems Associated with Water Ingress into Hard Rock Tunnels.* MSc (Applied Geosciences) dissertation, Hong Kong University, unpublished, 39p.

Building Research Establishment. 1993. Low-rise buildings on shrinkable clay soils: Part 1. *BRE Digest*, 240, London: CRC.

Buildings Department. 2017. *Code of Practice for Foundations 2017*, Hong Kong Government, 111p.

Bulletin of Atomic Scientists. 1976. *Preface to the Royal Commission on Environmental Pollution.* Sixth Report, Nuclear Power and the Environment.

Bunce, C.M., Cruden, D.M. and Morgenstern, N.R. 1997. Assessment of the hazard from rock-fall on a highway. *Canadian Geotechnical Journal*, 34, 344–356.

Burbidge, M.C &. Burland, J.B. 1985. Settlement of foundations on sand and gravel. *Proceedings of the Institution of Civil Engineers*, 78, 1325–1381.

Burland, J. 2007. Terzhagi: Back to the future. *Bulletin Engineering Geology and Environment*, 66, 29–33.

Burland, J.B. 1987. The teaching of soil mechanics: A personal view. *Proceedings of the Ninth European Conference on Soil Mechanics and Foundation Engineering*, Dublin, 1427–1447.

Burland, J.B. 1990. On the compressibility and shear strength of natural clays. *Géotechnique*, 40, 329–378.

Burland, J.B. 2008. Interlocking, and peak and design strengths. Discussion of Schofield, A.N., 2006, *Géotechnique*, 58, 527.

Burland, J.B. & Wroth, C.P. 1975. Settlement of buildings and associated damage. *Building Research Establishment Current Paper*, CP33/75, 44p.

Burnett, A.D., Brand, E.W. & Styles, K.A. 1985. Terrain classification mapping for a landslide inventory in Hong Kong. *Proceedings IVth International Conference and Field Workshop on Landslides*, Tokyo, 63–68.

Byerlee, J.D. 1978. Friction of rocks. *Pure Applied Geophysics*, 116, 615–626.

CALTRANS. 2010. *Soil and Rock Logging, Classification, and Presentation Manual*, 81p.

Carter, T.G., de Graaf, P., Booth, P., Barrett, S. & Pine, R. 2002. Integration of detailed field investigations and innovative design key factors to the successful widening of the Tuen Mun Highway. *Proceedings of the 22nd Annual Seminar*, organised by the Geotechnical Division of the Hong Kong Institution of Engineers, 8th May, 2002, 187–201.

Carter, T.G., Diederichs, M.S. and Carvalho, J.L. 2008. Application of modified Hoek-Brown transition relationships for assessing strength and post-yield behaviour at both ends of the rock competence scale. *Proc. SAIMM*, 108, 325–338.

Carvalho, J.L., Carter, T.G., and Diederichs, M.S. 2007. An approach for prediction of strength and post yield behaviour for rock masses of low intact strength. *Proceedings 1st*

Canadian-US Rock Mechanics Symposium. Meeting Society's Challenges and Demands, Vancouver, 249–257.

Casagrande, A. 1965. Classification and identification of soils. *Transactions American Society Civil Engineers*, 113, 901–992.

Cederstrom, M., Thorsall, P.-E., Hildenwall, B. & Westberg, S.-B. 2005. Incident with loss of seven post-tensioned 72 ton anchors in a dam. *Dam Safety 2005 Proceedings*, Association of State Dam Safety Officials.

Challinor, J.A. 1964. *A Dictionary of Geology*, 2nd edition. Cardiff: University of Wales Press, 289p.

Chandler, R.J. 1969. The effects of weathering on the shear strength properties of Keuper Marl. *Géotechnique*, 19, 321–334.

Chandler, R.J. 2000. Clay sediments in depositional basins: The geotechnical cycle. *Quarterly Journal Engineering Geology and Hydrogeology*, 33, 7–39.

Chaplow, R. 1996. The geology and hydrogeology of Sellafield: An overview. *Quarterly Journal of Engineering Geology*, 29, S1–S12.

Chen, T.-C., Lin, M.-L. & Hung, J.-J. 2003. Pseudostatic analysis of Tsao-Ling rockslide caused by Chi-Chi earthquake. *Engineering Geology*, 71, 31–47.

Cheng, Y. & Liu, S.C. 1993. Power caverns of Mingtan pumper storage project, Taiwan. *Comprehensive Rock Engineering*, 5, Surface and Underground Project Case Histories (Editor in Chief, J.A. Hudson; Volume Editor, E. Hoek), Pergamon Press, 111–132.

Chigira, M. & Yahi, H. 2006. Geological and geomorphological characteristics of landslides triggered by the 2004 Mid Niigata prefecture earthquake in Japan. *Engineering Geology*, 82, 202–221.

Chigira, M., Wang, W-N, Furya, T. & Kamai, T. 2003. Geological causes and geomorphological precursors of the Tsaoling landslide triggered by the 1999 Chi-Chi earthquake, Taiwan. *Engineering Geology*, 68, 259–273.

China Ministry of Construction. 2009. Code of Investigation of Geotechnical Engineering. National Standard of the People's Republic of China, GB5OO21-2001 (2009 edition), 269p.

Cho, S-M., Ahn, S-S., Lee, Y-K. & Oh, S-T. 2009a. Incheon Bridge: Facts and the state of the art. *Proceedings of the International Commemorative Symposium for the Incheon Bridge*, 18–28.

Cho, S-M., Kim, J-H., Lee, H-G., Kim, Z-C., Shin, S-H., Song, M-J., Kim, D-J. & Park, Y-H. 2009b. Pile load tests on the Incheon bridge project. *Proceedings of the International Commemorative Symposium for the Incheon Bridge*, 488–499.

Chung, S.G., Kim, G.J., Kim, M.S. and Ryu, C.K. 2007. Undrained shear strength from field vane test on Busan clay. *Marine Georesources & Geotechnology*, 25, 167–179.

Clark, C.D., Ely, J.C., Fabel, D. & Bradley, S. 2022. BRITICECHRONO maps and GIS data of the last British-Irish Ice Sheet 31 to15 ka, including model reconstruction, geochronometric age spreadsheet, palaeotopographies and coastline positions. Pangaea. https://doi.org/10.1594/PANGAEA.945729

Clayton, C.R.I, Simons, N.E. & Mathews, M.C. 1995. *Site Investigation. A Handbook for Engineers*, 2nd edition. Granada, 424p.

Clayton, C.R.I. 1995. The standard penetration test (SPT): Methods and use. *CIRIA Report*, 143, 144p.

Clayton, C.R.I. 2001. *Managing Geotechnical Risk: Improving Productivity in UK Building and Construction*. London: Thomas Telford, 80p.

Clayton, C.R.I. 2011. Stiffness at small strain: Research and practice. *Géotechnique*, 61, 5–37.

Clough, W., Sitar, N. & Bachus, R.C. 1981. Cemented sands under static loading. *ASCE, Journal of the Geotechnical Engineering Division*, 107,6, 799–817

Clover, A.W. 1986. Slope stability on a site in the volcanic rocks of Hong Kong. *Proceedings of the Conference on Rock Engineering and Excavation in an Urban Environment, Hong Kong*, 121–134.

Coburn, A. & Spence, R. 1992. *Earthquake Protection*. Wiley, 355p.

Cole, K.W. 1988. Foundations. ICE Works Construction Guides, Thomas Telford Ltd., 82p.

Collins, K. & McGowan A.M. 1974. The form and function of micro-fabric features in a variety of natural soils. *Géotechnique*, 24, 223–254.

Combault, J., Morand, P. & Pecker, A. 2000. Structural response of the Rion Antirion Bridge. *Proceedings of the 12th World Conference on Earthquake Engineering*. Auckland, Australia.

Combault, J., Morand, P. & Pecker, A. 2000. Structural response of the Rion-Antirion Bridge. *12th World Conference Earthquake Engineering*, Paper 1609, 7p.

Construction Industry Research and Information Association (CIRIA). 1995. *SP164, Vol III: site investigation & assessment* ISBN: 978-0-86017-398-4.

Cook, J.R. 2022 Reducing future climate impact on rural transport infrastructure in developing countries; The role of engineering geology. *Quarterly Journal of Engineering Geology and Hydrogeology*, 55, 2021–2052.

Cooper, A.H. & Waltham, A.C. 1999. Photographic feature: Subsidence caused by gypsum dissolution at Ripon, North Yorkshire. *Quarterly Journal of Engineering Geology*, 32, 4, 305–310.

Cosgrove, J. 2007. The use of shear zones and related structures as kinematic indicators: A review. *Geological Society of London Special Publications*, 272,1,59–74

Coulson, J.H. 1971. Shear strength of flat surfaces of rock. *Proceedings of the 13th US Symposium on Rock Mechanics*, Urbana, Illinois, 77–105.

Craig, R.F. 1992. *Soil Mechanics*, 5th edition. Chapman & Hall, 427p.

Crawford, A.M. and Curran, J.H. 1982. The influence of rate and displacement dependant, shear resistance on the response of rock slopes to seismic loads. *International Journal of Rock Mechanics & Mining Sciences and Geomechanics Abstracts*, 19, 1–8.

Cripps, J.C., Reid, J.M., Czerewko, M.A. & Longworth, T.I. 2019. Tacklingproblems in civil engineering caused by the presence of pyrite. *QuarterlyJournal of Engineering Geology and Hydrogeology*, 52, 481–500.

Culshaw, M.G. 2005. From concept towards reality: Developing the attributed 3D geological model of the shallow subsurface. *Quarterly Journal of Engineering Geology and Hydrogeology*, 38, 231–284.

Cunha, A.P. 1990. Scale effects in rock mechanics. *Scale Effects in Rock Masses (ed. Pinto da Cunha), Proceedings 1st International Workshop on Scale Effects in Rock Masses*, Löen, Norway, 3–27.

Darracott, B.W. & McCann, D.M. 1986. Planning engineering geophysical surveys. *S I Practice: Assessing BS5930, Geological Society, Engineering Geology Special Publication No.* 2, Hawkins, A.B (ed.), 85–90.

Darwin, C.R. 1842. *The Structure and Distribution of Coral Reefs*. Being the first part of the geology of the voyage of the Beagle, under the command of Capt. FitzRoy, R.N. during the years 1832 to 1836. London: Smith Elder and Co.

Das, B.M. & Sivakugan, N. 2007. Settlements of shallow foundations on granular soil–an overview. *International Journal of Geotechnical Engineering*, 1, 19–29.

Davies, R.J., Mathioas, S.A., Swarbrick, R.E. & Tingay, M.J. 2011. Probabilistic longevity estimate for the LUSI mud volcano, East Java. *Journal of the Geological Society, London*, 168, 517–523.

Davis, E.H. & Poulos, H.G. 1967. Laboratory investigations of the effects of sampling. *Civil Engineering Transactions (Australia)*, CE9, 86–94.

Davis, G.H. & Reynolds, S.J. 1996. *Structural Geology of Rocks and Regions*, 2nd edition. John Wiley & Sons, 776p.

Davis, J., Essex, R., Farooq, I. & Drake, A. 2023. *Geotechnical Baseline Reports: A Guide to Good Practice*, C807D, Ciria.

Davy, P., Bour, O., De Dreuzy, J.-R., & Darcel, C. 2006. Flow in multiscale fractal fracture networks. *Fractal Analysis for Natural Hazards*, Geological Society, London, Special Publications, 261, 31–35.

Daws, E. 2018. The impact of climate change on glaciers in the Chamonix Valley, French Alps. eGG: The e-journal for undergraduate research in Environment(e), Geography(G) and Geology(G),Volume2(2018):1–11

de Freitas, M.H. 2009. Geology; Its principles, practice and potential for Geotechnics. *Quarterly Journal of Engineering Geology and Hydrogeology*, 42, 397–441.

Dearman, W.R. & Coffey, J.R. 1981. Effects of evaporite removal on the mass properties of limestone. *Bulletin of the International Association of Engineering Geology*, 24, 91–96.

Dearman, W.R. & Fookes, P.G. 1974. Engineering geological mapping for civil engineering practice in the United Kingdom. *Quarterly Journal of Engineering Geology and Hydrogeology*, 7, 223–256.

Deere D.U. 1971. The foliation shear zone–an adverse engineering geologic feature of metamorphosed rocks. *Journal of the Boston Society of Civil Engineers*, 60, 4, 1163–1176.

Deere, D.U. & Deere, D.W. 1988. The RQD index in practice. *Proceedings Symposium on Rock Classification for Engineering Purposes*, ASTM Special Technical Publications 984, Philadelphia, 91–101.

Deere, D.U. 1968. Geological considerations. Chapter 1, *Rock Mechanics in Engineering Practice*, Stagg, K.G. & Zienkiewicz, O.C. (eds.), New York: Wiley, 1–20.

deFreitas, M.H. 1993. Discussion on session 3.2: Foundations and underground excavation. *The Engineering Geology of Weak Rock*, Cripps et al. (eds), Rotterdam: Balkema, 493.

DeGraff, J.M. & Aydin, A. 1987. Surface morphology of columnar joints and its significance to mechanics and direction of joint growth. *Geological Society of America Bulletin*, 99, 605–617.

Derbyshire, E. 2001. Geological hazards in loess terrain, with particular reference to the loess regions of China. *Earth-Science Reviews*, 54, 231–260.

Dering, C. 2003. Discussion. *Proceedings 14th SE Asian Geotechnical Conference*, 3, Hong Kong, 41–43.

Dershowitz, W. & LaPointe, P. 1994. Discrete fracture approaches for oil and gas applications. *Proceedings 1st North American Rock Mechanics Symposium, Rock Mechanics*, Nelson & Laubach (eds.), Rotterdam: Balkema, 19–30.

Dershowitz, W., Lee, G., Geier, J., Foxford, T., LaPointe, P. & Thomas, A. 1996. *FracMan Interactive Discrete Feature Data Analysis, Geometric Modeling, and Exploration Simulation, User Documentation, Version 2.5*. Redmond, Washington: Golder Associates.

Devonald, D.M., Thompson, J.A., Hencher, S.R. & Sun, H.W. 2009. Geomorphological landslide models for hazard assessment–a case study at Cloudy Hill. *Quarterly Journal Engineering Geology*, 42, 473–486.

Dobie, M.J.D. 1987. Slope instability in a profile of weathered norite. *Quarterly Journal of Engineering Geology and Hydrogeology*, 20, 279–286.

Domenico, P. A. & Schwartz, F. W. 1990. *Physical and Chemical Hydrogeology*. New York, Chichester, Brisbane, Toronto, Singapore: John Wiley & Sons, 824 p.

Döse, C. Stråhle, A., Rauséus, G., Samuelsson,E. & Olsson, O. 2008. *Revision of BIPS-Orientations for Geological Objects in Boreholes from Forsmark and Laxemar*. Svensk Kärnbränslehantering AB (SKB Report No. P-08-37), 97p.

Douglas, P.M. & Voight, B. 1969. Anisotropy of granites: A reflection of microscopic fabric, *Géotechnique*, 19, 376–398.

Dowding, C.H. 1985. *Blast Vibration Monitoring and Control.* Prentice-Hall Inc., 297p.

Dowrick, D.J. 1987. *Earthquake Resistant Design for Engineers and Architects*, 2nd edition. John Wiley & Sons.

Dumbleton, M.J. & West, G. 1970 *Air Photograph Interpretation for Road Engineers in Britain.* Ministry of Transport, RRL Report, LR369, Crowthorne: Road Research Laboratory.

Dunnicliff, J. 1993. *Geotechnical Instrumentation for Monitoring Field Performance.* New York: Wiley, 577p.

Dusseault, M.B. & Morgenstern, N.R. 1979. Locked sands. *Quarterly Journal of Engineering Geology*, 12, 117–131.

Early, K.R. & Skempton, A.W. 1972. Investigations of the landslide at Walton's wood, Staffordshire. *Quarterly Journal of Engineering Geology*, 5, 19–41.

Eberhardt, E., Stead, D. and Coggan, J.S. 2004. Numerical analysis of initiation and progressive failure in natural rock slopes–the 1991 Randa rockslide. *International Journal of Rock Mechanics and Mining Sciences*, 41, 69–87.

Eberl, D.D. 1984. Clay formation and transformation in rocks and soils. *Philosophical Transactions, Royal Society, London*, A311, 241–257.

Ebuk, E.J. 1991. *The Influence of Fabric on the Shear Strength Characteristics of Weathered Granites.* Unpublished PhD thesis, the University of Leeds, 486p.

Ebuk, E.J., Hencher, S.R. & Lumsden, A.C. 1993. The influence of structure on the shearing mechanism of weakly bonded soils. *Proceedings 26th Annual Conference of the Engineering Group of the Geological Society, The Engineering Geology of Weak Rock*, Leeds, 207–215.

Edwards, R.J.G., 1971. The practical application of rock bolting–Jeffrey's Mount rock cut on the M6. *Journal Institution Highway Engineers*, 17, 12, 21–27.

Egan, D. 2008. The ground: Clients remain exposed to unnecessary risk. Proceedings of the Institution of Civil Engineers, *Geotechnical Engineering*, 161, 189–195.

Eggers, M.J. 2016. Diversity in the science and practice of engineering geology. *Geological Society, London, Engineering Geology Special Publications*, 27, 1–18.

Ehlers J. & Gibbard P.L. 2007. The extent and chronology of Cenozoic global glaciation. *Quaternary International*, 164–165, 6–20.

Eisma, D. (ed.) 2006. *Dredging in Coastal Waters.* Taylor & Francis, 244p.

Elmo, D., Stead, D., Yang, B., Marcato, G. & Borgatti, L. 2022. A new approach to characterise the impact of rock bridges in stability analysis. *Rock Mechanics and Rock Engineering*, 55, 2551–2569.

Emerson, W.W. 1967. A classification of soil aggregates based on their coherence in water. *Australian Journal of Soil Research 1967*, 5, 47–57.

Emery, K.O. & Aubrey, D.G. 1991 *Sea Levels, Land Levels, and Tide Gauges.* New York: Springer-Verlag, 237p.

Engelder, T. 1985. Loading paths to joint propagation during a tectonic cycle: An example from the Appalachian Plateau, USA. *Journal of Structural Geology*, 7, 459–476.

Engelder, T. 1999. Transitional-tensile fracture propagation: A status report. *Journal of Structural Geology*, 21, 1049–1055.

Engelder, T. & Peacock, D. 2001. Joint development normal to regional compression during flexural-slow folding: The Listock buttress anticline, Somerset, England. *Journal of Structural Geology*, 23, 259–277.

Entwisle, D.C., Hobbs, P.R.N., Northmore, K.J., Skipper, J., Raines, M.R., Self, S.J., Ellison, R.A. & Jones, L.D. 2013. *Engineering geology of British rocks and Soils–Lambeth Group. British Geological Survey Open Report*, OR/13/006. 316p.

Erlich, P.R. 1969a. Discussion. *The Optimum Population for Britain. Proceedings of a Symposium at the Royal Geographical Society*, 25 and 26 September, 1969, London, 129p.

Erlich, P.R. 1969b. Population control or Hobson's choice. The optimum population for Britain. *Proceedings of a Symposium at the Royal Geographical Society*, 25 and 26 September, 1969, London, 151–162.

Essex, R.J. 2023. Geotechnical Baseline Reports: Suggested Guidelines: 154 (Manuals and Reports on Engineering Practice). Prepared by the Task Committee on Geotechnical Baseline Reports of the Construction Institute of ASCE Geotechnical Baseline Reports: Suggested Guidelines, MOP 154, 110p.

Ewing, M. & Donn, W.L. 1956. A theory of ice ages. *Science* 123 3207, 1061–1066.

Fecker, E. & Rengers, N. 1971. Measurement of large scale roughness of rock planes by means of profilograph and geological compass. *Proceedings Symposium on Rock Fracture*, Paper, France: Nancy, 1–18.

Fell, R., Ho, K.K.S., Lacasse, S. & Leroi, E. 2005. A framework for landslide risk assessment and management. *Landslide Risk Management*, Hung, Fell, Couture & Eberhardt (eds.), 3–25. Taylor & Francis.

Fletcher, C.J.N. 2004. *Geology of Site Investigation Boreholes from Hong Kong*. Applied Geoscience Centre, Hong Kong Construction Association and AGS Hong Kong, 132p.

Fletcher, C.J.N., Wightman, N.R. & Goodwin, C.R. 2000. Karst related deposits beneath Tung Chung new town: Implications for deep foundations. *Proceedings Conference on Engineering Geology HK 2000*, Hong Kong: Institution of Mining and Metallurgy, 139–149.

Fogg, G.E., Noyes, C.D. & Carle, S.F. 1998. Geologically based model of heterogeneous hydraulic conductivity in an alluvial setting. *Hydrogeology Journal*, 6, 1, 131–143.

Fookes P.G. 1997. Geology for engineers: The geological model, prediction and performance. *Quarterly Journal of Engineering Geology*, 30, 293–431.

Fookes, P.G. & Parrish, D.G. 1969. Observations on small-scale structural discontinuities in the London Clay and their relationship to regional geology. *Quarterly Journal of Engineering Geology*, 1, 217–240.

Fookes, P.G. & Hawkins, A.B. 1988. Limestone weathering: Its engineering significance and a proposed classification scheme. *Quarterly Journal of Engineering Geology*, 21, 7–31.

Fookes, P.G., Baynes, F.J. & Hutchinson, J.N. 2000. Total geological history: A model approach to the anticipation, observation and understanding of ground conditions. *GeoEng 2000*, 1, Melbourne, 370–460.

Fookes, P.G., Sweeney, M., Manby, C.N.D., & Martin, R.P. 1985. Geological and geotechnical engineering aspects of low-cost roads in mountainous terrain *Engineering Geology*, 21, 1–13, 17–53, 57–65, 69–85, 91–99, 102, 105–109, 115–131, 133, 135–152.

Fry, N. 1984. *The Field Description of Metamorphic Rocks*. Milton Keynes, UK: The Open University Press, 110p.

Fu, R., Shang, A. & Naudts, A. 2007. Pre-excavation grouting through water bearing zone under high pressure under extreme flow conditions. *Toronto, Rapid Excavation and Tunneling Conference*, June, 14p.

Fujii, Y., Takemura, T., Takahashi, M. & Lin, W. 2007. Surface features of uniaxial tensile fractures and their relation to rock anisotropy in Inada granite. *International Journal of Rock Mechanics and Mining Science*, 44, 98–107.

Geotechnical Control Office. 1979. *Geotechnical Manual for Slopes*, 1st edition. Hong Kong: Geotechnical Control Office, 242p.

Geotechnical Control Office. 1984a. *Geotechnical Manual for Slopes*, 2nd edition. Hong Kong: Geotechnical Control Office, 295p.

Geotechnical Control Office. 1984b. *Mid-Levels Study: Report on Geology, Hydrology and Soil Properties*. Hong Kong: Geotechnical Control Office, Public Works Department.

Geotechnical Control Office. 1987. *Guide to Site Investigation (Geoguide 2)*. Hong Kong Government, 359p.

Geotechnical Control Office. 1988. *Guide to Rock and Soil Descriptions (Geoguide 3)*. Hong Kong: Geotechnical Engineering Office, 189p.

Geotechnical Engineering Office. 1992. *Guide to Cavern Engineering (Geoguide 4)*. Hong Kong: Geotechnical Engineering Office, 156p.

Geotechnical Engineering Office. 1993. *Guide to Retaining Wall Design (Geoguide 1)*, 2nd edition. Hong Kong: Geotechnical Engineering Office.

Geotechnical Engineering Office. 2000a. *Highway Slope Manual*. Hong Kong: Geotechnical Engineering Office, 114p.

Geotechnical Engineering Office. 2000b. *Technical Guidelines on Landscape Treatment and Bio-engineering of Man-made Slopes and Retaining Walls*. GEO Publication No. 1/2000, Hong Kong: Geotechnical Engineering Office, 146p.

Geotechnical Engineering Office. 2005. *Methods Other than Recompaction for Upgrading Loose Fill Slopes*. GEO Report No. 162, Hong Kong: Geotechnical Engineering Office, 43p.

Geotechnical Engineering Office. 2006. Foundation Design and Construction. GEO Publication No. 1/2006, 348p.

Geotechnical Engineering Office. 2007. *Engineering Geological Practice in Hong Kong*, GEO Publication No. 1/2007, 278p.

Geotechnical Engineering Office. 2009. *Catalogue of Notable Tunnel Failure Case Histories (up to December 2008)*. Hong Kong: Geotechnical Engineering Office.

Gerstner, R., Fey, C., Kuschel, E., Voit, K., Valentin, G., & Zangerl, C. 2023. Polyphase rock slope failure controlled by pre-existing geological structures and rock bridges. *Bulletin of Engineering Geology and the Environment*, 82, 363, 25p.

Giles, D.P. 2020. Introduction to geological hazards in the UK: Their occurrence, monitoring and mitigation. *Geological Society, London, Engineering Geology Special Publications*, 29, 1–41.

Gioda, G. & Sakurai, S. 2005. Back analysis procedures for the interpretation of field measurements in geomechanics. *International Journal for Numerical and Analytical Methods in Geomechanics*, 11, 555–583.

Glossop, R. 1968. The rise of geotechnology and its influence on engineering practice. Eighth rankine lecture. *Géotechnique*, 18, 107–150.

Gnirk, P. 1993. OECD/NEA International Stripa Project Overview. Natural Barriers. SKB: Stockholm, Sweden.

Gomes, J.P., Batista A.L. & Oliveira, S. 2009. Damage-Chemo-Viscoelastic model on the analysis of concrete dams under swelling processes. *Proceedings International Conference on Long Term Behaviour of Dams (LTBD09)*, Graz.

González, J., González-Pastoriza, N., Castro, U., Alejano, L.R. & Muralha, J. 2014. Considerations on the laboratory estimate of the basic friction angle of rock joints. In *Rock Engineering and Rock Mechanics: Structures in and on Rock Masses–Proceedings of EUROCK 2014, ISRM European Regional Symposium*, Alejano, L.R., Perucho, A´., Olalla, C., and Jimenez, R.(eds), 199–204.

Goodman, R., D. Moye, A. Schalkwyk, and I. Javendel. 1965. Groundwater inflow during tunnel driving. *Bulletin Association Engineering Geology*, 2, 39–56.

Goodman, R.E. 1989. *Introduction to Rock Mechanics*. 2nd edition. New York: John Wiley & Sons, 562p.

Goodman, R.E. 1993. *Engineering Geology. Rock in Engineering Construction*, Wiley, 412p.

Goodman, R.E. 2002. Karl Terzaghi's legacy in geotechnical engineering. *Geo-Strata*, October, ASCE.

Goodman, R.E. 2003. Karl Terzaghi and engineering geology. *Geotechnical Engineering*,. Ho, K.K.S & Li, K.S. (eds.), A.A. Balkema, 115–122.

Gordon, T., Scott, M.J. & Statham, I. 1996. The identification of bedding shears and their implications for road cutting design and construction. *Prediction and Performance in Rock Mechanics and Rock Engineering, Eurock '96*, Torino, Italy, 597–603.

Graham, R.H. 1981. Gravity sliding in the Maritime Alps. *Geological Society, London, Special Publications*, 9, 335–352.

Green, P. & Western, R. 1994. Time to Face the Inevitable. A Submission from Friends of the Earth Ltd. To the UK Department of the Environment's Review of Radioactive Waste Management Policy, 72p.

Green, R.G. & Hawkins, A.B. 2005. Rock cut on the A5 at Glyn Bends, North Wales, UK. *Bulletin Engineering Geology and Environment*, 64, 95–109.

Grose, W.J. & Benton, L. (2005). Hull wastewater flow transfer tunnel: Tunnel collapse and causation investigation. *Geotechnical Engineering*, 158, 4, 179–185.

Hack, H.R.G.K. & Huisman, M. 2002. Estimating the intact rock strength of a rock mass by simple means. *9th congress of the International Association for Engineering Geology and the Environment (IAEG); Engineering geology for developing countries*, Durban, South Africa, 1971–1977.

Hack, H.R.G.K. 1998. *Slope Stability Probability Classification, SSPC*, 2nd edition. International Institute for Aerospace Survey and Earth Sciences (ITC), Publication No. 43, 258p.

Haimson, B.C. 1992. Designing pre-excavation stress measurements for meaningful rock characterization. *Rock Characterization, Eurock'92*, Hudson, J.A. (ed.), Thomas Telford, 221–226.

Halcrow Asia Partnership Ltd. 1998b. *Report on the Landslide at Ten Thousand Buddha's Monastery of 2 July 1997*. GEO Report No. 77, Hong Kong: Geotechnical Engineering Office, 96p. www.cedd.gov.hk/eng/publications/geo_reports/geo_rpt077.htm

Halcrow Asia Partnership. 1998a. Report on the Ching Cheung Road Landslide of 3rd August 1997. GEO Report No. 78, Hong Kong: Geotechnical Engineering Office, 47p.

Hancock, P.L. 1985. Brittle microtectonics: Principles and practice. *Journal of Structural Geology*, 7, 437–457.

Hancock, P.L. 1991. Determining contemporary stress directions from neotectonic joint systems. *Philosophical Transactions of the Royal Society Land A*, 337, 29–40.

Haneberg, W.C. 2008. Using close range terrestrial digital photogrammetry for 3-D rock slope modelling and discontinuity mapping in the United States. *Bulletin of Engineering Geology and the Environment*, 4, 457–469.

Hansen, A. 1984. Landslide hazard analysis. In *Slope Instability*, Brunsden, D & Prior, D.B. (eds), New York: John Wiley and Sons, 523–602.

Harris, D.I., Mair, R.J., Love, J.P., Taylor, R.N. & Henderson, T.O. 1994. Observations of ground and structure movements for compensation grouting during tunnel construction at Waterloo station. *Géotechnique*, 44, 691–713.

Hartwell, D.J. 2006. Discussion: Hull wastewater flow transfer tunnel: Tunnel collapse and causation investigation. *Geotechnical Engineering*, 2, 125–126.

Hashash, Y.M.A., Hook, J.J., Schmidt, B. & Yao, J.I.-C. 2001. Seismic design and analysis of underground structures. *Tunnelling and Underground Space Technology*, 16, 247–293.

Haszeldine, R.S. & Smythe, D.K. (eds.) 1996. *Radioactive Waste Disposal at Sellafield, UK. Site Selection, Geological and Engineering Problems*. University of Glasgow, 520p.

Hausfather,A. & Peters, G.P. 2020. Emissions – The "business as usual" story is misleading. *Nature*, 577, 618–620,

Hawkins, A.B. (ed.) 1986. *Site Investigation Practice: Assessing BS 5930*. Geological Society of London, Engineering Geology Special Publication No. 2, 423p.

Hawkins, A.B., Larnach, W.J., Lloyd, I.M. & Nash, D.F.T. 1989. Selecting the location, and the initial investigation of the SERC soft clay test bed site. *Quarterly Journal of Engineering Geology and Hydrogeology*, 22, 281–316.

Head, J.M. & Jardine, F.M. 1992. Ground-borne vibrations arising from piling. *Technical note 142*, UK: Construction Industry Research and Information Association (CIRIA).

Heald, M.T. & Larese, R.E. 1974. Influence of coatings on quartz cementation. *Journal Sedimentary Petrology*, 44, 1269–1274.

Health and Safety Executive. 1985. *The Abbeystead Explosion.* A report of the investigation by the Health and Safety Executive into the explosion on 23 May 1984 at the valve house of the Lune/Wyre Water Transfer Scheme at Abbeystead.

Health and Safety Executive. 2000. *The Collapse of NATM Tunnels at Heathrow Airport.* 110p. Transport Research Board, HSC Books.

Heathcote J.A., Jones M.A & Herbert A.W., 1996. Modelling the groundwater flow in the Sellafield Area, *Quarterly Journal of Engineering Geology*, 29, 559–581

Hencher, S.R. & Martin, R.P. 1982. The description and classification of weathered rocks in Hong Kong for engineering purposes. *Proceedings 7th Southeast Asian Geotechnical Conference*, 1, Hong Kong, 125–142.

Hencher, S.R. & McNicholl, D.P. 1985. Engineering in weathered rock. *Quarterly Journal of Engineering Geology*, 28, 253–266.

Hencher, S.R. & Richards, L.R. 1989. Laboratory direct shear testing of rock discontinuities. *Ground Engineering*, 22, 2, 24–31.

Hencher, S.R. & Richards, L.R. 2015. Assessing the shear strength of rock discontinuities at laboratory and field scales. *Rock Mechanics and Rock Engineering*, 48, 883–905.

Hencher, S.R. 1976. Correspondence: A simple sliding apparatus for the measurement of rock friction. *Géotechnique*, 26, 4, 641–644.

Hencher, S.R. 1977. *The Effect of Vibration on the Friction between Planar Rock Surfaces.* Unpublished PhD Thesis, Imperial College Science and Technology, London University.

Hencher, S.R. 1981a. Friction parameters for the design of rock slopes to withstand earthquake loading. *Proceedings of Conference on Dams and Earthquake, ICE*, London, 79–87.

Hencher, S.R. 1981b. *Report on Slope Failure at Yip Kan Street (11SW-D/C86) Aberdeen on 12th July 1981.* Geotechnical Control Office Report No. GCO 16/81, Hong Kong: Geotechnical Control Office, 26p.

Hencher, S.R. 1983a. *Landslide Studies 1982 Case Study No. 4 South Bay Close.* Special Project Report No. SPR 5/83, Hong Kong: Geotechnical Control Office, 38p.

Hencher, S.R. 1983b. *Landslide Studies 1982 Case Study No. 10 Ching Cheung Road.* Special Project Report No. SPR 11/83, Hong Kong: Geotechnical Control Office, 38p.

Hencher, S.R. 1983c. *Landslide Studies 1982 Case Study No. 1 Chai Wan Road.* Special Project Report No. SPR 2/83, Hong Kong: Geotechnical Control Office, 34p.

Hencher, S.R. 1983d. *Summary Report on Ten Major Landslides in 1982.* Special Project Report No. SPR 1/83, Hong Kong: Geotechnical Engineering Office, 28p.

Hencher, S.R. 1985. Limitations of stereographic projections for rock slope stability analysis. *Hong Kong Engineer*, 13, 7, 37–41.

Hencher, S.R. 1987. The implications of joints and structures for slope stability. *Slope Stability–Geotechnical Engineering and Geomorphology*, Anderson, M.G. & Richards, K.S. (eds.), UK: John Wiley & Sons Ltd, 145–186.

Hencher, S.R. 1995. Interpretation of direct shear tests on rock joints. *Proceedings 35th US Symposium on Rock Mechanics*, Deaman & Schultz (eds.), Balkema, Rotterdam, Lake Tahoe, 99–106.

Hencher, S.R. 1996a. Fracture flow modelling: Proof of evidence. *Radioactive waste disposal at Sellafield, UK: Site Selection, Geological and Engineering Problems*, Haszeldine, R.S. &. Smythe, D.K (eds.), published by Department of Geology and Applied Geology, University of Glasgow, 349–358. www.foe.co.uk/archive/nirex/sfoe6.html

Hencher, S.R. 1996b. Fracture flow modelling: Supplementary proof of evidence. *Radioactive waste disposal at Sellafield, UK: Site Selection, Geological and Engineering Problems* (edited by R S Haszeldine and D K Smythe) Published by Department of Geology and Applied Geology University of Glasgow 359–370.

Hencher, S.R. 2000. Engineering geological aspects of landslides. Keynote paper. Proceedings Conference on Engineering Geology HK 2000, Institution of Mining and Metallurgy, Hong Kong, 93–116.

Hencher, S.R. 2006. Weathering and erosion processes in rock–implications for geotechnical engineering. *Proceedings Symposium on Hong Kong Soils and Rocks*, March 2004, Institution of Mining, Metallurgy and Materials and Geological Society of London, 29–79.

Hencher, S.R. 2007. Hazardous ground conditions–reducing the risks. Keynote Paper. Engineering Geology in Geotechnical Risk Management, Hong Kong Regional Group of the Geological Society of London, 32–40.

Hencher, S.R. 2008. The "new" British and European standard guidance on rock description. A critique by Steve Hencher. *Ground Engineering*, 41, 7, 17–21.

Hencher, S.R. 2010. Preferential flow paths through soil and rock and their association with landslides. *Hydrological Processes*, 24, 1610–1630.

Hencher, S.R. 2013. Characterizing discontinuities in naturally fractured outcrop analogues and rock core: The need to consider fracture development over geological time. Geological society of London special publication, *Advances in the Study of Fractured Reservoirs*, 374, 113–123.

Hencher, S.R. 2015. *Practical Rock Mechanics*. Taylor & Francis, 356p.

Hencher, S.R. 2019. The Glendoe tunnel collapse in Scotland. *Rock Mechanics and Rock Engineering*, 52, 4033–4055.

Hencher, S.R., Massey, J.B. and Brand, E.W. 1985. Application of back analysis to some Hong Kong landslides. *Proceedings of the 4th International Symposium on Landslides*, 1, Toronto, 631–638.

Hencher, S.R. & Mallard, D.J. 1989. On the effects of sand grading on driven pile performance. *Proceedings of the International Conference on Piling and Deep Foundations*, London, 255–264.

Hencher, S.R., Toy, J.P. & Lumsden, A.C. 1993. Scale dependent shear strength of rock joints. *Scale Effects in Rock Masses 93*, Pinto da Cunha (ed.), Rotterdam: Balkema, 233–240.

Hencher, S.R. & Acar, I.A. 1995. The Erzincan earthquake, Friday, 13 March 1992. *Quarterly Journal of Engineering Geology*, 28, 313–316.

Hencher, S.R., Liao, Q.H. & Monighan, B. 1996. Modelling slope behaviour for open pits. *Transactions of the Institution of Mining & Metallurgy*, 105, A37–A47.

Hencher, S.R. & Daughton, G. 2000. Anticipating geological problems. *The Urban Geology of Hong Kong, Geological Society of Hong Kong Bulletin*, 6, 43–62.

Hencher, S.R., Tyson, J.T. & Hutchinson, P. 2005. Investigating substandard piles in Hong Kong. *Proceedings 3rd International Conference on Forensic Engineering*, London: Institution of Civil Engineers, 107–118.

Hencher, S.R., Anderson, M.G. & Martin, R.P. 2006. Hydrogeology of landslides in weathered profiles. *Proceedings of International Conference on Slopes*, Malaysia, 463–474.

Hencher, S.R. & Knipe, R.J. 2007. Development of rock joints with time and consequences for engineering. *Proceedings of the 11th Congress of the International Society for Rock Mechanics*, 1, 223–226.

Hencher, S.R., Sun, H.W. & Ho, K.K.S. 2008. The investigation of underground streams in a weathered granite terrain in Hong Kong. In *Geotechnical and Geophysical Site Characterization*, Huang & Mayne (eds.), London: Taylor & Francis Group, 601–607.

Hencher, S.R., Lee, S.G., Carter, T.G. & Richards, L.R. 2011. Sheeting joints–Characterisation, shear strength and engineering. *Rock Mechanics and Rock Engineering*, 44, 1–22.

Heuer, R.E. 1974. Important ground parameters in soft ground tunnelling. *Proceedings of Speciality Conference on Subsurface Exploration for Underground Excavation and Heavy Construction*, ASCE, 41–45.

Hight, D.W. 2009. Nicoll highway, Singapore: The post collapse investigations. Meeting Report by Cabarkapa, Z. *Ground Engineering*, October, 9–13.

Hight, D.W., Ellison, R.A. & Page, D.P. 2004. The engineering properties of the Lambeth group. Report RP576 *Construction Industry Research and Information Association (CIRIA)*, London.

Hill, S.J. & Wallace, M.I. 2001. Mass modulus of rock for use in the design of deep foundations. *Proceedings 14th Southeast Asian Geotechnical Conference*, 1,. Ho, K.K.S & Li, K.S (eds.), Hong Kong, 333–338.

HKIE 2004. *Code of Practice for Foundations*. Buildings Department, The Government of the Hong Kong Special Administrative Region, 57p.

HKIE 2017. *Code of Practice for Foundations* (2nd Edition). Buildings Department, The Government of the Hong Kong Special Administrative Region, 104p.

Ho, K.K.S., Sun, H.W. & Hui, T.H.H. 2003. *Enhancing the Reliability and Robustness of Engineered Slopes*. GEO Report No. 139, Hong Kong: Geotechnical Engineering Office.

Ho, H.Y., King, J.P. & Wallace, M.I. 2004. *A Basic Guide to Air Photo Interpretation in Hong Kong*. Applied Geoscience Centre, University of Hong Kong, 115p.

Hoek, E. 1968. Brittle failure of rock. In *Rock Mechanics in Engineering Practice*, Stagg & Zienkiewicz (eds.), London: Wiley and Sons, 99–124.

Hoek, E. 1999. Putting numbers to geology–an engineer's viewpoint. The 2nd Glossop lecture. *Quarterly Journal of Engineering Geology*, 32, 1–20.

Hoek, E. 2000. Big tunnels in bad rock. 2000 Terzhagi Lecture. *ASCE Journal of Geotechnical and Geoenvironmental Engineering*, 127, 726–740.

Hoek, E. & Bray, J.W. 1974. *Rock Slope Engineering*. Institution of Mining and Metallurgy, 309p.

Hoek, E. & Brown, E.T. 1980. *Underground Excavations in Rock*. London: Institution of mining and Metallurgy, 527p.

Hoek, E. & Moy, D. 1993. Design of large powerhouse caverns in weak rock. *Comprehensive Rock Engineering*, 5, Surface and Underground Project Case Histories (Editor in Chief, J.A. Hudson; Volume Editor, E. Hoek), Pergamon Press, 85–110.

Hoek, E., Kaiser, P.K. & Bawden, W.F. 1995. *Support of Underground Excavations in Rock*. Rotterdam: A.A. Balkema, 215p.

Hoek, E. & Marinos, P. 2000. Predicting tunnel squeezing. *Tunnels and Tunnelling International*, Part 1, 32/11, 45–51, November 2000; Part 2, 32/12, 33–36, December 2000.

Hoek, E. & Diederichs, M.S. 2006. Empirical estimation of rock mass modulus. *International Journal of Rock Mechanics & Mining Sciences*, 43, 203–215.

Hoek, E. & Brown, E.T. 2019. The Hoek–Brown failure criterion and GSI–2018 edition. *Journal of Rock Mechanics and Geotechnical Engineering*, 11, 3, 445–463.

Holmes, A. 1965. *Principles of Physical Geology*, 2nd edition. Nelson & Sons.

Holmøy, K.H. & Nilsen, B. 2014. Significance of geological parameters for predicting water inflow in hard rock tunnels. *Rock Mechanics & Rock Engineering*, 47, 853–868.

Holzhausen, G.R. 1989. Origin of sheet structure, 1. Morphology and boundary conditions. *Engineering Geology*, 27, 225–278.

Horn, H.M. & Deere, D.U. 1962. Frictional characteristics of minerals. *Géotechnique*, 12, 319–335.

Hoshino, K. 1993. Geological evolution from the soil to the rock: Mechanism of lithification and change of mechanical properties. *Geotechnical Engineering of Hard Soils–Soft Rocks*, Anagnostopoulos et al. (eds.), Balkema, 131–138.

Houghten, D.A. & Wong, C.M. 1990. Implications of the karst marble at Yuen Long for foundation investigation and design. *Hong Kong Engineer*, June, 19–27.

Houlsby, A.C. 1976. Routine interpretation of the Lugeon water–test. *Quarterly Journal of Engineering Geology*, 9, 303–313.

Hudson, J.A. 1989. *Rock Mechanics Principles in Engineering Practice*. CIRIA Ground Engineering Report: Underground Construction, Butterworths, 72p.

Hudson, J.A. 1992. *Rock Engineering Systems*. New York: Ellis Horwood, 185p.

Hudson, R.R. & Hencher, S.R. 1984. The delayed failure of a large cut slope in Hong Kong. *Proceedings of the International Conference on Case Histories in Geotechnical Engineering*, Missouri: St Louis, 679–682.

Hudson, J.A. & Harrison, J.P. 1992. A new approach to studying complete rock engineering problems. *Quarterly Journal of Engineering Geology and Hydrogeology*, 25, 93–105.

Hudson, J.A. & Harrison, J.P. 1997. *Engineering Rock Mechanics. An Introduction to the Principles*, Pergamon, 444p.

Hungr, O., Corominas, J. & Eberhardt, E. 2005a. Estimating landslide motion mechanism, travel distance and velocity. *Landslide Risk Management*, Hungr, Fell, Couture & Eberhardt (eds.), Taylor & Francis Group, 99–128.

Hungr, O., Fell, R., Couture, R. & Eberhardt, E. (eds.) 2005b. *Landslide Risk Management*. Taylor & Francis Group, 764p.

Hunt, R.E. 2005. *Geotechnical Engineering Investigation Handbook*, 2nd edition. Taylor & Francis, 1066p.

Hutchinson, J.N. 2001. Reading the ground: Morphology and geology in site appraisal. *Quarterly Journal of Engineering Geology and Hydrogeology*, 334, 7–50.

Institute of Geological Sciences. 1971. *British Regional Geology: Northern England*, 4th edition. HMSO, 125p.

Institution Civil Engineers. 1996. *Sprayed Concrete Linings (NATM) for Tunnels in Soft Ground*. Thomas Telford, 88p.

Institution Civil Engineers. 2005. *NEC3 Engineering and Construction Contract*. Thomas Telford.

International Society for Rock Mechanics. 1978. Suggested methods for the quantitative description of discontinuities in rock masses. *International Journal of Rock Mechanics and Mining Sciences & Geomechanics Abstracts*, 15, 319–368.

International Society for Rock Mechanics. 1981. Basic geotechnical description of rock masses. *International Journal of Rock Mechanics Mining Sciences and Geomechanics Abstracts*, 18, 85–110.

Irfan, T.Y. & Tang, K.Y. 1993. *Effect of the Coarse Fractions on the Shear Strength of Colluvium*. GEO Report No. 23, Hong Kong: Geotechnical Engineering Office, 232p.

Irfan, T.Y. 1996. Mineralogy, fabric properties and classification of weathered granites in Hong Kong. *Quarterly Journal of Engineering Geology*, 29, 5–35.

Itasca. 2004. *UDEC Version 4.0 User's Guide*, 2nd edition.

Iverson, R.M. 2000. Landslide triggering by rain infiltration. *Water Resources Research*, 36, 1897–1910.

Jahns, R.H. 1943. Sheet structure in granites, its origin and use as a measure of glacial erosion in New England. *Journal of Geology*, 51, 71–98.

Jaksa, M.B., Goldsworthy, J.S., Fenton, G.A., Kaggwa, W.S., Griffiths, D.V., Kuo, Y.L. & Poulos, H.G. 2005. Towards reliable and effective site investigations. *Géotechnique*, 55, 2, 109–121.

James, D. 2007. *Fractured Reservoirs, Geological Society Special Publication no. 270*. Lonergan, L., Jolly, R.J.H., Rawnsley, K., Sanderson, D.J. (eds.), London, Bath: Geological Society of London, 285p.

Janbu, N. 1973. Slope stability computations. *Embankment-Dam Engineering*, New York: John Wiley & Sons, Inc., 47–86.

Jardine, R.J., Symes, M.J. & Burland, J.B. 1984. The measurement of soil stiffness in the triaxial apparatus. *Géotechnique*, 34, 323–340.

Jenkins, G.J., Perry, M.C. & Prior, M.J.O. 2009. *The Climate of the United Kingdom and Recent Trends. UKCIP09*. Exeter: Met Office Hadley Centre.

Jiao, J.J. & Malone, A.W. 2000. An hypothesis concerning a confined groundwater zone in slopes of weathered igneous rocks. *Symposium on Slope Hazards and their Prevention*, Hong Kong, 165–170.

Jiao, J.J., Wang, X-S. & Nandy, S. 2005. Confined groundwater zone and slope instability in weathered igneous rocks in Hong Kong. *Engineering Geology*, 80, 71–92.

Jiao, J.J., Ding, G.P. & Leung, C.M. 2006. Confined groundwater near the rockhead in igneous rocks in the Mid-Levels area, Hong Kong, China. *Engineering Geology*, 84, 207–219.

Jones, A. 1971. Soil piping and stream channel initiation. *Water Resources Research*, 7, 3, 602–609.

Jones, L.D. & Jefferson, I. 2012. Expansive soils. In *ICE Manual of Geotechnical Engineering. Volume 1, Geotechnical Engineering Principles, Problematic Soils and Site Investigation*, Burland, J. (ed.), London, UK, ICE Publishing, 413–441.

Jones, L.D. 2006. Monitoring landslides in hazardous terrain using terrestrial LIDAR: an example from Montserrat. *Quarterly Journal of Engineering Geology and Hydrogeology*, 39, 371–373.

Jouzel, J. et al. 2007. Orbital and millennial Antarctic climate variability over the past 800,000 Years. *Science*, 317, 5839, 793–797.

Kamewada, S., Gi, H.S., Taniguchi, S. and Yoneda, H. 1990 Application of borehole image processing system to survey of tunnel. *Proceedings of the International Symposium on Rock Joints*, Barton, N.R & Stephansson, O. (eds.), Loen, Norway, Rotterdam: Balkema, 51–58.

Karakuş, M.M. & Fowell, R.J. 2004. An insight into the New Austrian Tunnelling Method (NATM). *ROCKMEC'2004-VIIth Regional Rock Mechanics Symposium*, Sivas, Turkey, 14p.

Karrow, P.F. & White, O.L. 2002. A history of neotectonic studies in Ontario. *Tectonophysics*, 353, 3–15.

Katsura, S., Kosugi, K., Mizutani, T., Okunaka, S. & Mizuyama, T. 2008. Effects of bedrock groundwater on spatial and temporal variations in soil mantle groundwater in a steep granitic headwater catchment. *Water Resources Research*, 44, W09430.

Keefer, D.K. 1984. Landslides caused by earthquakes, Geological *Society of America Bulletin*, 95, 406–421.

Keefer, D.K. 2002. Investigating landslides caused by earthquakes–a historical review. *Surveys in Geophysics*, 23, 473–510.

Khorasani, E., Amini, M., Hossaini, M.F., Medley, E. 2019 Statistical analysis of bimslope stability using physical and numerical models. *Engineering Geology*, 254, 13–24.

Khorasani, E., Amini, M., Medley, E. et al. 2022. A General solution for estimating the safety factor of bimslopes. *Rock Mechanics and Rock Engineering*, 55, 7675–7693.

Kikuchi, K. & Mito, Y. 1993. Characteristics of seepage flow through the actual rock joints. *Proceedings 2nd International Workshop on Scale Effects in Rock Masses*, Lisbon, 305–312.

Kim Y.I., Hencher, S.R., Yoon, Y.H. & Cho, S.G. 2001. Determination of the construction method for the Young dong tunnel by risk assessment. *Korea Society of Civil Engineers, Seminar*, Seoul, 200–206.

Knill, J.L. 1976. *Cow Green revisited*. Inaugural Lecture, Imperial College of Science and Technology, University of London.

Knill, J.L. 1978. Geology in the construction industry. *Industrial Geology*, Knill, J.L (ed.), Oxford University Press, 259–286.

Knill, J.L. 2003. Core values: The first Hans Cloos lecture. *Proceedings 9th AEG Congress*, Durban, 1–45.

Koe, A., Murphy., W. & Nicholson, R. 2018. *Rock Netting Systems–Design, Installation and Whole-Life Management (C775)*. CIRIA Report No. C775, 136p.

Kong, W.K. 2011. Water ingress assessment for rock tunnels: a tool for risk planning. *Rock Mechanics & Rock Engineering*, 755–765.

Kovacevic, N., Hight, D.W. & Potts, D.M. 2007. Predicting the stand-up time of temporary London Clay slopes at Terminal 5, Heathrow Airport. *Géotechnique*, 57, 63–74.

Krauskopf, K.B. 1988. *Radioactive Waste Disposal and Geology*. Topics in the Earth Sciences, Volume 1, Chapman & Hall, 145p.

Kromer, R.A., Hutchinson, D.J., Lato, M.J., Gauthier, D. & Edwards, T. 2015. Identifying rock slope failure precursors using LiDAR for transportation corridor hazard management. *Engineering Geology*, 195, 93–103.

Kruseman, G.P. & de Ridder, N.A. 2000. *Analysis and Evaluation of Pumping Test Data*. 2nd Edition. International Institute for Land Reclamation and Improvement, 372.

Krynine, D.P. & Judd, W.R. 1957. *Principles of Engineering Geology and Geotechnics*. McGraw-Hill, 730p.

Kulander, B.R. & Dean, S.L. 1995. Observations on fractography with laboratory experiments for geologists. In *Fractography: Fracture Topography as a Tool in Fracture Mechanics and Stress Analysis*, Ameen, M.S. (ed.), Geological Society London Special Publication, 92, 59–82.

Kulander, B.R., Dean, S.L. & Ward, B.J. 1990. *Fractured Core Analysis. Interpretation, Logging and Use of Natural and Induced Fractures in Core*. AAPG Methods in Exploration Series, No. 8, 88p.

Kwong, A.K.L. 2005. Drawdown and settlement measured at 1.5 km away from SSDS Stage 1 Tunnel C (a perspective from hydro-geological modelling). *Proceedings of K.W.Lo Symposium*, London, Ontario, Canada, Session D, 1–24.

Lambe, T.W. & Whitman, R.V. 1979. *Soil Mechanics*, John Wiley & Sons, 553p.

Lawn, B.R. & Wilshaw, T.R. 1975. *Fracture of Brittle Solids*, Cambridge University Press.

Leach, B. & Herbert, R. 1982. The genesis of a numerical model for the study of the hydrogeology of a steep hillside in Hong Kong. *Quarterly Journal of Engineering Geology*, 15, 243–259.

Lee, E.M. 2020. Statistical analysis of long-term trends in UK effective rainfall: Implications for deep-seated landsliding. *Quarterly Journal of Engineering Geology*, 53, 587–597.

Lee, K.C., Jeng, F.S., Huang, T.S., Hsieh, Y.M. & Orense, R.P. 2010. Can tilt test provide insight regarding frictional behaviour of sandstone under seismic excitation? *Proceedings New Zealand Society for Earthquake Engineering 2010 Conference*, Paper 19, 8p.

Lee, S.G. & Hencher, S.R. 2009. The repeated collapse of a slope despite continuous reassessment and remedial works. *Engineering Geology*, 107, 16–41.

Leeder, M. 1999. *Sedimentology and Sedimentary Basins*, Blackwell Publishing, 592p.

Legislative Council Panel on Environmental Affairs. 2000. Paper for discussion on 25 October 2000.

Lerouiel, S. & Tavernas, F. 1981. Pitfalls of back-analysis. *Proceedings 10th ICSMFE*, 1, Stockholm, 185–190.

Li, Z.H., Huang, H.W., Xue, Y.D. and Yin, J. 2009. Risk assessment of rockfall hazards on highways. *Georisk: Assessment and Management of Risk for Engineered Systems and Geohazards*, 3, 14–154.

Lindquist, E.S. & Goodman, R.E. 1994. Strength and deformation properties of a physical model melange. *Proceedings 1st North American Rock Mechanics Symposium*, Austin, Texas, 843–858.

Lisle, R.J. & Leyshon, R.J. 2004. *Stereographic Projection Techniques for Geologists and Civil Engineers*, 2nd edition. Cambridge University Press, 112p.

Lumb, P. 1962. Effect of rain storms on slope stability. *Proceedings of the Symposium on Hong Kong Soils*, Hong Kong, 73–87.

Lumb, P. 1976. Discussion on the assessment of landslide potential with recommendations for future research by A.A. Beattie & E.P.Y. Chau. *Hong Kong Engineer*, 4, 2, 55.

Lumsdaine, R.W. & Tang, K.Y. 1982. A comparison of slopes stability calculations. *Proceedings of the Seventh Southeast Asian Geotechnical Conference*, Hong Kong, 31–38.

MacGregor, F., Fell, R., Mostyn, G.R., Hocking, G. & McNally, G. 1994. The estimation of rock rippability. *Quarterly Journal of Engineering Geology*, 27, 123–144.

Magnus R., Teh C.I., Lau J.M. 2005. *Report of the Committee of Inquiry into the Incident at the MRT Circle Line Worksite that Led to the Collapse of Nicoll Highway on 20 April 2004*. R. Magnus, Er. Dr. Teh Cee Ing and Er. Lau Joo Ming. May 2005. Submitted to The Hon. Minister of Manpower. Singapore.

Maidl, B., Schmid, L., Ritz W. & Herrenknecht, M. 2008. *Hardrock Tunnel Boring Machines*, Berlin: Ernst and Young, 343p.

Malone, A.W. 1998. Slope movement and failure: Evidence from field observations of landslides associated with hillside cuttings in saprolites in Hong Kong. *Proceedings 13th Southeast Asian Geotechnical Conference*, 2, Taipei, 81–90.

Malone, A.W., Hansen, A., Hencher, S.R. & Fletcher, C.J.N. 2008. Post-failure movements of a large slow rock slide in schist near Pos Selim, Malaysia. *The 10th International Symposium on Landslides and Engineered Slopes*, Xian, 457–461.

Mandl, G. 2005. *Rock Joints. The Mechanical Genesis*. Springer, 221p.

Marcuson, W.F. III, Hynes, M.E. & Franklin, A.G. 1990. Evaluation and use of residual strength in seismic safety analysis of embankments. *Earthquake Spectra*, 6, 529–572.

Marinos, P. & Hoek, E. 2000. GSI–a geologically friendly tool for rock mass strength estimation. *Proceedings of GeoEng 2000*, Melbourne, Australia, 1422–1442.

Marshall, R.H. and Flanagan, R.F. 2007. Singapore's deep tunnel sewerage system–experiences and challenges. *RETC 2007*, Toronto, 1308–1319.

Martel, S.J. 2006. Effect of topographic curvature on near-surface stresses and application to sheeting joints. *Geophysical Research Letters*, 33, LO1308.

Martin, C.D., Davison, C.C. & Kozak, E.T. 1990. Characterising normal stiffness and hydraulic conductivity of a major shear zone in granite. *Proceedings International Symposium on Rock Joints*, Loen (Barton & Stephansson, eds.), Balkema: Rotterdam, 549–556.

Martin, D.C. 1994. Quantifying drilling-induced damage in samples of Lac du Bonnet granite. *Proceedings 1st North American Rock Mechanics Symposium*, Austin, Texas, 419–427.

Martin, D.C. & Kaiser, P.K. 1984. Analysis of a rock slope with internal dilation. *Canadian Geotechnical Journal*, 21, 605–620.

Martin, R.P. & Hencher, S.R. 1986. Principles for description and classification of weathered rocks for engineering purposes. *S I Practice: Assessing BS5930, Geological Society, Engineering Geology Special Publication No. 2*, Hawkins, A.B. (ed.), 299–308.

Masses, Lake Tahoe (Meyer, Cook, Goodman & Tsang, eds.), June 1992, Balkema: Rotterdam, 567–572.

Masset, O. & Loew, S. 2010. Hydraulic conductivity distribution in crystalline rocks, derived from inflows to tunnels and galleries in the Central Alps, Switzerland. *Hydrogeology Journal*, 18, 863–891.

Mathews, M.C., Clayton, C.R.I. & Owen, Y. 2000. The use of field geophysical techniques to determine ground stiffness profiles. *Proceedings Institution Civil Engineers, Geotechnical Engineering*, January, 31–42.

Matsuo, S. 1986. Tackling floods beneath the sea. *Tunnels & Tunnelling*, March, 42–45.

Maunsell Consultants Asia Ltd. 2000. *Investigation of Unusual Settlement in Tsuen Kwan O Town Centre*. Final Report, November 2000, Territory Development Department, Hong Kong Government.

Maury, V. 1993. An overview of tunnel, underground excavation and boreholes collapse mechanisms. *Comprehensive Rock Engineering*, 4 (ed. in chief J. Hudson), Pergamon Press, 369–412.

Mayne, P.W., Christopher, B.R. & deJong, J. 2001. *Manual on Subsurface Investigations*. National Highway Institute, Publication No. FHWA NHI-01-031 Federal Highway Administration, Washington, DC, 305p.

McCabe, B.A., Nimmons G.J. & Egan, D. 2009. A review of field performance of stone columns in soft soils. *Proceedings of the Institution of Civil Engineers: Geotechnical Engineering*, 162, 323–334.

McFeat-Smith, I., MacKean, R. & Waldmo, O. 1998. Water inflows in bored rock tunnels in Hong Kong: prediction, construction issues and control measures. *ICE Conference on Urban Ground Engineering*, Hong Kong.

McLearie, D.D., Forekman, W., Hansmire, W.H. & Tong, E.K.H. 2001. Hong Kong strategic sewage disposal scheme stage 1 deep tunnels. *2001 RETC Proceedings*, 487–498.

McMahon, B.K. 1985. Geotechnical design in the face of uncertainty. *Journal of the Australian Geomechanics Society*, 10, 7–19.

McNicholl, D.P., Pump, W.L. & Cho, G.W.F. 1986. Groundwater control in large scale slope excavations–five case histories from Hong Kong. *Groundwater in Engineering Geology*, Cripps, J.C et al. (eds.), Geological Society, Engineering Geology Special Publication no. 3, 561–576.

Mead, L. & Austin, G.S. 2005. *Dimension stone. Industrial Minerals and Rocks*, 7th edition. Littleton CO: AIME-Society of Mining Engineers, 907–923.

Megaw, T.M. & Bartlett, J.V. 1981. *Tunnels. Planning, Design, Construction, Volume 1*. Ellis Horwood Limited, 284p.

Menard, L. & Broise, Y. 1975. Theoretical and practical aspects of dynamic consolidation. *Géotechnique*, 25, 3–18.

Michie, U. 1996. The geological framework of the Sellafield area and its relationship to hydrogeology. *Quarterly Journal of Engineering Geology*, 29, S13–S28.

Mikkelsen, P.E. & Green, G.E. 2003. Piezometers in fully grouted boreholes. *Proceedings Symposium on Field Measurements in Geomechanics*, Oslo, 1–10.

Miller, K.G., Kominz, M.A., Browning, J.V., Wright, J.D., Mountain, G.S., Katz, M.E., Sugarman, P.J., Cramer, B.S., Christie-Blick, N. & Pekari, S.F. 2005. The Phanerozoic record of global sea-level change. *Science*, 310, 1293–1298.

Miller, K.G., Mountain, G.S., Wright, J.D. & Browning J.V. 2011. A 180-million-year record of sea level and ice volume variations from continental margin and deep-sea isotopic records. *Oceanography* 24(2): 40–53.

Miller, L. D., Goldfarb, R. J., Gehrais, G. E. & Snee, L. W. 1994. Genetic links among fluid cycling, vein formation, regional deformation and plutonism in the Juneau gold belt, southeastern Alaska. Geology, 22, 203–206.

Montgomery, D.R., Dietrich, W.E. & Heffner, J.T. 2002. Piezometric response in shallow bedrock at CB1: Implications for runoff generation and landsliding. *Water Resource Research*, 38, 12, 1274, 10-1–10-18.

Moore, R. & Brunsden, D. 1996. A physico-chemical mechanism of seasonal mudsliding. *Géotechnique*, 46, 259–278.

Moore, R., Hencher, S.R. & Evans, N.C. 2001. An approach for area and site-specific natural terrain hazard and risk assessment, Hong Kong. *Proc. 14th SEA Geotech. Conference*, Hong Kong, 155–60.

Morgenstern, N.R. & Cruden, D.M. 1977. Description and classification of geotechnical complexities. General Report, Session 2, *International Symposium of the Geotechnics of Structurally Complex Formations*, 24p.

Morgenstern, N.R. & Price, V.E. 1965. The analysis of the stability of general slip surfaces. *Géotechnique*, 15, 79–93.

Morgenstern, N.R. 2000. Common ground. *Proceedings of GeoEng 2000*, 1, Melbourne, Australia, 1–30.

Morton, K.L., Bouw, P.C. & Connelly, R.J. 1988. The prediction of minewater inflows. *Journal South African Institution Mining Metallurgy*, 88, July, 219–226.

Moye, D.G. 1955. Engineering geology for the snowy mountains scheme. *Journal of the Institution of Engineers*, 27, Australia, 287–298.

Muir, T.R.C., Smethurst, B.K. & Finn, R.P. 1986. Design and construction of high rock cuts for the Kornhill Development, Hong Kong. *Conference on Rock Engineering and Excavation in an Urban Environment*, Hong Kong, Institution of Mining and Metallurgy, 309–329.

Muir, Wood, A. 2000. *Tunnelling: Management by Design*. E & FN Spon, 320p.

Murray, A.D. & Gray, A.M. 1997. The Pergau hydroelectric project part 3: Civil engineering design. *Proceedings Institution Civil Engineers Water, Maritime & Energy*, 124, 173–188.

New Zealand Geotechnical Society. 2005. *Field Description of Soil and Rock*. 36p.

Newmark, N.M.1965. Effects of earthquakes on dams and embankments. *Géotechnique*, 2, 139–160.

Ng, H.Y. 2021 Revisiting lessons learned from the Nicoll Highway collapse. *Structure*, February 2021, 4p.

Nichols, T.C. Jr. 1980. Rebound, its nature and effect on engineering works. *Quarterly Journal of Engineering Geology*, 13, 133–152.

Nicholson, D.P. 1994. The observational method in geotechnical engineering–Preface. *Géotechnique*, XLIV, 4, 613–618.

Nicholson, D.T., Lumsden, A.C. & Hencher, S.R. 2000. Excavation-induced deterioration of rock slopes. *Proceedings of Conference on Landslides in Research, Theory and Practice*, 3, Cardiff: Thomas Telford, , 1105–1110.

Nilsen, B. 2014. Characteristics of water ingress in Norwegian subsea tunnels. *Rock Mechanics & Rock Engineering*, 47, 933–945.

Nirex 1996. *Nuclear Science and Technology Testing and Modelling of Thermal, Mechanical and Hydrogeological Properties of Host Rocks for Deep Geological Disposal of Radioactive Waste*. Proceedings of a Workshop held in Brussels, 12–13 January 1996, 127–140.

Nirex. 2007. *Geosphere Characterisation Report: Status Report*. October 2006, 198p.

Norbury, D. 2010. *Soil and Rock Description in Engineering Practice*. CRC Press, 288p.

Norbury, D. 2017. Standards and quality in ground investigation; squaring the circle. *Quarterly Journal of Engineering Geology and Hydrogeology*, 50 (3), 212–230.

North Wales Geology Association. 2006. *Newsletter*, 46, July, 1.

Note:–all the following GCO and GEO publications are available to download from: www.cedd.gov.hk/eng/publications/http://www.cedd.gov.hk/eng/publications/

Nowson, W.J.R. 1954. The history and construction of the foundations of the Asia insurance building, Singapore. *Proceedings Institution Civil Engineers*, Pt 1, 3, 407–443.

Odling, N.E. 1997. Scaling and connectivity of joint systems in sandstones from western Norway. *Journal of Structural Geology*, 19, 1257–1271.

OECD. 2009. *Considering Timescales in the Post-Closure Safety of Geological Disposal of Radioactive Waste*. Paris: OECD Publishing, Radioactive Waste Management.

Oldroyd, D.R. 2002. Nirex and the great denouement. Chapter 20. In *Earth, Water, Ice and Fire: Two Hundred Years of Geological Research in the English Lake District*, Geological Society Memoir No. 25. xvi, London, Bath: Geological Society of London, 328p.

Ollier, C.D. 1975. *Weathering*, 2nd impression. Longman Group Limited, 304p.

Ollier, C.D. 2010. Very deep weathering and related landslides. *Weathering as a Predisposing Factor to Slope Movements*, From Calceterra, D. & Parise, M. (eds.), London: Geological Society, Engineering Geology Special Publications, 23, 5–14.

Olsson, O. & Gale, J.E. 1995. Site assessment and characterization for high-level nuclear waste disposal: Results from the Stripa Project, Sweden. *Quarterly Journal of Engineering Geology and Hydrogeology*, 28, 1, S17–S30.

Olsson, R. & Barton, N. 2001. An improved model for hydromechanical coupling during shearing of rock joints. *International Journal of Rock Mechanics & Mining Sciences*, 38, 317–329.

Orr, W.E., Muir-Wood, A., Beaver, J.L., Ireland, R.J. & Beagley, D. 1991. Abbeystead Outfall Works: Background to repairs and modifications–and the lessons learned. *Journal of the Institution of Water and Environmental Management*, 5, 7–20.

Osipov, V.I. 1975. Structural bonds and the properties of clays. *Bulletin of the International Association of Engineering Geology*, 12, 13–20.

Palmström, A. 1982. The volumetric joint count–a useful and simple measure of the degree of rock jointing. *Proceedings 4th Congress International Association Engineering Geologists*, Delhi, 221–228.

Papaliangas, T., Lumsden, A.C., Manolopoulou, S. & Hencher, S.R. 1990. Shear strength of modelled filled rock joints. *Proceedings of the International Symposium on Rock Joints*, Loen, Norway, 275–282.

Papaliangas, T.T., Hencher, S.R. & Lumsden, A.C. 1994. Scale independent shear strength of rock joints. *Proceedings IV CSMR/ Integral Approach to Applied Rock Mechanics*, Santiago, Chile, 123–133.

Parks, C.D. 1991. A review of the mechanisms of cambering and valley bulging. *Quarternary Engineering Geology*, Forster, A., Culshaw, M.G., Cripps, J.A. Little, J.C. & Moon, C. (eds.), Geological Society Engineering Geology Special Publication No.7, 381–388.

Parry, S., Campbell, S.D.G. & Fletcher, C.J.N. 2000. Kaolin in Hong Kong saprolites–genesis and distribution. *Proceedings Conference Engineering Geology HK 2000*, IMM HK Branch, 63–70.

Patton, F.D. 1966. Multiple modes of shear failure through rock. *Proceedings 1st International Congress International Society Rock Mechanics*, 1, Lisbon, 509–513.

Peck, R.B. 1969. Advantages and limitations of the observational method in applied soil mechanics. *Géotechnique*, 19, 171–187.

Peck, R.B., Hanson, W.E. & Thornburn, T.H. 1974. *Foundation Engineering*, 2nd edition. New York: John Wiley and Sons, 514p.

Pells, P.J.N. 2004. Rock mass grouting to reduce inflow. *Tunnels and Tunnelling*, March, 34–37.

Petley, D. 2024 (ongoing). Dave's Landslide Blog. http://daveslandslideblog.blogspot.com/

Pettifer, G.S. & Fookes, P.G. 1994. A revision of the graphical method for assessing the excavatability of rock. *Quarterly Journal of Engineering Geology*, 27, 145–164.

Phillips, F.C. 1973. *The Use of Stereographic Projection in Structural Geology*, 3rd edition. Edward Arnold, 90p.

Phillipson, H.B. & Chipp, P.N. 1982. Air foam sampling of residual soils in Hong Kong. *Proceedings of the ASCE Specialty Conference on Engineering and Construction in Tropical and Residual Soils*, Honolulu, 339–356.

Picarelli, L. & Di Maio, C. 2010. Deterioration processes of hard clays and clay shales. *Weathering as a Predisposing Factor to Slope Movements*, From Calceterra, D. & Parise, M. (eds.), Geological Society, London, Engineering Geology Special Publications, 23, 15–32.

Pierson, T.C. 1983. Soil pipes and slope stability. *Quarterly Journal of Engineering Geology*, 16, 1–15.

Pine, R.J. & Roberds, W.J. 2005. A risk-based approach for the design of rock slopes subject to multiple failure modes–illustrated by a case study in Hong Kong. *International Journal of Rock Mechanics & Mining Sciences*, 42, 261–275.

Pinho, A., Rodrigues-Carvalho, J.A., Gomes, C.F. & Duarte, I.M. 2006. Overview of the evaluation of the state of rock weathering by visual inspection. *10th IAEG Congress, Engineering Geology for Tomorrow's Cities*, Nottingham, Paper 260.

Pirazzoli, P.A. 1996. *Sea-Level Changes. The Last 20000 Years*, John Wiley & Sons, 211p.

Piteau, D.R. 1973. Characterising and extrapolating rock joint properties in engineering practice. *Rock Mechanics*, 2, 2–31.

Pollard, D.D. & Aydin, A. 1988. Progress in understanding jointing over the past century. *Geological Society America Bulletin*, 100, 1181–1204.

Pomeroy, J.S. 1981. Storm-induced debris avalanching and related phenomena in the Johnstown area, Pennsylvania, with references to other studies in the Appalachians. *US Geological Survey Professional Paper*, pp 1191.

Pope, R.G., Weeks, R.C. & Chipp, P.N. 1982. Automatic recording of standpipe piezometers. *Proceedings of the 7th Southeast Asian Geotechnical Conference*, 1, Hong Kong, 77–89.

Poulos, H.G. 2005. Pile behavior–consequences of geological and construction imperfections. 40th Terzaghi Lecture, *Journal Geotechnical & Geoenvironmental Engineering*, 131, ASCE, 538–563.

Powderham, A.J. 1994. An overview of the observational method: Development in cut and cover and bored tunnelling projects. *Géotechnique*, 44, 619–636.

Power, C.M. & Hencher, S.R. 1996. A new experimental method for the study of real area of contact between joint walls during shear. *Rock Mechanics, Proceedings 2nd North* American Rock Mechanics Symposium, Montreal, 1217–1222.

Power, C.M. 1998. *Mechanics of Modelled Rock Joints under True Stress Conditions Determined by Electrical Measurements of Contact Area.* Unpublished PhD thesis, University of Leeds, 293p.

Power, M.S., Rosidi, D., Kaneshiro, J.Y. 1998. Seismic vulnerability of tunnels and underground structures revisited. *North American Tunneling*, Ozdemir, L. (ed.), Balkema, 243–250.

Preene, M. and Brassington, F.C. 2003. Potential groundwater impacts from civil engineering works. *Water & Environment Journal*, 17, 59–64.

Price, N.J. 1959. Mechanics of jointing in rocks. *Geological Magazine*, 96, 149–167.

Price, N.J. & Cosgrove, J. 1990. *Analysis of Geological Structures.* Cambridge University Press.

Priest, S.D. 1993. *Discontinuity Analysis for Rock Engineering.* Chapman & Hall, 473p.

Pugh D.T. 1987. *Tides, Surges and Mean Sea-Level–A Handbook for Engineers and Scientists.* Ed. John Wiley & Sons, 472p.

Puller, M. 2003. *Deep Excavations: A Practical Manual,* 2nd edition. Thomas Telford Ltd., 584p.

Pun, W.K., Cheung, W.M. & Shum, K.W. 2009. *Geoguide 7–Guide to Soil Nail Design and Construction.* Hong Kong: GEO, 100p.

Pye, K. & Miller, J.A. (1990). Chemical and biochemical weathering of pyritic mudrocks in a shale embankment. *Quarterly Journal of Engineering Geology*, 23, 365–382.

Quinn J.D., Philip L.K. & Murphy W. 2009. Understanding the recession of the Holderness Coast, East Yorkshire, UK: A new presentation of temporal and spatial patterns. *Quarterly Journal of Engineering Geology and Hydrogeology*, 42, 165–178.

Ramsay, J. G. & Huber, M. I. 1987. The Techniques of Modern Structural Geology: Volume 2, Folds and Fractures. Academic Press, London.

Randolph Glacier Inventory (RGI). 2017. *A Dataset of Global Glacier Outlines: Version 6.0*: Technical Report, Global Land Ice Measurements from Space, Colorado, USA. Digital Media. https://doi.org/ 10.7265/N5-RGI-60

Rawnsley, K.D. 1990. *The Influence of Joint Origin on Engineering Properties*. Unpublished PhD thesis, University of Leeds, 388p.

Rawnsley, K.D., Hencher, S.R. & Lumsden, A.C. 1990. Joint origin as a predictive tool for the estimation of geotechnical properties. *Proceedings of the International Symposium on Rock Joints*, Löen, Norway, 91–96.

Rawnsley, K.D., Rives, T., Petit, J.-P., Hencher, S.R. & Lumsden, A.C. 1992. Joint development in perturbed stress fields near faults. *Journal of Structural Geology*, 14, 939–951.

Raymer, J.H. 2001. Predicting groundwater inflow into hard-rock tunnels: estimating the high-end of the permeability distribution. In *Proceedings of Rapid Excavation and Tunnelling Conference*, 1027–1038.

Read, J. & Stacey, P. 2009. *Guidelines for Open Pit Slope Design*. CSIRO Publishing, 496p.

Reason, J. 1990. *Human Error*. New York: Cambridge University Press.

Reeve, D., Chadwick, A. & Fleming, C. 2018. *Coastal Engineering: Processes, Theory and Design Practice*, 3rd edition. CRC Press, 542p.

Reid, J.M. & Cripps, J.C. 2019. Geochemical lessons from Carsington Dam Failure of 1984 and reconstruction. *Quarterly Journal of Engineering Geology and Hydrogeology*, 52, 4, 2018–184.

Richards, K.S. and Reddy, K.R., 2007. Critical appraisal of piping phenomena in earth dams. *Bulletin Engineering Geology Environment*, 66, 381–402.

Richards, L.R. & Cowland, J.W. 1982. The effect of surface roughness on the field shear strength of sheeting joints in Hong Kong granite. *Hong Kong Engineer*, 10, 39–43.

Richards, L.R. & Cowland, J.W. 1986. Stability evaluation of some urban rock slopes in a transient groundwater regime. *Proceedings Conference on Rock Engineering and Excavation in an Urban Environment*, Hong Kong: IMM, 357–363 (Discussion 501–6).

Richards, L.R. & Read, S.A.L. 2007. New Zealand greywacke characteristics and influences on rock mass behaviour. *Proceedings 11th Congress of the International Society of Rock Mechanics*, 1, Lisbon, 359–364.

Richey, J.E. 1948. *Scotland. The Tertiary Volcanic Districts*, 2nd edition revised. British Regional Geology, Her Majesty's Stationery Office, 105p.

Rico, M., Benito, G., Salgueiro, A.R., Díez-Herrero, A. & Pereira, H.G. 2008. Reported tailings dam failures. A review of the European incidents in the worldwide context. *Journal of Hazardous Materials*, 152, 846–852.

RILEM. 2003. Recommended test method AAR-I, Detection of potential Alkali-reactivity of aggregates–petrographic method, *Materials & Structures*, 36, 480–496.

Rives, T., Rawnsley, K.D. & Petit, J.-P. 1994. Analogue simulation of natural orthogonal joint set formation in brittle varnish. *Journal of Structural Geology*, 16, 419–429.

Robertshaw, C. & Tam, T.K. 1999. Tai Po to butterfly valley treated water transfer. *World Tunnelling*, 12, 8, 383–385.

Robertson, P.K. 2009. Interpretation of cone penetration tests–a unified approach. *Canadian Geotechnical Journal*, 46, 1337–1355.

Rodin, S. 1979. Contribution to Session 3. *Sixth Asian Regional Conference, ISSMFE*, Singapore.

Rodriguez, C.E. 2001. *Hazard Assessment of Earthquake-Induced Landslides on Natural Slopes*. Unpublished PhD Thesis, Imperial College of Science and Technology, University of London, 367p.

Rogers, C.M. & Engelder, T. 2004. The feedback between joint-zone development and downward erosion of regularly spaced canyons in the Navajo Sandstone, Zion National Park, Utah. *The Initiation, Propagation and Arrest of Joints and Other Fractures*, 231, Cosgrove, J.W and Engelder, T. (eds.), Geological Society Publication, 49–71.

Romana, M. 1991. SMR Classification. *Proceedings 7th Congress Rock Mechanics*, 2, Aachen, Germany: ISRM., 955–960.

Roorda, J., Thompson, J.C. and White, O.L., 1982. The analysis and prediction of lateral instability in highly stressed, near-surface rock strata. *Canadian Geotechnical Journal*, 19, 4, 451–462.

Roscoe, K.H., Schofield, A.N. & Wroth, C.P. 1958. On the yielding of soils. *Géotechnique*, 8, 22–53.

Ross-Brown, D.M. & Walton, G. 1975. A portable shear box for testing rock joints. *Rock Mechanics*, 7, 129–153.

Rouleau, A. & Raven, K.G. 1995. Site specific simulation of groundwater flow and transport using a fracture network model. Proceedings of Conference on Fractured and Jointed Rock

Rowe, P.W. 1962. The stress dilatancy relation for static equilibrium of an assembly of particles in contact. *Proceedings of the Royal Society*, 269A, 500–527.

Ruxton, B.P & Berry, L. 1957 The weathering of granite and associated erosional features in Hong Kong. *Bulletin of the Geological Society of America*, 68, 1263–1292.

Sadiq, A.M., & Nasir, S.J. 2002. Middle Pleistocene karst evolution in the State of Qatar, Arabian Gulf. *Journal of Cave and Karst Studies*, 64, 132–139.

Salehy, M.R., Money, M.S. & Dearman, W.R. 1977. The occurrence and engineering properties of intra-formational shears in Carboniferous rocks. *Proceedings Conference on Rock Engineering*, Potts, E.L.J. & Attewell, P.B. (eds.) Newcastle upon Tyne, 311–328.

Santamarina, J.C., and Cho, G.C. 2004. Soil behaviour: The role of particle shape. *Advances in Geotechnical Engineering: The Skempton Conference*, 1, Jardine, R.J. et al. (eds.), March 2004, London: Thomas Telford, , 29–31, 604–617.

Sausse, J. and Genter, A. 2005. Types of permeable fractures in granite. *Geological Society, London, Special Publications* 2005, 240, 1–14.

Schmidt, B. 1966. Discussion on earth pressures at rest related to stress history. *Canadian Geotechnical Journal*, 3, 239–242.

Schneider, H.J. 1976. The friction and deformation behaviour of rock joints. *Rock Mechanics*, 8, 169–184.

Schofield, A.N. 2006. Interlocking, and peak and design strengths. *Géotechnique*, 56, 357–358.

Scholz, C.H. 1990. *The Mechanics of Earthquakes and Faulting*. Cambridge University Press, 439p.

Seed, H.B. & Idriss, I.M. 1982. *Ground Motions and Soil Liquefaction during Earthquakes*, Monograph No. 5, Berkeley, California: Earthquake Engineering Research Institute, 134p.

Seed, H.B., Idriss, I.M. & Kiefer, F.W. 1968. *Characteristics of Rock Motions during Earthquakes*, EERC 68 5, Berkeley: University of California, September.

Selby, M.J. 1993. *Hillslope Materials and Processes*, 2nd edition. Oxford University Press, 451p.

Seol, H & Jeong, S. 2009. Load-settlement behaviour of rock-socketed drilled shafts using Osterberg-Cell tests. *Computers and Geotechnics*, 36, 1134–1141.

Sewell, R.J., Campbell, S.D.G., Fletcher, C.J.N., Lai, K.W., & Kirk, P.A. 2000. *The Pre-Quaternary Geology of Hong Kong*. Geotechnical Engineering Office, Civil Engineering Department, Hong Kong SAR Government, 181p plus 4 maps.

Shang, J., Hencher, S.R., West, L.J. & Handley, K. 2017. Forensic excavation of rock masses: A technique to investigate discontinuity persistence. *Rock Mechanics and Rock Engineering*, 50, 2911–2928.

Shang, J., Hencher, S.R., West, L.J. 2016. Tensile strength of geological discontinuities including incipient bedding, rock joints and mineral veins. *Rock Mechanics and Rock Engineering*, 49, 4213–4225.

Shang, J., West, L.J., Hencher, S.R., & Z. Zhao 2018. Geological discontinuity persistence: Implications and quantification. *Engineering Geology*, 241, 41–54.

Shaw, R. & Owen, R.B. 2000. Geomorphology and ground investigation planning. *Proceedings Conference on Engineering Geology HK 2000*, Hong Kong: Institution of Mining and Metallurgy, 151–163.

Shibata, T. & Taparaska, W. 1988. Evaluation of liquefaction potential of soil using cone penetration tests. *Soils and Foundations*, 28, 49–60.

Shirlaw, J.N. 2006. Discussion: Hull wastewater flow transfer tunnel: Tunnel collapse and causation investigation. *Geotechnical Engineering*, 2, 126–127.

Shirlaw, J.N., Hencher, S.R. & Zhao, J. 2000. Design and construction issues for excavation and tunnelling in some tropically weathered rocks and soils. *Proceedings of GeoEng 2000*, 1, Melbourne, Australia, Invited Papers, 1286–1329.

Shirlaw, J.N., Broome, P.B., Chandrasegaran, S., Daley, J., Orihara, K., Raju, G.V.R., Tang, S.K., Wong, K.S. & Yu, K. 2003. The fort canning boulder bed. *Underground Singapore 2003. Engineering Geology Workshop*, 388–407.

Shirlaw, N., Ong, J.C.W., Rosser, H.B., Tan, C.G., Osborne, N.H. and Heslop, P.E. 2003. Local settlements and sinkholes due to EPB tunneling, *Proceedings ICE, Geotech. Engineering*, 156 (GE4), 193–211.

Simons, N., Menzies, B. & Mathews, M. 2001. *A Short Course in Soil and Rock Slope Engineering*. Thomas Telford, 432p.

Singapore Standard. 2003. *Code of Practice for Foundations*. CP4:2003, Singapore: Spring, 253p.

Skempton, A.W. 1953. The colloidal "activity" of clays. *Proceedings 3rd International Conference of Soil Mechanics and Foundation Engineering*, 1, 57–61.

Skempton, A.W. 1960. Effective stress in soils, concrete and rocks. *Conf. Pore Pressure and Suction in Soils*, 4–16.

Skempton, A.W. 1970. The consolidation of clays by gravitational compaction. *Quarterly Journal of Geology*, 125, 373–411.

Skempton, A.W. & Macdonald, D.H. 1956. The allowable settlement of buildings. *Proceedings of the Institution of Civil Engineers, Part 3*, 5, 727–784.

Skempton, A.W., Schuster, R.L. & Petley, D.J. 1969. Joints and fissures in the London clay at Wraysbury and Edgeware. *Géotechnique*, 19, 205–217.

Skempton, A.W., Leadbetter, A.D. & Chandler, R.J. 1989. The mam Tor landslide, North Derbyshire. *Philosophical Transactions Royal Society London, A*, 329, 503–547.

Skempton, A.W. & Vaughan, P.R. 1993. The failure of Carsington Dam. *Géotechnique*, 43, 1, 151–173.

Slob, S. 2010. *Automated Rock Mass Characterisation Using 3-D Terrestrial Laser Scanner*. Unpublished PhD Thesis, Technical University of Delft, 287p.

Smart, P.L. 1985. Applications of fluorescent dye tracers in the planning and hydrological appraisal of sanitary landfills. *Quarterly Journal of Engineering Geology*, 18, 275–286.

Smith, M.R. & Collis, L. (eds.) 2001. Aggregates. *Geological Society*, London, Engineering Geology Special Publications No. 17, 339p.

Snow, D.T. 1968. Rock fractures spacings, openings and porosities. *J. Soil Mech. Found. Div.*, ASCE, 94 (SMI), 73–91.

Spang, K., & Egger, P. 1990. Action of fully-grouted bolts in jointed rock and factors of influence, *Rock Mechanics and Rock Engineering*, 23, 201–229.

Spink, T. 1991. Periglacial discontinuities in Eocene clays near Denham, Buckinghamshire. *Quarternary Engineering Geology*, Forster, A., Culshaw, M.G., Cripps, J.C., Little, J.A. & Moon, C. (eds.), Geological Society Engineering Geology Special Publication No. 7, 389–396.

Starfield, A.M. & Cundall, P.A. 1988. Towards a methodology for rock mechanics modelling. *International Journal of Rock Mechanics and Mining Sciences & Geomechanics Abstracts*, 25, 99–106.

Starr, D., Dissanayake, A., Marks, D., Clements, J. & Wijeyakulasuriya, V. 2010. South West transport Corridor landslide. *Queensland Roads*, 8, 57–73.

Statham, I. & Baker, M. 1986. Foundation problems on limestone: A case history from the Carboniferous Limestone at Chepstow, Gwent. *Quarterly Journal of Engineering Geology*, 19, 191–201.

Stauffer, D. 1985. *Introduction to Percolation Theory*. Taylor and Francis, London.

Sterling, R.L. 1993. The expanding role of rock engineering in developing national and local infrastructures. *Comprehensive Rock Engineering*, 5. Surface and Underground Project Case Histories (Editor in Chief, J.A. Hudson; Volume Editor, E. Hoek), Pergamon Press, 1–27.

Stille, H. & Palmström, A. 2003. Classification as a tool in rock engineering. *Tunnelling and Underground Space Technology*, 18, 4, 331–345.

Stow, D.A.V. 2005. *Sedimentary Rocks in the Field. A Colour Guide*. Manson Publishing Ltd., 320p.

Streckeisen, A. 1974. Classification and nomenclature of plutonic rocks: IUGS Subcommission on the Systematics of Igneous Rocks. *Geologische Rundschau*, 63, 773–786.

Streckeisen, A. 1980. Classification and nomenclature of volcanic rocks, lamprophyres, carbonatites and melilitic rocks: IUGS Subcommission on the Systematics of Igneous Rocks. *Geologische Rundschau*, 69, 1, 194–207.

Stringer, C. 2006. *Homo Britannicus*. Allen Lane, 319p.

Strozzi, T., Delaloye, R., Poffet, D., Hansmann, J. & Loew, S. 2011. Surface subsidence and uplift above a headrace tunnel in metamorphic basement rocks of the Swiss Alps as detected by satellite SAR interferometry. *Remote Sensing of Environment*, 8p.

Styles, K.A. & Hansen, A. 1989. *Geotechnical Area Studies Programme: Territory of Hong Kong* (GASP Report No. XII). Hong Kong: Geotechnical Control Office, 356p.

Swannell, N.G. and Hencher, S.R. Cavern design using modern software. *Proceedings of 10th Australian Tunnelling Conference*, Melbourne, March 1999.

Sweeney, D.J. & Robertson, P.K. 1979. A fundamental approach to slope stability problems in Hong Kong. *Hong Kong Engineer*, 7, 10, 35–44.

Tada, R. & Siever, R. 1989. Pressure solution during diagenesis. *Annual Reviews Earth Planetary Sciences*, 17, 89–118.

Talbot, D.K., Hodkin, D.L. & Ball, T.K. (1997) Radon investigations for tunnelling projects: A case study from St. Helier, Jersey. *Quarterly Journal of Engineering Geology*, 30, 2, 115–122.

Tam, A. 2010. Lofty aspirations at destination West Kowloon. *Hong Kong Engineer*, May, 8–11.

Tanner, K. & Walsh, P. 1984. *Hallsands: A Pictorial History*. Kingsbridge: Tanner & Walsh, 32p.

Tapley, M.J., West, B.W., Yamamoto, S. & Sham, R. 2006. Challenges in construction of Stonecutters Bridge and progress update. *International Conference on Bridge Engineering*, Hong Kong.

Tarkoy, P.J. 1991. Appropriate support selection, *Tunnels and Tunneling*, 23 (October), 42–45.

Taylor, L.R. 1969. Preface. *The Optimum Population for Britain. Proceedings of a Symposium at the Royal Geographical Society*, London, 25 and 26 September, 1969, viip.

Terando, A., Reidmiller, D., Hostetler, S.W., Littell, J.S., Beard, T.D., Jr., Weiskopf, S.R., Belnap, J., and Plumlee, G.S. 2020. Using information from global climate models to inform policymaking – The role of the U.S. Geological survey: *U.S. Geological Survey Open-File Report 2020–1058*, 25p.

Terzaghi, K. 1923. *Die berechnung der durchlassigkeitzifer des tones aus dem verlauf der hydrodynamischen spannungserscheinungen, Mathematish-naturwissenschaftliche, Klasse*. Vienna: Akademie der Wissenschaften, 125–138.

Terzhagi, K. 1925. The physical causes of proportionality between pressure and frictional resistance. From *Erdmechanic*, translated by A. Casagrande in *From Theory to Practice in Soil Mechanics*, J. Wiley.

Terzhagi, K. 1929. The effect of minor geological details on the stability of dams. *American Institute of Mining and Metallurgical Engineers*, Technical Publication 215, 31–44.

Thomas, A.L. & La Pointe, P.R. 1995. Conductive fracture identification using neural networks. *Proceedings 35th US Symposium on Rock Mechanics*, Lake Tahoe, 627–632.

Thorpe, R.S. & Brown, G.C. 1985. *The Field Description of Igneous Rocks.* Milton Keynes, UK: The Open University Press, 154p.

Thrush, P.W. & The Staff of the US Bureau of Mines. 1968. *A Dictionary of Mining, Mineral and Related Terms.* US Department of the Interior, 1269p.

Tika, Th., Sarma, S.K., & Ambraseys, N. 1990. Earthquake induced displacements on pre-existing shear surfaces in cohesive soils. *Proceedings of the 9th European Conference on Earthquake Engineering*, 4A, 16–25.

Todd, D.K. 1980. *Groundwater Hydrology*, 2nd edition. Wiley, 535p.

Tomlinson, M.J. 2001. *Foundation Design and Construction*, 7th edition. Prentice Hall, 569p.

Tondel, M.& Lindahl, L. 2019. Intergenerational ethical issues and communication related to high-level nuclear waste repositories. *Current Environment Health Reports*, 6, 338–343.

Tottergill, B.W. 2006. *FIDIC Users' Guide: A Practical Guide To The 1999 Red And Yellow Books: Incorporating Changes And Additions To The 2005 MDB Harmonised Edition.* Thomas Telford.

Trenter, N. 2003. Understanding and containing geotechnical risk. *Proceedings of the Institution of Civil Engineers: Civil Engineering*, 156, 42–48.

Trotter, F.M., Hollingworth, S.E., Eastwood, T. & Rose, W.C.C. 1937. Gosforth District (One-inch Geological Sheet 37 New Series). In *Memoirs of the Geological Survey of Great Britain*, Department of Scientific and Industrial Research..

Trurnit, P. 1968. Pressure solution phenomena in detrital rocks. *Sedimentary Geology*, 2, 89–114.

Tsuji, H., Sawada, T. & Takizawa, M. 1996. Extraordinary inundation accidents in the Seikan undersea tunnel. *Proceedings Institution Civil Engineers, Geotechnical Engineering*, 119, 1–14.

Tucker, M.E. 1982. *The Field Description of Sedimentary Rocks.* Milton Keynes, UK: The Open University Press, 112p.

Turner, A.K. 2006. Challenges and trends for geological modelling and visualisation. *Bulletin Engineering Geology and Environment*, 65, 109–127.

Turner, J.P., Bell, T.W., Bowman, M.J. & Chapman, N.A. 2023. Role of the geosphere in deep nuclear waste disposal–an England and wales perspective. *Earth-Science Reviews*, 242, 1–24.

Twidale, C.R. 1973. On the origin of sheet jointing. *Rock Mechanics*, 5, 163–187.

Uchida, T., Kosugi, K. & Mizuyama, T. 2001. Effects of pipeflow on hydrological process and its relation to landslide: A review of pipeflow studies in forested headwater catchments. *Hydrological Processes*, 15, 2151–2174.

Ulusay, R., Hudson, J.A. (eds.) 2007. *The Complete ISRM Suggested Methods for Rock Characterisation, Testing and Monitoring*: 1974–2006: ISRM, Turkish National Group, www.isrm.net/gca/?id=305

van der Berg J.P., Clayton C.R.I., Powell D.B. 2003. Displacements ahead of an advancing NATM tunnel in the London clay. *Géotechnique*, 53, 767–784.

Varley, P.M. & Warren, C.D. 1996. History of the geological investigations for the Channel Tunnel. Chapter 2 In *Engineering Geology of the Channel Tunnel*,. Harris, C.S., Hart, M.B., Varley, P.M & Warren, C. (eds.), Thomas Telford, 5–18.

Vaughan, P.R. 1969. A note on sealing piezometers in boreholes. *Géotechnique*, 19, 405–413.

Vermeulen, B., Hoitink, A.J.F., van Berkum S.W., & Hidayat, H. 2014. Sharp bends associated with deep scours in a tropical river: The river Mahakam (East Kalimantan, Indonesia), *Journal of Geophysical Research: Earth Surface*, 119, 1441–1454.

Vernon, R.H. 2018. *A Practical Guide to Rock Microstructure*, 2nd edition. Cambridge University Press.

Vervoort, A. & De Wit, K. 1997. Correlation between dredgeability and mechanical properties of rock. *Engineering Geology*, 47, 259–267.

Vidal Romani, J.R. & Twidale, C.R. 1999. Sheet fractures, other stress forms and some engineering implications. *Geomorphology*, 31, 13–27.

Wakasa, S., Matsuzaki, H., Tanaka, Y. & Matskura, Y. 2006. Estimation of episodic exfoliation rates of rock sheets on a granite dome in Korea from cosmogenic nuclide analysis. *Earth Surface Processes and Landforms*, 31, 1246–1256.

Wallis, K. 2000. *Breakthroughs made on Hong Kong Sewer*. T & T International, December, 2000.

Wallis, S. 1999. Heathrow failures highlight NATM (abuse?) misunderstandings. *Tunnel*, 3, 99, 66–72.

Waltham, A.C. 2002. *Foundations of Engineering Geology*, 2nd edition. Spon Press, 92p.

Waltham, A.C. 2008. Lötschberg tunnel disaster, 100 years ago. *Quarterly Journal of Engineering Geology & Hydrogeology*, 41, 131–136.

Waltham, A.C. 2016. Control the drainage: The gospel accorded to sinkholes. *Quarterly Journal of Engineering Geology & Hydrogeology*, 49, 5–20.

Walton, J.G. 2007. *Unforeseen Ground Conditions and Allocation of Risk: Before the Roof Caved in.* Paper delivered at the first of two sessions organised by the Society of Construction Law on 1 May 2007.

Warner, J. 2004. *Practical Handbook of Grouting–Soil, Rock and Structures.* John Wiley & Sons.

Warren, J.E. & Root, P.J. 1963. The behaviour of naturally fractured reservoirs. *Society Petroleum Engineers Journal*, 3, 245–255.

Weltman, A.J. & Healy, P.R. 1978. *Piling in "Boulder Clay" and Other Glacial Tills.* DOE and CIRIA Piling Development Group, Report PG5, 78p.

Wentzinger, B., Starr, D., Fidler, S., Nguyen, Q. & Hencher, S.R. 2013. Stability analyses for a large landslide with complex geology and failure mechanism using numerical modelling. *Proceedings International Symposium on Slope Stability in Open Pit Mining and Civil Engineering*, Brisbane, Australia, 733–746.

West, G., Carter, P.G., Dumbleton, M.J. and Lake, L.M. 1981. Site investigation for tunnels, *International Journal Rock Mechanics Mining Science*, 18, 345–367.

West, L.J., Hencher, S.R. & Cousens, T.C. 1992. Assessing the stability of slopes in heterogeneously graded soils. *Proceedings 6th International Symposium on Landslides*, 1, Christchurch, New Zealand, 591–595.

Whitten, D.G.A. & Brooks, J.R.V. 1972. *The Penguin Dictionary of Geology*. Penguin Books, 495p plus appendices.

Wong, H.N. 2005. Landslide risk assessment for individual facilities. *Landslide Risk Management*, Hungr, Fell, Couture & Eberhardt (eds.), Taylor & Francis Group, 237–296.

Woodcock, N.H. & Mort, K. 2008. Classification of fault breccias and related fault rocks. *Geological Magazine* 145, 3, 435–440.

Woods, N, Karimi, M., Jafari, I. and Hinchliffe, R. 2022. Geological and geotechnical investigations for the Salang highway, Afghanistan. *Quarterly Journal of Engineering Geology and Hydrogeology*, 55, 3. https://doi.org/10.1144/qjegh2021-189

Woodworth, J.B. 1896. On the fracture system of joints, with remarks on certain great fractures. *Proceedings of the Boston Society of Natural History*, 27, 163–184.

Woodworth, P.L. 2018. Sea level change in great Britain between 1859 and the present. *Geophysical Journal International*, 213, 222–236.

Wyllie, D.C. 1999. *Foundations on Rock: Engineering Practice*, 2nd edition. E & FN Spon.

Wyllie, D.C. & Mah, C.W. 2004. *Rock Slope Engineering*, 4th edition. E & FN Spon, 431p.

Yin, Y., Wang, F. & Sun, P. 2008. Landslide hazards triggered by the 12 May 2008 Wenchuan Earthquake, Sichuan, China, 2008, *Proceedings First World Landslide Forum*, Tokyo, 1–17.

Young, G.M. 2008. Origin of enigmatic structures: Field and geochemical investigation of columnar joints in sandstones, Island of Bute, Scotland. *The Journal of Geology*, 116, 527–536.

Younger, P.L. & Manning, D.A.C. 2010. Hyper-permeable granite: Lessons from test pumping in the Eastgate Geothermal Borehole, Weardale, UK. *Quarterly Journal of Engineering Geology & Hydrogeology*, 43, 5–10.

Yu, Y.F., Siu, C.K. & Pun, W.K. 2005. *Guidelines on the Use of Prescriptive Measures for Rock Slopes*. GEO Report 161, 34p.

Yu, Y.S. & Coates, D.F. 1970. *Analysis of rock slopes using the finite element method*. Department of Energy, Mines and Resources Mines Branch, Mining Research Centre, Research Report, R229 (Ottawa).

Zambri, B., Robock, A., Mills, M.J., & Schmidt, A. 2019. Modeling the 1783–1784 Laki eruption in Iceland: 2. Climate impacts. *Journal of Geophysical Research: Atmospheres*, 124, 6770–6790.

Zare, S. & Bruland, A. 2006. Comparison of tunnel blast design models. *Tunnelling and Underground Space Technology*, 21, 533–541.

Zhang, L & Einstein, H.H. 2010. The planar shape of rock joints. *Rock Mechanics & Rock Engineering*, 43, 55–68.

Index

Data Analytics in the AWS Cloud